The Ascent of Humanity

Charles Eisenstein

Panenthea Press

The Ascent of Humanity

ISBN 978-0-9776222-0-7

Library of Congress cataloging-in-publication data: contact publisher

Panenthea Press, www.panenthea.com
266 F, N. Arlington Ave., Harrisburg, PA, 17109

Printed in the United States

First Printing: March, 2007

Dedicated to the more beautiful world our hearts tell us is possible

And they said, "Come, let us build us a city and a tower, whose top may reach unto heaven."

Contents

Chapter V: The World Under Control

Chapter VI: The Crumbling of Certainty

Chapter VII: The Age of Reunion

Chapter VIII: Self and Cosmos

Introduction

More than any other species, human beings are gifted with the power to manipulate their environment and the ability to accumulate and transmit knowledge across generations. The first of these gifts we call technology; the other we call culture. They are central to our humanity.

Accumulating over thousands of years, culture and technology have brought us into a separate human realm. We live, more than any animal, surrounded by our own artifacts. Among these are works of surpassing beauty, complexity, and power, human creations that could not have existed—could not even have been conceived—in the times of our forebears. Seldom do we pause to appreciate the audacity of our achievements: objects as mundane as a compact disc, a video cellphone, an airplane would have seemed fantastical only a few centuries ago. We have created a realm of magic and miracles.

At the same time, it is quite easy to see technology and culture not as a gift but as a curse. After millennia of development, the power to manipulate the environment has become the power to destroy it, while the ability to transmit knowledge transmits as well a legacy of hatred, injustice, and violence. Today, as both the destruction and the violence reach a feverish crescendo, few can deny that the world is in a state of crisis. Opinions vary as to its exact nature: some people say it is primarily ecological; others say it is a moral crisis, a social, economic, or political crisis, a health crisis, even a spiritual crisis. There is, however, little disagreement that the crisis is of human origin. Hence, despair: Is the present ruination of the world built into our humanity?

Is genocide and ecocide the inevitable price of civilization's magnificence? Need the most sublime achievements of art, music, literature, science, and technology be built upon the wreckage of the natural world and the misery of its inhabitants? Can the microchip come without the oil slick, the strip mine, the toxic waste dump? Under the shadow of every Chartres Cathedral, must there be women burning at the stake? In other words, can the gift of technology and culture somehow be separated from the curse?

The dashed Utopian dreams of the last few centuries leave little hope. Despite the miracles we have produced, people across the ideological spectrum, from Christian fundamentalists to environmental activists, share a foreboding that the world is in grave and growing peril. Temporary, localized improvements cannot hide the ambient wrongness that pervades the warp and woof of modern society, and often our personal lives as well. We might manage each immediate problem and control every foreseeable risk, but an underlying disquiet remains. I am referring simply to the feeling, "Something is wrong around here." Something so fundamentally wrong that centuries of our best and brightest efforts to create a better world have failed or even backfired. As this realization sinks in, we respond with despair, cynicism, numbness, or detachment.

Yet no matter how complete the despair, no matter how bitter the cynicism, a possibility beckons of a world more beautiful and a life more magnificent than what we know today. Though we may rationalize it, it is not rational. We become aware of it in moments, gaps in the rush and press of modern life. These moments come to us alone in nature, or with a baby, making love, playing with children, caring for a dying person, making music for the sake of music or beauty for the sake of beauty. At such times, a simple and easy joy shows us the futility of the vast, life-consuming program of management and control.

We intuit that something similar is possible collectively. Some of may have experienced it when we find ourselves cooperating naturally and effortlessly, instruments of a purpose greater than ourselves that, paradoxically, makes us individually more and not less when we abandon ourselves to it. It is what musicians are referring to when they say, "The music played the band."

Another way of being is possible, and it is right in front of us, closer than close. That much is transparently certain. Yet it slips away so easily that we hardly believe it could be the foundation of life; so we relegate it to an afterlife and call it Heaven, or we relegate it to the future and call it Utopia. (When nanotechnology solves all our problems… when we all learn to be nice to each other… when finally I'm not so busy…) Either way, we set it apart from this world and this life, and thereby deny its practicality and its reality in the here-and-now. Yet the knowledge that life is more than Just This cannot be suppressed, not forever.

I share with dreamers, Utopians, and teenagers an unreasonable intuition of a magnificent potential, that life and the world can be more than we have made of them.

What error, then, what delusion has led us to accept the lesser lives and the lesser world we find ourselves in today? What has rendered us helpless to resist the ugliness, pollution, injustice, and downright horror that has risen to engulf the planet in the last few centuries? What calamity has so resigned us to it, that we call this the human condition? Those moments of love, freedom, serenity, play—what power has made us believe these are but respites from real life?

Inspired by such moments, I have spent the last ten years trying to understand what keeps us—and what keeps me—from the better world that our hearts tell us must exist. To my endless amazement, I keep discovering a common root underneath all the diverse crises of the modern age. Underlying the vast swath of ruin our civilization has carved is not human nature, but the opposite: human nature denied. This denial of human nature rests in turn upon an illusion, a misconception of self and world. We have defined ourselves as other than what we are, as discrete subjects separate from each other and separate from the world around us. In a way this is good news. In this book I will describe the profound changes that will flow, and are already flowing, from the reconception of the self that is underway. The bad news is that our present conception of self is so deeply woven into our civilization—into our technology and culture—that its abandonment can only come with the collapse of much that is familiar. This is what the present convergence of crises portends.

Everything I wrote in the preceding paragraph about our civilization also applies to each of us individually. Saints and mystics have tried for thousands of years to teach us how we are trapped in a delusion about who we are. This delusion inevitably brings about suffering, and eventually a crisis that can only be resolved through a collapse, a surrender, and an opening to a state of being beyond previous self-limitation. You are not, they tell us, a "skin-encapsulated ego", and lasting happiness can never result from pursuing that ego's agenda. These spiritual teachings have helped me realize, at least part way, my intuitions of what work, love, human relationship, and health can be. They are not the main subject of this book, however, nor do I claim to exemplify them in my own life. Nonetheless, the shift in our collective self-conception is intimately related to a parallel shift in our individual self-conception. In other words, there is a spiritual dimension to the planetary crisis.

As this planetary crisis invades our individual lives, unavoidably, neither the personal nor the collective misconception of who we are will remain tenable. Each mirrors the other: in its origin, its consequences, and its resolution. That is why this book interweaves the story of humanity's separation from nature with the story of our individual alienation from life, nature, spirit, and self.

* * *

Despite my faith that life is meant to be more, little voices whisper in my ear that I am crazy. Nothing is amiss, they say, this is just the way things are. The rising tide of human misery and ecological destruction, as old as civilization, is simply the human condition, an inevitable result of built-in human flaws like selfishness and laziness. Since you can't change it, be thankful for your good fortune in avoiding it. The misery of much of the planet is a warning, say the voices, to protect me and mine, impelling me to maximize my security.

Besides, it couldn't be as bad as I think. If all that stuff were true—about the ecological destruction, the genocide, the starving children, and the whole litany of impending crises—then wouldn't everyone be in an uproar about it? The normalcy of the routines surrounding me here in America tells me, "It couldn't be that bad." That little voice echoes throughout the culture. Every advertising flyer, every celebrity news item, every product catalog, every hyped-up sports event, carries the subtext, "You can afford to care about this." A man in a burning house wouldn't care about these things; that our culture does care about them, almost exclusively, implies that our house is not burning down. The forests are not dying. The deserts are not spreading. The atmosphere is not heating. Children are not starving. Torturers are not going free. Whole ethnicities are not being exterminated. These crimes against humanity and crimes against nature couldn't really be happening. Probably they have been exaggerated; in any event, they are happening somewhere else. Our society will figure out solutions before the calamities of the Third World affect me. See, no one else is worried, are they? Life hums on as usual.

As for my intuition of magnificent possibilities for my own life, well, my expectations are too high. Grow up, the voices say, life is just like this. What right have I to expect the unreasonable magnificence whose possibility certain moments have shown me? No, it is my intuitions that are not to be trusted. The examples of what life is surround me and define what is normal. Do I see anyone around me whose work is their joy, whose time is their own, whose love is their passion? It can't happen. Be thankful, say the voices, that my job is reasonably stimulating, that I feel "in love" at least once in a while, that the pain is manageable and life's uncertainties under control. Let good enough be good enough. Sure, life can be a drag, but at least I can afford to escape it sometimes. Life is about work, self-discipline, responsibility, but if I get these out of the way quickly and efficiently, I can enjoy vacations, entertainment, weekends, maybe even early retirement. Listening to these voices, is it any wonder that for many years, I devoted most of my energy and vitality to the escapes from life? Is it any wonder that so many of my students at Penn State look forward already, at age 21, to retirement?

If life and the world are Just This, we are left no choice but to make the best of it: to be more efficient, to achieve better security, to get life's uncertainties under control. There are voices that speak to this too. They are the evangelists of technology and self-improvement, who urge us to improve the human condition basically by trying harder. My inner evangelist tells me to get my life under control, to work out every day, to organize my time more efficiently, to watch my diet, to be more disciplined, to try harder to be a good person. On the collective level, the same attitude says that perhaps the next generation of material and social technologies—new medicines, better laws, faster computers, solar power, nanotechnology—will finally succeed in improving our lot. We will be more efficient, more intelligent, more capable, and finally have the capacity to solve humanity's age-old problems.

For more and more people today, these voices ring hollow. Words like "high-tech" and "modern" lose their cachet as a multiplicity of crises converge upon our planet. If we are fortunate, we might, for a time, prevent these crises from invading our personal lives. Yet as the environment continues to deteriorate, as job security evaporates, as the international situation worsens, as new incurable diseases appear, as the pace of change accelerates, it seems impossible to rest at ease. The world grows more competitive, more dangerous, less hospitable to easy living, and security comes with greater and greater effort. And even when temporary security is won, a latent anxiety lurks within the fortress walls, a mute unease in the background of modern life. It pervades technological society, and only intensifies as the pace of technology quickens. We begin to grow hopeless as our solutions—new technologies, new laws, more education, trying harder—only seem to worsen our problems. For many activists, hopelessness gives way to despair as catastrophe looms ever closer despite their best efforts.

This book explains why trying harder can never work. Our "best efforts" are grounded in the same mode of being that is responsible for the crisis in the first place. As Audre Lord put it, "The master's tools will never dismantle the master's house." Soon, though, this mode of being will come to an end, to be replaced by a profoundly different sense of the self, and a profoundly different relationship between human and nature. This book is about the gathering revolution in human beingness.

* * *

When we say that the planetary crisis is of human (and not natural) origin, what do we mean? Human beings are mammals, after all, biological creatures no less natural than any other. In a sense, there can be no distinction between human and nature, because human beings are a part of nature and everything we do is therefore "natural". However, we *do* distinguish. We recognize in nature a kind of harmony, balance, authenticity, and beauty lacking in the world of technology—think of the connotations of the word "artificial". Whether in fact or in perception, we modern humans live in a way that is no longer natural.

At the crux of the human-nature distinction is technology, the product of the human hand. While other animals do make and use tools, no other species has our capacity to remake or destroy the physical environment, to control nature's processes or transcend nature's limitations. In the mental and spiritual realm, the counterpart of technology is culture, which modifies and even supersedes human nature in the same way technology modifies physical nature. In thus mastering nature with technology, and mastering human nature with culture, we distinguish ourselves from the rest of life, establishing a separate human realm. Believing this to be a good thing, we think of this separation as an ascent in which we have risen above our animal origins. That is why we naturally refer to the millennia-long accumulation of culture and technology as "progress".

It is separation, then, in the form of technology and culture, that defines us as human. As well, it is separation that has generated the converging crises of today's world. People of a religious persuasion might attribute the fundamental crisis to a separation from God; people of an ecological persuasion, to a separation from nature. People engaged in social activism might focus on the dissolution of community (which is a separation from each other). We might also investigate the psychological dimension, of separation from lost parts of ourselves. For good or ill, it is separation that has made us what we are.

Through long and tortuous pathways, these forms of separation have created the world we know today. Our intuitions that life and the world are meant to be more reflect the ultimate illusoriness of that separation. But it is a powerful illusion, having generated the converging crises we see today in politics, the environment, medicine, education, the economy, religion, and many other realms. In this book I will trace the pathways to these crises. Constantly I am amazed how the same fundamental misconception of self underlies phenomena as apparently unconnected as the war in Iraq, intellectual property, antibiotic resistance, acid rain, ethnic cleansing, junk mail, suburban sprawl, and declining U.S. literacy. (No, I'm not going to blame it all on "capitalism", for our economic system too is more a symptom than a cause of separation.)

The root and the epitome of separation is the discrete, isolated self of modern perception: the "I am" of Descartes, the "economic man" of Adam Smith, the individual phenotype of Darwinian competition for resources, the skin-encapsulated ego of Alan Watts. It is a self conditionally dependent on, but fundamentally separate from, the Other: from nature and other people. Seeing ourselves as discrete and separate beings, we naturally seek to manipulate the not-self to our best advantage. Technology in particular is predicated on some kind of individuation or conceptual separation from the environment, because it takes the physical world as its object of manipulation and control. Technology, in effect, says, "Let us make the world better."

If, as I wrote above, our self-conception as discrete and separate beings is an illusion, then the whole ascent of humanity, the species of culture and technology, is based on an illusion as well. That is why the implications of our present reconceiving of ourselves are so profound, promising no less than a radical redefinition of what it is to be human, how we relate to one another, and how we relate to the world.

Not only is technology based on a conceptual separation from nature, but it also reinforces that separation. Technology distances us from nature and insulates us from her rhythms. For example, most Americans' lives are little affected by the seasons of the year. We eat the same food year round, shipped in from California; air conditioning keeps us cool in the summer and heating warm in the winter. Natural physical limitations of muscle and bone no longer limit how far we can travel, how high we can build, or the distance at which we can communicate. Each advance in technology distances us from nature, but it also frees us from natural limitations. That's why we so naturally think of human development as an ascent. But how can all these improvements add up to the world we find ourselves in today?

We are faced with a paradox. On the one hand, technology and culture are fundamental to the separation of humans from nature, a separation that is at the root of the converging crises of the present age. On the other hand, technology and culture explicitly seek to improve on nature: to make life easier, safer, and more comfortable. Who could deny that the first digging stick was an improvement over hands and fingernails; who could deny that fire keeps us warmer and medicine healthier than the primitive living in a state of nature? At least, that is what these technologies intend. But have we actually made the world better? If not, why has technology not achieved its intended purpose? Again, how can a series of incremental improvements add up to crisis?

Chapter One begins to answer these questions by describing a grave constitutional flaw in the very premise of technology, and beyond that, in technology's generalization as a program of control. By considering it through the lens of addiction, we will see that the despair mentioned above is justified, that our entire approach to problem-solving renders us helpless to do anything but worsen the gathering crisis. Like an animal caught in quicksand, the harder we struggle the faster we sink.

Chapter Two describes how we got into this quagmire to begin with. It digs far beneath the usual culprits of industry and agriculture to identify the origins of separation in everything that makes us human: language, art, measure, religion, and technology, even Stone Age technology. These have built upon each other, converging into the tidal wave of alienation and misery that engulfs the planet today. Nonetheless, by tracing the separation responsible for our current crisis to prehistoric or even prehuman times, we begin to see separation not as a "monstrously wrong turn" (to use John Zerzan's words) but as an organic inevitability leading, perhaps, to a new phase of human and natural development.

It was with the Scientific Revolution of Galileo, Newton, Bacon, and Descartes that the ideology of separation received its full articulation. We call this articulation "science". Chapter Three describes how the conceptual distinction between self and world is built into our very vocabulary of thought. The methods and techniques of modern science, along with that entire mode of thought we call rational, objective, or scientific, reinforce the regime of separation even when we try to ameliorate it. The master's tools will never dismantle the master's house. An example of this is the urge to "save the environment" or "conserve natural resources", locutions that reaffirm an external environment, fundamentally separate from ourselves, upon which we are only conditionally dependent. This echoes the classical scientific cosmology which, though obsolete, still forms the basis of our intuitions: we are isolated, separate beings gazing out upon an objective universe of impersonal forces and generic masses.

Religion, too, is shown to be complicit in the despiritualization of the world that we associate with science. By retreating into an ever-shrinking non-material realm of the spirit, or by flagrantly denying elementary scientific observations, religion has effectively ceded the material world to the science of Newton and Descartes. With spirit separate from matter and God separate from Creation, we are left impotent and alone in Fritjof Capra's "Newtonian World-machine".

After language and measure have labeled and quantified the world, and science made it an object, the next step is to turn it into a commodity. Chapter Four describes the vast consequences of the conversion of all wealth—social, cultural, natural, and spiritual—into money. Phenomena as diverse as the dissolution of community, the weakening of friendship, the rise of intellectual property, the shortening of attention spans, the professionalization of music and art, and the destruction of the environment have a common source in our system of money and property. This system, in turn, arises from (and reinforces) our self-conception as discrete and separate beings in an objective universe of others. And this self-conception manifests as usury. To simply try to stop being so greedy will never be enough, because selfishness is built in at an impossibly deep level. This selfishness, however, is not "human nature", but rather human nature denied, human nature contorted by our misconception of who we are.

The consequences of our fundamental misunderstanding of self and world, introduced in Chapter One, are portrayed in full flower in Chapter Five. Our opposition to nature and human nature, implicit in technology's mission to improve them, can only result in a "world under control." Manifesting in every realm, from religion to law to education to medicine, we maintain the world under control only at an ever-greater price. Helplessly, we respond to each failure of control with more of it, postponing but ultimately intensifying the eventual day of reckoning. As the social, cultural, natural, and spiritual capital of Chapter Four is exhausted, as our technology proves helpless to avert the impending crises, the collapse of the world under control looms closer. It is this collapse, which the present convergence of crises portends, that will set the stage for the Age of Reunion described in Chapter Seven.

While classical science presents the illusion of separation as fact, scientific developments of the last century have rendered the Newtonian world-machine obsolete. Chapter Six describes how the crumbling of the objective, reductionistic, deterministic worldview opens the door not just to a new mode of technology, but also to a spirituality that sees sacredness, purpose, and meaning as fundamental properties of matter. Part of our separation has been to see spirit as distinct from matter, either imposed from the outside by an extra-natural God, or a mere figment of our imagination. Assiduously avoiding New Age clichés about quantum mechanics, Chapter Six draws on recent developments in physics, yes, but also evolutionary biology, ecology, mathematics, and genetics. It lays the scientific groundwork for a reuniting of matter and spirit, as well the reuniting of man and nature, self and other, work and play, and all the other dualisms of the Age of Separation.

We are witnessing in our time the intensification of separation to its breaking point—the convergence of crises mentioned above that is birthing a new era. I call it the Age of Reunion. Chapter Seven portrays what life might look like no longer founded on the illusion of the discrete and separate self. Drawing on the new scientific paradigms of Chapter Six, it describes a system of money, economics, medicine, education, science and technology that seeks not the control or transcendence of nature, but our fuller participation in nature. Yet it is not a return to the past, nor a divestiture of the gifts of hand and mind that make us human. The Age of Reunion is rather a new human estate, a return to the harmony and wholeness of the hunter-gatherer but at a higher level of organization and a higher level of consciousness. It does not reverse but rather integrates the entire course of separation, which we may begin to see as an adventure of self-discovery instead of a terrible blunder.

Although I affirm the general, growing premonition of our civilization's impending crash, the enormous misery and ruination we have wrought is not in vain. Look at the New York City skyline, or a closeup of an integrated circuit board: Could it all be for nought? Could the incredible complexity, furious activity, and vast scientific knowledge of our civilization be merely, to paraphrase Shakespeare, "a sound and a fury, signifying nothing"? Following my intuition to the contrary, in Chapter Eight I describe what I believe to be the cosmic purpose of our "ascent" to the furthest reaches of separation. Drawing on religious, mythological, and cosmological metaphors, Chapter Eight puts the tides of separation and reunion into a vast context in which none of our efforts to create a world of wholeness and beauty, however doomed they seem right now, are futile, foolish, or insignificant.

Even in the darkest days, everyone senses a higher possibility, a world that was meant to be, life as we were meant to live it. Glimpses of this world of wholeness and beauty have inspired idealists for thousands of years, and echo in our collective psyche as notions of Heaven, an Age of Aquarius, or Eden: a once and future Golden Age. As mystics have taught throughout the ages, such a world is closer than close, "within us and among us". Yet as well it is impossibly far off, forever inaccessible to any effort arising from our present self-conception. To reach it, our present self-conception and the relationship to the world it implies must collapse, so that we might discover our true selves, and therefore our true role, function, and relationship to the universe.

This book exposes the futility, the fraudulence, and ultimately the baselessness of the program to control the world, to label it and number it, to categorize it and own it, to transcend nature and human nature. Thus exposed, that program will loosen its grip on us, so that we may let go of it before it consumes every last vestige of life and beauty on earth. The extensive scientific chapters are there to persuade you that the mechanistic, objective world of the discrete and separate self is not reality but a projection, merely the image of our own confusion.

The Ascent of Humanity is not merely another critique of modern society, and the solutions I explore are not along the lines of "we should do this" and "we shouldn't do that." Who the hell is "we"? You and I are just you and I. That is why so much political discourse (about what "we" must do) is so disheartening; that is why so many activists experience such despair, such despondency. You and I, no matter how much we agree with each other, are not the "we" of collective action, as in "we need to live more sustainably" or "we need to pursue diplomatic options." I find many people resonating with my intuition of a wrongness about life and the world as we know it, but their response is not empowered indignation, it is despair, helplessness, impotence. What can one person do? These emotions too are symptoms of the same separation behind all of our crises. When I am a discrete and separate individual, whatever I do makes little difference. But this logic is founded upon an illusion. We—you and I—are actually powerful beyond imagining.

Because the illusion of separateness is crumbling, the alternative I offer is practical, natural, and indeed inevitable. The ruin and violence of the present age do not typify an immutable "human condition". They originate in a confusion about self and world, a confusion embodied in our fundamental scientific and religious principles and applied in every aspect of modern life, from politics and economics to medicine and education. Social and environmental destruction is an inevitable consequence of this worldview, just as rejuvenation and wholeness have been, and will be, the consequence of a different worldview, one that has roots in primitive culture and religion, and that is the inescapable yet heretofore generally unrealized implication of 20th century science.

Our current self/world distinction, and its consequent parsing of all the world into discrete entities, has run the course of its usefulness as the dominant paradigm. Our individuation, as individuals and as a species separate from nature, is complete; in fact it is over-complete. What started with agriculture and even before, with pre-human gropings toward the technologies of stone and fire, has reached its outer limit. It has taken us far, this separation; it has fueled the creation of wonders. To the extent that the separation is an illusion and that we too are part of nature, that illusion has unleashed a new force of nature that has transformed the planet. But if the human gifts of hand and mind are natural too, then what happened to the "harmony, beauty, and authenticity" whose absence everyone can feel in the world of technology? Can we ever attain that human condition that we sense is possible in those moments of spiritual connection? This book will explore the extremes of separation we have reached, as well as the potential reunion that lies in the fulfillment, and not the abandonment, of the gifts that make us human.

CHAPTER I

The Triumph of Technology

Gee Whiz—The Future!

For at least 200 years now, futurists have been predicting the imminent rise of a technological Utopia, drawing on the premise that technology will free humankind from labor, suffering, disease, and possibly even death. Underlying this view is a defining story of our civilization: that science has brought us from a state of ignorance to an increasing understanding of the physical universe, and that technology has brought us from a state of dependency on nature's whims to an increasing mastery of the material world. Someday in the future, goes the story, our understanding and control will be complete.

At the dawn of the Industrial Revolution, it seemed obvious that the Age of Coal would usher in a new era of leisure. In one industry after another, a machine was able to "do the work of a thousand men". Soon the day would come when all work was mechanized: if a machine could do the work of a thousand men, then it stood to reason that each man would have only to work one-thousandth as hard.

As the Industrial Revolution progressed it soon became apparent that most people were doing more work, not less. True, the spinning jenny and power loom freed millions of women from the tedium of spinning their own thread and weaving their own cloth, but replaced that tedium with the horrors of the textile mill. Similarly, the steel foundry replaced the blacksmith's shop, the railroad car replaced the horse and cart, the steam shovel replaced the pick and spade. Yet in terms of working hours, working conditions, danger and monotony, the Industrial Revolution had not lived up to the promise encoded in the term "labor-saving device". The Age of Leisure, where coal-powered machines would do the work while people looked on and reaped the benefits, was going to arrive later than expected.

The futurists did not give up hope though—maybe they had only been premature. They hadn't realized that coal wasn't enough—it was the Age of Electricity that would finally usher in technotopia. Modern man would live in a paradise of electrified comfort. The spate of inventions that followed the harnessing of electricity made it obvious that we had the power to eliminate most forms of work (still largely associated with physical labor) and bring unprecedented leisure to the masses.

Almost no one doubted the power, the inevitability, and the desirability of technological transcendence of our natural limitations. The slogan of the 1933 World's Fair exemplified this attitude: Science Invents; Industry Applies; Man Conforms. The ascent of technology carries an aura of inevitability, destiny, and triumph. As John von Neumann put it, "Technological possibilities are irresistible to man. If man *can* go to the moon, he *will*. If he can control the climate, he will."[1] What fool would doubt it or stand in the way of progress?

In the decades after World War II, all signs pointed toward the impending triumph of technology. The 1940s and 1950s witnessed revolutionary innovations in medicine, including antibiotics and vaccines that (apparently) brought an end to the mass killers that had haunted civilization for centuries. Flush with victory, medical researchers confidently predicted the imminent end of all disease. Surely cancer, heart disease, and arthritis would succumb to modern medicine just as polio, smallpox, cholera, and plague already had. In agriculture, chemical fertilizers brought record harvests and the seeming promise of an unlimited cornucopia in the future, which would be protected from insect depredation by the new classes of pesticides such as DDT, lauded as nothing short of miraculous. Soon, it seemed, agriculture would no longer depend on nature at all,[2] as modern chemistry improved on the soil and modern breeding improved on the organism. Also around this time, atomic power offered the potential of virtually unlimited energy, electricity "too cheap to meter." Just as oil and coal had supplanted animal power, so would atomic energy increase our energy supply by several more orders of magnitude. And as the 60's drew to a close, space—the final frontier—also succumbed to human conquest, first with the orbiting of the earth and culminating with the moon landing of 1969.

Atomic Energy Commissioner Lewis Strauss summed up the vision nicely in 1954:

> It is not too much to expect that our children will enjoy in their homes electrical energy too cheap to meter, will know of great periodic regional famines in the world only as matters of history, will travel effortlessly over the seas and under them and through the air with a minimum of danger and at great speeds, and will experience a lifespan far longer than ours as disease yields and man comes to understand what causes him to age.[3]

Meanwhile, the horrors of the Industrial Revolution seemed to be in retreat—its hellish slums, child labor, disease epidemics, 16-hour workdays, and starvation wages. The blossoming new sciences of economics, psychology, and sociology promised to bring the same wonders to the social universe that the hard sciences had brought to the physical universe. The goal of a rational society, engineered for maximum happiness just as a machine is engineered for maximum efficiency, was just around the corner.

So we cannot blame ourselves for believing that technology would indeed usher in the Golden Age of humanity, would make us finally independent of nature, independent of suffering, independent, perhaps, even of death. All that was needed was to extend our victories a little farther, to make our understanding and control of nature just a little more precise. And perhaps, the faithful opine, nanotechnology and genetic engineering will finally allow us to achieve that precision, to control nature on the molecular level in the same way we already (ostensibly) control it on the macro level. As one technology evangelist puts it, "We would have an army of molecular robots and nanodevices that would allow us to completely dominate Nature. We now dominate it at a macroscopic level; we would then dominate it at a microscopic level too."[4]

The paradigm of ever-ascending understanding and control represents a fundamental myth of our culture, which I call the ascent of humanity. Its culmination would be the totalization of that understanding and control, the complete mastery of nature. The myth goes something like this: whereas in the beginning we were fully at the mercy of natural forces, someday we will transcend nature completely. We will control the weather; conquer old age, disease, and death; improve upon the cell and the gene; augment or replace the body with mechanical parts; download our consciousness onto computers; even leave nature behind entirely by colonizing space. Consider the following futurist ravings:

> The systematic application of nanotechnology, self-reproducing micro-miniaturised robots armed with supercomputer processing power, and ultra-sophisticated genetic engineering, perhaps using retro-viral vectors, will cure the root of all evil in its naturalistic guise throughout the living world. And once the pain has gone, with the right genes and designer drugs there's no reason why life shouldn't just get better and better...."[5]

> In the near future, a team of scientists will succeed in constructing the first nano-sized robot capable of self-replication. Within a few short years, and five billion trillion nano-robots later, virtually all present industrial processes will be obsolete as well as our contemporary concept of labor. Consumer goods will become plentiful, inexpensive, smart, and durable. Medicine will take a quantum leap forward. Space travel and colonization will become safe and affordable.[6]

The above quotes are from the far margins of futuristic thinking, but the underlying attitudes are alive and well, to wit: (1) that the answer to our problems lies in new technology; (2) that progress consists of increasing our control over nature; and (3) that someday our control over nature will be complete, or at least far greater than it is today, enabling the conquest of disease, reduction of work, lengthening of life-span, space travel, and so forth. As recently as the 1970s and 80s, futurists like Alvin Toffler were writing that the greatest challenge facing society in the year 2000 would be how to use all of our leisure time. Today, analyses of the future of retirement routinely assume that people will be living longer and, thanks to medical technology, will enjoy greater health into their later years. Every day we hear about "advances" and "progress", and although these words no longer bear the magical cachet they once did, we still wonder with anticipation what the next revolution in medical, information, or entertainment technology will be. Especially pronounced in magazines like *Wired!*, *Discover*, and *Scientific American*, the "Gee Whiz!" attitude about the future is everywhere, an ideology of progress written into our fundamental beliefs. What will the next wonder be? Where will Moore's Law take us next?[7] Naïve on the surface, the extreme opinions quoted above are merely distillations of a pervasive cultural myth: that we are on the way toward fulfilling our destiny of rising above nature.

That the words "The Ascent of Humanity" reverberate with a religious connotation is not surprising. Where else do we find the idea that our present age of suffering is only a temporary stage on the way to some perfect state of future existence? The myth of technological Utopia is uncannily congruent to the religious doctrine of Heaven, with technology as our savior. Thanks to the god Technology, we will leave behind all vestiges of mortality and enter a realm without toil or travail and beyond death and pain. Omnipotent, technology will repair the mess we have made of this world; it will cure all our social, medical, and environmental ills, just as we escape the consequences of our sins of this life when we ascend to Heaven.

This, in a nutshell, is the ascent of humanity that Jacob Bronowski was referring to in his classic *The Ascent of Man*, after which the present volume is ironically named. It is an ascent from the depths of superstition and ignorance into the light of scientific reason; an ascent from fear and powerlessness in the face of natural forces to the mastery of those forces. A myth is a story that provides a template for understanding ourselves and our world; as well it is a program that guides our choices and priorities. Accordingly, I will distinguish the myth of ascent into two aspects: the Scientific Program of complete understanding and the Technological Program of complete control.

Together, the Scientific Program and the Technological Program form a defining myth of our civilization. The two are intimately related: technology, the way we control the world, arises from science, the means by which we understand and explain it. Technology in turn provides the means for science to probe even more deeply into the remaining mysteries of the universe. Technology also proves the validity of science—if our scientific understanding of the world were no better than myth and superstition, then the technology based on that science wouldn't work.

Philosophers of science will protest that it is already well-established, even in conventional circles, that perfect knowledge and perfect control of the universe is probably impossible (due to such things as mathematical incompleteness, quantum indeterminacy, and sensitive dependence on initial conditions). Be that as it may, this information has yet to filter down to the level of popular consciousness, even among scientists. What I am talking about is the faith encapsulated in the saying, "Science will surely explain it someday." It is the faith that the answer is there, the answer is accessible to science, and that science itself is well-grounded in its primary principles and methods. The technological corollary to this faith in science is our faith in the technological fix. Whatever the problem, the solution lies in technology—finding a way to solve the problem. Science will find an answer. Technology will find a way.

The Myth of Ascent

The Technological Program	The Scientific Program
Starting with stone tools and fire, technology gives us increasing control over nature, insulates us from her whims, and provides us with safety and comfort.	By making methodical observations of the universe and creating and testing theories, we replace myth and superstition with a growing body of objective knowledge.
Past	**Past**
We had very little control over the physical environment. We were at the mercy of nature, barely surviving before technology came to the rescue.	We had very little understanding of the laws of the universe, so we resorted to myth and superstition in a vain attempt to explain the world.
Present	**Present**
Although there are still many problems to be solved, we have made great strides in our ability to engineer and control nature. We have conquered many illnesses, reduced the hardship of survival, moved mountains and drained lakes, augmented the processing power of our own brains with computers.	While there are still many things about the universe that we do not understand, we have discovered at least the basic framework of how the universe works: the laws of gravity, quantum mechanics, evolution, and so forth. We can explain most of the phenomena we observe, and we have plausible theories about the rest. Myth and superstition have no place.
Future	**Future**
Someday our control over nature will be complete. We will prolong human life indefinitely, eliminate pain and suffering, eliminate labor, travel to the stars and leave earth behind entirely.	Someday our understanding of nature will be complete. We will formulate a "Theory of Everything" that combines relativity and quantum mechanics into a single equation, and apply that theory to explain any and all observed phenomena. There will be no more mysteries—even the workings of the human brain will be fully understood according to scientific principles.

Underlying the Technological Program is a kind of arrogance, that that we can control, manage, and improve on nature. Many of the dreams of Gee Whiz technology are based on this. Control the weather! Conquer death! Download your consciousness onto a computer! Onward to space! All of these goals involve controlling or transcending nature, being independent of the earth, independent of the body. Nanotechnology will allow us to design new molecules and build them atom by atom. Perhaps someday we will even engineer the laws of physics itself. From an initial status of subordination to nature, the Technological Program aims to give us mastery over it, an ambition with deep cultural foundations. Descartes' aspiration that science would make us the "lords and possessors of nature" merely restated an age-old ambition: "And God said to them, Be fruitful and multiply, and fill the earth and subdue it; and have dominion over the fish of the sea and over the birds of the air and over every living thing that moves upon the earth" (Genesis 1:28).

Yet a contrary thread runs concurrently through the world's religious traditions, a recognition of the hubris of our attempt to improve on nature. Greek mythology has given us the figure of Daedelus, who arrogated to himself the power of flight in violation of ordinary mortal limitations. The power to transcend nature's limitations is for the gods alone, and for his temerity Daedelus was punished when his son, Icarus, soared too high in his desire to attain to the heavens. In the Bible we find a similar warning in the Tower of Babel, a metaphor for the futility of reaching the infinite through finite means. Have we not, through our technology, attempted to rise above nature—sickness, uncertainty, death, and physical limitation—to attain to an immortal estate?

Utopia Postponed

The 1960s were in many ways the summit of our civilization. We had beaten polio, smallpox and plague. Surely cancer and the rest would succumb in due course. We had beaten the Nazis. Surely the Commies were next to go. Social problems like poverty, racism, illiteracy, crime, and mental illness would be engineered out of existence. Everything pointed to unlimited growth and continued triumph: atomic power, robots, space, artificial intelligence, maybe even immortality. But in the words of Patrick Farley, the future has been running a little behind schedule.[8]

Hints that technology was not the vehicle of Utopia began to emerge early in the Industrial Revolution, but its successes were so spectacular that it was easy to believe that social and environmental problems were merely temporary impediments, engineering challenges we would overcome through the same methods, mindsets, and techniques that had solved previous problems: more technology, more control. Today the successes are less spectacular, the crises harder to deny, the promise of Utopia "just around the corner" more hollow, but we still act as though more control were the answer.

For example, the medical establishment is having more and more trouble hiding the fact that, with the sole exception of emergency medicine, the last forty years of "advances" have had little impact on human health and mortality. Consider the overall effect of the successes. Organ transplants were a real breakthrough, but their effect is limited to a few thousand patients annually. Premature babies have much higher rates of survival—but far more babies are born prematurely.[9] Most of the new pharmaceuticals merely control symptoms, often with severe side-effects. Hormone replacement therapy is turning out to be a disaster; the same is true for cholesterol-lowering medication, anti-depressants, and many over-the-counter drugs. When a new medicine is unveiled, a "twenty percent improvement in outcomes among a significant group of patients" is considered a great success. Despite decades of huge investment, it appears that the age of dramatic cures is over. There has been no "cure" for any of the poster-child diseases such as muscular dystrophy and breast cancer. Certainly no major disease has been wiped out since we conquered the great killers of the nineteenth century. Coronary artery disease has retreated little, if at all, in thirty years. Cancer is doing just fine, thank you. Arthritis is just as devastating as ever, strokes nearly as common, Alzheimer's disease on the rise. Meanwhile a host of formerly uncommon conditions, for which conventional medicine can offer palliative remedies at best, have grown into epidemics: diabetes, autism, allergies, multiple sclerosis, lupus, obesity, chronic fatigue, fibromyalgia, multiple chemical sensitivities, inflammatory bowel syndrome, chronic fungal infections, and many others. Not only have formerly rare diseases become epidemic, but entirely new diseases such as AIDS have appeared seemingly out of nowhere. Finally, to add insult to injury, some of the "conquered" diseases of the past such as tuberculosis seem to be making a comeback, usually due to antibiotic resistance. This state of affairs constitutes a great unspoken crisis in medicine. Despite unprecedented billions of dollars in pharmaceutical research, medicine seems to be losing ground in the "battle against disease." Typically, the response is more technology, more precise control at the genetic and molecular level. A continued search for the "cure".

Life expectancy has similarly failed to live up to predictions. For half a century now, futurists have been predicting dramatic increases in life expectancy: there is no reason why 120 years shouldn't be common; perhaps with gene therapy this could be extended indefinitely. A glance at the statistics, however, shows that the most dramatic gains in life expectancy all occurred in the first half of the 20th century, not the second. From 1900 to 1950, life expectancy at birth rose by an impressive 21 years; since then it has risen only 9 years.[10] Moreover, much of this improvement must be due to lower infant mortality and emergency life-saving procedures, because if we look at life expectancy at age 65, we find a paltry 4-year increase over the past half-century.[11]

Gains in life expectancy show the same familiar pattern of diminishing marginal returns that we see in the agricultural application of fertilizer. The first application of technology (fertilizer) brings dramatic results, but subsequent applications have less and less benefit until enormous amounts are eventually required to boost yields a tiny bit, or even to prevent yields from falling. As burgeoning healthcare expenditures demonstrate, we are presently pouring enormous technological effort into healthcare with only tiny gains compared to the dramatic improvements that accompanied the comparatively modest expenditures of the early 20th century. We can expect that unless the fundamental direction of medical technology changes, life expectancy will stagnate and probably even begin to fall within ten years.[12]

Nor has technology lived up to its promise to usher in an age of leisure. In the United States, leisure time did seem to be increasing throughout the 20th century until about 1973, when it began a gradual, sustained decline. Most researchers agree that leisure time has decreased in the thirty years since then: we are spending more time working, more time commuting, more time running errands, more time meeting the obligations of life.[13] The computer, trumpeted as the final key technology that would do for the drudgery of mental labor what machines had (supposedly) done for physical labor, has brought about the opposite: more time spent in offices, at desks, at keyboards. By now it is apparent that the computer has not eliminated the drudgery of office work, any more than the steam engine eliminated the ordeal of physical labor. Despite the "information revolution", few people would argue that office work has become more intellectually stimulating or meaningful in the last thirty years. The solution? Again it is more technology, more labor-saving devices, greater efficiency, better "time management".

Technology has similarly failed to bring about a world of plenty. While the food supply has indeed grown enough to feed a doubled world population, hunger and famine are no less prevalent, and for the same old political and environmental reasons: war, repression, and drought. Moreover, vast areas of land that were once agriculturally productive have turned to desert, so that we face the prospect of a food crisis, not a cornucopia. (At the present writing, world grain stocks are in decline,[14] and most of the great fisheries nearly depleted.) The last forty years of "development" in the third world have not brought the promised prosperity. To the contrary: the gap between rich and poor has widened globally and within third-world countries themselves. Travel in the Third World and you can easily see for yourself that the destitution, disease, and dislocation of our own Industrial Revolution are still common today. And the justification is the same too: It is just a temporary phase before you become a "developed" nation like us. We had to go through it, and so do you; it is the natural order of things. The prescription, then, is again more of the same, further "advances" in agriculture and more vigorous economic development. Now, people around the world are starting to disbelieve this dogma, simply because the decline in third-world living standards has continued unabated for too long. That wasn't supposed to happen. In many Latin American countries, the middle class has almost disappeared altogether.

In space, the triumphs of the 1960s and 1970s never led to the space colonies, Mars landings, and interstellar space travel that were confidently predicted by the year 2000. There have essentially been no significant advances in propulsion technology since the rocket, developed some seventy years ago. When I was growing up in the early 1970s, space fever gripped the minds of all my contemporaries: we had space board games, space lunch boxes, even, I recall, rocket-shaped shampoo bottles. We landed on the moon; then we did it again. And again. We have not returned to the moon since the 1970s, however, and there is little enthusiasm for such a mission today. Been there, done that... where has it led? At the time of this writing, President Bush has just proposed a new drive to establish a permanent moon base and a manned mission to Mars, yet there is not even a shadow of the excitement that enthralled the nation in the days of my early childhood.

The age of leisure and easy plenty, technotopia, is forever just around the corner. First it was the Age of Coal that was supposed to free us from labor: in the dawning Golden Age of the 19th century, coal-fired steam-driven machines would do all the work. Instead we got the sweat-shop, the coal mine, the foundry, the Satanic mills and Stygian forges (no idle metaphor, these), the eighty-hour work week, child labor, industrial accidents, starvation wages, fabulous wealth alongside wretched slums, childhoods spent in coal mines, horrific pollution, shattered communities and ruined lives. But not to worry! The Golden Age was just around the corner, thanks to electricity! Chemistry! The automobile! Nuclear power! Rockets! Computers! Genetic engineering! Nanotechnology! Unfortunately, none of these ever quite lived up to their promise.

And now we are in the 21st century, which was supposed to have been the Age of Leisure, the Information Age, the Knowledge Economy. In the latter phrase some key prejudices of the ascent myth are laid bare. It implies a progression from the industrial age, mired in materiality, to a separate, exclusively human realm of pure knowledge. The base concerns of material production were to be left to less advanced countries; our society was to have risen above that to deal in the products of the mind. Eventually, with the perfection of robotics, all societies were to follow us there.

The aspiration to rise above materiality defines modern religion as much as it does modern economics and modern technology. This is no accident. All arise from a common source that I will discuss throughout this volume. All are variations on the theme of ascent, the ascent of humanity. "Materiality", after all, is just a pejorative word for nature, and we identify the ascent of humanity with a progressive transcendence of nature. Once nature's slave, now its master. So of course it is higher, better, more *ascended* to be in the realm of the mind than the realm of base material production.

This is why occupations such as "executive" and "consultant" carry a cachet absent from "industrial engineer" and "plumber". For the last twenty years or more, young people have aspired to such roles without even caring what their actual subject matter is. They major in business, marketing, and finance, hoping to be an "executive" somewhere, anywhere. Part of the reason is the wealth and status that accord to such occupations, but a deeper principle is at work, too: the separation of spirit and matter, mind and body, human and nature that is as old as civilization. From the first social division of labor, prestige went to those whose hands were not sullied by the dirt—the soil of farming at first, but eventually the entire material world. Thus it was that ancient kings' feet were not allowed to touch the earth. Today's knowledge worker was supposed to be the consummation and democratization of that trend. Every man a king.

The bankruptcy of the ambition encoded in the words "knowledge economy" is now becoming plain. Office work is no less tedious than that of the assembly line or vegetable monofarm—and for the same systemic reasons of standardization and mass scale I will describe in Chapter Two. Much of today's knowledge economy consists of data input. Furthermore, the recent migration of "knowledge-intensive" jobs such as engineering and computer programming to new industrial powerhouses such as India and China demonstrates that the realms of mind and matter are not so separate as we might wish to think.

The promise of Utopia just around the corner to justify today's sacrifices is a common thread connecting every application of the Technological Program. We saw it in the Age of Coal, we see it in the Computer Revolution today: We must undertake the vast project of inputting all the data; then computers will run everything much more efficiently. We see it in the Third World in the IMF's austerity programs, which call for sacrifice today to bring prosperity tomorrow. IMF policies are often criticized as instruments of globalization that benefit the already-wealthy, but their systemic necessity springs from a much deeper source than that. Sacrifice is a built-in feature of any capitalist system based on interest-bearing money: sacrifice now to scrape together the money-that-breeds-money. Even more fundamentally, it is a defining mindset of agriculture, in which we must sow today in order to reap tomorrow. The same mentality affects religion, which calls upon us to sacrifice worldly pleasures for the sake of a hypothetical future Heaven. The problem with all this is that, whether in the Third World or in the endless task of data input, the sacrifice seems to be perpetual. Heaven never comes. Speaking from the bowels of the Industrial Revolution, William Wordsworth said it best:

> With you I grieve, when on the darker side
> Of this great change I look; and there behold
> Such outrage done to nature as compels
> The indignant power to justify herself....

Then, in full many a region, once like this
The assured domain of calm simplicity
And pensive quiet, an unnatural light
Prepared for never-resting Labour's eyes
Breaks from a many-windowed fabric huge;
And at the appointed hour a bell is heard—
. . . A local summons to unceasing toil!. . . .
Men, maidens, youths,
Mother and little children, boys and girls,
Enter, and each the wonted task resumes
Within this temple, where is offered up
To Gain, the master idol of the realm,
Perpetual sacrifice.[15]

Perpetual sacrifice. It is an ideology that invades nearly every aspect of our lives. What is being sacrificed? What is the common thread? Most fundamentally, it is a sacrifice of the present for the future. Cut back today so you will have enough for tomorrow. Work comes before play. No pain, no gain. Control yourself. Whether it is in diet, education, or personal development, we find the same sad prescription. Why is it that for so many people, the Heaven of physical fitness, or financial independence, or cessation of an addiction remains forever just as distant as technological Utopia? How long do your New Year's resolutions last? Well, try harder. It is like the man who decided to walk to the horizon, and failing to get there, concluded that he needed to run instead. This book will uncover the origins and evolution of the regime of perpetual sacrifice that we have endured in our attempt to build a tower to Heaven.

Because the exhilarating "Gee Whiz!" aspect of technology has failed to deliver the futuristic wonderama we all expected in the 1960s, the dark side of technology has become more difficult to ignore. Certainly there has been ample evidence for centuries that technology is not an unqualified good, but until the twentieth century the ideology of progress dominated all but the most independent thinkers. The horrific conditions of the Industrial Revolution could be explained as merely a temporary sacrifice on the way to Utopia. Only a few romantics had the vision to resist this ideology. People like Wordsworth, William Blake, Lord Byron, Henry David Thoreau, and Mary Shelley saw the ruination within mass industrial society not as a temporary phase or an engineering challenge, but as its fundamental character.

All this started to change in 1914, when the world finally got to see the result of industrialization applied to warfare: battlefield carnage on the mass scale of industry, a whole generation of young men decimated. Twenty-five years later, the carnage returned to encompass entire civilian populations in the conflagration of total war, ending in the first application of the century's greatest scientific triumph: the atomic bomb. At the same time, the organizational principles of the industrial revolution, based on the same scientific tools of analysis and control, reason, logic, and efficiency, were applied to the purposeful mass extermination of innocent people under Hitler, Stalin, and their imitators.

Ironically, it was precisely these principles of logic, reason, and efficiency that were supposed to elevate humanity to a more noble state, just as the technologies of physical and chemical engineering—used in the world wars—were supposed to elevate humanity to a new level of material comfort, health, and security. The irony was not lost on artists, writers, and other cultural sensitives, who have been grappling with the resulting feelings of betrayal and despair ever since.

From Plato onward, Utopian philosophers thought that reason, planning, and method would bring the same progress to the social realm as material technology brought to the physical. Social planning would conquer the wilderness of human nature, just as technology subdued the wilderness of physical nature. The failure of both is seen merely as evidence that we need more of the same. The ambition of nanotechnology, to extend physical control to a new level of microscopic precision, parallels the social technologies of education and law as they strive toward ever-finer regulation of human behavior.

Underlying both material technology and sociopolitical methods of control we find the same conceptual foundation. Is it mere accident that from this foundation, the same decimation has visited both the human and natural realm? There is a flaw in the common position that technology is neutral, up to us to use for good or for evil. The pogroms and the genocides, the ethnic cleansings and wars of extermination, the despoliation of the planet and the wrecking of indigenous cultures, all these are attributed to the misuse of technology, not technology itself. But perhaps this position is mistaken. Perhaps something basic to the very mindset of technology has generated the twin crises in the social and environmental realm.

In some quarters the faith in technology continues. Ozone layer destruction? We'll make new ozone. Soil erosion? We'll find a way to grow food without soil; maybe we'll just synthesize it. Total environmental collapse? No matter, we'll colonize the stars. Who needs nature anymore? The can-do spirit that has brought us this far will surely overcome any future obstacles. Human ingenuity is unlimited. If things seem to be getting worse, not better, if people seem to be getting sicker, busier, and more anxious, if life seems more stressful and the environment less healthy—rest assured! This is a temporary sacrifice, one step backward necessary to take a giant leap forward.

Today, though, the rhetoric of progress is wearing thin. It looks as though the future, always just around the corner, is never going to come. Since the mid-20th century, that feeling of betrayal and despair has spread beyond artists and intellectuals to engulf the general population. Superficially, many people still affirm that the onward march of technology will someday render all our present problems obsolete, but on a deeper level they have lost confidence in both science and technology. The long-promised marvels, the next step in our transcendence of nature, have failed to materialize, while new and unforeseen problems multiply faster than we can solve them. Gone is the Sixties optimism that sparked the War on Poverty, the War on Cancer, the Conquest of Space. Now we hope merely to stave off the problems that threaten to overwhelm us: the convergence of crises in the environment, health, education, the economy, and politics.

The Addiction to Control

Whereas technology once promised a grand future of leisure and security, today we need intensifying doses of it merely to keep the world from falling apart. A pattern of diminishing marginal returns seems to have infiltrated all areas of technology, whether material or social. Early in the twentieth century, modest expenditures in medical research brought enormous improvements in lifespan; today vast outlays barely succeed in maintaining present standards. In agriculture, small amounts of chemical fertilizers and pesticides once brought huge increases in crop yields; today, ever-greater chemical input can hardly prevent yields from falling, despite "improved" varieties. In daily life, inventions such as cellular phones, personal digital assistants, convenience foods, and the Internet barely enable us to keep pace with the ever-quickening pace of modern life.

Recently I had a conversation with a long-time Washington D.C.

native who was recalling the building of the Beltway back in the 1960s. Everyone was excited because you'd be able to circumnavigate the whole city in just an hour. D.C. was starting to have traffic jams, and the Beltway would usher in a new era of ease and convenience. Well, everyone knows what happened. The new road facilitated new real estate development and encouraged people to use cars rather than public transport. Soon the beltway was jammed. The solution? Make it wider and add even more roads. Of course that caused even more development and congestion. The immediate engineering solution—more roads to accommodate an excessive car-to-road ratio—worsens the problem in the long run. That is a classic example of a technological fix. Technology usually has unintended consequences, often including, as in this case, a worsening of the problem the technology was supposed to solve. Generally speaking, unintended consequences are not the result of sloppy engineering, lazy planning, or lack of diligence; they cannot be eliminated through tighter control; rather, they are built into the very attempt at control.

By now this pattern of escalating dosage for a diminishing effect may remind you of another meaning of the word "fix"—a drug fix. Our dependency on technology shares many features in common with drug addiction. Returning to the example of agriculture, once we've killed the natural predators, lost the topsoil, and depleted the minerals, we cannot grow crops at all without repeated applications of more and more technology. Each fix brings some temporary improvement, but then crop yields start falling and we need another fix. At this point we're hooked: if we go back to zero fertilizer, crop yields fall way below the original pre-fertilizer level. Eventually, the soil is so damaged that no amount of fertilizer can coax life from it. The parallel with the course of addiction is uncanny: escalating dosage to get a less and less intense high, followed ultimately by complete desolation.

The history of life expectancy is another example. The "dosage" of technology must go up and up, at a greater and greater expense to the rest of life, in order to achieve diminishing returns. Eventually, addicts tell us, huge doses of the drug are needed to even feel just normal. In parallel, huge medical expenditures are needed to keep people functioning at all. Half of American adults take some form of prescription medication[16]; the average senior citizen takes between two and seven medications per day.[17]

At the beginning of Terry Gilliam's dark science-fiction film, *Brazil*, the main character's aunt has gotten some minor plastic surgery to fix a blemish on her face. We see her with a little bandage. The next time we see her, there are two or three larger bandages, because there were complications from the initial surgery. The next time, bandages cover most of her face, because she had new surgery to fix the complications of the second surgery, which was to fix the complications of the first. By the end, her whole head is swaddled in bandages. Each time she says something like, "It's almost perfect" or, "The doctors tell me it will just be a matter of one or two more procedures." A series of incremental improvements ends up in total ruin.

Why is the technological fix so attractive? Because from the short-term perspective, it really does work. The first digging stick really did make it easier to obtain roots. A cup of coffee really does make us feel energized. A good stiff drink really does make the pain go away. Air conditioning makes us feel cooler on a hot day. Cars get us there faster. Fertilizer boosts the yield. With each stage of construction, the Tower rises higher. See, it's working! We're getting closer to the sky.

Invisible at first is the fact that the fix is a trap. At the end of the day, the coffee exhausts our adrenal glands and makes us more tired, not less. The air conditioning habituates us to a narrow range of comfort, trapping us indoors. Cars inevitably bring more roads, more cars, and more time in transit. Food production technology brings population increases, and eventually less security and more anxiety.

Ultimately, even the immediate efficacy of the fix is diminished. The problems it once ameliorated grow to overwhelming proportions. Today, new technology can barely keep pace with the acceleration of modern life, the proliferation of new threats, new diseases, and new uncertainties. Eventually, the alcoholic becomes so sick that each drink causes more pain than it removes.

The principle of diminishing marginal returns that characterizes the technological fix was explored by archaeologist Joseph Tainter in his classic work, *The Collapse of Complex Societies.*[18] Tainter says that a society's investments in complexity bring fewer and fewer benefits, until its maintenance alone consumes all resources. Bureaucracies, legal systems, technological systems, and complex divisions of labor solve a society's immediate problems and achieve dramatic initial returns, but come with hidden costs. These costs may be exported into the future, delayed in their manifestation through growth and conquest. Eventually, though, in a pattern that has repeated itself from ancient Sumer to Rome and now to the American Empire, the society collapses under the weight of the structure it has erected. First the burden grows heavy; then one crisis after another is barely averted. Wars over resources break out, the leadership degenerates into corruption, the environment deteriorates, and finally one or another crisis, something the civilization could have easily overcome in its youth, deals the final blow. The society "collapses" to a state of much diminished complexity.

The Babel story offers an allegory for this process. The organizational overhead required to manage an increasingly complex project manifests as a growing confusion, an inability to communicate across the vast range of specializations and subsystems that need to be coordinated. In the Bible story, the builders find themselves speaking different languages, unable to communicate, unable to unite in a common task—a situation eerily prescient of the specialized jargon that separates each scientific and professional field and stalls meaningful progress. In the story, the Tower is eventually abandoned. In my mind's eye I picture its abandonment preceded by frantic attempts to shore it up, to repair the proliferation of cracks and cave-ins that foretell its ultimate collapse. After the initial rapid progress slows and eventually stalls, the ambition of reaching the sky becomes a mere dogma, an ideology that no one believes. Such is our attitude toward the technological Utopia. No one believes it anymore. Indeed, it consumes all of our effort to even maintain the Tower at its present height. Even as we make an addition here and there, other parts crumble, and a spreading infirmity undermines its very foundation.

The parallel to the life of an addict is uncanny. Easy to maintain at the beginning, the addiction soon demands increasingly complex structures to support it. The addict sacrifices long-term for short-term benefits, establishes webs of deceit that must eventually fail, and devotes more and more resources to maintaining the addiction. Get my fix today; deal with the consequences tomorrow. The consequences build and build, the burden on life grows, and eventually the whole fragile structure collapses. Just as the immediate cause of the collapse varies from person to person, so we must look beyond the proximate causes for the collapse of civilizations as well. On one level, yes, it will be the energy crisis, or an economic depression, or a military defeat, or an environmental crisis, or a combination of these, or something totally different that will end our civilization. The immediate cause is impossible to predict, but the end result is inescapable.

While hints of the built-in failure of the Technological Program have been nagging civilization for thousands of years, it is only in the present era that they are becoming undeniable and inescapable. In the past, the effects of ecological destruction were localized: the rich and the lucky could always move somewhere else (which is, in itself, a kind of temporary fix as well). Today, as ecosystem collapse becomes global, there is no longer a "somewhere else". There is nowhere to go. Even when people retreat into a fortress mentality, our systemic social and environmental problems find a way in.

In any addiction, the fix appears to work beautifully at first: a servant of life, an easer of pain, coming at a manageable cost. At first the sacrifices seem worth it, cast into some corner to be dealt with later. But sooner or later the cost grows to such proportions as to engulf the whole of life, even as its power to numb the pain diminishes.

The technological fix puts off the problem to the future, just as a drinking binge puts off until tomorrow the problems of life. No longer. The future is now, and it will not be put off much longer. "The future" of the Technological Program is one where all the problems are solved once and for all. Here and now, though, we are waking up to another kind of future, and with a hangover to boot: vomit on the floor, the apartment trashed, the world a mess.

Just as any personal addiction inexorably unravels the fabric of the addict's family, friendships, work and indeed all relationships to the world, so has our technological addiction progressively destroyed our natural and social environment. And as with other addictions, before technology's glittering promise began to fade, such destruction was easy to ignore. The appalling pollution of the nineteenth century was actually more destructive to human quality of life (though more localized) than pollution today, but was easily dismissed as a temporary problem, a cost of progress that would inevitably be solved through more technology. Today, at least among the affluent, the effects of pollution are more distant, more subtle, and certainly less easily attributable to a single specific cause, but also more systemic and more a threat to the whole of the planet. From ozone layer destruction to global warming to the PCBs in every living cell[19], the destruction today is pervasive, inescapable.

The inescapability of the present crises is demolishing the fundamental illusion beneath the course of separation. As long as we believe ourselves to be discrete beings fundamentally separate from the environment, then in principle there is no limit to our ability to insulate ourselves from the degeneration of the social and natural environment. The world is an Other, and its suffering has nothing to do with me, provided

I am skillful enough in insulating myself. Today, as the wreckage proliferates, its effects become increasingly difficult to manage. The habitual response is to try harder: to invent new technology to clean up the problems of the old, to insulate ourselves still more skillfully from the mess. But as this becomes impossible, as burgeoning crises overwhelm us, another possibility emerges: to abandon the program of insulation and control, and the conception of the separate self on which it rests.

The process of addiction recovery described by the Twelve Steps program offers a parallel. The first three steps boil down to something like this: "We realized that we were powerless over our addiction, that our lives had become unmanageable. Therefore we made a decision to give our lives and our will over to a higher power." In the context of technology, the first sentence above amounts to an admission of the failure of the Technological Program. It is the realization that the more we try to manage and control nature, the more unmanageable and uncontrollable our problems become. The second sentence is a statement of surrender to and trust in that which is beyond ourselves. The religious content of the Twelve Steps testimony translates in this context to a transcendence of the limiting and delusionary conception of self implicit in our physics and biology, economics and politics, philosophy and religion.

The way we relate to the world is written into our most basic mythos, our cosmology, our ontology—belief systems that *underlie* the superstructures of science and religion. It is our fundamental beliefs about who we are and about the nature of the universe that have generated human life as we know it, and the world as we experience it. If these beliefs remain unchanged, then unchanged as well will be the direction they take us. Our despair, then, is justified. Technology as we know it, and with it the program of control, will never fulfill the promise of the ascent of humanity. But herein also lies a great hope, because from despair comes surrender, and from surrender comes an opening to new beliefs, a new conception of self and world. From this might come a new way of relating to the world, a new mode of technology no longer dedicated to the objectification, control, and eventual transcendence of nature.

The collapse we are facing is of more than "our civilization" but of *civilization* itself, civilization as we know it. It is a collapse of a whole way of relating to the world, a whole way of being, a whole definition of self. For at the root of the technological addiction is our own off-separation from the universe, our self-conception as discrete and separate beings that goads us toward control. The disintegration of historical civilizations was a mere preview, the diffraction back onto history of the archetypal collapse that is overtaking us today.

What drives our addiction to technology? Underneath all addictions there is an authentic need that the addiction promises to meet. The narcotic says, "I will kill the pain." But of course, the promise is a lie that leaves the true need unmet. The same goes for technology, driven by the imperative to control nature, which itself comes as well from an unmet need. It is a need that we all feel in different ways: as an anxiety endemic to modern life, as a near-universal feeling of meaninglessness, as a relentless ennui from which we can only ever be temporarily distracted, as a pervasive superficiality and phoniness. It is a feeling that something is missing. Some people call it a hole in the soul. What we are seeking in our technological addiction is nothing less than our lost wholeness, and its recovery is what lies on the other side of the imminent collapse of the regime of separation.

From Separation to Boredom

The ascent of humanity has come at a price, and I am not speaking here merely of the destruction of the ecological basis of human civilization. Our separation-fueled ascent exacts its toll not just on the losers, the victims of our wars, industry, and ecocide, but on the winners as well. It is the highest of all possible prices: it comes out of our very being. For all we have built on the outside, we have diminished our souls.

When we separate ourselves from nature as we have done with technology, when we replace interdependency with "security" and trust with control, we separate ourselves as well from part of ourselves. Nature, internal and external, is not a gratuitous though practically necessary other, but an inseparable part of ourselves. To attempt its separation creates a wound no less severe than to rip off an arm or a leg. Indeed, more severe. Under the delusion of the discrete and separate self, we see our relationships as extrinsic to who we are on the deepest level; we see relationships as associations of discrete individuals. But in fact, our relationships—with other people and all life—define who we are, and by impoverishing these relationships we diminish ourselves. We *are* our relationships.

"Interdependency", which implies a conditional relationship, is far too weak a word for this non-separation of self and other. My claim is much stronger: that the self is not absolute or discrete but contingent, relationally-defined, and blurrily demarcated. There is no self except in relationship to the other. The economic man, the rational actor, the Cartesian "I am" is a delusion that cuts us off from most of what we are, leaving us lonely and small.

Stephen Buhner calls this cleavage the "interior wound" of separation. Because it is woven into our very self-definition, it is inescapable except through temporary distraction, during which it festers inside, awaiting the opportunity to burst into consciousness. The wound of separation expresses itself in many guises, ranging from petty but persistent dissatisfactions that, when resolved, quickly morph into other, equally petty dissatisfactions in an endless treadmill of discontent, to the devastating phthisis of despair that consumes vitality and spirit.

Riding any vehicle it can, the pain from the interior wound manifests in a million ways: an omnipresent loneliness, an unreasonable sadness, an undirected rage, a gnawing discontent, a seething resentment. Unaware of its true source, we assign it to one or another object, one or another imperfection in the outside world. We then seek to forestall the pain by suppressing its vehicles: getting life under control. In a personalized version of the Technological Program, we identify happiness with the maximum possible insulation from danger, dirt, and discomfort. But of course, this insulation cuts us off even further from the world and, so, exacerbates the separation that is the actual source of the pain.

A saying goes, "Seek not to cover the world in leather—just wear shoes." It is a spiritual cliché that happiness is not to be found by engineering the world to make everything goes your way: such happiness is transient, doomed. But that's the way we act, culturally and individually, much of the time. Someday, everything will be perfect and we'll be able to relax and be happy forever.

The futility of the personal and collective Technological Program of complete control finds incontrovertible demonstration in the phenomenon of boredom, which shows us the human condition when the Technological Program *succeeds*. What is the ground state, the default state of the human being when everything is under control, when no personal calamity imminently threatens? What happens if we just sit here, with nothing to do and nothing that needs to be done?

Boredom is so endemic to our culture, particularly among youth, that we imagine it to be a near-universal default state of human existence. In the absence of outside stimuli we are bored. Yet, as Ziauddin Sardar observes, boredom is virtually unique to Western culture (and by extension to the global culture it increasingly dominates). "Bedouins," he writes, "can sit for hours in the desert, feeling the ripples of time, without being bored."[20]

Whence comes this feeling we call boredom, the discomfort of having nothing to occupy our minds? Boredom—nothing to do—is intolerable because it puts us face to face with the wound of separation. Boredom, that yearning for stimulation and distraction, for *something* to pass the time, is simply how we experience any pause in the program of control that seeks to deny pain. I am not suggesting that we ignore the causes of pain. Pain is a messenger that tells us, "Don't do that," and we are wise to heed it. But we step far beyond that when we suppose, even when the wound has been inflicted and the consequent pain written into reality, that we can still somehow avoid feeling it. A saying of Chinese Buddhism goes, "A Bodhisattva avoids the causes; the ordinary person tries to avoid the results."

Apparently, boredom was not even a concept before the word was invented around 1760, along with the word "interesting".[21] The tide of boredom that has risen ever since coincides with the progress of the Industrial Revolution, hinting at a reason why it has, until recently, been an exclusively Western phenomenon. The reality that the factory system created was a mass-produced reality, a generic reality of standardized products, standardized roles, standardized tasks, and standardized lives. The more we came to live in that artificial reality, the more separate we became from the inherently fascinating realm of nature and community. Today, in a familiar pattern, we apply further technology to relieve the boredom that results from our immersion in a world of technology. We call it entertainment. Have you ever thought about that word? To entertain a guest means to bring him into your house; to entertain a thought means to bring it into your mind. To be entertained means to be brought into the television, the game, the movie. It means to be removed from your self and the real world. When a television show does this successfully, we applaud it as entertaining. Our craving for entertainment points to the impoverishment of our reality.

All the causes of boredom are permutations of the interior wound of separation. Aside from the impoverishment of our reality, we are uncomfortable doing nothing because of the relentless anxiety that dominates modern life. This in turn arises from the paradigm of competition that underlies our socioeconomic structures, which (as I will explain in Chapter Four) is written into our conception of self. Second, we desire constant stimulation and entertainment because in their absence, we are left alone with ourselves with nothing to distract us from the pain of the wound of separation. Finally, technology contributes directly to boredom by bombarding us with a constant barrage of intense stimuli, habituating our brains to a high level of stimulation. When it is removed, we suffer withdrawal. We are addicted to the artificial human realm we have created with technology. Now we are condemned to maintain it.

That we have unprocessed pain inside us, waiting for any empty moment so that it may assert itself and be felt, is not so surprising given that a main imperative of technology is to maximize pleasure, comfort, and security, and to prevent pain. The urge to make life easier, safer, more convenient, and more comfortable has motivated technology from its inception. When the inventor of the Levallois flint-working technique produced his first spearhead, his contemporaries enthusiastically adopted it because it made life easier: "Not nearly so much work, now, to produce each spearhead." The new technique was so much more efficient. Life got easier. Need I cite more examples? Today we go to the pharmacy cabinet to apply technology to the alleviation of any discomfort, no matter how minor. Have a hangover? Take an aspirin. Have a runny nose? Take a cold medicine. Depressed? Have a drink. The underlying assumption is that *pain is something that need not be felt.* And the ultimate fulfillment of technology would be to discover the means to eliminate pain and suffering forever.

Maximizing pleasure and eliminating pain is the goal of the Technological Program taken to its logical extreme. An articulation of this goal in fairly pure form is David Pearce's "Hedonistic Imperative," which advocates the total elimination of suffering through genetic engineering, nanotechnology, and neurochemistry by disabling pain receptors, stimulating pleasure centers, and so on, as foreshadowed by today's happy drugs but also by the entire medical apparatus that seeks to remove or palliate symptoms. The mood-altering drugs, most notably the "selective" serotonin uptake inhibitors, are applied on the premise that the real cause of mental anguish is low levels of serotonin and norepinephrine in the brain. Raise levels of these neurotransmitters and the anguish goes away. The treatment is a success!

Underneath the assumption "the pain need not be felt" lie some even deeper assumptions. One of these is disconnection. The low serotonin levels are viewed in isolation from a patient's whole being, like a car with a broken part. This mechanistic paradigm denies the organic nature of a body, in which the health of any part reflects the health of the whole. It denies that there are reasons for the low serotonin, and reasons for the reasons, and reasons for those, spreading out to encompass the patient's whole being.

Related to disconnection is a further assumption, that we live in a dead and purposeless universe. Events happen essentially at random; there is no orchestrating purpose to make each event significant and right. Depression did not a serve a higher purpose because there is no such thing as a higher purpose, no reason except the identifiable, mechanistic reason, and therefore no cause to expect the pain will return in another form when this avenue is blocked. Reality is infinitely manageable.

If, however, we see technology (both on the personal and the collective level) as a means not to eliminate pain but to defer it, then it stands to reason that it will be waiting for us in any empty moment. All the more so if the very effort to defer pain generates new pain: the new problems caused by the previous technology, the symptoms caused by the drug itself.

In a connected, purposeful universe, managing the pain is like patching a leaky pipe when the water pressure is too high. Fixing one leak ensures another will spring elsewhere. Meanwhile, the pressure keeps rising. The apparatus of civilization springs one leak after another, as frantically we try to seal the spreading cracks.

It has been said in a Judaic-Christian-Islamic context that separation from God, the Fall, is the source of all suffering. Buddhism names attachment as the cause of suffering, but careful examination reveals its teaching to be nearly identical to that of esoteric Western religion. Attachment, to the impermanent, delusory ego self and all those things that reinforce it, maintains a separation from the rest of the universe from which we are not actually separate. Attachment *is* separation. As for separation from God, what is God but that which transcends our separate selves and interpenetrates all being? On the origin of suffering, Eastern and (esoteric) Western religion are in fundamental agreement.[22]

In everyday human life, happiness and security come from strong connections—to family, community, nature, place, spirit, and self—and not from "independence" whether psychological or financial. Because the story of technology is one long saga of widening separation from nature, widening separation from community (because of specialization and the mass scale of society), widening separation from place (because of our highly mobile and indoor-centered lifestyles), and widening separation from spirit (because of the dominant scientific paradigms of the Newtonian World-machine), it is no wonder that the pain of the human condition has only grown throughout the modern era. Even as outright physical hardship has declined, psychological suffering in the form of loneliness, despair, depression, anxiety, angst, and anger has grown to epidemic proportions. Even when our technology succeeds in holding off the external consequences of separation, we still internalize it as a wound, a separation from our own souls.

A final indication of the nature of the wound lies in the phenomenon of greed. When I ask my students the source of global problems such as pollution, they invariably cite greed, which they see as a fundamental characteristic of human nature that can be controlled but never eliminated. But greed like boredom is absent in most hunter-gatherer cultures based on a more open conception of self. Acquisitiveness is merely another attempt to fill the void and assuage the ache of separation, as if the accretion of more and more self, in the form of possessions, could compensate for the profound denial of self that is separation. Tellingly, we often use acquisitive metaphors for the ways we distract ourselves from the existential unease we call boredom: have a cigarette, have a drink, have something to do. It is by *having* as well that we strive for security, whether material—having possessions—or interpersonal, even to the extent of "having sex". But of course, no matter how much accrues to the discrete and separate self, that self is still fundamentally alone in the universe.

From Affluence to Anxiety

He must be cut off from the past… because it is necessary for him to believe that he is better off than his ancestors and that the average level of material comfort is constantly rising.

--George Orwell, *1984*

So complete is our identification of boredom as the default state of human existence that when asked to define it, most people say, "Boredom is when there is nothing to do." That this is an unpleasant state is by no means a logical necessity. Not only pre-modern people, but the entire animal kingdom seems to be perfectly fine with inactivity. This observation calls into question one of the fundamental tenets of the conventional explanation of the history of technology, which Stephen Buhner names the "anxiety theory" in the context of the invention of brewing.[23] More broadly, the concept is that human technological progress in general is driven by the struggle to survive, and that this struggle, this precariousness of existence, expresses itself in the human organism's physiology and psychology as anxiety, which is eased by creating better means to survive. Anxiety, then, is the way that threats to survival are translated into action to mitigate those threats. We can restate the anxiety theory as follows: (1) life is dangerous and survival difficult; (2) this makes us feel anxious; (3) the unpleasantness of this feeling drives us to control the dangerous/difficult circumstances, for example through technology; (4) we now feel less anxious.

On an individual level, the anxiety theory purports to explain boredom as follows: we really cannot afford to sit there and do nothing. If life is a competition for survival, then our genes should drive us to make the best possible use of each moment to augment our chances of survival and reproduction. Sitting around doing nothing goes against our genetic programming, which generates feelings of discomfort that impel us to do something productive. Certainly this is what many people feel during empty moments or deliberate experiments at meditation: a churning unease that says, "I should be doing *something*." This cultural compulsion is so strong that even spiritual practices such as meditation and prayer are easily converted into just another thing to do, moments mortgaged to the campaign of improving life.

Is the anxiety theory true? Ask some random people on the street and you will find that most would not want to go back to a primitive life before technology. We assume a dark picture of the primitive life as an uncomfortable, never-ending struggle for existence. This assumption is at the root of our cultural belief that technology has rescued us from the caprices of nature and enabled us to develop our higher potential. Here we have, in a nutshell, "the ascent of humanity."

The main problem with this view is that life in the Stone Age was not necessarily "nasty, brutish, and short" at all. Ethnographic studies of isolated Stone Age hunter-gatherers and pre-modern agriculturalists suggest that "primitive" peoples, far from being driven by anxiety, lived lives of relative leisure and affluence. An oft-cited example is the !Kung of the

Kalahari Desert in southern Africa, who were studied by the anthropologist Richard Lee.[24] He followed them around for four weeks, kept a log of all their activities, and calculated an average workweek of approximately twenty hours spent in subsistence activities. This figure was confirmed by subsequent studies by Lee and other researchers in the same region. In one of the harshest climates in the world, the !Kung enjoyed a leisurely life with high nutritional intake. This compares to the modern standard of forty hours of work per week. If we add in commuting time, shopping, housework, cooking and so forth, the typical American spends about eighty hours per week aside from leisure time, eating, and sleep. The comparable figure for the !Kung is forty hours including such necessary activities as making tools and clothes.

Other studies worldwide, as well as common sense, suggest that the !Kung were not exceptional. In more lush areas life was probably even easier. Moreover, much of the "work" spent on these twenty hours of subsistence activities was by no means strenuous or burdensome. Most of the men's subsistence hours were spent hunting, something we do for recreation today, while gathering work was occasion for banter and frequent breaks.

Primitive small-scale agriculturalists enjoyed a similar unhurried pace of life. Consider Helena Norberg-Hodge's description of pre-modern Ladakh, a region in the Indian portion of the Tibetan Plateau.[25] Despite a growing season only four months long, Ladakh enjoyed regular food surpluses, long and frequent festivals and celebrations, and ample leisure time (especially in winter when there was little field work to do). This, despite the harsh climate and the (proportionately) enormous population of non-working Buddhist monks in that country's numerous monasteries! More powerfully than any statistic, Norberg-Hodge's video documentary *Ancient Futures* conveys a sense of the leisurely pace of life there: villagers chat or sing as they work, taking plenty of long breaks even at the busiest time of the year. As the narrator says, "work and leisure are one."

Living in today's depleted world, it is hard to imagine its original bounty:

> Early European accounts of this continent's opulence border on the unbelievable. Time and again we read of "goodly woods, full of Deere, Conies, Hares, and Fowle, even in the middest of Summer, in incredible aboundance," of islands "as completely covered with birds, which nest there, as a field is covered with grass," of rivers so full of salmon that "at

> night one is unable to sleep, so greate is the noise they make"... They describe rivers so thick with fish that they "could be taken not only with a net but in baskets let down [and weighted with] a stone."[26]

These and other wonders—flocks of passenger pigeons and Eskimo curlews (both now extinct) that darkened the sky for days—comprised the provenance of this continent's native inhabitants. How much of a struggle could life have been? Note as well that this cornucopia existed despite humans having inhabited the continent for at least 12,000 years. It was not as if the Native Americans hadn't sufficient time to deplete nature's resources. We cannot conclude that their attitude of easy abundance was a temporary consequence of rich natural capital; their relationship with nature also preserved and sustained that abundance.

More significant than the actual time spent on subsistence was the hunter-gatherer's attitude toward subsistence, which was generally relaxed and nonchalant. As Marshall Sahlins describes:

> [The hunter] adopts a studied unconcern, which expresses itself in two complementary economic inclinations. The first, prodigality: the propensity to eat right through all the food in the camp, even during objectively difficult times, "as if", Lillian said of the Montagnais, "the game they were to hunt was shut up in a stable". Basedow wrote of native Australians, their motto "might be interpreted in words to the effect that while there is plenty for today never care about tomorrow. On this account an Aboriginal inclined to make one feast of his supplies, in preference to a modest meal now and another by and by."
>
> .... A second and complementary inclination is merely prodigality's negative side: the failure to put by food surpluses, to develop food storage. For many hunters and gatherers, it appears, food storage cannot be proved technically impossible, nor is it certain that the people are unaware of the possibility. One must investigate instead what in the situation precludes the attempt. Gusinde asked this question, and for the Yahgan found the answer in the self same justifiable optimism. Storage would be "superfluous", "because through the entire year and with almost limitless generosity the she puts all kinds of animals at the disposal of the man who hunts and the woman who gathers. Storm or accident will deprive a family of these things for no more than a few days. Generally no one need reckon with the danger of hunger, and everyone almost anywhere finds an abundance of what he needs. Why then should anyone worry about food for the future... Basically our Fuegians know that they need not fear for the future, hence they do not pile up supplies. Year in and year out they can look forward to the next day, free of care...."[27]

Significantly, aboriginal peoples typically refer to food as a "gift" from the land, the forest, or the sea. To us moderns it is a charming metaphor; to pre-agricultural people the providence of the earth was a living reality. The land provides all things—plants grow, animals are born—without the necessity of human effort or planning. Gifts are not something that must be earned. To see life in terms of receiving gifts bespeaks an attitude of abundance and naturally fosters a mentality of gratitude. Only with agriculture did the freely received gifts of the land become objects of exchange, first an exchange of work for harvest, and eventually the objects of commerce. In contrast, the mentality of the gift corresponds to the forager's nonchalance, which makes sense when the necessities of life are *provided* and not extracted.

Maybe we can still rescue the anxiety theory—what about disease? When I ask students to identify the most valuable achievements of modern technology, they invariably point to medicine, which they claim has given us levels of health, security, and longevity unprecedented in history. Such a view, however, fails to recognize the power and sophistication of traditional herbal medicine for curing the wounds and diseases common in those times. It also must contend with the observations of Weston Price, an American dentist who lived in the early twentieth century.[28] Price was curious about the decline of dental health he had seen over the decades of his practice, and hypothesized that the rapid increase in the prevalence of tooth decay, crowded dentition, and a host of other, formerly rare, non-dental maladies had something to do with our diets. He quit his practice and spent many years traveling to remote corners of the world where people still lived without modern foods. The societies he visited weren't all Stone Age, but they were primitive by our standards. He went to remote Swiss villages accessible only by mule, and to the outer islands of Scotland; he lived with the Masai in Africa, the Inuit in Alaska, the aborigines in Australia, Polynesians in the Pacific. In all these places he found almost no tooth decay, no obesity, no heart disease, and no cancer. Instead he observed magnificent physical stamina, easy childbirth, and broad jaws with all 32 teeth. The diets were different everywhere but there were some things in common. People ate very few refined carbohydrates, plenty of live fermented food, and substantial quantities of fats and organ meats. Their vitamin intake was many times greater than the norm today. Price's work lends support to the contention that at least in some respects, primitive people enjoyed better health than is the norm today, even without the modern medicine that we *think* keeps us healthy.

I do not mean to idealize life before modern technology. Certainly, we were more subject to the elements: heat, cold, rain, and wind. In Ladakh, people were cold a lot in the wintertime. For the !Kung at certain times of year the best foods were hard to come by and people typically would lose a few pounds over the dry season. Sometimes people would be hungry. While infectious diseases were rare in the days before high concentrations of population, and degenerative diseases rare before the advent of industrial food processing, other threats to life and limb abounded. Sometimes a child would be killed by a lion or a hyena. The !Kung, a peaceful and sharing people, even experienced occasional homicide, usually sparked by sexual jealousy, resulting in intermittent blood feuds spanning decades.[29]

In other parts of the globe, hunter-gatherer tribes lived in a state of constant low-level warfare with occasional outbreaks of horrendous violence—or so it is claimed. Most famous are tribes of the New Guinea highlands, with reported adult male mortality rates from violence of 20-30%[30], and the Yanomamo Indians of the Amazon, immortalized by Napoleon Chagnon in his book, *Yanomamo: The Fierce People*. Chagnon claims that these "living ancestors" lived in a state of perpetual warfare in which some 44% of adult males have killed.[31] Others maintain that his figures are greatly exaggerated.[32] A lot is at stake: perhaps violence is coded into our genes. Chagnon's mentor, geneticist James Neel, "thought that modern culture, with its supportive interventions on behalf of the weak, was 'dysgenic.' It had strayed too far from humankind's original 'population structures': small, relatively isolated tribal groups where men competed with one another—violently—for access to women. In these societies, Neel assumed, the best fighters would have the most wives and children, and pass on more of their genetic 'index of innate ability' to the next generation, leading to a continual upgrading of the quality of the gene pool."[33]

Have we ascended from a Hobbesian past of violence and fear? How else can we interpret the warlike nature of so many of the primitive peoples we have encountered? Is our species consigned to perpetual war until we have, through culture, overcome the genetic programming of the dominating "demonic males"?

In fact, in a pattern that is typical, the violence Chagnon encountered sprung in large part from the disruptions initiated by Western contact and, ironically, from his own presence. Investigative journalist Patrick

Tierney writes, "Kenneth Good, who worked with Chagnon while researching his Ph.D., has lived among the Yanomami for twelve years—longer than any other American anthropologist. Good calls Chagnon 'a hit-and-run anthropologist who comes into villages with armloads of machetes to purchase cooperation for his research. Unfortunately, he creates conflict and division wherever he goes.'"[34] Tierney continues:

> In 1995, Brian Ferguson, an anthropologist at Rutgers University, published a book entitled "Yanomami Warfare: A Political History," which challenged the sociobiological theories drawn from "The Fierce People" and other studies by Chagnon. Ferguson, whose book analyzes hundreds of sources, wrote that most of the Yanomami wars on record were caused by outside disturbances, particularly by the introduction of steel goods and new diseases. Ferguson noted that axes and machetes became highly coveted among the Yanomami as agricultural tools and as commodities for trade. In his account, evangelical missionaries, who arrived in Yanomami territory during the fifties, inadvertently plunged the region into war when they disbursed axes and machetes to win converts. In time, some of the missions became centers of stability and sources of much needed medicine. But Chagnon, whose study of Yanomami mortality rates took him from village to village, dispensed steel goods in order to persuade the people to give him the names of their dead relatives—a violation of tribal taboos.… these methods destabilized the region—in effect, promoted the sort of warfare that Chagnon attributed to the Yanomami's ferocity.

This may be an extreme example, but the principle is clear. It is very difficult to know what any society was like "pre-contact". The effects of Western technology, germs, and commerce typically precede the first anthropologists to even the most remote regions, initiating social breakdown. The same goes for non-humans. As primatologist Margaret Power demonstrates, the murderous behavior of chimpanzees in the wild, cited as evidence of our innate badness, emerges only in disturbed populations (which, strictly speaking, are the only ones accessible to researchers).[35] Specifically, the methods that researchers use reduce mobility and generate conflict.

When we see the warlike nature of primates and primitives, we may be seeing mostly our own shadow.

The debate over primitive savagery goes back at least to Rousseau's "noble savage" versus Hobbes' "nasty, brutish, and short" state of nature. It rages on today. I leave the question to anthropologists, but one thing is clear: Death, whether by human or natural agency, was a very visible part of life in primitive times. Death and discomfort are less visible today, but this does not mean we have conquered them. We have only hidden them away. Perhaps, even if Hobbes was right, our ascent is an illusion.

Nature, too, can be cruel, at least from our present perspective. What of those millions of tadpoles devoured by fish before they can ever enjoy the pleasures of froghood? Most animals have natural predators, except for a few top carnivores and the hugest herbivores, who must face their own brand of uncertainty. The Technological Program notwithstanding, life at all levels is inherently uncertain. Yet somehow, the rest of the animal kingdom does not seem beset by anxiety. Animals spend lots of time grooming, playing, and just loafing around. Do birds really need to spend all that time singing to find a mate and establish territory? Even bees, the epitome of busyness, spend lots of time in the hive apparently doing nothing at all.[36]

The unnaturalness of constant anxiety is also written into our physiology, which is not designed to handle constant stimulation of the sympathetic nervous system and its stress hormones. We have evolved to handle conditions of general leisure and relaxation, punctuated by the occasional burst of emergency overdrive. Many physiological functions, such as digestion, tissue-building, and immunity, operate only under conditions of relaxation. The stress we consider normal interferes with them and damages our health.

A final indication that anxiety is not, in fact, the default state of human existence lies in the relative lack of anxiety among cultures today that, while certainly not primitive, are nonetheless incompletely integrated into the Western social model. Go to nearly any Third World country and you will find that, in the absence of outright war or intense civil unrest, people are generally more laid back, less anxious, less driven, and less competitive than they are here. As the old joke goes, in Mexico everything is done *mañana.* In Taiwan, where modernity has come so fast that the old-time agrarian society is still visible in the person of the eldest generation, shoulder to shoulder with the youth culture of cell phones and fast food, traces of the slower pace of life that once existed are still visible. Whereas Chinese New Year is now a five-day, or even a three-day, holiday, in the old days it lasted two weeks. Other festivals were similarly extended, and required lengthy preparation of costumes and foods that are simply purchased in stores today. And every day, an extended siesta broke apart the workday. Whether in Taiwan or anywhere in the world, the pace of life in the more traditional parts of the country—the American South, for example—is much slower, less pressured, more leisurely. Extrapolating backward, we may surmise that this, and not anxiety, represents the actual "default state" of human existence. Witness the following description of the work attitudes of a group of South American hunter-gatherers:

> The Yamana are not capable of continuous, daily hard labor, much to the chagrin of European farmers and employers for whom they often work. Their work is more a matter of fits and starts, and in these occasional efforts they can develop considerable energy for a certain time. After that, however, they show a desire for an incalculably long rest period during which they lie about doing nothing, without showing great fatigue… It is obvious that repeated irregularities of this kind make the European employer despair, but the Indian cannot help it. It is his natural disposition.[37]

A racist interpretation of this passage is easily remedied by acknowledging the sneaking suspicion that *this is our natural disposition too.* "Their work is a matter of fits and starts..." Does that not describe a child as well? Reader, do you sometimes feel the desire to "rest" even when you're not actually tired? That our actual behavior contradicts our "natural disposition" testifies to the power of our acculturation. We have been convinced that we cannot afford to live like that, and so we condition ourselves and our children to override their "natural disposition" and work hard. Just as technology seeks to improve upon nature, so does culture seek to improve upon human nature.

The denial of Stone Age affluence is ideologically necessary, else the myth of ascent would lose its foundation. The Hobbesian view of the state of nature—nasty, brutish, and short—motivates and justifies the entire Technological Program. It is implicit in the myth of progress and the ideology of ascent. That is also why many in the opposing camp, who affirm Stone Age affluence, see technology and culture as a long series of blunders, a Fall, a descent. Another view is possible, however. Perhaps our eons-long accumulation of technology and culture has a different purpose entirely—neither the minimization of suffering nor the completion of control—that we have yet to recognize.

Another reason why we assume that life in the past was a struggle for survival, ruled by anxiety, is that we project our own experience onto the past. As Stephen Buhner points out, any of us would certainly be quite anxious, and have to struggle to survive, if we were plunked down suddenly into primitive conditions. But there is a deeper aspect of projection: We believe that anxiety ruled their lives, because anxiety rules *our* lives. *We* are the ones who feel that life is a struggle for survival, not they.

Consider our economic paradigm. Whereas in primitive societies cooperation was the rule, in our society it is competition. More for you means less for me. I've got to stake out my territory and protect my interests. Even education is based on competition for grades (which are implicitly associated with eventual success in life; i.e., survival). Both our ontology and our economics set us in competition with one another and therefore generate anxiety. However, the best way to see the powerful role of anxiety in our lives is on a personal level, by examining the emotions and considerations that determine important life decisions.

Every semester at Penn State I take a poll of my students, and ask them to complete the following sentence: "I am at Penn State…" (a) to get a degree so I can get a good job; (b) because my parents expect me to and I don't want to let them down; (c) I don't know, college comes after high school; (d) because this is where I can satisfy my thirst for knowledge. Semester after semester, a consistent 70-90% of the students choose "a". "B" and "c" typically draw 5-10% of responses, while "d" averages 2-5%. In other words, most of the students are at Penn State because they feel that they *have* to be here—have to be here in order to get a degree, which means a secure job, which means money, which we need for the basic necessities of survival: food, shelter, and clothing. "In other words," I tell them, "you are here at Penn State, at least in large part, due to survival anxiety. Hey, it's a beautiful day! Why don't you spend the afternoon playing Frisbee? Why don't you go hang out with your friends? Why don't you play your guitar on Old Main lawn? Is it because you love your classes and studies so much you cannot tear yourself away from them? Hey, you are young. Why don't you travel the world?" It is because they feel they "can't afford to", that it isn't practical, that it would somehow interfere with their ability to achieve financial security. And even these are mere rationalizations for an ambient dread and guilt that informs their every moment of leisure. I gave a class assignment requiring students to go home and spend 15 minutes doing absolutely nothing. One student wrote, "Pretty much the whole 15 minutes all I could think about was what work I could have been getting done." This is a typical response.

Because our culture so closely associates money with survival, the refrain "I cannot afford to" gives us a glimpse of the survival anxiety that underlies so many of our life decisions, large and small. "I cannot afford to" is certainly not confined to contexts involving purchases. It points to the monetization of all life. As the sphere of monetized human activity grows, so also grows the pervasiveness of the anxiety arising from a scarcity- and competition-inducing money system. To choose based on what we can afford is to choose from a position of lack. The mechanics of interest-based money, which I describe in Chapter Four, ensure that we never have enough.

It is because anxiety is such a powerful force in our lives, that we project it onto primitive life and assume that it, too, was driven by anxiety.

We also project our own anxiety onto biology when we see it primarily as a competition driven by the imperatives of survival and reproduction. The anxiety theory is essentially a restatement of Darwinism applied to human technological development. The genes of any organism will program it to do everything it can to negate threats to its survival; any gene that did not do this would surely exit the gene pool. Anxiety is one of these programs (terror is another). Technological progress, then, is viewed as an expression of the Darwinian drive to survive.

As in economics, biology posits discrete individual actors, i.e. Genes, behaving to maximize their self-interest, the means to survive and reproduce. Our very understanding of biology, i.e. of life, and in particular of progress in biology, i.e. of evolution, rests on a foundation of competition for survival. It is no wonder that we see human life and human progress in the same terms. The anxiety that defines so much of modern life is built into our understanding of what it is to be alive and what it is to be human.

The view of life as a struggle for survival is woven into our worldview on a much deeper level than Darwinism. In fact, our guiding scientific paradigms can admit no alternative. Competition is implicit in our culture's very conception of the self as an independent entity, distinct and separate from the environment and from other beings. This conception reached its fully developed form with Descartes, who identified the self as a discrete point of conscious awareness, a non-material soul separate from material reality, and with Francis Bacon, who enunciated the ideal of objectivity in science and the independence of the observer from factual reality. The foundations of science entail separation. When the definition of the self (and more generally, of an organism) is exclusive and discrete, any interdependency is therefore contingent on circumstances and can in principle be eliminated. This is known as "independence" or "security"—not to depend on others. Beings are naturally set in competition with one another, because more for me is less for you.

The other key feature of Darwinism that conforms to our basic scientific ideology is its purposelessness or randomness, features which comprise yet another source of our anxiety. Darwinism represents a valiant attempt to reconcile life's order and spontaneity with the mechanical, deterministic laws of (classical) physics. In the words of one of Darwinism's most eloquent exponents, Richard Dawkins, "The universe we observe has precisely the properties we should expect if there is, at bottom, no design, no purpose, no evil and no good, nothing but blind, pitiless indifference."[38]

The classical understanding of the universe, in which all things are composed of atoms[39] and void, gives rise to a further level of anxiety, one that the reader may have felt upon reading the Dawkins quote above. This is, to quote Robert Lenoble, the "anxiety of modern man" that comes from the recognition that we, too, are composed of nothing more nor less than any other object in the universe: atoms and void. Perhaps underneath the arbitrarily discrete beingness we have assigned ourselves lurks a kind of existential panic, the suspicion that perhaps, in some fundamental sense, we do not exist at all. Herein also lies an alienation of the human spirit from the cold, deterministic, impersonal laws of physics, a sense that something essential is left out.

Sigmund Freud is often quoted as saying, "The goal of psychoanalysis is to convert neurotic misery into ordinary unhappiness." Actually he has been misquoted and taken out of context,[40] but the very fact of the misquotation's tenacity points to the impossibility of finding real happiness within our present worldview. We can distract ourselves from the misery of purposelessness, emptiness, and meaninglessness, but they are always there waiting for us.

Anxiety and boredom flow from a common confluence of sources. Technology has separated us from each other, from nature, and from ourselves, inflicting the interior wound of separation. Secondly, the definition of the self as a discrete entity, fundamentally separate from other beings and the environment, contributes to our psychological loneliness. Thirdly, the competitive view of the world that is inseparable from the edifice of science weaves anxiety into the very fabric of life, which becomes a competition for survival. Finally, the belief that the universe at its most fundamental level consists of atomic particles interacting according to impersonal forces creates an existential insecurity, an alienation from the living, enspirited world and selves we intuitively sense.

Our society is based upon competition and anxiety in part because these are implicit in our basic understanding of the universe. To forge a new psychology—and, collectively, a new society—that is not underpinned by anxiety, will therefore require a new conception of self and life, and therefore of science and the universe. Other societies, fast disappearing under the deluge of Western culture, were remarkably free from the ambient anxiety we know today. It is no coincidence that their social systems were based on cooperation and that their self-definitions were not atomistic like ours are, but relativistic: defined in relationship to a greater whole such as family, village, forest, nature.

It is a primary goal of this book to establish a different conception of life and self, founded on both scientific and psychological reasoning, from which a different sort of society might naturally grow. When our basic ontology and self-definition changes, everything else will change with it. How will this happen? What might it change into? To answer these questions, in the next chapter I will discuss how the ascent of separation got started in the first place, tracing it back to well before the beginning of what we call "technology". Knowing the state we have departed, we may better envision the state to which we might attain. Understanding the dynamics of separation as an historical process, we may know better how to fulfill that process and grow into a new stage of human development.

CHAPTER II

The Origins of Separation

Happy the age, happy the time, to which the ancients gave the name of golden, not because in that fortunate age the gold so coveted in this our iron one was gained without toil, but because they that lived in it knew not the two words 'mine' and 'thine'!
-- Miguel de Cervantes, *Don Quixote*, Part I, Chapter XI

The Origins of Self

Technology is both a cause and a result of our separation from and objectification of nature. It distances us from nature, as today's artificial environments, reliance on machinery, and processed foods exemplify. On the other hand it is precisely our conceptual distancing from nature that encourages us to apply technology to it as an object of manipulation and control. How did this chicken-and-egg scenario get started? What initiated our separation from nature, of which technology is one aspect?

Jared Diamond has written compellingly of the cascade of changes that followed the adoption of agriculture some ten thousand years ago: writing, math, calendars, division of labor, wars of conquest, private property and its accumulation, money, epidemics, slavery, famines, and so on, making it, in his words, "the worst mistake in the history of the human race."[1] He also is one of many authorities to point out that early farmers were "smaller and less well nourished, suffered from more serious diseases, and died on the average at a younger age than the hunter-gatherers they replaced."[2] Nor did they enjoy easier lives: as mentioned before, hunter-gatherers also enjoy much more leisure than agriculturalists. So how did agriculture get started? Was it indeed a "mistake", a bad choice, or was it an unfolding of an inevitable process? We might also ask the same question of technology in general: Was the entire "ascent of humanity" as a technological species predestined, in the context of anthropology and even evolutionary biology? Is it built into who we are? Or is technology a mistake repeated again and again, from its earliest pre-agricultural beginnings through agriculture, industry, and information technology?

Agriculture was not the first technology. Before it we had tools of stone and other materials, we had fire, and we had language. There is no such thing as a pre-technological human being, a pre-technological *Homo sapiens*. In fact, the earliest representative of our genus, *Homo habilis*, whose fossils date back 2.4 million years, was already a fashioner of stone tools. Its successor *Homo erectus* had gained mastery over fire as well by about 1.5 million years ago.[3] Our hands have since evolved to use tools; our jaws have evolved to eat a diet of softer, cooked foods; our digestive systems have likewise developed the enzymes necessary to digest cooked food; our bodies have lost most of the hair necessary for survival without clothing; our brains have evolved to handle language. While it is certainly possible for a man or woman dropped naked into the wilderness to survive (provided the right training), the way he or she does so is to create tools and build a fire. As these tool-making skills must be taught, not genetically inherited, they surely fit anyone's definition of technology. Humans are by their very nature technological animals.

Insofar as technology comprises learned (as opposed to genetically encoded) skills for manipulating the physical environment, humans are not even the only animals to use technology. Most mammals and birds learn behaviors for survival from their parents; some even use tools. Chimpanzees not only use sticks as tools, but also select and alter them for purposes of climbing, catching ants and termites, digging roots, withdrawing honey, and as levers.[4] The Egyptian vulture will hold a rock in its beak as a tool for breaking open ostrich eggs.[5] A Galapagos finch uses cactus spines to pry insects from crevices, and a species of crow in New Caledonia goes a step farther by actually creating tools—shaping leaves into blades.[6] If a tool is considered an extension of the body for the purpose of manipulating the environment, we might even have to consider as tools the calcite shells of microscopic coccolithophores, made from environmental calcium. Where does the body end and the "extension of the body" begin? Here again we run into a perhaps arbitrary or blurry distinction between self and not-self. Our customary self/not-self distinction, a characteristic of our worldview projected onto biology, breaks down under close examination.

The same considerations apply to culture. Birds and mammals learn key behaviors from their mothers, through play if not through conscious imitation. At the very least, mothers (and sometimes fathers) provide triggering stimuli for neurological development. Extragenetic information is transmitted, even if the ability to use this information may be "genetically programmed". Our own genes also provide us with the physical structures to learn language. Culture, like technology, is incipient in prehuman biology.

Few people would consider a coccolithophore's shell to be an example of tool use, and even tool use in birds is usually dismissed as "instinctive." Likewise, some authorities dismiss the tool use of wild chimpanzees, bonobos, and hooded monkeys with the claim that they don't really "understand" their tools, that their behavior is somehow automatic, learned through unthinking imitation. I do not really "understand" how my computer works either! The point is that the learned use of extrasomatic objects for the purpose of manipulating or altering the environment began long before human beings walked the earth. We cannot blame technology on an unfortunate decision. What *Homo sapiens* have wrought is merely an acceleration of what has been going on for billions of years.

Some philosophers distinguish between human and animal technology by observing that humans are the only animals to use tools to make more tools. Does this distinction represent a qualitative cognitive difference between animals and people? Either way, it does get at an important characteristic of technology: its cumulative nature, the fact that once started it naturally builds upon itself, progressively distancing its users from their naked origins.

Animal technology is rudimentary at best, but so is their degree of individuation from nature, their consciousness of themselves as separate beings. Consider the possibility that our individuation, our separation from nature, was not a choice or a mistake but an inevitability set in motion before we even became human. Its culmination, then, is to take individuation to the absolute limit implicit in the Newtonian-Cartesian-

Darwinian worldview.

Since many animals can be considered to use and even manufacture tools, we might expect that the separation of the individual from nature implicit in technology applies to some degree to animals as well. And if animals, why not plants? Fungi? Bacteria? We commonly look upon primitive humans as being "in harmony with Nature" if not "one with Nature". All the more for non-human species. Certainly, they are closer to such a state than we are, but even in the simplest of organisms there are intimations of separation, hints of what was to come. Consider that the current age of separation started eons ago, written into the future through the very dynamics of biological evolution. It did not spring from a blunder, a distinct note of discord that one or another group of hominids introduced into nature's grand symphony. Separation, rather, is an inevitable unfolding of a cosmic process.

On the most basic level, by maintaining a constant internal environment removed from chemical and thermal equilibrium, all living creatures create a distinction between themselves and the outside world. All modern definitions of life draw on this concept, which is called homeostasis.[7] Because homeostasis entails a localized arrest or reversal of entropy, it demands an energy source, for example the sun, through which entropy is in effect exported to the outside environment. By definition, then, life creates a dualism, an inside and an outside, and what's more requires an irreversible *taking* from the environment. Speaking in terms of thermodynamics, life exists only at an (entropic) cost to the environment.[8]

Life, then, not only requires but actually *is* a separating off of a temporarily self-maintaining part of the universe. Contrary to our fundamental cultural assumptions, though, this separation is neither permanent nor absolute, but admits to degrees. Separation inexorably builds upon itself and has done so for hundreds of millions of years on earth, first through a biological phase and then through a technological phase.

The distinction between self and environment is minimal among the earliest form of life, the bacteria, which blur the self/other distinction with their fluid sharing of genetic material. Even higher animals and plants, however, rely upon one another for the cocreation of the internal and external environments essential to their mutual existence. No plant or animal is a completely individuated, separate, distinct being. As we shall see in Chapter Six, there is no clear-cut, absolutist definition of the self or the organism; our belief to the contrary is only a projection of our mistaken view of our own selves. Nonetheless, the evolution of life from bacteria to higher life forms did set the stage for the radical acceleration of individuation that was to follow.

For a long time it was assumed that because there is no shuffling of the chromosomes in asexual reproduction, that evolution must happen very slowly among asexual organisms such as bacteria. As it turns out, bacteria are actually far *more* genetically promiscuous than other organisms, to the point where the very concepts of species and individual hardly apply to them at all. Bacteria regularly exchange genetic material through a variety of means: via bacteriophages, by emitting plasmids and other DNA fragments into the environment for other bacteria to take up, and even by joining together and directly exchanging genetic material in a kind of asexual bacterial "sex".[9]

The next major advance in the development of individuation came with the nucleated cell and sexual reproduction. A nucleus insulates the genetic material from the environment and allows for a more discrete, more rigidly demarcated self. Sex replaces the ubiquitous genetic promiscuity of the bacterial world with a severely circumscribed realm of genetic mixing. To the extent to which the genes define the organism (which is vastly overstated), the restriction of gene mixing to the act of sex made the organism more discrete. Given the fluid genetic sharing of prokaryotic bacteria, sexual reproduction is not a innovation bringing organisms closer together, but a circumscription of a prior openness, a sharper demarcation of boundaries. Sexual genetic mixing channeled into a separate category an interchange that previously happened all the time.

Indulge me while I speculate, half-seriously, that bacterial life is one of near-constant bliss, akin to a perpetual state of sexual union with the universe. When we humans engage in sexual intercourse we recover, for a few moments, a state of being that was once the baseline of existence in a time of greater union and less separation. When we "make love" we let down our boundaries on many levels. The euphemism is appropriate, love being nothing other than a release of the boundaries that separate us from another being. Since bacteria maintain such boundaries far less vigilantly than nucleated organisms, they could be said to be that much more in love with the world. All the more blissful would be the state that admits no boundaries whatsoever, not even the homeostatic boundaries of a bacterial membrane, which goes by the name of cosmic consciousness, oneness with God, or universal love. Might we view evolution, by which the self divides from the whole and competes against other selves for survival, as a progressive distancing from that state? In this case we can do "future primitive" radicals such as John Zerzan one better: instead of a mere return to a pre-technological hunter-gatherer state of oneness with nature, let us undo evolution as well as technology—let us overthrow the tyranny of the eukaryotes! It wasn't with agriculture that we went wrong, nor even with representational language, but with the cell nucleus!

The insulation of genetic material in the cell nucleus effectively aligned the survival interests of the organismal self with those of the DNA. Let's pretend you and I are bacteria with common ancestry. From the perspective of the DNA, since reproduction happens only asexually to produce an exact clone, my survival is no more important than your survival, and there is no evolutionary impediment to altruism. Indeed, from the genetic perspective an entire bacterial species could be considered a single, widely distributed individual. The frequency of horizontal genetic transfer among bacteria further illustrates the relative unimportance of the bacterium *qua* individual from the perspective of genetic replication.

Contrast this to sexually reproducing taxa, in which each individual is genetically unique. Because you and I (no longer bacteria) have many genes which we do not share, and because we cannot transfer genetic material to each other, it behooves our genes to program us to enact behavior that maximizes our personal survival and reproduction, if even at others' expense. Animals, because their reproduction is almost exclusively sexual, have the most to gain, genetically, from this type of selfishness. The more highly differentiated from other species, the greater the genetic incentive for selfishness. Separation between the individual and the rest of her species, as well as the rest of nature, has an increasing genetic basis as one moves further down the line of descent.

Let me hasten to add that the above analysis is somewhat misleading. The fact is that competition is much less a determinant of behavior and evolution than commonly supposed, and our view of nature as "red in tooth and claw" is mostly a projection of our own cultural prejudices. We find what we look for. Secondly, the idea that genes "program" behavior and serve as the blueprint of physiognomy is also wrong, a product of our mechanistic worldview. Evidence is emerging that the environment triggers and even alters DNA to serve purposes that transcend the individual. Thirdly, the genetic integrity of higher organisms is not absolute as commonly supposed: plants, fungi, and even animals share the genetic fluidity of the bacteria in previously unsuspected ways. Fourthly, the leap from anuclear bacteria to eukaryotic cells, like other macroevolutionary jumps, happened through a symbiotic merger of simpler organisms. Co-operation, not competition, is the primary basis of life and the primary engine of evolution. The individuality of modern single-celled eukaryotes and the multicellular higher plants, animals, and fungi derived from them is built upon a merger of simpler individuals. As Alfred Ziegler puts it in *Archetypal Medicine*, life is a chimera.

DNA aside, we are all of course permeable to our environment as we routinely exchange materials with the world. We are semi-permanent patterns of flux with an existence independent of the specific material substances that compose us, just as an ocean wave only temporarily comprises a certain collection of water molecules. The molecules simply bob up and down as the wave moves forward onto new ones. Similarly, even though the matter of the universe cycles through each of us at varying rates and in a unique way, we share this matter and, in our relationships, co-determine each other's ever-mutating patterns of flux. Neither the matter nor its patterning constitute autonomous, independent units. The self has only a conditional reality.

To the extent gene-based competition does determine organisms' behavior, the separation of self from environment has an evolutionary, not merely a technological, basis. Whether through proto-tools or not, living creatures, especially animals, manifest a nascent dualism through their manipulation of the environment for selfish purposes. Even if it together comprises a unified whole, life embodies at least a conditional separation, a fracturing of perspective into mine and yours. Separation began long before humans walked the earth.

Fire and Stone

Bridging the gap between the incipient separation of prehuman biology and the relatively recent invention of agriculture are earlier technologies of stone and fire, language, counting, religion, time, and representational art. By objectifying Nature and humanizing the Wild, by converting the world into an object of management and control, and by interposing representational systems between observer and reality, the above technologies initiated the process of separation hundreds of thousands, if not millions, of years ago.

In a remarkable series of essays, John Zerzan makes a compelling argument for the pre-agricultural origins of separation. He sees the institution of linear time measurement, language, number, art and so on as elements of a kind of original crime, seeds from which have sprouted forth all the noxious weeds of the ecocidal, genocidal, suicidal modern world. Addressed primarily to Marxists and anarchists who see in the overthrow of capitalism the solution to the world's problems, his book *Elements of Refusal* establishes that the Revolution must go much deeper than that, for the root of civilization's ruin goes as deeply as the all the devices which enable our conceptual separation from the world.[10] Along with related ideas of Daniel Quinn and Derrick Jensen, Zerzan's critique demands to be addressed by any serious philosopher of civilization.

I will recast the phenomena Zerzan describes as a natural progression of an imperative of separation going back even further to deep prehuman times. Separation was not a crime or a blunder, but an inevitability. The age of separation whose apogee we are experiencing today was, as we have seen, written into the laws of biology, even into the laws of physics, from the very beginning, just as the widening of separation that came with agriculture grew naturally from who we were before that.

Let us first ask, in what sense does the use of the first technology—the stone tools that date back over two million years—represent a distancing from nature? Stones, after all, are natural objects (as is petroleum). What separates the tool-making H. *habilis* from the non-tool making *Australopithicus* that preceded him? What separates them from other animals? For that matter, what separates *us* from other animals?

The key distinguishing factor between human and animal technology is innovation—not just tool use, but cumulative improvements in tools. And even this is a quantitative, not a qualitative, distinction. Animals do learn and pass on new "technologies". With humanity, the accumulation of new technologies has greatly accelerated, but it is nothing new.

Whether human or animal, each new or improved tool changes that species' relationship with the environment by altering its ecological niche. For early humans, inventions like digging sticks gave access to new food sources and raised the effective carrying capacity of the land. Each improvement or invention was a step away from the "natural" state preceding it, and, foreshadowing the present addiction to technology, each step was irreversible once the population rose to the new carrying capacity.

With the accumulation of technology, human beings became increasingly dependent on skills that had to be learned, not genetically programmed. This does not constitute a qualitative difference between humans and other mammals and birds. All rely on knowledge transmitted through extragenetic channels. Early tool-using humans merely took an earlier step of separation to a new level.

Tool innovation was slow at first. Hundreds of thousands of years would pass without significant improvements. But as the culturally-transmitted knowledge accumulated, its sum total came to comprise a separate human realm. The difference between an acculturated human and a feral human widened. A new stage in our conceptual distancing from nature was underway.

Soon, the emergence of a separate human realm manifested in our physiology. Each technological innovation represents an alteration in the environment that exerts new selection pressures. Technological evolution thus feeds back into biological evolution. The post-tool *H. habilis* was no longer the same species as before; by changing his environment he changed himself. Hands, eyes, and postures all changed to facilitate tool use. Each great leap in protohuman evolution was precipitated or accompanied by a leap in technology. *Homo erectus* emerged from the very outset with marked improvements in technology over *Homo habilis*. Perhaps, indeed, such improvements were a key part of the speciation process. If so, we may view the early stages of technological development not as a series of inventions by a single species, but as a coupled biological/technological evolution of a phenotype extending farther than ever before out into the realm of inorganic matter. We would also then expect that as the pace of technological development has accelerated, that speciation would accelerate as well. We might also speculate on whether a new speciation is in the offing, or perhaps even underway.

As the human realm grew and separated off from the natural, technology came to represent the manipulation and control of the world, the subordination of nature to human intentions and purposes. A self manipulates that which is outside the self. Inherent in technology is the division of the world into self and other, me and environment.

Perhaps no other technology exemplifies this division better than fire, the next great step toward separation. Like the other steps, mastery of fire came about gradually, not as a distinct, deluded human decision to choose technology instead of trusting in Nature to provide. *Homo erectus* probably used it without knowing how to make it for hundreds of thousands of years. Eventually fire came to define human beings as unique among animals. Its use in cooking changed the human digestive system forever. Its use for warmth and protection allowed the habitation of whole new ecosystems. Ultimately, of course, fire led to ceramics and metals, engines and factories, chemistry and electronics, and the whole edifice of the artificial modern world. But that is getting ahead of ourselves.

From the very beginning, fire reinforced the concept of a separate human realm. The circle of the campfire divided the world into two parts: the safe, domestic part, and the Wild. Here was the hearth, the center of the circle of domesticity. Here was warmth, keeping the cold world at a distance. Here was safety, keeping predators at bay. Here was light, defining a human realm but making the night beyond all the deeper, all the more alien. Outside the circle of firelight was the other, the wild, the unknown.

Today, as fire-based technology covers the globe and the lights of civilization penetrate into the planet's few remaining dark places, we easily imagine that our conquest of the world is nearly complete: the domestication of all the wild, the bringing of the world under human control. Similarly, we imagine the light of science illuminating the few remaining mysteries of the universe, converting the unknown into the known, and subjecting the mysterious to the structures of human understanding and measurement. Consider though, that just as a campfire deepens the shadows beyond its circle of light, perhaps our science succeeds in illuminating only that which is within its purview, which we have deluded ourselves into thinking is the whole of reality, while making the vast beyond even more impenetrable. We have convinced ourselves that the world outside the campfire's circle does not even exist, or is not important, or will succumb to light as we build the fire higher and higher, consuming in the process every available bit of fuel.

With fire, the separate human realm began to take on a new character—linearity. Linearity is at the root of the unsustainability of the present system, which assumes an infinite reservoir of inputs and limitless capacity for waste. Fire is a fitting metaphor for such a system, for it involves a one-way conversion of matter from one form to another, liberating energy—heat and light—in the process. Just as our economy is burning through all forms of stored cultural and natural wealth to liberate energy in the form of money, so does our industry burn up stored fossil fuels to liberate the energy that powers our technology. Both generate heat for awhile, but also increasing amounts of cold, dead, toxic ash and pollution, whether the ash heap of wasted human lives or the strip mine pits and toxic waste dumps of industry.

It is not that fire is unnatural. Fire, along with its biological counterpart of oxidation, is a stage of a natural cycle. Our delusional folly is to act as if that stage of the cycle could exist permanently and independently. Only someone who cannot see the whole of reality would say, "Of course we can keep the fire burning forever. When it burns low we'll just add more fuel." To believe that a larger and larger fire can be sustained forever is transparently absurd, ignorant, and delusional. While fuel is plentiful, perhaps the delusion might be sustained. But today it is increasingly evident that we are running out of fuel—both social capital and natural "resources"—even as we suffocate in the ash.

The original technologies of fire mostly employed wood, thereby removing it from the normal biological cycle and preempting the natural flow of matter and energy. No longer did it nourish generations of insects, fungi, and soil. This arrogation of wood's oxidative energy to human purposes defined very early on the dominating relationship that technology embodies. Today, the same logic sees all the materials of the world as "resources", classifying them according to their usefulness to man.

The domination of nature that fire represents manifested in two of the earliest fire-based technologies: metal-working and ceramics. Both involve a transformation of the substances of the earth. Fire abetted the development of a separate human realm by converting the substances of nature—clay and ore—into the substances of man—ceramic and metal. Fire, the defining human technology, brings things over from the natural realm into the human.

If fire consumes the basis of oxidative life, then it is no wonder that the modern technologies of fire are themselves life-consuming, both in the literal sense of ecological destruction and in the figurative sense of their depletion of cultural, social, and spiritual wealth. For modern society is based primarily on the technologies of fire. It is fire that powers our automobiles and airplanes at supra-biological speeds; it is fire that enables us to smelt metal and etch silicon; it is fire that powers our electrical grid and communications system; it is fire that allows us to distill or synthesize chemicals that do not exist in pure form in nature; it is fire that powers the quarrying of limestone and the crushing of rock to build roads and skyscrapers. Even objects as "environmental" as a bicycle utilize the technologies of fire. We even, unlike any other animal, apply fire to our food in the process known as "cooking".

Fire-based technology epitomizes the Technological Program of controlling and improving upon nature, usurping the oxidation of stored energy for purposes we deem superior. Not coincidentally, these purposes themselves involve the further abrogation of natural cycles. The wholesale disruption of nature and reengineering of the physical landscape would be impossible without fire-based technology. From the building of superhighways and dams to the clearcutting of forests, nearly all large-scale domination of nature depends on fire technologies such as the internal combustion engine and the coal- or oil-fired turbine. However, let us not forget that the initial clearcutting of the entire Northeast was accomplished with hand-axes and saws alone (also fire-based insofar as they are made of metal). I doubt a Stone Age culture could accomplish this even if it tried. But as soon as fire-based technology gains ascendency, such projects as clearcutting become not just technologically feasible but morally conceivable, as the Epic of Gilgamesh testifies to as far back as the time of ancient Sumer.

The reader might protest that most fire technology is based on fossil fuels, whose burning does not, strictly speaking, "usurp" stored energy that would otherwise feed life processes (though it does diminish life in other ways). I will offer some speculations on the Gaian significance of such deep-storage of energy in the last chapter. The point for now is that, whether wood or oil, the mentality of burning is the same: the arrogation of stored energy to human purposes of control, accompanied by the degradation of other phases of the cycle in an unsustainable pretense of eternal linear growth.

Our age is so defined by the technologies of fire that we sometimes forget the possibility of other realms of technology. Other humans in other times were actually more highly advanced than ourselves in plant-based, earth-based, body-based, and mind-based technologies. Many of the practices that we dismiss as magic or superstition actually represented modes of mind-body development whose possibility and power we do not even suspect today. Their inaccessibility is not due to historical accident, nor to willful ignorance, nor to any *intentional* campaign to eradicate competitors to the dominant fire-based technology,[11] but rather to their incompatibility with our fundamental self-definition in a dualistic cosmology of self and other. Today, as our division of ourselves from the universe becomes increasingly untenable, a new understanding of self is beginning to emerge that will naturally foster these near-forgotten or yet-to-be-discovered realms of technology. We cannot understand or utilize them operating from our current dualistic ontology of the discrete, separate self.

Language and Label

As a separate human realm coalesced around the technologies of fire and stone, another even more powerful technology grew alongside them—the technology of mind we call language. Consisting of symbols that are connected only arbitrarily to the objects, attributes, and processes they name, language is indeed a separate human realm, a human-created map or representation of reality.

Language is prior to any technology requiring the accumulation of knowledge and the coordination of human activity. Anything human civilization has ever created, from the pyramids to the space station, rests ultimately upon a foundation of symbols. Without blueprints, instructions, specifications, guidelines, computer programs, money, science texts, laws, contracts, schedules, and databases, could anyone build a microchip, a hydrogen bomb, or a radio telescope? Could anyone operate an airport or a concentration camp?

Referring to technology, I asked in the Introduction, "Can the gift be separated from the curse?" As the above examples make clear, we might ask the same of language. Language is the foundation of the separate human realm, and from the very beginning it has borne a destructive as well as a creative power.

The destructive potential of language is contained within the very nature of representation. Words, particularly nouns, force an infinity of unique objects and processes into a finite number of categories. Words deny the uniqueness of each moment and each experience, reducing it to a "this" or a "that". They grant us the power to manipulate and control (with logic) the things they refer to, but at the price of immediacy. Something is lost, the essence of a thing. By generalizing particulars into categories, words render invisible the differences among them. By labeling both A and B a tree, and conditioning ourselves to that label, we become blind to the differences between A and B. The label affects our perception of reality and the way we interact with it.

Hunter-gatherers, who were closer to a time before generic labels, were animists who believed in the unique sacred spirit of each animal, plant, object, and process. I can imagine a time when a tree was not a tree, but a distinct individual. If it is just a tree, one among a whole forest of trees, it is no great matter to chop it down. Nothing unique is being removed from the world. But if we see it as a unique individual, sacred and irreplaceable, then we would chop it down only with great circumspection. We might, as many indigenous peoples do, meditate and pray before committing an act of such enormity. It would be an occasion for solemn ritual. Only a very worthy purpose would justify it. Now, having converted all of these unique, divine beings into just so many trees, we level entire forests with hardly a second thought.

The same goes, of course, for human beings. The distancing effect of language facilitates exploitation, cruelty, murder, and genocide. When the other party to a relationship is a mere member of a generic category, be it "customer", "terrorist", or "employee", exploitation or murder comes much more easily. Racial epithets serve the same purpose: we call it "dehumanizing the victim". Yet the dehumanization begins with any categorization, even the word "human". This is not to advocate the abolition of nouns, only to be mindful of their relative unreality. It is when we get lost in the manmade realm of abstractions—statistics, names of countries, figures in accounting ledgers—and believe them to be real that we end up perpetrating violence.

When we knew every face intimately, there was no need to generalize into "people." Our ancestors experienced a richness of intimacy that we can hardly imagine today, living as we do among strangers. It is not only social richness that is muffled underneath our words, it is the entirety of sensual experience. Margaret Mead once observed, "For those who have grown up to believe that blue and green are different colours it is hard even to think how any one would look at the two colours if they were not differentiated, or how it would be to think of colours only in terms of intensity and not of hue."[12] And if we had no words for color at all, might we not see a world painted in the tens of millions of colors that the human eye is capable of discerning? How much richer and more alive such a world would be. Each moment a visual feast. Perhaps it is the increasing abstraction of ourselves from the world, to which language contributes, that explains why "fifteen years ago people could distinguish 300,000 sounds; today many children can't go beyond 100,000 and the average is 180,000. Twenty years ago the average subject could detect 350 shades of a particular color. Today the number is 130."[13] By naming the world, abstracting it and reducing it, we impoverish our perception of it. Language is the basis and the model for the standardization, generalization, and abstraction that underlie present-day science and industry. In science, it is the assumption of universal laws applying generally to a featureless substrate of fundamental particles. In industry, it is the standardization of parts and processes. And the price we pay is a loss of the original richness of the ground of being.

Occasionally one may be fortunate enough to catch a momentary glimpse of perception unmediated by language and other representational systems. The world vibrates with an unspeakable richness of sound and color. As soon as we try to explain, interpret, or exploit that state, we distance ourselves from immediate reality and the experience vanishes. Habitually interpreting the world second-hand through symbolic representations keeps us distanced from the glory of reality all the time.

The realization that language can distance us from reality goes back thousands of years, at least to the time of Lao-tze, who opened the *Tao Te Ching* with the words, "The Tao that can be spoken is not the true Tao; the name that can be named is not the true name."[14] The first line of one of the world's greatest classics of spiritual scripture is a disclaimer, an admonition about the insufficiency of language to represent truth.

In the Heart Sutra as well, one of the most important works in the Buddhist canon, we have a similar warning of the "emptiness of all teachings." The truth is not to be found in the words of the teachings; it is a mistake to assume that the words themselves contain the truth.

On the other hand, the ancients recognized a creative aspect to language alongside its tendency to distance and delude. There is a mythological thread hinting at the existence long ago of an Original Language, a true language that somehow did not symbolize and abstract from reality, but that was itself part of reality. Perhaps this language is what Derrick Jensen calls "a language older than words," akin to the vocalizations of wild animals. This language is almost wholly lost to us today, except in a few surviving exclamations having primal reverberations in the body and psyche—words like "Tada!" "Yahoo!" "Wow!" "Amen!" "Ahh," and "Oooh". Some of these words derive directly from Sanskrit roots. Indeed, there are those who claim for Sanskrit a special status of being closer than any other language to the Original Language of reality. As anyone who has experienced Hindu chanting can attest, Sanskrit words and phrases often have an emotional resonance that may be quite distinct from their semantic denotation. Listeners with no knowledge of Sanskrit may be strongly affected. Words like "Om," "Ah," "Ram" and others are considered not to denote or represent the divine, but to actually *be* aspects of the divine. This point is very difficult for the dualistic mind to grasp.

The same resonance can be found in other antique languages. In Taoism as in Hinduism, certain sounds are invested with a psycho-spiritual power quite apart from their semantic denotation.[15] The correct pronunciation of these words is considered extremely important in certain *qigong* exercises. It is not enough to know their meanings, if indeed the sounds have meanings at all in the conventional sense. Like "yippee!" and "wow!", the sound *is* the meaning. In Judaism as well, the sacred power of certain words is considered to arise from their sound. To merely hear them uncomprehendingly is enough, it is claimed, to induce psychological changes in the listener.

Similar claims have been advanced for the indigenous languages of North America. Joseph Epes Brown observes, "Among all Native American and Inuit languages, there is a blending of rich verbal and non-verbal expressions."[16] That is, the distinction between sound and word is not so clearly conceived as it is in modern languages. Moreover, "Spoken words or names are not understood symbolically or dualistically, as they are in English.... Such separation [between sound and meaning] is not possible in Native American languages, in which a mysterious identity between sound and meaning exists."[17] Because they are not merely labels, names and nouns in such a language are an intrinsic and inseparable aspect of the being named: "To name a being, or any aspect or function of creation, actualizes that reality."[18]

Traditional Native Americans will therefore use the real names of things only with great circumspection, for to name the bear, for example, will actually invoke its presence. The creative power of speech is again difficult for the dualistic mind to understand—just talking about something won't actually change it, right?—but we can see vestigial traces of this understanding in certain "superstitions" that survive to the present day. Chinese culture has strong taboos against speaking aloud dark possibilities, lest it bring them into reality. Even in America, we still knock on wood.

That words are not arbitrary labels affixed to an objective reality, but have creative force, echoes the Hindu association of certain sounds with divine forces and the Biblical equation of the Word with God, as well as the near-universal identity of breath and spirit.[19] For what is a word but a special kind of breath? Word is an intentional breath, a meaning-carrying breath, a creative breath because it infuses meaning into a world that otherwise just is. Out of the raw material of nature, we speak a human realm into existence, just as the God of the Book of Genesis spoke into existence the material world. Like God, in whose image we are made, we speak worlds into existence.

Why, then, does present-day language seem so impotent, so ineffectual? Why has talk become cheap? What has happened to that original language with its creative power? How has the creative breath devolved into the ubiquitous matrix of lies we find ourselves in today?

In the beginning, there were no words as we know them today, no representational sounds, only the cries of the human animal. What was this Original Language like? In fact, we can still access it today. Because it is not conventional but is part of reality, the Original Language can never be irretrievably lost, but only temporarily forgotten. It is locked deep inside all of us, ready to emerge whenever we shed the inhibitions of civilization. One such occasion is, of course, love-making. The vocalizations of passionate sexual abandon are nothing other than the Original Language remembered. These utterances do not have meaning in the way ordinary words do, but they cannot be considered meaningless either; they are vectors of a communication far more honest and intimate than any semantic exchange. Taoist and Tantric tradition has apparently made a study of these utterances, although I am not aware of anything more than passing references and superficial descriptions in the open literature. However, the contemporary psychologist Jack Johnston has developed a powerful system of sexual healing through higher-level orgasm utilizing a key sound which is difficult to transcribe but goes something like ahhh-*ahhh*, with a rolling quality in the middle. Significantly, Johnston discovered this sound through an "intuitive search"; he did not invent it but unearthed a latent capability intrinsic to what it is to be human.[20] This is a perfect example of the technology, or anti-technology, of the Age of Reunion, which is not based on control or separation.

Any intensely emotional experience may also elicit utterances of the Original Language—spontaneous vocalizations of ecstasy, lamentation, glee, fear, rage, and so forth, as well as the cooing noises we make at infants. They come out when words are simply insufficient to express ourselves, and when our emotions overpower the inhibitions of culture; that is, when we go wild. They are not really words. They are sounds, cries, the calls of the human animal. They do not derive their meaning from a grammar, nor are they subject to convention.

Nor has the Original Language entirely disappeared from ordinary speech; it interpenetrates modern language and could be called the voice behind the words. In his technical work, *Languages Within Language: An Evolutive Approach*, linguist Ivan Fonagy has even taken a step toward describing it. Fonagy's achievement was to develop a statistical approach demonstrating a correspondence across languages between sounds and meanings. For example, he found that whether in English, French, or Hungarian, front vowels are more prevalent in words for concepts like light, above, cheerful, and pretty, than for their opposites. Soft consonants predominate in the words "love", "tender", "soft", "good", and "sweet", while hard consonants predominate in "anger", "wild", "hard", "bad", and "bitter". The individual pronunciations are otherwise unrelated, but in these statistical commonalities we can see a glimpse of a mode of vocal communication prior to language. He also catalogs a number of changes in the articulatory organs that are common across numerous languages in the expression of various emotions: the lips are protruded and rounded in displays of tenderness; the tongue is withdrawn in the expression of hatred and anger, which are also characterized by pharyngeal contraction and reduced acoustic intensity relative to the expiratory effort.[21]

The semantic meanings of our words obscure the intonations that communicate our real state of being, and we have learned to listen to the words and not the voice. Yet part of us, the deep primal part usually beneath consciousness, still tunes into the voice, which communicates far more honestly than words can. The simplest example lies in emotional exclamations. Fonagy comments, "The effect produced by 'emotive phonemes' might be attributed to their 'strangeness' due to the violation of phonemic rules. I would suspect instead that their impact is due to the fact that these sound gestures are not devalued, they escape the general rule of arbitrariness; thus they can be freely enjoyed. Moreover, they are 'meaningful', linked by natural (more or less narrow) ties with real (non-verbal) physical or mental phenomena."[22] We delude ourselves when we suppose that the main impact of speech lies in the words (as opposed to the voice), just as we delude ourselves when we cite logical reasons, which are actually rationalizations or justifications, for our decisions. (This link between logic and language is embodied in the Greek root logos, which means logic, law, and language, something imposed from the outside, in contrast to voice, which comes from within and is actually a form of breathing; i.e., spirit.)

Like logic, law, and technology, the control implicit in language is a façade. We carefully label and categorize the whole world, hoping thereby to impose order upon it, to domesticate the wild, but we delude ourselves to think that the wild respects our boundaries any more than a squirrel respects a "no trespassing" sign. To this day, it is the voice that communicates more than the speech.

This Original Language was the subject of a misguided search by the philosophers and linguists of the Age of Reason, who referred to it as the *lingua adamica.* Unable to see beyond language as a system of symbols, Leibnitz and others sought to reinvent a language that would correspond perfectly to reality, in which truth could be discerned through grammar. Their program failed miserably, of course, because they did not understand that the Original Language was non-representational rather than a system of perfect representation. Nonetheless, Leibnitz' program continues on in the ever-finer labeling of the world through the lingo and jargon of science. It is another version of the Tower of Babel, a man-made edifice that seeks to rival the infinity of the real world.

Ivan Fonagy exemplifies the projection of our ontological assumptions onto primitive language when he observes: "The far-reaching parallels in unrelated languages in the expression of emotions at all levels of sound-making clearly show that the basic tendencies which appear in emotive vocal behavior are not language-specific. They seem to be governed by a *paralinguistic semiotic system.*" [emphasis in the original].

Fonagy adopts the conventional dualistic interpretation of language in assuming that these commonalities across languages are a system of signs in parallel with the usual semantic one. But perhaps what Fonagy calls "natural languages" are not semiotic systems at all. He interprets, for example, the spasmodic tongue movements of anger and hatred to be part of another system of signs, in parallel to the semantic meanings of the words thus articulated, that *represent* anger. However, these expressions are really not "representations" at all, they do not represent anger, they *are* anger. They are part of the corporeal state which equally includes hormonal releases, vascular dilation, elevated heart rate and breathing, and so on. Unlike semiotic language, these vocalizations do not distance us from the emotions being expressed. Or as Thoreau put it, "Most cry better than they speak, and you can get more nature out of them by pinching than by addressing them."[23]

John Zerzan writes, "As soon as a human spoke, he or she was separated. This rupture is the moment of dissolution of the original unity between humanity and nature." He implies a catastrophic moment of separation, a blunder, a Fall. But we have always vocalized, as do most mammals, and birds, and even many reptiles and insects; and it would be arrogant indeed to assume these animals sounds are devoid of meaning. There are stories of Native American trackers whose ability to interpret animal calls borders on the magical, and legends abound in every culture attributing to the ancients the ability to talk to animals.

As the human realm gradually separated from the natural, the original vocabulary of human utterances became insufficient. New objects, new distinctions, and new processes came into being, as well as a new-found objective relationship to nature. Slowly, gradually, language accompanied being into a widening dualism: self and other, human and nature, name and thing.

The ascent of humanity is a descent into a language of conventional symbols, representations of reality instead of the integrated vocal dimension of reality. This gradual distancing, in which and through which language assumed a mediatory function, paralleled, contributed to, and resulted from the generalized separation of man and nature. It is the discrete and separate self that desires to name the things of nature, or that could even conceive of so doing. To name is to dominate, to categorize, to subjugate and, quite literally, to objectify. No wonder in Genesis, Adam's first act in confirmation of his God-given dominion over the animals is to name them. Before the conception of self that enabled dominion, there was no naming—none of the original vocalizations were nouns.

Fascinatingly, ancient languages were far less dominated by nouns than modern languages: from the ancient nounless original language, it is claimed, by Neolithic times only half of all words were verbs, declining to less than ten percent of words in modern English.[24] The trend continues to this day, with the growth of passive and intransitive uses of verbs that objectify and abstract reality by saying, in effect, A is B. Language has evolved toward an infinite regression of symbols, words defined in terms of each other, that distances us from the world. Significantly, some indigenous languages apparently lack a word for "is", as the shaman Martin Prechtel claims for at least two Native American languages.[25] I have also noticed that Taiwanese, an ancient Chinese dialect firmly based in a preindustrial society, has an amazing profusion of descriptive action words that do not exist in or have disappeared from modern Mandarin and English. In English the same tendency manifests as a gradual supplanting of the simple present by the present progressive ("I am walking" instead of "I walk").

A few modern thinkers have sought to reverse or undo this trend. Alfred Korzybski, in his monumental tome, *Science and Sanity*, spends over a thousand pages reproving us for our wanton use of the "is" of identity, which reduces things to other things, proposing what he believes is a new "non-Aristotelean" mode of thought. He was apparently unaware that numerous mystics (such as Lao Tze) preceded him in this insight by thousands of years. Nonetheless, writing in the 1920s, Korzybski was ahead of his time, and helped to launch the movement known as neurolinguistic programming that seeks to induce mental health (sanity) through new language patterns. More recently, the physicist-sage David Bohm has proposed a new mode of language he calls the rheomode, aimed specifically at recovering the dwindling verb form and thereby fostering an understanding of the universe in terms of process rather than thing. "The Rheomode" is the first chapter of his book *Wholeness and the Implicate Order*, in which Bohm attempts to introduce his interpretation of quantum mechanics. We might understand him to imply that the rheomode is the only way of speaking that is consistent with the true nature of physical reality, which is a fundamentally unified and interconnected whole. In Bohm's view, the artificial division of the world into subject and object is, at bottom, incoherent. I am not a separate I, I am the universe "Charles-ing".

We will probably never know when the descent into representational language began. Citing anatomical evidence such as the hyoid bone, enlarged thoracic spinal cord, and enlarged orifice to carry the hypoglossal nerve to the tongue, paleontologists date the origin of language back to the Neanderthals, certainly, and probably back to *Homo erectus* or even further.[26] This contrasts with the views of theorists such as Noam Chomsky, Stephen Pinker, and Julian Jaynes, who fix the date much later in the Upper Paleolithic, 30,000-50,000 years ago. Their view is based on the association of language with the cognitive development implied by concurrent developments in technology, art, and so forth. Both camps, however, see language as a "symbolic coded lexicon and syntax";[27] that is, a representational system.

In this regard we might ask, what was there to talk about in Stone Age society? Some researchers propose that speech was necessary to teach the two-hundred-plus different blows required for the production of middle-Paleolithic blades, but such skills are better learned by observation and imitation, not description. Others claim that hunting, which began only when we developed weapons, requires speech to coordinate hunters' movements. But here again silence usually benefits the human hunter more than speech; besides, wolves and other pack animals seem to hunt just fine without language. But let's not fall into the trap of trying to explain everything based on how it might aid survival. Might there be other reasons for speech in hunter-gatherer times?

Could it be that speech did not arise out of necessity at all? An important and ancient function of speech is to play, to joke, to tell stories. Perhaps these were the origins of language. Perhaps its function as an instrument of separation grew gradually, in tandem with other alienating developments in culture and technology.

Until fairly recently, human beings lived in kin bands of usually no more than twenty people, loosely associated into tribes of perhaps a few hundred. Open to nature and each other, they knew each other more intimately than we can imagine today. Speech may have been superfluous, as it often is between lovers, or between mother and baby. When we know someone that well, we know without asking what they are thinking and feeling. All the more in prelinguistic times, when our empathetic faculties were yet unclouded by the mediatory apparatus of language. Spend some time alone with a person or small group in silence, and observe whether, after just a few days or even hours, you feel more intimately connected with them than if you'd been talking. The empathy and intuitive understanding of others that develops in such circumstances is amazing.[28]

We might therefore speculate that language only becomes *necessary* when other forms of separation deaden our intuitive connections and, at the same time, demand a more complex coordination of human activity. Especially relevant is the division of labor, incipient if not already underway in the late Stone Age, which brought "a standardizing of things and events and the effective power of specialists over others... Division of labor necessitates a relatively complex control of group action; in effect it demands that the whole community be organized and directed."[29] Standardization of things accords naturally with their abstraction and naming. It is part and parcel of the separate human realm that grew up around technology in general; it both arises from that separation and reinforces it. Language cannot be considered in isolation from all the other elements of separation I describe in this chapter, but only as part of a vast, comprehensive pattern.

Spoken language was only the beginning of this division. Voice inevitably lives on in spoken words, though masked more deeply the more controlled and refined the speech. The invention of writing, therefore, was another huge step away from the Original Language and toward the complete replacement of direct communication by arbitrary, abstract symbols. The divorce between written words and concrete objects and processes was gradual, progressing from the first representational hieroglyphs to increasingly abstract forms, and eventually to the alphabet, which is wholly non-representational. Alphabets changed our way of thinking in subtle but far-reaching ways. "The alphabet codified nature into something abstract, to be cut and controlled impersonally."[30] Unlike a pictogram, an alphabetic word can be figured out through analysis, by breaking it down into parts; pictograms derive their meaning through resemblance to the real world. Alphabets therefore encourage an atomistic conception of meaning and, by extension, of the universe.

In writing the voice is gone, replaced by the apparent objectivity of ink on paper divorced from any tangible speaker. Written words exist as independent entities unto themselves, no longer addressed to a specific listener. Written words foster the illusion that they have objective meanings—definitions—not contingent on the state of speaker and listener. The apparent objectivity of written words explains why people tend to believe what they read more than what they hear. The written word seems more authoritative. Dictionaries, a comparatively recent phenomenon (the first significant Western dictionaries were compiled in the seventeenth and eighteenth centuries),[31] further substantiate the illusion that words have fixed, objective meanings apart from the interaction of speaker and listener. In the same vein, books concretize the belief that knowledge is to be found outside the individual. Non-literate societies may have been more apt to seek it within.

Printing and electronic media take the divorce between meaning and speaker to an even further extreme, for if handwritten words lack voice, they at least have "hand". Each hand is unique and conveys to the attentive observer the emotional and spiritual state of the writer. Typeface replaces this hand with a mass-produced one, leaving very little room for the Original Language to creep in. Yet still it does, irrepressible, in the sub-semantic idiosyncrasies of style that we persist, following some unconscious wisdom, in calling the "writer's voice". We can therefore see the standardization of grammar and usage, the descent into jargon and formulaic locutions, and the general blanding of public speech such as corporate and political press releases to be the final stage in excising voice from language. The goal would seem to be to pretend that the words had no human author at all, existing as purely objective facts. Indeed, use of the first person is considered bad form in academic writing—a convention the author of the present work finds ridiculous!

Words defined in terms of other words in a system of abstract representation maroon us in a factitious, anthropized, domesticated, and finite world, and render us susceptible to the illusion that we can manipulate and control reality in the same way we can manipulate and control its symbolic representation. But because the map is necessarily a partial and distorted version of the mapped, our manipulations based on that map invariably produce a profusion of unforeseen results: the unintended consequences of technology. When we mistake words for reality, or even assume a full one-to-one, linear correspondence between symbol and reality, the symbols assume a reified, objective status that invests them with an unwarranted authority (particularly when they are written and thus divorced from a specific speaker). The proliferation of the passive voice exacerbates this tendency. The speaker disappears; process becomes thing, becoming becomes being, impersonal forces act upon inert objects. The parallel to classical physics is quite striking. The notion that words have objective meanings, independent of speaker and listener, reader and writer, is completely consistent with and accessory to the Newtonian-Cartesian universe of independently-existing "objects" possessing a reality independent of the observer. John Zerzan puts it this way: "Like ideology, language creates false separations and objectifications through its symbolizing power. This falsification is made possible by concealing, and ultimately vitiating, the participation of the subject in the physical world."[32] The world becomes an object.

The fallacy of objective meaning is widely recognized, from Lao Tze to the post-modern deconstructionists; Thoreau said, "It takes two to speak the truth: one to speak, and another to hear."[33] Only recently, however, has this fallacy begun to enter the general consciousness, resulting in a generalized breakdown in linguistic meaning. Increasingly, words don't mean anything anymore. In politics, campaigning candidates can increasingly get away with saying words that flatly contradict their actions and policies, and no one seems to object or even care. It is not the routine dissembling of political figures that is striking, but rather our nearly complete indifference to it. We are as well almost completely inured to the vacuity of advertising copy, the words of which increasingly mean nothing at all to the reader. Does anyone really believe that GE "brings good things to life?" That a housing development I passed today, "Walnut Crossing", actually has any walnut trees or crossings? From brand names to PR slogans to political codewords, the language of the media that inundates modern life consists almost wholly of subtle lies, misdirection, and manipulation. No wonder we thirst so much for "authenticity".

People everywhere are talking about a search for meaning, recognizing that it is not to be found in words. Perhaps herein lies the reason for the astounding decline in U.S. literacy over the last half-century. What is typically seen as a failure of education and a symptom of social breakdown, might be, at least in part, a form of rebellion. Frustration with language could also be the reason for the much-maligned proliferation of "like" and "you know" in the speech of young people. A more charitable view is that by using "like", we deny the false identity inherent in "is". As for "you know", could that be a groping toward a more intuitive mode of communication? Even though the listener may not understand the meaning of the words, if she listens to the voice behind them, she does indeed already know.

Another symptom of the breakdown of semantic meaning is the routine use of words like "awesome," "amazing," and "incredible" to describe what is actually trivial, boring, and mundane. We are running out of words, or words are running out of meaning, forcing us into increasingly exaggerated elocutions to communicate at all.

Like all our other technologies, language is not working so well any more. It has failed to live up to the promise, echoed in the Technological Program to control nature, of providing a fully rational, objective, logical system of representation, the rigorous use of which will bring us to accurate knowledge of reality. Just as any technological fix always neglects some variable that generates unexpected outcomes and new problems, so also is any language, any system of signs, a distortion of reality riddled with blind spots that unavoidably generates error and misunderstanding. The attempt to control the world is futile. For too long now, we have sought to remedy the consequences of failed control by imposing even more control, more technological solutions; in language this equates to more rigor, more definitions, more names, an ever-finer categorization of reality. In our era, we are finally witnessing the collapse of the technological program of language.

The increasing obviousness of the corruption of the language is a blessing in disguise, for it makes all the clearer the authenticity of non-verbal modes of communication based on immediate experience rather than representation. These modes of communication, in contrast to the distancing implicit in the abstraction, naming, and symbolizing of the universe, demand a letting go of the barriers between self and world. When we gaze into a lover's eyes, the most authentic communication happens when both people drop their masks and pretenses, stop contriving to send a message, and simply open themselves up to the other. When we finally let go of the enormous effort to hold ourselves separate and apart from other people and the world, words will become less necessary.

Less necessary, but not obsolete. The development of language was not a mistake, an original blunder, but like technology developed in a gradual and inevitable evolution from animal origins. The descent into representation was foreordained. If so, let us consider whether it might harbor a purpose outside its function as an instrument of separation. What will be the purpose of language in a healed world? It will be what it always has been—to tell stories. This is no trivial function. Our entire civilization is built on a story, a story of self. The separate human realm is not in fact separate—just look at how it has altered the planet. In the future we will wield the world-creating power of word consciously, to tell a new story, and thus usher in a consciously creative phase of human development.

Mathematics and Measure

The earliest form of mathematics was no doubt counting; i.e., the invention of numbers. Like nouns, numbers are an abstraction of reality, a reduction of the infinite variability of nature to a collection of standard things. To say there are five of anything presupposes that there could possibly be more than one of any given object, thereby denying the particularity of each being in the universe. When your family sits down to dinner, you don't have to count them to make sure everyone is there. In a society where every person is known as an individual, and where every thing is perceived in its unmediated uniqueness, number would be an absurdity. One could imagine a Paleolithic philosopher protesting, "How can you say those are three? They are one and one and one, each occupying a unique place in the world."

It is therefore unsurprising that numerous hunter-gatherer societies have been discovered, in remote areas, who do not have words for numbers other than "one," "two," and "many."[34] The modern mind typically interprets this as evidence of their childlike simplicity or their lack of cognitive development. But perhaps they simply do not have the need. They live in a concrete world. This does not mean that they cannot distinguish between five and six—reportedly even crows can perform this feat. It just means that these amounts were not subject to abstraction.

Numbering is one of the most primitive forms of measurement, which is none other than the conversion of quality into quantity, the conversion of the specific and unique into the standard and general. In numbering things, we implicitly perform an abstraction by turning a multiplicity of unique objects into just so many uniform *ones*. We realize today that we cannot add apples and oranges. In an earlier world that appreciated the individuality of all things and moments, we realized as well that one cannot even add apples and apples.

Numbers, even more than words, remove objects from their original cross-referential matrix and imply that they are discrete, uniform things. Eventually, as the symbolic, representational world separated us increasingly from the real, immediate world, we took the final step of assigning to numbers an ontological status more real than the things from which they were originally abstracted. Pythagoras, and Plato after him, reversed the original order of abstraction to assign primary reality to the abstractions themselves. Aristotle describes the Pythagorean view that "They supposed the elements of numbers to be the elements of all things, and the whole heaven to be a musical scale and a number."[35] His critique highlights the danger in basing knowledge on the manipulation of abstractions:

> And all the properties of numbers and scales which they could show to agree with the attributes and parts and the whole arrangement of the heavens, they collected and fitted into their scheme; and if there was a gap anywhere, they readily made additions so as to make their theory coherent, e.g., as the number ten is thought to be perfect, they say that the bodies which move through the heavens are ten, but as the visible bodies are only nine, to meet this they invent a tenth.

Already, in the sixth century BCE, scientists were selectively gathering and omitting data to prove their assumptions!

Astonishingly, Pythagoras' elevation of abstraction apparently occurred before the Greeks had even begun using numerals.[36] His mathematics was based purely on geometry and proportion, and therefore still tangibly linked to concrete pebbles and lines in the dirt. Each new advance in mathematics widened the abstraction, made our mode of thought more symbolic, and removed it further from the reality symbolized. The concept of numeral was one such advance, the invention of decimal numbering and the zero another. With the zero, for the first time something stood for nothing—an apt statement of the generalized divorce between our symbols and the reality they purport to represent.

It is no wonder that, because number reduces the variability of concrete reality, neglected variables creep back in to wreak havoc on our attempt to control the world by extending its measurement. Since the time of Galileo, the goal and *modus operandi* of science has essentially been to convert the entire world of observed phenomena into numbers. Measurement converts things to numbers; then the equations of science convert these numbers into other numbers in a burgeoning tower of abstraction. The assumption seems to be that if one day we could measure everything, perfect understanding and thus perfect control would be ours. Indeed, today the hard sciences and even the social sciences offer us "data", i.e., numbers, purporting to encapsulate just about every observable phenomenon. Yet as the failed promise of "Gee whiz! The future!" demonstrates, perfect control remains elusive no matter how much of the world we quantify. We seem to have forgotten that mathematics, and therefore the science and technology built on its scaffold, *by its nature as abstraction* leaves something out. So far our response has been that of the technical fix: to extend measurement still further to encompass those things left out—to remedy its failures with more of the same. On a conceptual level, this program hit a brick wall in the twentieth century with the development of quantum mechanics and chaos theory. On a practical level, we have so far failed to appreciate the lesson in the repeated failure of the program to better manage reality by reducing it to numbers. Instead we call for more numbers, more data.

Mathematics and measure are objective, in the sense that they vitiate objects of the particularity which resides in the interaction of observer and observed. They are consistent with separately existing objects that are "out there", external to our subjectivity, denying a principle common to ancient mysticism and modern physics that "existence" is a two-place predicate, an *interaction*. Today the concept of objectivity is central to our worldview that includes ourselves as separate, discrete individuals. It also underlies classical physics and the Scientific Method, and it informs what we mean by the very adjective "scientific". To see how deeply it has influenced our perceptions, visualize something just "existing". Is your picture that of something floating by itself, alone? No wonder we feel so alone ourselves. To be is to be separate. In this book, I am calling for a revolution, the deepest possible revolution—the replacement of our conception of beingness with a new equation: being equals relationship.

As with language, the abstraction of reality inherent in number has horrifying consequences. As Derrick Jensen says, "It's easier to kill a number than an individual, whether we're talking about so many tons of fish, so many board feet of timber, or so many boxcars of *untermenschen*."[37] The logic and processes of the machine can equally accept as inputs anything that can be quantified and measured. The part left out by number's reduction of reality does not enter into the calculations, even if that part is someone's home, someone's livelihood, or someone's life. Hence the refrain, "I am not a casualty figure, I am a human being." I don't think the cruelty of today's world could exist without the distancing effects of language and measure. Few people can bring themselves to harm a baby, but, distanced by the statistics and data of national policy-making, our leaders do just that, on a mass scale, with hardly a thought.

In the most extreme application, number and language combine in the ultimate expression of objectification and abstraction in the locution "one" to mean "I". Herein, the vitiation of the particular extends even to the self, which is generalized, depersonalized, and made interchangeable, its individuality denied. The generality of this usage of "one" turns all other people into a collection of identical "ones" as well, like so many interchangeable uniform parts of a vast world-machine.

From the very beginning, the concept of number implied an objectification of the universe and a subjugation of the world to human manipulation. Tellingly, the very word number has root meanings of grasping, taking, and seizing,[38] just as the word "digit" means a finger. It is likely, therefore, that the concept of number only arose when the other forces described in this chapter—technology, language, division of labor, and most importantly agriculture—turned the world into an object of manipulation. Number and commodity are highly interdependent concepts that contributed to the replacement of sharing with exchange, commerce, and money. "This ox" became "one ox"; number became abstracted from specific objects. "Number and commodity were now cut apart forever and, as a result, most significantly, numbers could now be applied to quantify any thing around in the world. We were now able to think of the world as something which could, like grain or sheep, be inventoried, controlled, and redistributed."[39]

The conversion of all the world to number implicit in science (and laid out explicitly by Galileo, Leibnitz, and Kant) is inseparable from the program to bring the world under complete control. It is no accident that shortly after the Enlightenment scientists' articulation of the program to convert nature into number, reformers sought to objectify measure as well, converting the old units of weight, length, and volume into new, "rational" units. The metric system replaces the human scale with a scale based on objective features of the observable universe. The old Fahrenheit scale correlates naturally to human experience, wherein zero is about as cold as it ever gets and a hundred is about as hot as it ever gets, whereas the Celsius scale is based on the freezing and boiling point of water at a given pressure. Compare also the foot with the meter, originally defined "objectively" as one forty-millionth of the earth's circumference, and now in terms of the wavelength of a certain frequency of light.[40] Measure has been removed from its original source in the human body and everyday experience.

In the last century, the reduction of the world to numbers has only accelerated, especially in the computer age. Music is a prime example. While musical notation had set the stage long before, the mathematization of music intensified with Bach, because of whom, "The individual voice lost its independence and tone was no longer understood as sung but as a mechanical conception. Bach, treating music as a sort of math, moved it out of the stage of vocal polyphony to that of instrumental harmony, based always on a single, autonomous tone fixed by instruments, instead of somewhat variable with human voices."[41] Today, with the digitization of music, its transformation into just so many bits of data is complete: music is, like any other "digital content", nothing but a series of numbers. Here is a good example of the reduction that quantification entails: contrary to popular belief, the standard CD-audio format is discernibly different from analogue, particularly at the higher frequencies. Some of the original richness is gone forever; music connoisseurs sometimes speak of the "warmth" of vinyl as compared to the coldness of digital sound. The technological solution, of course, is to increase the sampling rate to the point where the lost data is below human powers of auditory discernment. The infinite will still have been made finite, the continuous made discrete, but at least, it is hoped, this semblance of reality will be "good enough".

Digitization is applied to images as well, and potentially to anything capable of "analysis." The motions of the human body, for example, can be converted into a set of numerical coordinates in 3-dimensional space; the sounds of human speech to just so many sine waves. The supposedly inherent reducibility (and notice the naturalness of the word "reducibility") of the world into numbers, and the assumption that either nothing significant is lost, or that we'll never notice what is lost provided enough numbers are used, motivates the ultimate technological separation of humans from reality called "virtual reality" (VR). VR is the penultimate step in the substitution of manufactured reality for natural reality. With VR, the separate human realm will be complete except for one, final step. If, as science implies, the entire universe is reducible to numbers, then we too are so reducible; hence the science-fiction scenarios of one day achieving immortality by uploading our consciousness onto a computer, where we could enjoy the best, most pleasurable of artificial experiences forever.[42] The Technological Program of complete control over the world we experience would be fulfilled. Or so goes the fantasy.

Science fiction writers such as Neal Stephenson and Vernor Vinge have described futures in which people live almost entirely in digital representations of reality, or in wholly constructed realities. Such scenarios are already taking shape in the individual RFID labeling of all products sold, setting the stage for the Internet to become nothing less than a virtual copy of the entire planet:

> The present IP, offering 32-bit data labels, can now offer every living human a unique online address, limiting direct access to something like 10 billion Web pages or specific computers. In contrast, IPv6 will use 128 bits. This will allow the virtual tagging of every cubic centimeter of the earth's surface, from sea level to mountaintop, spreading a multidimensional data overlay across the planet. Every tagged or manmade object may participate, from your wristwatch to a nearby lamppost, vending machine or trash can—even most of the discarded contents of the trash can.[43]

As presaged by naming and number, the potential exists for us to attempt the conversion of every object and person in the world into a dataset.

Whether in the digitization of music or the quantization of reality implicit in VR, the insanity of our ferocious strivings to manufacture a reality almost as good as the real thing should be plain. We are exerting tremendous effort to make an inferior version of a freely available original, similar to our fevered attempt to recreate the original affluence, in which "work" was not yet a concept, through ever more efficient labor-saving devices. That manufactured reality is inferior is demonstrated by the increasing intensity of our simulations, which seek to compensate for the lost richness and intensity of unmediated experience. No matter how large a set of numbers we produce to describe reality, something of its infinitude is lost. Instead we must settle for "close enough", the lesser lives we live in the age of separation. The systems of representation—number, language, image, and so forth—we interpose between ourselves and reality are always a reduction of what they purport to represent. The Scientific Program of a complete mathematization of the universe is yet another Tower of Babel, aiming to attain the Infinite by finite means.

Keeping Time

The ultimate and perhaps most significant conversion of reality into numbers is the measurement of time. Clocks do to time what name and number do to the material world: they reduce it, make it finite. And what is time, but life itself? Time is experience, process, the flow of being. By measuring time, by converting it into numbers, we rob it of its infinitude and uniqueness in precisely the same way that nouns and numbers reduce the physical world. Time measurement turns a succession of unique moments into just so many seconds, minutes, and hours, and denies the particularity of each person's subjective experience of them.

The keeping of time began in Neolithic times with the calendar, used to manage the planting of crops. Since calendars are based on natural cycles of the sun, moon, and seasons, their distancing effect is minimal, just as early agriculturalists were still tightly wedded to Nature. Because the measure of time was cyclical and not linear, early calendars did not have the effect of binding time, creating history, and numbering the years of one's life. Soon, however, with the rise of long-distance commerce and hierarchical government, it became necessary to keep records over a span of years. In Egypt, Mesopotamia, India, China, and Central America, people began to number the years, e.g. from the start of a dynasty, thereby introducing linearity to time and divorcing it from the cycles of Nature. The artificial division of the day into hours (curiously, both the Babylonians and ancient Chinese used twelve) and the Hebrew invention of seven-day weeks only deepened this divorce, which has culminated in the replacement of circular clocks with digital clocks, obliterating the last remaining link between measured time and the cyclical processes of nature.

Crude division of the day into hours was sufficient for the demands of the Iron Age, but industry requires a far more precise coordination of human activity. The development of mechanical clocks in the late Middle Ages set the stage for the Industrial Revolution. As Lewis Mumford put it, "The clock, not the steam engine, is the key machine of the industrial age."[44] The more finely we divided and measured time, first into hours, then minutes and seconds, the less we seemed to have of it and the more the clock encroached upon and usurped sovereignty over life, until today we are all "on the clock."

To be punctual is the onus of a slave toward a master or a subject toward a king. Today we are all subject to schedules imposed by the machine requirements of precision, regularity, and standardization. We think of machines as our servants, but our constant rush to be on time says otherwise.

Immersed in linear time measurement, it is hard to appreciate the audacity of dividing up the day into standard units, manmade hours, minutes, and seconds, that are deliberately unconnected to natural processes and therefore "objective." The idea, to paraphrase Thomas Pynchon, that every second is of equal length and irrevocable is only as recent as the clock.[45] Or as Paul Campos puts it, "Until very recently there was no such thing as '6:17 a.m.'"[46]

The clock translates heavenly movement into earthly routine. Time measurement profoundly accelerated human separation from nature. Mumford comments:

> By its essential nature, [the clock] dissociated time from human events and helped create the belief in an independent world of mathematically measurable sequences: the special world of science. There is relatively little foundation for this belief in common human experience: throughout the year the days are of uneven duration, and not merely does the relation between day and night steadily change, but a slight journey from East to West alters astronomical time by a certain number of minutes. In terms of the human organism itself, mechanical time is even more foreign: while human life has regularities of its own, the beat

> of the pulse, the breathing of the lungs, these change from hour to hour with mood and action, and in the longers span of days, time is measured not by the calendar but by the events that occupy it. [47]

In effect, clocks turn time into another standardized, interchangeable part of the World Machine, facilitating the engineering of the world. Only time thus devalued is a conceivable object of commerce. Otherwise, who would sell their moments, each infinitely precious, for a wage? Who would reduce time, i.e. life, to mere money? Leibnitz' merciless phrase, "Time is money," encapsulates a profound reduction of the world and enslavement of the spirit.

It is not surprising that the revolutionaries of Paris's 1830 July Revolution went around the city smashing its clocks.[48] The fundamental purpose of clocks is not to measure time, it is to coordinate human activity. Aside from that it is a fiction, a pretense: as Thoreau said, "Time measures nothing but itself."[49] Smashing the clocks represents a refusal to sell one's time, a refusal to schedule one's life or to bring it into conformity with the needs of specialized mass society. Further, it represents a declaration that "I will live my own life," establishing the ascendancy of now.

The scheduled and hurried life is the life of a slave, whose life is not his own. A fundamental power over another is to compel him to appear when beckoned: "When I say come, you will come." To rule a person's time is to rule his life. In modern society we are chronically busy, too busy to do the things we want to, too busy to stop and smell the roses, too busy to spend an hour looking at clouds, too busy to play games with children, too busy to spend more time on anything than is necessary.

As John Zerzan so poignantly observes, the clock makes "time scarce and life short"; hence the compulsive obsession with speed, efficiency, and convenience in modern technological society. Why else would we seek to get there faster, do it faster, have it faster, except for the belief that our days are numbered? The anxiety of modern society comes in large part from the feeling that there is not enough time. Daniel Greenberg explains, "You've always got to be doing something useful. You have to account for every minute of the day in a productive way. If, when you go to sleep at night, you can't really say that you have used every minute of your time productively, then a piece of your life has flitted by, never to return again. You've just squandered it."[50] After all, any moment could be used to exercise more control over the world, to enhance survival and comfort. Maybe, after we have maximized the possibility of all these things, then we *can afford* some leisure, play, recreation.

Afford? That is a financial metaphor, is it not? Time is money.

To be "at leisure" originally meant not to be subject to the constraints of time; today we schedule in leisure along with everything else, and the freedom to linger at whim by the roadside until good and ready to leave seems a rare luxury. Our leisure is more akin to a prisoner's furlough. We have lost the primal right to our own time.

The pace of modern life continues to accelerate. In business we have "just-in-time" inventory management, "instantaneous communications," "same-day turnaround." We schedule our days more and more tightly, down to the minute, even as we extend the regime of the schedule further and further into childhood, starting with the imposition of hospital schedules on the newborn. "Time management" and "multitasking" have become essential skills in coping with the onrushing deluge of modern life. They are, along with devices such as cell phones and personal digital assistants, technological fixes that apply yet more control to deal with the problems caused by control.

Even as adult life marches to the ever-accelerating beat of the machine, so have the endless afternoons of childhood given way to the scheduled confines of school and other programmed activities. For the first time in history, children are too busy to play.

Consider the tragedy of that statement, let it reverberate in your chest: *children are too busy to play*. The reason comes down again to survival anxiety. Play is a luxury, a frivolity relegated to the cracks within the schedule of productive, educational, and developmental activities. The competitive demands of adulthood today dictate that no time be wasted in play, because every moment at play is a moment where your child could be getting ahead in life, preparing for the future. After all, play in adulthood is limited to our "time off," and childhood is preparation for adulthood, is it not? So we seek to instill good "study habits" and a strong work ethic in our children, a sense of responsibility lest they learn to put play, pleasure, and joy first. What kind of adult would they become then? Probably an undisciplined adult who cannot hold a nine-to-five job and has little patience with boring, demeaning, or unpleasant work—the kind of work most of the population accepts as a grim necessity. So school comes first, then homework, then piano practice, then little league soccer. Then, if there is any time left, they can play.

Several years ago I noticed that when I raised my voice at my children, it was usually because of time pressure. Perhaps we had to be somewhere at a certain time and they were not cooperating. More often, rather than a specific scheduled obligation, it was a non-specific feeling of anxiety, of time being short, of needing to move on to the next thing. Time scarcity becomes a habit of thought, a way of being.

To force spontaneous, uninhibited little children to conform to adult schedules requires just that: force. My children resisted scheduling. Whatever they wanted to do, they wanted to do it *now*, and for as long as it took. If we were never in a hurry we would never lose our patience. As I live my life less by the clock, I find I rarely yell at my children or lose my patience with them, not because I've become a saintly person, but simply because there is no reason to. Let them take as long as they want. Okay Matthew, go ahead and play with your socks for half an hour instead of putting them on. Why would I not let a little boy play with his socks? Only if we were "running late" as we chronically are in modern society. Always, the next obligation hangs over our head, keeping us from full devotion to what we are doing. The constant interruption in the natural rhythm of children's play, which must steal whatever moments it can from the cracks in the adult-imposed schedule, trains us for an adulthood of furtive, hurried pleasures.

As alluded to in the opening chapter, the endemic busy-ness of modern life is one of its defining features, and certainly not a temporary aberration to be abolished with the next generation of futuristic "labor-saving" devices. To be busy is to be unfree; it is to have one's time constrained. It is to be subject to the priorities of necessity. It is the natural result of a childhood acculturation that yokes our lives to the omnipresent threat of deprivation—of affection, approval, or even of physical comfort. We are by adulthood deeply conditioned against play.

When we say we are too busy, what do we mean? We mean that we have to do other things, priorities dictated by survival; we mean we are not at liberty to do what we want to. "It just would not be practical," we believe, "to put play before work." We envision losing our jobs, going bankrupt, ending up on the street. That play can actually be productive without consciously directing it at productivity rarely occurs to us, and when it does we assign it to the province of those lucky few, artists and geniuses, who get to do what they love. But actually the logic is backwards. Genius is the result of doing what you love, not a prerequisite for it. The problem, of course, is discovering what that is. That is what childhood is supposed to be for, but our culture has turned it into the opposite. When we are so thoroughly broken that we know not what we love, the only way out is first to stop doing what we do not love, to do nothing for a while. This is the message encoded in the Biblical story of Exodus, in which the children of Israel, after fleeing slavery, had to wander the desert for forty years before they found the land of milk and honey. Similarly, we must overthrow the dictatorship of busyness and allow ourselves to wander for a while, in order to discover our bliss.

The deepest irony of all, and the gravest indication of our servitude, is seen in our aversion to long expanses of empty time. The true slave is made to fear freedom. Thus we fill up our empty moments with "pastimes"; we seek to be "entertained"; that is, to be taken away from ourselves. The underlying anxiety of modern life has robbed us of our moments and bound us to ceaseless doing.

The measurement of time, especially its linear measurement, has among its consequences the concept of an abstract future, another fundamental source of anxiety and insecurity. There is always something to prepare for, always a reason to be incompletely immersed in the now. When we mortgage the present to the future it is usually in the interests of the same practicality we invoke when we "cannot afford to." Compare our survival anxiety to Marshall Sahlins' description of hunter-gatherers:

> A more serious issue is presented by the frequent and exasperated observation of a certain "lack of foresight" among hunters and gatherers. Orientated forever in the present, without the slightest thought of, or care for, what the morrow may bring, the hunter seems unwilling to husband supplies, incapable of a planned response to the doom surely awaiting him. He adopts instead a studied unconcern… [51]

In Sahlins' "original affluent society," where nature provides everything in easy abundance, there is no need to plan for the future. In an agricultural society, though, such nonchalance can be fatal, so it gave way to the time-bound mentality of sowing and reaping that still governs the modern mind. Always, we can be doing something to enhance our rational self-interest, to improve our position in the world, to increase our future security. Things are never okay just as they are when life is at root a struggle for survival. Thus the present becomes slave to the future. That is the defining mentality of agriculture, in which today's labor brings tomorrow's harvest. In the eternal present, would work ever supersede play?

A similar phenomenon obtains at the collective level: without the abstractions of future and past there would be no such thing as progress. Conversely, without a concept of progress there is little use for time. The accumulation of culture and technology defines an arrow of time by reference to simpler days. Early on in this exponential ascent, when change was slow, only the most rudimentary consciousness of time existed. As change built upon itself, the awareness of time crystallized as well.

Time was not an invention that could have been refused had only we been wiser. Rather, it was an inevitable product of the progression of language, number, and technology, each of which built upon itself and the others. Written language, for example, which was originally used for record-keeping, allowed the linear binding of time and the conception of history, as words were no longer confined to a moment.

Because it is inextricably tied up with speed, convenience, efficiency, and progress, technology as we know it is rooted in reified, linear time. This invites the question as to whether any other conception of technology is possible. It is difficult to imagine. It would seem that any technology which builds upon itself thereby defines progress and therefore linear time. However, progress need not be unnatural or destructive. Life, after all, has evolved into forms of increasing complexity over four billion years, only veering into planetary crisis in the last few thousand. This suggests a different mode of technology, which seeks explicitly to recover and harmonize with nature's patterns. Such technology actually has a precedent in the magical practices of primitive peoples, which sought, in John Zerzan's words, "the regularity, not the supercession, of the processes of nature." Post-technology technology, if I may use such a phrase, will take as its model the cycles of nature and in particular, the "magical" practices of ancient people. It will seek attunement and not conquest, and it will be occupied not with control but with beauty. This mode of technology, which I will describe later in the book, will not be a separation from nature at all, but rather an organic extension of nature that will return us to a timeless life in which linear time measurement is but a plaything.

Images of Perfection

The separate human realm that grew up around stone and fire, language, number, and time-binding finds explicit reflection in the world of images. Representational art—pictures of things—is quite literally a human-made copy of the world, the world interpreted through human eyes. Extending now far beyond art, the separate realm of images has increasingly taken on a life of its own, diverging from the reality it once represented. The fake images of politicians and corporations, along with our own self-images that we project, are part of a phony world, a realm of appearances. As these appearances diverge more and more from reality, so also grows our intuition of an inauthenticity to life. This section explores the evolution of the world of images that immerses us today.

Living in the Age of Separation, we tend to see and understand art either as a depiction (of things, ideas, or emotions) mediating human and nature, or as an appendage to life offering aesthetic pleasure: representation or decoration. But could images serve another purpose?

Representational paintings and sculpture are quite new on the face of the earth. Neanderthals did not make them, though they apparently did adopt the use of pigments, possibly by diffusion from modern humans.[52] The earliest known representational paintings were found in a Namibian cave dating back some 59,000 years, followed by an abundance of paintings throughout Africa, Europe, and Australia starting around 30,000 years ago. Most depict animals and are conventionally interpreted as attempts to bring luck to the hunt through "sympathetic magic". Similarly, magical or ritual uses are ascribed to the earliest sculptures, which also date to this period and which appear to be fertility symbols.

In other words, scholars explain Paleolithic art as an attempt to affect the real world by manipulating its representation. It was a bumbling attempt, though—the kind of magic we have replaced with technology that actually works. Ha ha! Those poor primitives actually thought pictures on a wall would influence the events those pictures depicted. Consider, however, that these scholars' interpretation is yet another projection, a projection of our own anxiety and separation onto primitive artists. Perhaps it is we, and not they, who manipulate a matrix of appearances under the delusion that it is the real thing.

Let us not blithely assume that hunter-gatherer art was representational in the sense we understand it. That a picture of an animal is separate and distinct from the real animal seems obvious to us, but not to the primitive mind. As Joseph Epes Brown observes, "Symbols in traditional Native American art… instead of simply pointing to what they represent actually become what they represent."[53] From the modern perspective we tend to wink patronizingly at such superstitious primitivism, at those poor fools who actually thought their paintings would bring luck to the hunt, and for that matter at all the rituals of every "primitive" society that has ever existed on earth, as if we were past all that now, as if technology has replaced the illusory control of nature that subsists in ritual with real control. But consider: if every human culture up until recent times believed in the power of magic ritual, is it not conceivable that there was something to it? This belief was *universal* among pre-agricultural peoples. Perhaps it is we moderns who have forgotten the truth, and lost touch with basic principles of the universe that every other culture knew. In any event, ostensibly representational art need not indicate a dualistic mindset; more likely, the dualism grew over time as other developments contributed to a great forgetting of the original magical significance of art.

Significantly, as technology and agriculture furthered the distance between human beings and nature, the depictions of artists became more and more stylized, standardized, and abstract, losing the vibrant aliveness of the Lascaux and Chauvet cave paintings, which seemed somehow to capture the spirit of the animals depicted. Compare these to the well-known stylized figures of the agricultural society of ancient Egypt. It is as if the farther the artist is removed from nature, the less able he or she is to be a direct channel for the infinity of the real world that somehow impregnates the truly great works of art, and the more necessary it becomes to control and confine the wild through its representation. As John Zerzan puts it, "art turns the subject into an object," confining the infinity of the real world within the conceptual and perceptual framework of a human artist.

There remains the possibility, even for a modern artist, of achieving a momentary freedom from culture and domestication, but this requires tapping into her own inner infinity: the spontaneous, undomesticated spirit that lies deeply buried under the vast weight of culture. Such art is described in the Gurdjieff tradition as "objective art," not because it reifies representational forms, but precisely because, like Paleolithic images, its primary existence is not representation but rather a thing in and of itself; that is, it is real. Such art has genuine power. Its presence moves us and can indeed change the world.

As human beings turned from the Original Religion toward the dualistic religions of agriculture, the magical import of art transformed from an inherency in the object itself to an inherency in its representational power. Pre-alienation peoples (animists) believed that the entire universe, animate and inanimate, is living spirit manifest. As separation grew, human beings separated spirit out from matter and conceived for the first time that some things might be more spiritual than others. Images and by extension rituals became, as John Raulston Saul puts it, "a magic trap."[54] Its power no longer resided in what it was, but in what it symbolized, the spirit it *contained*. Much like a soul inhabiting, but separate from, a non-spiritual body, the magical item became a *vessel* not magical in and of itself. It became magical only because of what was invested in it. Content became separate from form, exactly paralleling the development of semantic meanings for the sounds of the *lingua adamica*. Animistic art admits no such separation. Shaker furniture is a good example: separate neither from form nor function, it lies in the perfection of both.

As the magical power of art came to be understood in terms of its representational content, artists would pursue perfection in their images, which would be tantamount to perfect control over the world. Yet when Raphael, Michelangelo, and Leonardo da Vinci made the final technical breakthroughs in perspective around the year 1500, nothing magical happened. Saul writes, "In the West the painter's and sculptor's job has been to design the perfect trap for human immortality."[55] The ambition was to make representation real, the same ambition that lives on in the latest version of the trap for immortality: the software simulation of consciousness in a VR world. We now live almost wholly in a manufactured reality, a world of images, but really nothing has changed. Using finite methods to approximate the infinite, using representation to approximate reality, we succeed only in building a taller Tower. Can it ever reach the sky?

The solution of the puzzle of perspective was not just a technical breakthrough, but a conceptual shift as well. Perspective freezes objects in time by assuming a single viewpoint at a single moment. "Before the fourteenth century there was no attempt at perspective because the painter attempted to record things as they are, not as they look."[56] Perspective painting records objects as they look *at a given moment of time*, thereby implicitly subordinating reality to the human observer and affirming the fundamental dualism of self and world. Here is the separate "I am" of Descartes, a discrete point of perception gazing out upon the Other. Perhaps the Medieval painters were not awaiting a technical breakthrough, but were simply not interested in attempting perspective at a time when the modern conception of self had not fully matured, and when measured time had not fully supplanted the rhythms of nature.

There is an element of hubris in the capturing of reality that is the perfect image. Perhaps this explains the deep suspicion many religious traditions have of images of the divine or even of humans, animals, and objects. Judaism and Islam do without them altogether. So did Buddhism until Alexander the Great brought Greek culture, with its obsession with mortality and, not coincidentally, with images, to India.[57] There are also many cultures in which people resist having their photographs taken, intuiting perhaps the hubris and futility of the pseudo-immortality it suggests, or the Faustian exchange of a moment in time, which is real, for the frozen representation of a moment, which is not. Indeed I have noticed that cameras and—even worse—videotaping tend to detract from the happy events they are supposed to preserve on film. Taking photographs at birthday parties and weddings to preserve those happy memories substitutes an image for an experience, and can sometimes imbue the event with a staged feel, as if it were not real, as if it were an enactment to be enjoyed later. It is as if we are uncomfortable with real moments and prefer to experience them from a distance, second-hand. In the most extreme case, the photographic or video recording of the event comes to completely define the event itself. This is certainly the case in the realm of public relations and politics.

A small but significant way to reduce the alienation of modern life is to put down the camera and participate fully in the moment, rather than trying, futilely, to preserve the moment on film. The compulsion to record everything bespeaks the underlying anxiety of modern life, the conviction, stemming from measured time, that our lives are slipping away from us day by day, hour by hour, moment by moment. Perhaps if I photograph them, those precious moments of my children's childhood will always be there, preserved for eternity. I have noticed, though, that when I look at my sons' baby pictures my main emotion is wistfulness, a regret that I did not truly and fully appreciate those precious, unique times. I can seldom look at my most treasured photographs without feeling sadness and regret. The very effort to possess and preserve those moments diminishes them, just as technology in general leaves us alienated from and more afraid of the very world it attempts to control.

It is much better to enjoy each beautiful moment in the serene knowledge that an infinity of equally yet differently beautiful moments await. At the same time, the awareness of each moment's transience helps us appreciate it all the more, if only we don't succumb to the illusion, offered for example by photography, that it can be made permanent. That illusion robs life of its urgency and intensity, substituting for it an insipid complacency that conceals our buried unmet hunger for real experience. And that unmet hunger, in turn, fuels an endless appetite for the vicarious imitation experiences to be found in television, movies, amusement parks, spectator sports, and—the last gasp—reality TV.

Buddhism (and, arguably, the esoteric teachings of all religions) recognizes the suffering implicit in the attempt to make permanent that which is intrinsically impermanent. The beautiful sand paintings made by Tibetan monks and Navajo Indians, which by the nature of their medium last a very short time, demonstrate an important principle: the value of beauty does not depend on its preservation. The modern mind tends to think of their creation as a waste of time—creating something beautiful only to destroy it again—and wants to preserve it in a museum, derive some "benefit" from it. This way of thinking, in which we mortgage the present moment to future moments, is precisely the mentality of agriculture, in which we must sow in order to reap, in which the future motivates and justifies the labor of the present. When we photograph, record, and archive the present, we are driven by the same anxiety as the agriculturalist who knows that unless he stores up grain now, there will be scarcity in the future. Just as the agriculturalist no longer trusts (as hunter-gatherers do) in Providence, the easy bounty of nature, so also are we compelled to save up beautiful moments as if their supply were limited.

When nothing magical happened when the painted image was perfected, it did not take long for disappointment and eventually desperation to infect the world of art. Saul writes, "For some twenty years after Raphael's discovery, craftsmen celebrated their triumph with an outpouring of genius. But gradually the subconscious failure beneath this conscious success began to slow them and to darken their perspective. The viewer has only to watch Titian's opulence and sensuous joy gradually turn tragic. With no room left for progress, the image turned and dodged and circled back and buried itself like an animal chained to its own impossible promise, searching for some way to get beyond the mortality of the real."[58] But try as it might, through all the contortions of the successive waves of artistic "movements," from Impressionism's demonstration of the facticity of the image to Cubism's shattering of perspective, the image cannot escape from the reality that it is nothing more than paint on canvas.

Further perfections of the image only reinforced the disappointment. Photography, then motion pictures, then holography, similarly failed to produce magical results; that is, actual control of reality via control over its representation. Yet, unwilling to admit defeat, we press on with "virtual reality," a fitting metaphor for the dead end to which our separation of self has brought us. The separate human realm that originated with the circle of the campfire is nearly complete now, a wholly artificial reality. We have arrived, only to find ourselves feeling more lost than ever before.

The second-hand pseudo-experiences that today's entertainment media provide assuage our hunger for real experience temporarily, but in the end only intensify that hunger. Like any object of addiction, they are a counterfeit that leaves the real need unmet, a shortcoming that is temporarily disguisable by increasing the dose. Movies and music, for example, have become progressively more intense, louder, and faster-paced over recent decades. Computer 3-D animation has brought to fruition the painters' old dream of the perfect image, even a moving image, yet the inconvenient fact remains that however perfect, the image can never be real, and the virtual world can never be a real world. And this phenomenon is not limited to the entertainment media. John Zerzan writes, "Everyone can feel the nothingness, the void, just beneath the surface of everyday routines and securities." We live in a world of images, of representations, that separate us from real experience; we then try to meet the hunger for reality with yet more images. The further intensification of dosage is therefore an inevitability—we can never get enough. It is like trying to assuage hunger by chewing gum.

The 2004 Presidential campaign offers an example of just how far we have become lost in a world of images disconnected from reality. In the words of journalist Matt Taibbi, "The whole thing could easily be done in a movie studio."[59] It is a world in which nothing is real and everything is staged, in which every action is calculated and every encounter scripted according to how it will look. The inauthenticity is palpable as candidates everywhere obsess about their image. Speaking of the Kerry campaign, Taibbi explains, "All they want to do is hit those dial survey words they know are going to impress voters. Change. Leadership. Strength. These are literally infantile concepts that they want to communicate to you." The delusion that image trumps reality has even extended to official U.S. foreign policy: the "brand America" campaign to improve America's "image" in the Middle East.

The world of images of which I speak is not limited to corporate identity systems and the posturing of politicians. Even on an individual level, we tend to be obsessed with appearances, with projecting an acceptable image of ourselves to the world. I'm not talking about the superficialities of fashion, but of the subtle masks and poses we wear as shields in a world filled with strangers. The unreality of the modern world projects onto our faces, into our voices, into our thoughts; hence the mounting desire to "get real", the search for authenticity.

Just as words are losing their meaning, so also have images lost their power and mystique. Leibnitz and other thinkers in the Age of Reason believed in the possibility of inventing a perfect language so congruent to reality that no false utterances would even be possible. They did not realize that the object of their search existed prior to language and could never be reached by increased precision of representation, but only by transcending representation altogether. No language has ever before had the number of words that exist in English, the profusion of precise technical terms for increasingly fine divisions of reality, yet true meaning recedes all the further. In the same way, images of reality proliferate toward ubiquity, to the point where the cellular phones have digital cameras built in and video cameras record our every movement in public space, but they don't seem to mean anything anymore. The most graphic cinematic images of violence leave us unmoved, because we know that despite their near-perfect verisimilitude, they are not real. Nothing is, in a world of images.

What power images still bear is rapidly fading as it becomes easier and easier to manufacture perfect images unconnected to reality. Shortly before the Abu Ghraib prison abuse scandal, a photograph circulated of a smiling GI and an Iraqi boy next to him cheerfully holding up a sign that said "This man raped my mother and killed my father." The sick joke was that the boy didn't understand English and was unwittingly humiliating himself. But then other versions of the photo circulated with a different message on the sign, and it was impossible to tell which was authentic.

Yet even when their authenticity is not in question, somehow photographs are losing their power to shock. The muted American response to the Abu Ghraib images testifies to this. Where is the outrage, the disgust, the national shame? Perhaps because of the increasing ubiquity of depictions of graphic violence and diabolical cruelty in the entertainment media, we have seemingly become inured to violent or tragic images even when they are not fictitious. We treat them as just so many pixels. Does not life go on as usual anyway?

A further symptom and consequence of the divorce between reality and image is the gradual migration of art to the position of irrelevance that the dualistic distinction between art and craft implies. Art is concerned with aesthetics, craft with function. In medieval times, painters were considered to be craftsmen performing a useful social, political, and religious function. It was not until the 18th century that a distinction between fine arts and useful arts was conceived. As Saul wryly observes, "Put another way, by the eighteenth century society was beginning consciously to doubt that art was useful."[60] By the next century the first museums for purely aesthetic purposes were established. The aesthetic function of art, today accepted almost universally, virtually consigns it to irrelevance, a pretty flourish divorced from function. As Zerzan puts it, "Kierkegaard found the defining trait of the aesthetic outlook to be its hospitable reconciliation of all points of view and its evasion of choice. This can be seen in the perpetual compromise that at once valorizes art only to repudiate its intent and contents with 'well, after all, it is only art.'"[61]

Maybe this is why I, along with all little children, detest art museums. Ordinarily I can walk ten miles without undue discomfort, but after about twenty minutes in a museum my feet get tired and I start whining for someone to buy me an ice cream cone. Museums reaffirm that art is just for looking at. Their walls and glass cases physically separate it from the rest of the world. Don't touch! Despite their great efforts to preserve art, which ought to validate its importance, by their very nature museums affirm art's irrelevancy and separateness.

Going back a little further, we can call into question the distinction not just between art and craft but between art and life as well. According to Joseph Epes Brown, Native American languages have no term for art at all. He quotes James Houston as follows: "I believe that Eskimos do not have a satisfactory word for art because they have never felt the need for such a term. Like most other hunting societies, they have thought of the whole act of living in harmony with nature as their art."[62] Later he writes, "To dismiss utilitarian items as only 'crafts' in this way has contributed to the tragic separation of art from life. Within created traditional Native American forms, however, there can be no such dichotomy, because art is not only the particular created form but also the inner principle from which the outer form comes into being. An art form is often seen as beautiful not just in terms of aesthetics but also because of its usefulness and the degree to which it serves its purpose."[63]

Art can be a way of life that encompasses all its dimensions. Echoing Brown, Wendell Berry writes,

> The possibility of an entirely secular art and of works of art that are spiritless or ugly or useless is not a possibility that has been among us for very long. Traditionally, the arts have been ways of making that have placed a just value on their materials or subjects, on the uses and the users of the things made by art, and on the artists themselves. They have, that is, been ways of giving honor to the works of God.... There is not material or subject in Creation that in using, we are excused from using well; there is no work in which we are excused from being able and responsible artists.[64]

Such honoring of Creation—that is, the world—is utterly inconsistent with the logic of modern biology and economics. Why do anything better than it needs to be done? The dictates of biological survival and reproduction, or of economic competition, reward a certain level of excellence, but nothing beyond that, nothing motivated internally by the worker and her materials, but only externally, by the environment or the marketplace. Good enough, in other words, is good enough. I see the results of this all the time in education, in which learning is motivated by the grade. Why learn more than what is necessary for the "A"? Under the logic of the discrete and separate self, any reason to do more than good enough is always an illusion. Our very world-view has sundered art from life, rendering the former frivolous or irrelevant, and the latter a dispirited caricature of itself, an empty shell. The reuniting of these two categories, art and life, is central to the healing that will occur with the collapse of the Age of Separation.

The Marvelous Piraha

Some months after writing the rest of this chapter I came across the work of Daniel Everett, a linguist who has spent more than a decade studying the language and culture of the Piraha.[65] The Piraha, a small tribe of hunter-gatherers in Brazil, have resisted, with breathtaking consistency, all the developments in linguistic abstraction, representational art, number, and time described above.

While this tribe has been in contact with other Brazilians for two centuries, for some reason they have maintained an extreme degree of linguistic and cultural integrity, remaining monolingual to this day. Significantly, in not just one but *all the areas* described in this chapter, they exhibit very little of the separation implicit in modern symbolic culture.

They do not impose linearity onto time. They do not abstract the specific into the generic through numbering. They do not usually genericize individual human beings through pronouns. They do not freeze time into representation through drawing. They do not reduce the continuum of color to a discrete finitude by naming colors. They have little independent concept of fingers, the basis for number, grasping, and controlling; nor do they use fingers to point.

Most strikingly, the Piraha are unable to count.[66] Not only do they have no words for numbers, their language also lacks any quantifiers such as "many", "some", or "all". Even more amazing, they apparently are incapable of even *learning* to count. Despite eight months of sustained efforts, speech pathologist Peter Gordon failed to teach them, even with the Piraha's enthusiastic cooperation. They cannot mimic a series of knocks because they cannot keep count of how many there have been.[67]

The Piraha language is nearly devoid of any sort of abstraction. There is no semantic embedding, as in locutions like "I think she wants to come." ("She wants to come" is a nominalized phrase embedded in "I think [X]"). The lack of nominalized phrases means that words are not abstracted from reality to be conceived as things-in-themselves. Grammar is not an infinitely extendible template that can generate meaning abstractly through mere syntax. Words are only used in concrete reference to objects of direct experience. There are, for example, no myths of any sort in Piraha, nor do the Piraha tell fictional stories. This absence of abstraction also explains the lack of terms for numbers.

Even colors do not exist in the abstract for Piraha. While they are clearly able to discern colors and to use words like "blood" or "dirt" as modifiers to describe colored objects, these words do not refer to any color in the abstract. One cannot say, for example, "I like red things, " or "Do not eat red things in the jungle" in Piraha.[68]

Even the very idea of abstract representation is apparently impossible to explain to the Piraha. Everett describes his own attempt:

> If one tries to suggest, as we originally did, in a math class, for example, that there is actually a preferred response to a specific question, this is unwelcome and will likely mean changing subjects and/or irritation. As a further example of this, consider the fact that Pirahas will 'write stories' on paper I give them, which are just random marks, then 'read' the stories back to me, i.e. just telling me something random about their day, etc. which they claim to be reading from their marks. They may even make marks on paper and say random Portugeuse numbers, while holding the paper for me to see. They do not understand at all that such

> symbols should be precise (demonstrated when I ask them about them or ask them to draw a symbol twice, in which case it is never replicated) and consider their 'writing' as exactly the same as the marks that I make. [69]

Abstraction is also absent from their art. The Piraha do not draw representational figures at all, except for crude stick figures used to explain to the anthropologist the spirit world , of which they claim direct experience. They cannot even draw straight lines. As Everett continues from above, "In literacy classes, however, we were never able to train a Piraha to even draw a straight line without serious 'coaching' and they are never able to repeat the feat in subsequent trials without more coaching." This is highly significant, given that the straight line is itself an abstraction, being absent from nature. It is an abstraction, moreover, fraught with powerful cultural and psychological implications. At the most literal level, the Piraha do not engage in linear thinking.

This absence of linear thinking comes out in the language, which lacks tenses, and not only in the morphosyntactical sense of conjugations and tense markers. There is simply no verbal way to fix an event at a specific point in the past or future, for Piraha doesn't have words for tomorrow, yesterday, next month, or last year. The sentence, "Let us meet here in three days"—or even, "Let us meet here tomorrow"—is inexpressible in Piraha. Piraha has only twelve time words at all, such as day, night, full moon, high water, low water, already, now, early morning, and another day. None of them allow the establishment of a time line. Accordingly, the Piraha have no sense of history, no stories that reach back before living memory, and no creation myths. "Pirahas say, when pressed about creation, for example, simply 'Everything is the same', meaning nothing changes, nothing is created."[70] They often do not know the names of their deceased grandparents; their kinship terms do not apply to dead people. Theirs is a timeless world. The past, after all, is just another abstraction as soon as it extends back before living memory.

The Piraha similarly abstain from projection into the future, sharing with other hunter-gatherers the nonchalance and disdain for food storage described in Chapter One. They are aware of food storage methods such as drying, salting, and so forth, but only use these techniques to make items for barter. For themselves they store no food, explaining to Everett, "I store meat in the belly of my brother". In other words, says Everett, "They share with those who need meat, never storing for the future." A further level of interpretation of this statement is also possible, however: taken literally, it suggests a different conception of self-interest and therefore a different conception of self. To help another is to help oneself. We are not separate.

Like other hunter-gatherers, the Piraha have few material possessions, and those they do possess are very impermanent: baskets that last a day or two, dwellings that last until the next storm. Their material culture makes no provision for security in the future, no provision for progress, betterment, or accumulation.

A final refusal of separation lies in the Piraha's incapacity to form an abstract concept of value. Unable to understand money, they rely on barter for trade, and in these transactions tend to be painfully ingenuous. They present what they have to offer (Brazil nuts, raw rubber, and so forth) and point to items in the trader's boat until the trader says they are paid in full. "There is little connection, however, between the amount of what they bring to trade and the amount of what they ask for," Everett observes. "For example, someone can ask for an entire roll of hard tobacco in exchange for a small sack of nuts or a small piece of tobacco for a large sack." Yet the Piraha are intelligent people, skillful hunters and fishers, with a well-developed sense of humor.

The main thrust of Daniel Everett's paper is to refute a widely accepted hypothesis in linguistics, "Hockett's design features for human language," which lays out a dozen or so characteristics of human language it claims are universal. Piraha, says Everett, defies at least three of those characteristics. I believe that Hockett's design features only appear universal to us because of our present perspective of separation. Contrary to Hockett, Piraha is severely limited in its ability to speak of events removed in time and space from the act of communication (Displacement) and in its ability to generate new meaning via grammatical patterning (Productivity). All of this stems from the resistance of the Piraha to the distancing from reality that we call abstraction.

From numbers to colors to time, there is much that the Piraha language is constitutionally unable to express, inviting the conclusion that the Piraha are a cognitively impoverished and socially isolated people. Moreover, theirs happens to have the fewest phonemes of any known language: just three vowel sounds and seven consonants for women, eight for men. Accustomed as we are to associating communication with semantic meaning, we can only conclude that the Piraha suffer an extreme poverty of communication.

Of great significance, then, is Everett's observation that the Piraha communicate almost as much by singing, whistling, and humming—non-symbolic modes of voice communication—as they do by speaking. A rich prosody enhances their verbal communication as well. Could it be that they are closer to the Original Language than the rest of the world, mired as it is in representation and abstraction? Perhaps it is we, not they, who suffer a poverty of communication.

Cultivation and Culture

Separation from Nature, and the technological program to control the world, did not originate with agriculture, despite the eloquent arguments of Daniel Quinn and others who associate the expulsion from the Garden of Eden into the world of toil with the transition from a hunter-gatherer to an agricultural mode of existence. Agriculture, rather, marked an epochal acceleration of a pre-established trend, an inevitable expression of a long-gathering latency.

With agriculture, the separate human realm expanded into radically new territory to include the various animals, plants, and other parts of nature that we made ours. No longer was domesticity limited to the campfire circle. With agriculture, we began to domesticate the whole world.

Because it was agriculture that launched the ascent of humanity into its present phase, the question of how and why agriculture began is critical. Many of the theories in the literature are unconvincing. The fallacy of the "nasty, brutish, and short" assumption of anxiety theory casts doubt on theories depending on population pressure and food shortage (food production can sustain far more people per square mile than food gathering). Hunter-gatherers had the means to regulate their population levels and in many places did so successfully for thousands of years; population grew dramatically as a result of agriculture more than as a cause. Another theory is that climate change or increased CO_2 levels at the end of the Ice Age rendered old lifestyles untenable and made new plants available[71]. However, this ignores the rapid adaptability of hunter-gatherers, who inhabited a wide variety of ecosystems even before the end of the Ice Age, and in any event, the transition from one to another environment would seem much less difficult than the transition from foraging to farming. Most ridiculous are explanations that imply we only figured out the idea of planting seeds quite recently: foraging cultures have a highly sophisticated understanding of plant reproduction and the conditions for their growth.

The fact that agriculture arose independently in several locations around the globe points to a natural progression from earlier technology and mindsets, and not an accident (whether mistake or glorious invention) that could just as well not have happened. Agriculture arose independently in Mesopotamia, in China (possibly in two locations), in South America, in Central America, in the Eastern United States, and perhaps in New Guinea and sub-Saharan Africa.[72] Indeed, most of the places where agriculture failed to develop were where there was a dearth of easily domesticable plants and animals. With few exceptions, wherever we could develop agriculture, we eventually did. Somehow, an agricultural future was built in to who we were in the late Upper Paleolithic.

Inevitable or not, agriculture was not a sudden invention but the cumulative consequence of a series of incremental developments that marked a gradual shift in human attitudes toward nature. Although in our customary dualistic mindset we may be tempted to see agriculture as an invention, a distinct epochal transition, Jared Diamond plausibly describes the origin of agriculture as a gradual step-by-step transition from the hunter-gatherer lifestyle. At first, perhaps, nomadic hunter-gatherers merely followed the herds of the ancestors of modern cattle, sheep, and so on. Over many generations, these nascent herders began to provide food and protection at key moments, upon which the animals came increasingly to depend. Crop planting may have started as a wider scattering of seed or removal of competing plants to give favorite foods a head start, followed perhaps by months of nomadic foraging for wild foods. Eventually, these plants came also to depend on the assistance of the planters, whether through deliberate breeding or unconscious coevolution. In any event, the domestic corn plant cannot reproduce without human assistance; nor does a domestic chicken stand much chance of survival in the wild.

Once domestication began, the much larger population density it permitted meant there was no going back. Agriculture, the archetype of human control over nature, induces dependency and the need for ever-increasing control—over land, people, plants and animals—as the population continues to grow.

Along with the gradual shift to agriculture came a transformation in human attitudes toward nature. Hunting accords with a view of other animals as equals. After all, nature works that way—some eat and some are eaten—and the human hunter is doing nothing different from animal hunters. Domestication imposes a hierarchy onto the interspecies relationship, as man becomes lord and master of the animals. Understandably, this relationship is then projected onto the whole of nature, which becomes in its entirety the object of domestication and control. Yet we must also consider that the innovation of animal domestication could perhaps not have happened in the first place unless nature were first objectified conceptually. The solution to this chicken-and-egg problem lies in the embryonic self-other separation embodied in all life forms, going back to prehuman times. Domestication merely represents its crystallization into a new phase: a slow-motion gestalt which also included all the other elements of separation detailed in the foregoing sections.

The farmer's new relationship with nature engendered a new conception of the divine. As agriculture and other technology removed humans from nature, so also did the gods become supernatural rather than natural beings. The process was a gradual one, starting with ancient pantheons closely identified with natural forces. Gradually, identity evolved into rulership as the gods were abstracted out of nature, eventually resulting in the Newtonian watchmaker God completely separate from the earthly (the natural) realm. At the same time, as we lost touch with nature's harmonies and cycles, the gods took on the capricious character exemplified by the Greek pantheon and the Old Testament. Accordingly, the gods must be propitiated, kept happy through the offering of sacrifices, a practice found in most ancient farming and herding cultures but not among hunters.

The angry God that arose in early civilizations is also linked to the concept of good and evil and the concept of sin. The corn is good, the weeds are bad. The bees are good, the locusts bad. The sheep are good, the wolves bad. Technology overcomes nature by promoting the good and controlling the bad. As for nature, so also for human nature. The self is divided into two parts, a good part and a bad part, the latter of which we overcome with the controlling technologies of culture.

Whereas hunter-gatherers could easily adapt to all the vicissitudes of the local climate, farmers were at the mercy of drought, hail, locusts, and other threats to a successful harvest. While the resources of hunter-gatherers were virtually unlimited and their population fairly stable, agricultural civilizations experienced famines, epidemics, and wars that decimated whole populations and defied any attempt at prevention. Here was a source of constant, inescapable anxiety woven into the fabric of life itself—no matter how successful this year's harvest, what of next year?—as well as a motivation for the increased understanding and control represented, respectively, in science and technology. Scarcity and the threat of scarcity is implicit in the attempted mastery of nature. Jockeying for position in the face of scarcity, we endure an endlessly intensifying competitiveness that is built into our system of money, our understanding of biology, and our assumptions about human nature.

Paradoxically, while agriculture raised nature's productivity of food (for humans), it also introduced the contemporary concept of labor. Food was at once more abundant but also harder to get. With agriculture we had to work today to obtain food tomorrow—a primary example of the paradox of technology, which has brought us to the brink of catastrophe despite its motivating goals of ease, comfort, and security.

Agriculture, because it involves keeping nature from its rest state, necessarily involves effort. I don't need to do any work to grow thistles, burdock, and crack grass in my garden, because that is what those thousand square feet of land naturally tend to. But to grow cabbages, kale, and garlic I have to do all kinds of work—pulling up the plants that crowd them out, erecting fences to keep out the rabbits and woodchucks, etc. The truism that we reap only what we sow only goes back as far as agriculture. Before then, we could reap without sowing: nature was fundamentally provident. For the hunter-gatherer, the providence of nature requires little labor or planning, but only an understanding of nature's patterns. Primitive survival is a matter of intimacy and not control.

The advent of agriculture accelerated the demise of the gift mentality that characterizes hunter-gatherer societies. Whereas hunter-gatherers see game and food plants as gifts of the earth, a farmer tends to see them as items of exchange for labor, and it is always his goal to tilt that exchange constantly to his benefit. No longer is sustenance something that the world freely provides. Whereas the hunter-gatherer is part of the gift network that we call an ecology, the farmer separates himself from that network and seeks to extract what he needs from it. Thus Daniel Quinn names the hunter-gatherers "leavers" and the agricultural societies "takers",[73] though perhaps "givers" would be a better word for the former. Eventually, the new relationship of taking and exchanging manifested among humans as well, setting the stage for the rise of money and property.

When we need to apply effort to coax a livelihood from the land, humanity's relationship with nature tends to become adversarial. The land naturally drifts towards weeds, pests, and in general a less productive default state. With the technologies of agriculture we seek to prevent this from happening. The battle lines are drawn. Today we seek to live more sustainably, yet the oppositional view of nature that environmentalists lament as the most destructive force on our planet is built into the very origin of civilization—agriculture. What else can we expect from a technology founded on arresting or reversing the processes of nature? Consequently, the end of humanity's war against nature must involve a wholly different approach to technology, and not merely better planning, fewer accidents, more foresight, and tighter control.

Agriculture inaugurated our conception of the earth as a resource or asset, defined primarily by its productivity. The land gradually lost its intrinsic value—its sacredness—and assumed an extrinsic, conditional value based on what it could produce. For the first time there was good land and bad land. The transition was a slow one, advancing each time new technology separated us a step further from original natural cycles. The primitive farmer is still very close to the land, even if less completely embosomed by it than the hunter-gatherer. Each new technological advance freed us from one or another natural limitation, culminating in modern industrial monoculture in which given the right inputs, almost anything can be made to grow on any land. Even a desert can be made to bloom.

It is becoming increasingly apparent, though, that natural cycles can be ignored only temporarily. Their disruption bears consequences that can be postponed by a series of technical fixes, but never permanently denied. Deserts can be made to bloom, yes, but only at an increasing cost and not forever. Someday they will return to their natural state. Moreover, the consequences of the disruption of natural cycles intensify the longer it is sustained. The present accelerating desertification of the world's agricultural areas testifies to the impossibility of forestalling natural processes forever, and to the severity of the consequences of trying. It is as if the desert cannot be denied.

The valuing of land according to its productive function and not its innate sacredness projects onto human society in the division of labor. With agriculture, human beings began to be distinguished by function—farmer, soldier, metalsmith, builder, priest, king—in a way they never had before. True, there are chiefs and shamans in pre-agricultural societies, but with rare exceptions they were not exempt from hunting and gathering food. They did not specialize by trading food for their services. In agricultural societies people came to be defined more and more according to generic functional classifications, a trend that drew as well from the anonymizing effect of the vast increases in population density that agriculture permitted.

With agriculture, a new category of being came into existence: the stranger. Before then, humans lived in tribes of at most 500 people, comprising bands of about 15-20 people each. It is not difficult to know 500 people by name and face, especially after a lifetime of frequent association, but beyond that the identifying structures of kith and kin become tenuous and some people necessarily fall into the category of "other".

In a hunter-gatherer band or tribe, and even in a Neolithic village, we were intimately known by virtually everyone we ever interacted with. Our acquaintances collectively embodied a tightly integrated web of relationships from which we derived our identity, our sense of self. We answered the question, "Who am I" through relationships with people who knew us very well, as unique individuals. But as the scale of society expanded, these personal relationships gave way to generic ones governed by commerce, law, and religion. Accordingly, the sense of self came to depend on these structures as well, which are by their nature anonymous and impersonal.

Relationships in primitive societies are guided by kin structures that provide each person a place relative to each other person. When society expands in scale to the point where two people are strangers, unable to place each other in their respective constellations of self, then there is a serious potential for conflict. Some kind of impersonal governance is required in the absence of structures of known relationship. After all, when someone is not "self" then he is a potential competitor whose interests might be at odds with ours. Practically speaking, if someone is a stranger there is no rational reason not to cheat him. Since he is not linked to your own social network, the consequences need never come back to haunt you. Hence the need for some kind of regulatory structure imposed from above.

When hunter-gatherers from different bands run into each other in the bush, they immediately begin an urgent and often very long conversation about who they know in each other's band, seeking to identify their relationship. Eventually, they establish that one is the cousin of the sister-in-law of the nephew of the other one's brother-in-law, effectively bringing each other into the same constellation of self. Vestiges of this behavior are apparent today when two strangers talk: "You were in Taiwan? Hey, I had a classmate from Taiwan—do you know so-and-so?"

When encounters between strangers are common, then some kind of governance is necessary based not on their unique relationship as individuals, but on generic principles: "All are equal under the law." Laws in the form of explicit codes are never found in pre-civilized peoples, nor are they necessary. It is no accident that as modern society grows increasingly anonymous, and as we pay strangers to perform more and more life functions, that the reach of the law extends further and further into every corner of life. Disputes that were settled informally a generation ago are today routinely administered according to written rules. Indeed, without some kind of formal standard we would feel insecure, for we would literally be at the mercy of strangers. This trend is a necessary consequence of the alienation and depersonalization that began with agriculture.

The division of labor introduced a new kind of anxiety into human life rooted in the idea that "You have to work in order to survive," a concept apparent today in the locution "to make a living." Foraging peoples engaged in various arts and crafts beyond what was necessary for survival—the fashioning of musical instruments, for example—and there was surely some differentiation of skills and talents among them. However, food was always there for the taking, readily available regardless of prior planning or its lack. We typically applaud the agricultural surpluses that "freed the non-farmer to specialize in other skills," not realizing that the non-farmer, lacking the means to obtain food on his own, was thereby enslaved to his specialization. *Freedom is slavery.* Art became profession and its product became commodity. Work is nothing other than art diminished, degraded, and debased. Driven by economic necessity (a code phrase for survival), no longer could we work in "fits and starts, and in these occasional efforts... develop considerable energy for a certain time."[74] Even worse, art created in the interests of economics is no longer art, for good enough is good enough. Why make it any better when it is to be exchanged anonymously for food or money, when it is to be given over unto the Other?

In the early days such exchanges were not completely anonymous. Money only partially replaced other forms of reciprocity, and most human interaction was not with strangers. Moreover, the life of the subsistence farmer is still intimately involved in the cycles of nature, wedded to the soil, and sustained only through a knowledge and respect for natural laws rivaling a hunter-gatherer's. Indeed, for us moderns gardening takes us much closer to nature, not away from it. However, once it started the ascent of agriculture built upon itself: technology advanced, the population grew, the regime of control intensified, and the conceptual dichotomy of human and nature widened. Finally, the institution of the Machine that culminated in the Industrial Revolution completed the degradation of work from its original identity with art to its present condition of slavery.

The Machine

Lewis Mumford defines a machine as a "combination of resistant parts, each specialized in function, operating under human control, to utilize energy and to perform work,"[75] and makes a compelling case that the first machines were built not of metal or wood but of human beings. Though this was an "invisible" machine because of the spatial separation of its human components, it was in all the above respects the model for every machine built since, as these components, "though made of human bone, nerve, and muscle, were reduced to their bare mechanical elements and rigidly standardized for the performance of their limited tasks." The products of these proto-machines are still visible today, most famously in the pyramids of Egypt.

The Machine enabled a new and profound expansion of the human ambition to dominate, subjugate, and eventually transcend nature. Agriculture set the stage for this ambition, it is true, but at the scale of the primitive farming village or herding tribe, humans were not really producing anything new under the sun. Ants, after all, engage in farming, and all creatures develop symbiotic partnerships with other species. Agriculture fostered the mentality of domination and laid the foundation for the division of labor, and it is these developments that transformed humanity into a new force of nature, or perhaps, not of nature. In the pyramids was evidence that humans could perform superhuman—that is, godlike—feats such as raising a mountain. And better than a mountain—a geometric shape of perfect precision.

New-age pyramidologists who think that the ancient megaliths must have required the technology of an extraterrestrial or Atlantean civilization are right about one thing: only through the mindset and method of machine civilization could they have been built. Yes, the pyramids were built by machines—machines built not of metal or wood but of human beings, fueled by the surpluses of riparian agriculture and molded by the forms of civilization into the standardized parts any machine requires. The specialization of labor in ancient Egypt was impressive even by modern standards: mining expeditions alone employed over fifty different qualities and grades of officials and laborers.[76]

Truly, only a machine could build a structure nearly five hundred feet tall out of fifty-ton blocks transported from distant quarries. "Blocks of stone were set together with seams of considerable length, showing joints of one ten-thousandth of an inch; while the dimensions of the sides at the base differ by only 7.9 inches in a structure that covers acres."[77] The base of the Great Pyramid is just half an inch from true level, and the sides are in near-perfect alignment with true north, south, east and west. "In short, fine measurement, undeviating mechanical precision, and flawless perfection are no monopolies of the modern age."[78] The tremendous physical energy to accomplish this feat, the specialization of function, the coordination of parts, and the requisite socio-technological infrastructure are the hallmark of the machine.

Because the operator of such a machine could perform godlike feats, it is no wonder that in ancient civilizations from Egypt to China to Mesoamerica, the king was accorded divine status. The ancient megaliths were proof that the king was not subject to the ordinary constraints of nature. Who else but a god could raise a mountain or change the course of a river?

Of course, as long as human beings formed the parts of the machine, these divine powers, the power to transcend nature, accorded only to its operator, the king. But later, as new machines were built along the *same principles and logic* as the prototypical labor machines, a new possibility arose in the minds of philosophers: maybe someday everyone will be a king, with the equivalent of hundreds or thousands of manpower at our command. At the same time, as one after another natural limit was transcended, the idea also emerged that we could all in some sense be gods. It is amazing how the characteristics of the gods of the Greek pantheon so closely resemble the ambitions of technological Utopia: they possessed immortality, eternal youth, and flawless physical beauty; they could travel at incredible speeds and fly through the air; and they possessed lordship over the processes of nature. As presaged by the semi-divinity of the Chinese emperors and Egyptian pharaohs, today we aspire through our technology to the status of gods.

As we have seen, this ambition was implicit already in the drive to create the perfect image, as well as to reduce the world to representations—names and numbers—susceptible to control by finitary means. The machine vastly accelerated these embryonic strivings, which were nonetheless crucial prerequisites for the development of mechanical technology. Prior to language, number, time, and other forms of symbolic representation, no one could even have conceived of a machine, which requires as a conceptual fundament—and then vastly accelerates—the objectification of the world.

The regularity, standardization, and functional specialization of these flesh-and-blood machines constituted a huge step away from nature, which admits to none of these qualities. The world-wide age of the builders confirmed in the human mind that we were something different under the sun, producing works and utilizing methods found nowhere else in nature. True, this paradigm shift started with the first stone hand-axe, which we recognize today as being in some sense an unnatural object (thereby confirming the prejudice that human is separate from nature), but machines take the out-separation of human beings to a qualitatively different plane by producing artifacts beyond individual human capacity.

Ensuing millennia saw a gradual progression of the division of labor, and thus in the complexity of the megamachine perfected in ancient times. Animal power, pulleys, screws and wheels, iron and steel all enhanced our ability to dominate, cut, and control nature. But none of these developments in any way reduced the mechanical structure of society itself, in which each individual is but a "resistant part," "specialized in function." "Resistant" here means discrete, and "specialized" refers to the division of labor initiated by agriculture. For a machine to run smoothly and predictably, its parts must be standard and hence replaceable, features which contribute, respectively, to modern depersonalization and anxiety. "Each standardized component... was only part of a man, condemned to work at only part of a job and live only part of a life."[79] And as the machine realm—those aspects of life subject to specialized division of labor—grows, so the depersonalization and anxiety intensifies.

Today the realm of the machine has expanded to include almost everything. The factory system, with its emphasis on standardization of parts and mass production as a path toward efficiency, is applied far beyond manufacturing. In schools, for instance, the standardized curricula, trained operators, classification of product via "grading" are all reminiscent of the factory. The resemblance is not accidental—schools were designed by some of the same efficiency experts who designed factories—and the dehumanization is the same as well. Here the process begins of assigning to each person more and more numbers, which eventually come to define us as mere sets of data. Meanwhile in agriculture, the logic and methods of industry have been applied to the land itself, subjecting it to the same imperative of efficiency and, in effect, reconceiving the earth as itself a factory.

The Industrial Revolution that promised to make each human a king or a god also exacerbated human separation from nature. Whereas the ancient megamachine allowed works beyond individual human capacity, the steam engine allowed works beyond human *biological* capacity, reinforcing the idea that human beings are not actually part of nature, or perhaps that our destiny is to rise above nature. The remaining "unconquered" natural domains of old age, death, social ills, and so forth would fall before the juggernaut of science, technology, and industry, just as all the other limitations had fallen already. This, the Technological Program, received its greatest conceptual impetus from the manifestly supranatural works of industrial technology. Kirkpatrick Sale foresees grim consequences:

> Imagine what happens to a culture when it becomes based on the idea of transcending limits... and enshrines that as the purpose of its near-global civilization. Predictably it will live in the grip of the technological imperative, devoted unceasingly to providing machinery to attack the possible.... Imagine then what happens to a culture when it actually develops the means to transcend limits, making it possible and therefore right to destroy custom and community, to create new rules of employment and obligation, to magnify production and consumption, to impose new means and ways of work, and to control or ignore the central forces of nature. It would no doubt exist for quite a long time, powerful and expansionary and prideful, before it had to face up to the truths that it was founded upon an illusion and that there are real limits in an ordered world, social and economic as well as natural, that ought not be transgressed, limits more important than their conquest.[80]

The Machine, then, gave us both the confidence and the wherewithal to attempt the transcendence of natural limitations, to ascend, like the Babelians, to Heaven and assume the powers of a god. Prophets, poets, and Luddites aside, it is only recently that we have begun to doubt this program, and now only because it is so evidently stalling. One reason it is stalling is foretold by the Babel story, in which a babble of mutually unintelligible languages made it impossible to coordinate the tower-building. In parallel, the fine division of labor that makes the entire technological project possible eventually generates such difficulty in managing that labor, so much chaos, that the effort collapses under its own weight. We see this too in science, where hyper-specialization renders various fields inaccessible to each other. Communication among fields becomes impossible. Each progresses toward solving its own narrowly defined problems, but systemic problems become increasingly hopeless.

The immediate result of steam-driven industry was not to make each man a god, but a slave. For the worker, the transcendence of biological limitations meant the subordination of the rhythms of flesh and blood to the rhythm of the machine: never tiring, never requiring food, sleep, or rest. Kirkpatrick Sale describes it thus: "The task for the factory owner was to make sure that workers would be disciplined to serve the needs of the machines—'in training human beings,' [Andrew] Ure said, 'to renounce their desultory habits of work and to identify with the unvarying regularity of the complex automaton.'"[81] The factory marked, if not the start, a key quickening of the modern conception of work, something that we must discipline ourselves to do in denial of "desultory" biological impulses. Work became toil, of necessity repetitive and unvarying like the machine functions it served.

As it was for the labor, so it was for the products. If the early cultural technologies of label and number suggested the genericness of their objects, industry attempted to realize it through its emphasis on standardization and uniformity of product. With few exceptions, natural objects are unique, variable. Industry sought the opposite, the unnatural uniformity of its products further reinforcing the divorce between human and nature. Along with this mass scale uniformity came the disintegration of local differences, as people everywhere began to eat, wear, and use identical products and perform identical labor. We see this process culminating in the homogeneity of the early 21st-century American landscape: the generic roads, housing developments, franchises, superstores, and strip malls that comprise, in Jane Holtz Kay's phrase, the "landscape of the exit ramp".[82]

Finally, the standardization, homogeneity, and specialization of function that characterize the factory equally characterize the inhabitants of machine society. What began with Mumford's megamachine has never truly disappeared, only worked itself deeper. Our specializations define us and validate our existence. Lumped together as "consumers" or "the workforce", classified statistically into the categories of pollsters and social scientists, we humans have been made generic too, robbed of the individuality that once derived from a unique set of relationships to nature and kin.

Industry took to its conclusion the reduction of nature that started on the psychological level with symbolic culture and was projected onto the land with agriculture. Whereas these earlier developments reduced world to object, industry turned object into commodity, time into money, and human being into consumer. All the world, in other words, is being converted into money—the ultimate in anonymity, abstraction, and genericness—a story I take up in Chapter Four.

Is there then no hope, other than to undo the entire course of specialization and return to a society without division of labor? Is there no choice but to abolish all the technology that depends on it? I don't want to leave you on a note of despair, so let me offer a sneak preview of a later chapter. In fact, there is a system in which specialization leads not to the reduction of the individual, but to her fulfillment. The model for such a system is sitting in your chair—it is the society of cells that constitutes a human body. The organs of such a society are growing already, and it is this organic or ecological society that will soon supplant the dying civilization of the Machine.

Religion and Ritual

Ritual as we understand it today employs symbolism, in which representations of objects stand in for real objects, and ritual enactments of events stand in for real events. However, like language and art, symbolic ritual evolved gradually from a time before separation, when symbol and object were one. Very likely, the rituals of deep antiquity were an outgrowth of animal rituals of mating, dominance, and so forth, and were not symbolic at all, or at least no more symbolic than the songs of birds.

Rituals only became symbolic when spirit became abstracted from physicality. That only happened when divinity became separated from nature, and that only happened when technology and culture created a separate human realm, especially when agriculture placed nature in an adversarial role. Wearing dualistic blinders, it is hard for us to imagine any ritual that were not symbolic, a ritual that is a real thing and not the representation of a story, principle, or event. So accustomed are we to asking what it means, what it symbolizes, that we can hardly conceive that the appropriate inquiry might be to ask what it *is*. We can hardly conceive of *understanding* something apart from interposing another level of interpretation between us and it.

Symbolic religious ritual is itself a mediation between two separately-conceived realms, the human and the divine. In the original religion, animism, there was no such distinction. Often misunderstood as the belief that everything has spirit, animism actually holds that everything *is* spirit, that everything is sentient, sacred, special. Therefore, the abstraction of the particular embodied in language and number is, from the animistic point of view, a sacrilege, turning a continuum of unique places and moments into just so many things. Probably, even the purest animistic religions studied by anthropologists are already degraded versions of true animism, existing as they do in cultures that have already adopted representational language.

We look upon the Neanderthal custom of burying their dead (with artifacts) as a sign of their cognitive or spiritual development, as it implies a belief in an afterlife and therefore a concept of soul separate from body. I do not know if the Neanderthal already had a dualistic religion, but the above interpretation lays bare a deep cultural prejudice: that matters of spirit are in a realm separate from life in this world, separate from the here and now. What, may we ask, is the religion of animals? Religion, which literally means "that which ties back" or "that which reconnects us" is only necessary when there is separation. A being constantly in touch with the fullness, sumptuousness, and indeed the infinity of sheer existence would have no need for religion. We may therefore see religion as a symptom of separation.

But that does not mean religion is a mistake! More than a symptom of separation, religion is also a response to it, a manifestation of the primal urge to reunite with all that we have separated from. Perhaps religion originated as a calling back to an earlier time of wholeness.

Let us revisit this "earlier time of wholeness" to see how religion might have developed. An assumption rarely questioned in paleontology is that earlier species of humans were in important respects inferior to modern *Homo sapiens*. The evidence would seem incontrovertible: not only did all other *Homo* species become extinct, but their tools were much simpler, and they left little or no evidence of art, religion, and all the other creations of modern intelligence.

We must be careful, however, not to project modern-day prejudices onto the physical evidence of paleontology and archeology. We tend to interpret burial of the dead as evidence of belief in an afterlife and therefore a non-material soul, cave paintings as evidence of magico-religious rituals aimed at controlling events, and technological advances as motivated by the struggle to survive. All of these interpretations may be projections of our own dualism and anxiety. Moreover, the tendency to equate the developments of language, technology, and so forth with "progress" or "advancement" in culture and cognition depends on the self-assumed superiority of our own culture. However, in light of the negative effects of language, number, tools, agriculture, and time measurement discussed above, it would behoove us to reexamine the supposed inferiority of those who refused them.

Recent evidence, rigorously laid out in Stephen Oppenheimer's 2003 book, *Out of Eden*, has shattered the myth that modern human cognition developed in a gene-fueled European cultural explosion some 30,000 or 40,000 years ago. The most significant development in stone tool technology occurred some 300,000 years ago with the development of flake tools chipped off a specially-prepared stone core, "a multi-stage process requiring that the final product be fixed in the maker's mind throughout."[83] Neanderthals and their Cro-magnon cousins had almost identical technology up until 50,000 years ago.

The disappearance of European Neanderthals is something of a paleontologic enigma, especially given recent mitochondrial evidence that no interbreeding occurred between the Neanderthal and Cro-magnon humans that simultaneously inhabited Europe some 28,000-40,000 years ago. The usual survival-of-the-fittest, scarcity-based thinking assumes that the latter either exterminated the former directly, thanks perhaps to superior weaponry, intelligence, and social organization, or simply outcompeted them for habitat. In any event, a dramatic cultural diversion separated the two species at an accelerating rate starting around 40,000 years ago, and apparently resulted in the Neanderthal's extinction by 28,000 years BCE.

Although some exceptions have been claimed, it is generally thought that Neanderthals lacked art, trade, tools of bone, shell, and antler, and burial of the dead. Neanderthal primitivism is usually taken either (1) to imply that Neanderthals were insufficiently intelligent to develop the technology of modern humans, or (2) as evidence of culture's cumulative nature; that is to say, Neanderthals were as intelligent as we are, but the technology simply had not time to develop. The first position is increasingly untenable given the parallel development of Stone Age technology up until 50,000 years ago, not to mention Neanderthals' slightly larger brain size. Either way, they apparently suffered the same fate as most of the world's indigenous peoples did upon encountering a technologically more advanced culture. A third possibility is rarely discussed: that Neanderthals consciously rejected the innovations that led to our widening separation from nature.

It would not be the only time such a rejection occurred. The more recent history of technology is not without examples of technology rejected or even abandoned. Perhaps the Neanderthals had the anatomical and cognitive capacity to proceed on the accelerating arc of the ascent of humanity, but simply refused to do so. Perhaps they refused the distancing, alienating technologies of semiotic language, number, art, time, and standardized stone tools,[84] intuiting as have shamans and religious mystics ever since that they separate us from nature, spirit, and joy. Perhaps they recognized the idolatry implicit in representational images, the reduction implicit in symbolic language, the suffering implicit in separation of self from environment. Perhaps they thought that separation had gone far enough, and knew that its continued ascent could lead to one place only.

Consider the megafauna extinctions of the northern hemisphere. These happened remarkably recently, after the Neanderthals were gone and modern humans fully established, and are usually attributed to our superior technology and, by implication, superior intelligence. Pause for a moment to think about this—major ecosystem disruption is taken as a sign of superior intelligence! We assume that it is human nature to take as much as possible. Let us discard that assumption for a moment and suppose that the Neanderthals and other pre-modern humans had the intelligence but not the desire. Perhaps they had the wisdom to avoid practices that would disrupt the balance of nature. Later cultures, more distant from nature, nonetheless still understood the importance of maintaining that dwindling connection. They used religious ritual and magic to reaffirm and renew it, relying on an ancient lineage of shamans and stories going back to their original teachers, from a time when harmony with nature did not need these artificial means of reconnection. This idea finds support in certain indigenous myths and legends. Here is a particularly striking one, courtesy of Joseph Epes Brown:

> The Yurok of Northern California believe that, in the beginning, the world was inhabited by the *wo'gey*, or Immortals, who knew how to live in harmony with the earth. The *wo'gey* departed when the humans arrived. Yet, because they knew that humans did not always follow the laws of the world, they taught them how to perform ceremonies that could restore the earth's balance.[85]

The correspondence here with the actual historical departure of Neanderthals and other human species with the coming of *Homo sapiens* is quite remarkable. *The wo'gey departed when the humans arrived.* In Europe from 40,000-28,000 BCE, the overlap between Neanderthal and modern human was very brief; the former's retreat before the latter was almost instantaneous. Nowhere did the two coexist for very long. The story may have been the same with other human species across Asia.

Perhaps the Neanderthals and the other human groups that our ancestors replaced were not lesser humans, lower on the evolutionary ladder, but actually more evolved in thought and spirit than we are. I like to speculate that they were even our teachers, exemplars of a mode of being that already, 40,000 years ago, we had begun to forget, but which has been carried forward to the present time encoded in myth and ritual, preserved in fragments by lineages of shamans, Sufis, storytellers, Taoists, yogis, and mystics, and revived from time to time by artists, poets, and lovers, so that, like a spore or a seed, it might blossom forth again when the Age of Separation has run its course.

As separation accelerated through the rise of agriculture, the gap over which religion was required to "tie us back" widened. Humans departed further from "the laws of the world", and the old ceremonies became impotent to restore balance. Slowly, nature lost its inherent divinity in human eyes and became, progressively, just a thing. To be sure, nature-as-thing was never anything more than an ideology, and an ideology that direct experience invariably contradicts when we open ourselves to it. Nevertheless, that ideology was (and still is) powerful enough to direct and justify a millennia-long course of domination and destruction, the subordination and conquest of nature. As described in a previous section, agriculture gradually took the gods from identity with natural forces to lordship over them, in parallel to the human abstraction of ourselves from nature. Whereas the ancient kings and pharaohs were divine, starting in Mesopotamia at around 2000 B.C.E. kings became mere emissaries or representatives of divinity, which was elevated to a celestial realm.

The association of the divine with the celestial—which removes God from *within* nature to an estate *above* nature—is itself another consequence of agricultural and machine thinking. Ancient astronomers, concerned with the measurement of time and the making of calendars, observed regularities in the motions of the planets, impersonal cycles removed from the chaotic irregularity of nature. Which would the engineer—the fashioner and operator of a machine—prefer? A higher, more perfect law, it was thought, governed the heavens. This split between the order and perfection of the skies and the messy chaos of the living earth was only resolved in the seventeenth century with Isaac Newton's unification of all motion under a single law, which had the effect of abstracting God still further.

With divinity accorded a non-earthly status, human beings (who are after all of this earth) lost their innate divinity to become mere servants of God. Around 2000 BCE, "Mesopotamian myths began to appear of men created by gods to be their slaves. Men had become the mere servants, the gods, absolute masters. Man was no longer in any sense an incarnation of divine life, but of another nature entirely, an earthly, mortal nature. And the earth itself was now clay. Matter and spirit had begun to separate."[86] Accordingly, we have in the Jewish, Christian, and Islamic religions that originated in this region a concept of sinning against God that is entirely separate from violation of nature's order or harmony. And not only separate, but often directly opposed. Spirituality came to mean conquering the flesh, whose desires were opposed to the elevation of the spirit. The parallel with technology and the "ascent" of humanity is clear. Civilization's technological quest to overcome nature with the order and regularity of the machine projects into religion as a program to overcome our unruly inner nature, or "human nature."

Religious institutions thereby came to represent the precise opposite of what the original rites, myths, and teachings intended. The original purpose of religion implied in the Yurok legend is to bring us back to a state of harmony with the natural order, and in particular with our natural selves, our true being. Despite the relentless attempts of social institutions to coopt religion to its purposes of control, the original intent and message of religion lives on, often buried beneath layers of dogma and interpretation. Every once in a while reformers see through the dogma and remind us of the principles of the Original Religion, animism. A few examples: George Fox: "Look to that of God in everyone." Hallaj: "I and the Beloved are one." Jesus: "I and the Father are one." The latter two were both crucified, the first merely beaten, pilloried, and imprisoned; all denied the reigning doctrine that human and divine are separate realms. In Jesus's case, the teaching was almost immediately turned into its exact opposite: "I am God and so are you" was turned into "Jesus is God and you are not," recreating the same division and duality that Jesus taught against.

In Oriental religions the dualism between human and divine is less developed, and many teachings to the contrary remain preserved in scripture. Taoism in particular emphasizes the identity of the spiritual way and the way of nature. Buddhism is rife with admonitions to the effect that Buddhas are no different from ordinary people (yet completely different), and that everything has Buddha-nature. There are even Zen koans directed at deconstructing the implicit dualism in the statement "Everything has Buddha-nature," as if it were a separate thing to be "had". Similarly for Hinduism: the Bhagavad Gita states, "The Supreme Self, which dwells in all bodies, can never be slain... Eternal, universal, unchanging, immovable, the Self is the same forever." The more extreme separateness from the sacred embodied in Occidental religion fits into the same mindset of objectification and control that also characterizes technology. Perhaps it is no accident that modern technology, too, arose in the West.

The perversion of religion (tying back) into its opposite has taken different forms across the globe. In the East, concepts of karma were perverted into duty, and cosmic order into a blind servitude to the temporal order of feudal society. The result in India was the caste system, which minutely prescribed the role of each individual. Similarly, the doctrine of the illusoriness of the finite self was coopted to encourage the submergence of individual will into the social mass. In the Far East, the Tao degenerated from a principle of organic divinity immanent in nature and the universe, into a justification for a rigid social hierarchy.

Meanwhile in the West, the differentiation of nature and the divine went a step further. Beginning as names for aspects of nature, graduating into separate representatives of those aspects, the gods of ancient Greece and the God of the Levant eventually came to a position of capricious overlordship. They became rulers, and not aspects, of nature. "Whereas in the older view... the god is simply a sort of cosmic bureaucrat, and the great natural laws of the universe govern all that he is and does and must do, we have now a god who himself determines what laws are to operate; who says, 'Let such-and-such come to pass!' and it comes to pass."[87] The parallel with the mentality of technology is obvious. God was no longer at the mercy of nature; by understanding the principles of God's creation, then, neither need be man.

A corollary of the capricious celestial overlord in the human realm is free will. Modeled after our concept of God, we saw ourselves as separate manipulators of the rest of the world, and subject to a very different set of laws than God has ordained for "the beasts". We could choose to obey, or not to obey, and we could manipulate the world as we saw fit, rather than merely playing a role in a preexisting harmony. The Hebrew, Christian, and Moslem God (though not of these religions' esoteric traditions) provides the model for the ultimate separation of ourselves from the universe, the alienation of the discrete self in a world of objects. Eventually, of course, we dispensed with God altogether when science began explaining the workings of the world without appeal to an extra-natural mover. Another way to look at it, though, is that we *displaced* God. Through science and technology we ourselves presumed to the functions previously ascribed to God: the director of nature, the Free Will that has the power to say, "Let such-and-such come to pass!" If we don't like what is, we can change it, we can make something else come to pass. The Babelian logic of the Technological Program is that our power to do this is unlimited.

The ascent of science described in the next chapter can be seen as the final phase in the progression of religion away from its holistic, animistic roots. It is also a necessary phase, the culmination of an age-old process. Before I lay out The Way of the World according to science, I would like to offer another interpretation of the incredible descent-that-masquerades-as-an-ascent whose story I have told.

The Playful Universe

Eternity is in love with the productions of time.
-- William Blake

John Zerzan has written, "We have taken a monstrously wrong turn with symbolic culture and division of labor, from a place of enchantment, understanding and wholeness to the absence we find at the heart of the doctrine of progress. Empty and emptying, the logic of domestication with its demand to control everything now shows us the ruin of the civilization that ruins the rest. Assuming the inferiority of nature enables the domination of cultural systems that soon will make the very earth uninhabitable."[88]

Zerzan's eloquent lament is correct in every detail but one: symbolic culture and the division of labor was not a "monstrously wrong turn" but rather, as I have argued throughout this chapter, the direction we have been headed all along. And that would seem even worse! If not a bad choice that is in principle remediable, is this long fall from "enchantment, understanding, and wholeness" just the way of the universe? Are we destined for ruination, desolation, and extinction? Or is this descent a phase of a larger pattern or process?

Back in 1938, the historian Johan Huizinga advanced the concept of "*Homo ludens*", the playful human, in direct contradistinction to contemporary anthropology which with near unanimity ascribed early human behavior and development to a struggle for survival. Huizinga suggested that play, not struggle, was the formative element in cultural development, for it was the creative inner world of make-believe through which we rehearsed for the subsequent transformation of the external environment.

He could have broadened the concept of *Homo ludens* beyond the human realm to include all of life, because play is by no means exclusively human, a fact obvious to anyone who has ever raised a dog or a cat. And it is not just domestic animals that exhibit playfulness but wild ones too—hence the phrase, "monkeying around". (I refer the reader to Tom Brown Jr.'s captivating accounts of playing with wild animals in his books *Tracker* and *The Search*.)

The function of play is quite difficult to explain in Darwinian terms, as it would seem to consume energy that could be directed toward maximizing survival and reproduction. Individuals genetically programmed to play would be at a competitive disadvantage against those who devote their full energy toward food-gathering, mating, and so forth. Thus, when play is addressed at all, we would expect to find rather forced attempts to explain it (or rather explain it away) in terms of mating rituals, dominance, or practice of hunting or other skills. The other alternative, that life is naturally playful, leisurely, and fun, simply does not fit into our basic conceptions of what motivates animal behavior.

In our own culture, play is typically conceived to be the realm of children. The Puritan streak in our culture views it as a luxury, an indulgence, which is okay to allot to children in small amounts as long as they have finished their "work" (schoolwork, homework, housework, etc.) On the more tolerant side of the spectrum, play is okay as long as it is "educational": hence the numerous toys and games directed at young children which seek to smuggle in the alphabet, numbers, or other "cognitive skills". To echo the Darwinian explanation of animal play as a means to hone hunting skills, play is good because it is practice for life. Play purely for play's sake is a waste of time, a view based on the purpose-of-life-is-to-survive assumption that underlies modern science and economics. After all, every minute spent playing is a lost opportunity to get ahead in life.

Either way, eventually we grow up and there is no more time for play. Now, the grim business of real life begins. Oh sure, maybe we can "afford" to play in our "time off"; that is, the time left remaining to us after we have met the demands of survival. But unless we are extremely wealthy, we believe, the bulk of our time and energy must go toward work.

Perhaps the truth is something quite different. Instead of youth being the time for play, maybe it is play that keeps us youthful. Perhaps the boundless free flow of creative expression is what keeps us physically and mentally supple, as a child. When we attempt to control it, limit it, mortgage it to the acceptable and safe, then the bounds of that safety project themselves onto body and mind, subjecting both to a severely limited range of motion that hardens over time.

Let us question the work-play dualism. Consider the possibility that childhood play is practice, yes—but practice for adult play, not adult work! For in fact, the same qualities that characterize childhood play apply equally to the most creative, productive activities of the adult. Childhood play is practice in the exploration of limits, the loosening of inhibitions to creativity, the creative dialogue with the environment, the reimagining of the world presented us. Play is not enslaved to a preset end, but allows the end to emerge spontaneously through the process itself. Play does not require willpower to stay focused and overcome our natural desires; it *is* natural desire manifest. When we play, we are willing to try things without guarantee of their eventual usefulness or value; yet paradoxically, it is precisely when we let go of such motivations that we produce the things of greatest use. In writing this book, for example, when I steel myself to cover certain material necessary to the book's logical framework, my words come out pedantic and uninspired; my best writing comes when I'm "playing around" with the ideas, and in this play a logic and structure emerges that is far more potent than anything I could have thought up beforehand. I imagine Thomas Edison doing the same kind of thing, puttering around in his lab, trying this and that with no guarantee of success but in the process thinking of a new idea, trying it out. I imagine Albert Einstein, trying out,, just for his fun and delight, ideas in physics that must have seemed crazy at first, but not caring, exploring them anyway. I don't mean to compare myself to these geniuses, only to illustrate a principle of creativity: forgetting about what has been done before, what will work, what brings secure results, and trying something else for the fun of it.

Because the creativity of play is spontaneous, unbidden, and impervious to any rote formulization, we must consider that it comes from a source beyond ourselves. We are the universe's channel for play, an aspect of a universal playfulness expressed through our minds and bodies, employing our mental skills of reason and expression but originating beyond them.

We might then see the hallmarks of our humanity—language, math, art, technology—as originating in play, and indeed being new, highly developed means for the implementation of universal playfulness. The phenomenon of language is a case in point. In the toddler, words are a key element in the development of the imagination and the ability to create a world of make-believe. The same abstracting quality of language that distances us from reality also allows us to create and play with an inner reality, an ability which, honed through a period of mental play, we eventually reapply to the external world, creating novelty with physical things just as we did before with mental word-associated images. Joseph Chilton Pearce speaks of the importance of storytelling to children, who, as they listen with rapt attention, flesh out the stories with sequences of associated images so vivid that it is as if the experience were their own.[89] This capacity, developed in childhood, can then turn to the envisioning in adulthood of *what might be*. To envision what might be is fundamental both to play ("let's pretend I'm a bear") as well as to creativity, which, if authentic, requires the imagining of a possibility that did not exist before. In modern times, when prefabricated images accompany our stories (in the form of storybooks and television) the image-forming capacity never gets a chance to fully develop, and we are rendered incapable of conceiving anything other than the life that is presented us.

I propose, then, that language originated not to coordinate hunting, tool manufacture, or for any other survival purpose, nor, as John Zerzan and the linguist E.H. Sturtevant imply, for purposes of deception. Representational language arose as a form of play, a game of associating sounds to objects and actions. That is why primitive cultures recognized the relative unreality and unimportance of words as compared to sounds, voice, song, and silence; that underneath the conventional and contingent names of people and things lay a mystical True Name that, as I described in discussing the *lingua adamica*, was not a representation but rather an aspect of the thing in itself. Just as a child at play knows that the rag on a stick is not really a baby, yet makes it so for the purpose of a game, so also did the original users of language playfully abstract words from things for the purpose of creativity—storytelling and make-believe—in precise parallel to the cognitive function of language in a toddler.

This explains the apparent paradox that, even though words are on the one hand symbols, they are also endued with real generative power. As Joseph Epes Brown puts it, "… language has creative force. Words are not merely symbols that point to things; they call forth the reality and power of the being mentioned."[90] Modern thinking understands the symbolic meaning of words to be separate from their base reality as just objects, mere combinations of sound waves or shapes on a page (not so surprising, really, when the things they refer to are equally mere objects of the Cartesian universe). Nonetheless, the knowledge that words are not just representational but creative too can be found in the magical and religious traditions of our own culture. "In the beginning was the Word, and the Word was with God, and the Word *was* God."[91] Here in the Bible the word is identified as the wellspring of creativity, the Godhead. Is there, then, a nascent separation implicit in the very existence of a physical universe? Could we see cosmic processes also as a manifestation of play?

Another apparent paradox lies in the sacredness ascribed to language which made profanity impossible in native languages, the reverence and awe attached to the names of people, animals, and places. How is sacredness, reverence, and awe compatible with play? It seems, rather, deadly serious. The paradox is resolved when we realize that play *is* serious! Our culture assumes play to be puerile or frivolous, undeserving of the attention of serious, adult matters. Yet when we observe children at play we find in them the whole range of human emotions and attachments: laughter, yes, but solemnity too, and passion. Children, at least, take their play very seriously. The forms of recreation that masquerade these days as adult play are but pale imitations of it, sops to the spirit. They are not creative but dissipative; they don't engage us more fully in life but rather, in the guise of "entertainment", remove us from life. Whether sports cars, sailboats, or video games, we don't genuinely believe in our toys.

If we are to relearn from our children how to play, then we must not limit ourselves to the frivolous and puerile. We can, like them, have occasion to play very solemnly, with great dedication and commitment, with laughter or with tears; in short, we can play our way through the full cornucopia of life's experiences.

If language, technology, and the other elements of separation originated as play, then how did they become something else? In a sense they never did. We are still at play, but immersed neck-deep in a game gone wrong, a game from which we are unable to extricate ourselves. After all, the separation from nature, spirit, self, and other that I write about in this book is not real; it too is a play, a dance of energy and information. Our present loss of all the characteristics of play—spontaneity, fearlessness, spirit of exploration, creativity, willingness to test limits, non-attachment to results—is itself part of a larger game of individuation. In the current age, as the dance of separation becomes increasingly intolerable, as crises mount throughout the world, we are beginning to realize that the time has come to stop playing this game and begin another one. The game of "let's pretend we are discrete, isolated beings in an objective universe" with all that it entails has served its purpose.

Another way to look at it is that we never stopped playing, but we have forgotten that we are playing. Every once in a while we run into people who are still firmly embedded in modern technological society yet who have cast off its hurried, anxious, alienated mindsets and adopted a more easy-going, playful attitude toward life. Often this happens after a major illness or other personal calamity reveals the vacuity of former ambitions and preoccupations. They say, "I stopped taking life so seriously," or "I came back to what's really important." Yet despite not taking it so seriously, they are not detached from life. They are not passionless or indifferent; if anything they are more involved and more fully engaged in the moment, because it is precisely the anxiety of modern life that removes us from the moment, that makes us feel we cannot afford to "be here now" or to fully engage in whatever we are doing.

Another thing these people say is, "I stopped taking *myself* so seriously," pointing toward the original source of modern ugliness, suffering, and anxiety in our misconception of self. The forms and structures of our society conspire to instill a false conception of ourselves. They hide from us who we truly are, but we still have the power to disbelieve what we are told and to reclaim our birthright of play. Whereas primitive societies naturally fostered adults secure enough and connected enough to live playfully, our own holds us hostage through subtle, omnipresent threats to our survival. Survival anxiety is what the phrase "cannot afford to" encodes. The sense of being threatened is so subtle and so deeply woven into the fabric of our existence that we are barely aware of it, like the threat of thunder on a cloudless, humid summer afternoon. It begins in early childhood, when through punishment, shaming, and conditional approval our parents wield the greatest threat of all over our heads, indeed the archetypal threat, which is abandonment by the parent. Eventually we internalize it as an unremitting sense of disquiet that renders us unable to fully concentrate or fully relax (the two are closely related) in the absence of powerful stimuli such as drugs, movies, and other thrills.

Despite my acute awareness of the vast panorama of atrocity and suffering that comprises the history of humanity's alienation, I do not say that it was all a mistake. As I have argued, the extreme of separation that we have explored in recent centuries was written into the future long, long ago. But more than a neutral inevitability, perhaps the current condition of alienation is necessary for our future development, a possibility I will explore in a later chapter. In any event, it is time for the present game of "let's pretend" to end. I say this not as an exhortation but as a simple statement of fact. Just as the Age of Separation was inevitable from the inception of agriculture or even before, its ending is equally inevitable. It is inevitable because it is untenable, unsustainable, or, to be more precise, sustainable only at a higher and higher price. Our reunion with the other, in the form of nature, other people, and lost parts of ourselves, will happen when the price of separateness becomes intolerable. I say this as certainly as I can say to an alcoholic, "Your addiction will not last forever." The addiction generates its own demise. As the gathering crises of the world visit themselves personally upon more and more of us, as it becomes more difficult to insulate ourselves from their effects no matter how wealthy we are or how skillful we are in exercising control over the world, we are collectively "hitting bottom". The effect on our society will be similar to the effect on the individual of a close brush with death.

When recovering addicts share stories of incredible lost opportunities, torn families, ruined lives, and the destruction left in their wake, their regret is tempered by the knowledge that they have emerged somehow wiser. They tell of a breakthrough to self-forgiveness, which is nothing less than the knowledge that given who they were, they could have done no other than what they did. This parallels the inevitability I speak of. The horrifying course of human history was built into who we were, and through its passage we are becoming something different and greater. Collectively and as individuals, we are being born into a new self. The good news is that this will happen no matter what, as inexorably as a baby is born when gestation is complete. The bad news is that we nonetheless have the power to delay our birthing indefinitely, even to the point where both mother and child must perish. The purpose of my writing is simply to tell everyone not to resist the inevitable! The gig is up. The game is over. It is time to wake up and play something else.

CHAPTER III

The Way of the World

The Scientific Method

Starting around the 16th century, our "ascent" to a separate human realm underwent a dramatic acceleration. Language, technology, number, image, and time each became the object of an ambition of Babelian audacity: to extend their demesnes to encompass the whole of reality. Although hints of this program can be found in ancient times, in Greek and Biblical urges to exert dominion over the world, it was only with the Scientific Revolution that we began to envision a plausible means to actually achieve it.

Four centuries later, we see a world utterly transformed. Miracles and magic, the province of the gods, now operate on a daily basis. Instantaneous communication across continents, travel through the air, entire books at the press of a button, perfect moving images, and much more are now commonplace—thanks to science. It is to science that we owe civilization's ascent. It is science, we believe, that has lifted us above primitive superstition to obtain verifiable, objective knowledge. Science, the crowning achievement of modern man. Science, unlocking the deepest secrets of the universe. Science, destined to bring the whole of the universe into the human realm of understanding and control.

The very adjective "scientific" implies deep suppositions about the nature of reality and our relationship to reality. Science offers prescriptions on how to live life and how to organize society, how to understand the world and how to pursue knowledge; it tells us who we are, how we came to be, where we are going. Speaking of another culture, we might describe these prescriptions and these stories about the way of the world as a religion. For ourselves, we call them truth, fact, science—fundamentally different from other cultures' myths. But why?

Our culture is not alone in believing its myths and stories to be special. We think that ours are true for real, while other cultures merely *believed* theirs true. What are our justifications? Two stand out, one theoretical and one practical: the doctrine of objectivity, and the power of technology.

On the practical level, we believe in the validity of our science because of the great and demonstrable power it has given us, through technology, to manipulate the material environment. The world of our experience—an artificial world, a human world—is science materialized. The very existence of the world we live in is proof of the validity of science. The god Science has given us the ability to reshape the very landscape, alter the code of life, and enact the "magic and miracles" enumerated above. Having given us such tangible power, how could it be a false god?

Yet it is not hard to imagine other cultures that would dismiss our power over the physical environment as inconsequential, an unimportant aspect of life, or that would even deny that our power is really so great. Do we not eat, sleep, pass waste, make love, grow old, grow sick, and die as all other human beings? Do we not experience the same gamut of emotions as human beings everywhere and everywhen? Henry Miller said,

> We devise astounding means of communication, but do we communicate with one another? We move our bodies to and fro at incredible speeds, but do we really leave the spot we started from? Mentally, morally, spiritually, we are fettered. What have we achieved in mowing down mountain ranges, harnessing the energy of mighty rivers, or moving whole populations about like chess pieces, if we ourselves remain the same restless, miserable, frustrated creatures we were before? To call such activity progress is utter delusion. We may succeed in altering the face of the earth until it is unrecognizable even to the Creator, but if we are unaffected wherein lies the meaning?[1]

Yes, we can move mountains and build skyscrapers and talk with people on the other side of the world, but perhaps the importance we place on these things as validations of our science says more about our values and emphasis than it does about the ultimate validity of science.

In other words, we have a highly developed science of (certain aspects of) the material world, and we cite our power over exactly those aspects of the material world as proof of the validity of our science. The logic is circular. Another culture might have a highly developed science of aspects of the world that we do not even recognize, or that we consider unimportant. An Australian aborigine might consider us hopelessly primitive in our understanding and use of dreams; a traditional Chinese doctor might find us laughably ignorant of the energetics of plants and the human body.

This leads to the second justification for the belief system we call science. We accord a privileged status to our stories because we think that the Scientific Method ensures objectivity. Ours is more than a mere religion, we think, because unlike all before it, it rests on verifiable, objective truth. Science is not just another alternative; it encompasses and supersedes all other approaches to knowledge. We can examine dreams or Chinese medicine scientifically. We can perform measurements, we can run double-blind studies, we can test the claims of these other systems of knowledge under controlled conditions. The Scientific Method, we believe, has eliminated cultural bias in prescribing an impartial, reliable way to derive truth from observation. As physicist Jose Wudka puts it, "The scientific method is the best way yet discovered for winnowing the truth from lies and delusion."[2] This belief is essential to the ideology of science: that it has escaped the bounds of culture and, by insisting on replicability and logic, freed knowledge from the yoke of subjectivity.

The Scientific Method, through its testing of hypotheses, invites a kind of certainty absent from other approaches to knowledge, a universal validity that is not culture-bound. Anyone from any culture can perform the same experiment and get identical results. As long as we abide rigorously by the Scientific Method, we have a reliable way to distinguish fact from superstition, an intellectual razor that cuts through layers of cultural belief to get at the objective truth underneath. Finally, we are free from the bonds of subjectivity, the personal and cultural limits to understanding.

But are we? Or could it be that the Scientific Method is not a supracultural royal road to truth, but itself embodies our own cultural presuppositions about the universe? Could it be that science itself is a vast elaboration of our society's more general beliefs about the nature of reality? Could it be that the entire edifice of science merely projects our culture onto the universe, a projection which we then validate and reinforce through selective observation and facile interpretation? Could it be, in other words, that we too have constructed a myth?

Perhaps we have simply done as all other cultures have done. Those observations that fit into our basic mythology, we accept as fact. Those interpretations that fit into our conception of self and world, we accept as candidates for scientific legitimacy. Those that do not fit, we hardly bother to consider or verify, prove or disprove, dismissing them as absurdities unworthy even of consideration: "It isn't true because it couldn't be true." It was in that spirit that Galileo's scholarly contemporaries refused to look through his telescope, because they *knew* Jupiter couldn't have moons.

History has shown that scientists are no less subject than anyone else to peer pressure, self-deception, institutional blindness, and tunnel vision. Our culture is not alone in believing itself to possess the truth; nor is it alone in the certainty of its belief. However, the problem goes much deeper than the abuse and manipulation of the Scientific Method. More significant are the Method's inherent limitations, which spring from hidden assumptions woven so seamlessly into our world-view that we rarely question them; indeed, we rarely notice them at all. They imbue our common sense about how to live life and how to organize society, how to understand the world and how to pursue knowledge. Yes, the Scientific Method can be twisted to serve cultural or institutional biases; what is less obvious is that the Scientific Method *itself represents* a very deep and subtle presumption about the nature of reality. Rather than freeing us from cultural prejudice, it draws us in even deeper.

Do you have any idea what I'm talking about? The scientific reader might think I'm babbling. What presumptions could there be, when the Scientific Method says in effect: "We will accept nothing on faith. We will test every hypothesis to see whether it is really true."

Here is a central assumption of the Scientific Method that may seem so obvious as to be beyond dispute: if two people perform an identical experiment, they will get the same results. This requires (1) determinism: that the same initial conditions will result in the same final conditions, and (2) objectivity: that the experimenter can be separated out from the experiment. These two assumptions are intertwined. If we include the experimenter as part of the "initial conditions", then they are never really identical—not even if the experimenter is the same person performing it at a different point in space and time.

At bottom, the Scientific Method assumes that there is an objective universe "out there" that we can query experimentally, thus ascertaining the truth or falsity of our theories. Without this assumption, indeed, the whole concept of a "fact" becomes elusive, perhaps even incoherent. (Significantly, the root of the word is the Latin *factio*, a making or a doing,[3] hinting perhaps at a former ambiguity between existence and perception, being and doing; what is, and what is made. Perhaps facts, like artifacts and manufactures, are made by us.)

The universe "out there" is in principle unconnected to one or another observer; hence the replicability of scientific experiment. If you and I query the universe with an identical experiment, we arrive at an identical result. So blinded are we by our ontology that we see this not as an assumption, but a logical necessity. We can hardly imagine a cogent system of thought that doesn't embody objectivity. Neither can we imagine a system of thought that dispenses with determinism, which encodes the modern notion of causality. These we see as basic principles of logic, not the conditional cultural assumptions that they are.

The unfortunate fact that the whole of 20th century physics invalidates precisely these principles of objectivity and determinism has not yet sunk into our intuitions. The classical Newtonian world-view has been obsolete for a hundred years, but we have still not absorbed the revolutionary implications of the quantum mechanics that replaced it. Amazingly, eighty years after its mathematical formalization, quantum mechanics defies interpretation. Today some five or six major interpretations of quantum theory, along with countless variations, boast adherents not just among amateur philosophers and new-age seekers but among mainstream academic physicists, many of whom eschew interpretation altogether and use the mathematics of the theory in apparent disregard of its ontological significance. Either they cannot agree on the interpretation of the theory or they have given up even trying, because no interpretation is compatible with our fundamental cultural assumptions about the nature of reality.

The worldview of classical science I describe in this chapter, obsolete though it may be, still informs the dominant beliefs and intuitions of our culture. Science is a vast and elaborate articulation of the defining myth of our civilization: that we are discrete and separate selves, living in an objective universe of others. Science presupposes, embodies, and reinforces that myth, blinding us to other ways of thinking, living, and being.

Both justifications for the religion of science engender the same kind of limitation. The power of technology confirms, with circular logic, the practical truth of our science in precisely those areas in which it applies, yet it limits us to those areas. Meanwhile, the Scientific Method by its very nature excludes whole classes of possible phenomena from ever being established. It is constitutionally incapable of apprehending phenomena that are not objective or deterministic. Believing in the Scientific Method as the "best way yet discovered for winnowing the truth from lies and delusion," we therefore conclude that such phenomena don't really exist: they are lies, delusions, hoaxes, superstitions. Our ontology and method is fundamentally unable to countenance such possibilities as "The unicorn was there for me and not for you" except by explaining them away along the lines of "it was there but you didn't see it." Existence, being "there", is assumed to have an absolute, objective reality, independent of whether or not anyone observes it. More precisely, we naively associate existence with some event in an absolute Cartesian coordinate system: if the unicorn was at point X,Y,Z at time T, then it exists.

Go ahead, try it! Close your eyes and visualize something, a fork, say, *merely existing*. Do you see a disembodied fork floating alone in space? Separation is woven into our conception of being. Existence happens in isolation, not in relationship. To be is to occupy a discrete point in space and time.

That would be fine if the universe were "really" like that. However, ancient ways of thinking and 20th-century physics both agree that it is not. The absolute Cartesian universe is at best an approximation, a mathematical tool useful for solving a very narrow range of problems. Yet we have attempted to force the whole of reality into its mold. In elevating the Scientific Method to a defining test for truth, we implicitly decree, by fiat, the very assumptions upon which it rests.

The Scientific Method relies for its supra-cultural validity on principles that are themselves among its own assumptions. The logic of its justification is circular. A parallel would be an aborigine insisting, "Okay, let's settle this question of whether scientific experiment or dreaming is the way to true knowledge once and for all… Let's settle it by entering the dreamtime and asking the ancestors." The principal assumptions of objectivity and determinism that underlie the Scientific Method are by no means shared by all the world's traditions of thought. A non-objective, non-deterministic, yet coherent system of thought is possible. It is more than possible: it is necessary given the impending collapse of the world of the discrete and separate self that we have wrought. It is also necessary in light of the new scientific revolution of the last hundred years. Our ways of thinking and being are not working anymore.

Science is the intensification of trends of self-conception going back thousands of years. Objectivity and determinism reflect profoundly the way we understand ourselves in relation to the world, infiltrating at the deepest levels our thought, language, and reason. Witness the above phrase, "… if the universe were 'really' like that." What is this "really"? It means something like, "Not just in the opinion of some, but in actual fact." And what is this "actual fact"? Non-objective thinking is exceedingly difficult to communicate, when the assumption of objectivity is built into the language of that communication. Again, the master's tools will never dismantle the master's house. Objectivity and determinism are woven into our very self-definition. That is why the new sciences of the 20th century have been so difficult to integrate into our general understanding of the universe. That is why the findings of quantum mechanics seem so counterintuitive, so weird.

We sense ourselves as discrete beings, separate from the universe around us. Accordingly, the Scientific Method is unable to handle phenomena of which the experimenter is an inseparable aspect. Suppose something like telepathy works if and only if the experimenter genuinely believes it will. If this is the case, telepathy is inherently beyond the reach of the Scientific Method, because the experiment is not freely replicable. A skeptical experimenter cannot confirm it by precisely replicating the "initial conditions", because the initial conditions bring different results for different experimenters. This failure of determinism—identical initial conditions bringing different results—can only be resolved by recognizing the experimenter as part of those conditions, thereby invalidating the principle of objectivity.

I am not advocating the abolition of the Scientific Method. I am simply illuminating its inherent limitations as well as the type of knowledge it is constitutionally able to produce. The Scientific Method has a place—a very important place—in the new science that will emerge as our society evolves. At its heart, the Scientific Method rests upon a beautiful impulse, an ideal of humility and intellectual non-attachment that would serve any system of belief in good stead. In Chapter Six I will describe a different approach to experimental science, a play with nature and not a Baconian interrogation, that preserves the ideal of humility without objectivity's alienation.

Generally speaking, the Scientific Method fails to deliver certainty whenever the very act of formulating and testing a theory about a supposed reality "out there" creates or alters that reality. A model for understanding how objectivity in science can fail is journalism, which clearly illustrates the necessary codependence of observer and reality. People behave differently when they know a journalist is present; moreover, by reporting an event we change its significance, to the point where some events are newsworthy only because there are journalists present. Further compromising objectivity is the unavoidable projection of a news organization's values and priorities onto its criteria for newsworthiness. Nonetheless, journalists still pretend to an objectivity which is not only an impossible ideal but, even worse, an incoherent concept and a dangerous trap.

Yet objectivity remains a near-universal standard of intellectual probity in our culture. The ideal of the detached scientist eliminating personal prejudice in a quest for pure objective truth projects into other fields like journalism, law, and even personal life, in which a "rational" decision is one made free from emotional attachment. A little reflection reveals that such objectivity is impossible; moreover, its blind pursuit limits what we can possibly apprehend and erects a dispiriting wall of separation between ourselves and the universe.

Not only scientific objectivity but reason itself expresses and reinforces our gradual abstraction from nature. As David Bohm explains, reason is essentially the application of an abstracted relationship onto something new. We observe the relationship between A and B, and say that the relationship between C and D will be like that too. For example: "All humans die; I am a human; therefore I will die." A is to B and C is to D. Or, A:B::C:D. That is a ratio; it is rational. Reason is the recognition and application of abstracted regularities. The criticism of professional skeptics on such forums as reason-dot-com, that New Age spirituality and other denials of objectivity-based science are *irrational*, has merit. Reason as they understand it is contingent upon, and not prior to, the assumption of an objective reality from which we can abstract. If this assumption is untrue, then other forms of cognition are valid, and reason fraught with peril.

The mathematical connotations of the word "rational", containing as it does the word "ratio", suggest as well that rational thought embodies the rigor and certainty of mathematics. And unlike a mere number, which (originally at least) is associated with concrete objects and therefore attached to units, a ratio cancels out units to arrive at a pure abstraction, disconnected from any specific experience. This contributes to our elevation of reason to a supernatural province abstracted from concrete reality—abstracted, that is, from nature. As the faculty of reason is unique to human beings, we consider it proof of our ascent above nature. And the more we rely on reason, the farther above nature we ascend; hence the dream of the perfectly rational society as the pinnacle of human development.

Science and indeed reason itself are based on the discovery of regularities in nature. The possibility that we are not observing but rather *creating* these regularities through our beliefs presents a profound challenge to the very validity of science as we know it, implying that we are merely observing our own reflection. Beyond these two opposed possibilities lies a third, dialectic, possibility of mutual co-creation of belief and reality that affirms the fundamental non-separation of subject and object. The deep presumption underlying the Scientific Method and the whole edifice of scientific reason is that there is a reality "out there", independent of me, waiting to be discovered. It is the exact counterpart of our foundational cultural assumption of the separate self in a world of discrete others. It is yet another flavor of the basic dualism of subject and object.

This is the cultural assumption that grew through the ascent of humanity whose origins I explored in the previous chapter, which motivated technology and was in turn motivated by technology in a self-reinforcing loop. Science has merely taken it to its logical extreme, to its fulfillment. Science, therefore, is not *prior* to our understanding of self and universe, it is an outgrowth of this understanding. Science is an ideology.

Like other cultures before us, we have created a mythology, a constellation of stories to explain The Way of the World. It includes the forces of nature, the forces of human nature, the story of our origins, and an account of our role and function in the universe. Like those of all cultures, our mythology is not wholly fabricated but a window on the truth, paned, however, with the distorting lens of our culture's prejudices. Science is this mythology's consummation. The study of science, therefore, reveals as much about ourselves as it does about the world. This chapter explores the scientific life, personal and collective, that we have made for ourselves.

My Personal Age of Reason

Rational linear proof is what the mind demands. The heart's way begins when one lays one's head on a person's chest and drifts into the answer.
— Coleman Barks

What is it to be "scientific"? What does that word mean in the context of day-to-day life? Since the 17th century the man of reason has been elevated above all others, and reason assigned the highest status among all forms of cognition. Reason is, after all, the unique province of humankind, distinguishing us from the animals, elevating us past our biological heritage of instinct and emotion. To be a creature of reason, then, is to be more ascended; it is to be more a man and less an animal.

I use the word "man" for "human" advisedly here, because cultural prejudice has long held reason to be a distinguishing quality of the male sex. Men were supposed to be more rational; women more emotional and more intuitive. Women were thought closer to their biology, more subject to periodic fluctuations of hormones. Tied up in our elevation of reason, then, is the elevation of the male sex above the female.

Reason is a male-associated trait not because men are more capable of it, but because they apply it more broadly. Generally speaking, the men of our culture deploy reason in more situations than women—even and especially in situations in which reason is inappropriate. If we see reason as a hallmark of separation, then the male predilection for reason is evidence for the more extreme separation of the male sex. Women (and more generally, the feminine, including the inner feminine of both sexes) offer us men a way to reconnect with the biology, emotion, and intuition that reason has separated us from.

Like the mythological Greek hero Theseus, we wander in a labyrinth, a labyrinth of rationality and ego in which lurks a monster that can devour us. The Minotaur is a beast with a human head,[4] the face of reason atop the drives of biology. To it we sacrifice, as did the Athenians, the best and brightest of our youth. Once inside the Labyrinth, the only way out is by following the lifeline that Woman alone offers.

I am intimately familiar with the labyrinth of rationality, because I lived in it for many years. Applying logic first to any situation, suppressing emotion and deploying reason, I thought myself more cognitively evolved, more intelligent than most other people. A higher sort of man, more scientific, more rational—in a word, better. I acted from the mind, which was higher, rather than the emotions, which were lower. In applying reason to my own life, I echoed the ideals of Enlightenment philosophers who thought that a new Age of Reason would usher in a perfect society. Reason was to be an omnipotent tool that would solve all the world's problems.

You can probably imagine some of the calamities that befell me as I pursued my personal Age of Reason. Numbing down my emotions, I could make decisions only by comparing one list of reasons (the pros) to another (the cons). I tried very hard to "figure things out", but either way, my decisions lacked the certitude that accompanies a heart-based choice. Doubt plagued me, and I was unable to follow through on any decision wholeheartedly. I became paralyzed with indecision before, and doubt after, any decision I made. To compensate for the numbness, I would create vast, elaborate networks of reasons and justifications to bolster my decisions, but no matter how hard I tried, I felt hopelessly adrift.

Much later I understood that all my reasons were actually rationalizations for choices already made, non-rationally, based on something unconscious and unfelt. To bring to light the hidden determinants of my choices became an urgent priority. If not for reasons, then why do I do the things that I do?

Notice the parallels between this and our own collective Age of Reason. For all our clever elaborations, are we too not adrift in a sea of impotent analysis and explanation, propelled headlong toward calamity by forces we are only dimly aware of? Does not our collective history repeat itself, just as I helplessly recreated the same patterns in my life again and again? Do we not still look to reason—embodied in science—to save us, just as I tried repeatedly to figure out a solution to my problems?

Whether on a personal or planetary level, the same conceit rules: that if we only try hard enough, we can figure everything out and live happily ever after.

When did I finally abandon my personal version of the Scientific Program? Only when a series of crises ripped apart the fabric of my life in a way that made it abundantly clear that the program of management and control was doomed. Look at the world around us. Is there any doubt that collectively, we too are approaching such a moment?

I am not advocating the abolition of reason. As with the Scientific Method, reason has its proper place. Operating within their proper spheres, both are tools for the creation of wonders. The problem comes when they exceed their proper bounds, as they did in my own life, and seek to bring the whole of reality under their sway. The consequences of this happening are escalating damage and depletion of the social and physical environment. In the personal example, other people get hurt. In the collective example, it is whole cultures, ecosystems, and the planet itself. Eventually, because the self-other distinction is not fundamentally valid, the damage and depletion circles back to affect the self—inevitably. Managing, fixing, and controlling these consequences can only work temporarily, and at an escalating price, because it is in fact the mindset of separation behind these responses that caused the damage in the first place.

The anthroposophical physician Tom Cowan offers an interesting metaphor for this process, drawing on the Medieval alchemical tradition.[5] The alchemists understood human beings as consisting of three parts: the head, the thoracic region of the heart and lungs, and the viscera. The head, symbolized by silver, is cool, static, and reflective; the heart, symbolized by gold, warm and rhythmic; the viscera, symbolized by sulfur, hot and transformative. In this philosophy, the faculty of knowledge resides not in the head, but in the heart. The heart is for knowing, the head for reflection. Cowan explains that when the head function invading the rest of the body, its stillness manifests as concrescences and sclerosis: stones in the organs, plaques in the heart and arteries, and tumors all over.

Is it mere coincidence that the technological artifacts of the Age of Reason are also predominantly hard? Pavement, buildings, metal and plastic, all are designed to resist the rhythms and transformative processes of nature. Looking down from an airplane, the burgeoning suburbs look like metastasizing tumors radiating out from the urban hubs. The earth's softness turned hard.

In the personal story I have related, and in the whole culture of science, the head has usurped the heart's function as the organ of knowing. We intuit this when we associate the alchemical head-quality of coolness with the exercise of reason—"coldly rational", we say, or "cool-headed". The head is meant to reflect, to consider, to explore, but it is the heart that is meant to know and, therefore, to choose.

The idea of choice not compelled by reasons runs counter to a fundamental principle of classical physics: determinism. A mass in a Newtonian system has no choice; its motion forever into the future is wholly determined by the forces acting upon it. When we see ourselves in the same way, science's promise of control comes at an enormous price—an equivalent feeling of helplessness. We, like a Newtonian mass, are wholly determined by the totality of forces acting upon us.

The universe of classical science is a universe of force. While the achievements of the Scientific Revolution gave explicit form to this principle, it actually goes back much farther, back to the very origins of separation. It harks back to the ancient farmer, applying force to keep the land from reverting to its wild state; it harks back to the builder societies, replacing natural forms with human creations through human effort. The conquest of nature and human nature requires, like all military endeavors, force. The history of technology is a history of humanity putting greater and greater energies at its disposal, graduating from human power, to animal, to water and wind, to steam, to oil, gas and electricity, to nuclear power. And since energy, in physics, is nothing but force integrated over a distance, the ascent of humanity amounts to the exercise of more and more force. It is the bringing of force under human control. The fulfillment of human destiny, then, would be the harnessing of all the forces of nature, to bring them wholly into the human realm.

Forces in physics are the counterpart of reasons in our own lives. Just as we attempt to understand events by looking into the reasons that *caused* them to happen, so also do we understand the behavior of a physical system by adding up all the forces acting upon it. My personal Age of Reason had all the hallmarks of the classical scientific worldview. Steeped in the ideology of science, any other approach to knowledge seems nonsensical.

It is ironic indeed that my own attempt to live a scientific life governed by reason, which was motivated by a desire for control and certainty, generated instead crisis and uncertainty. The same has happened globally. The Scientific Method really comes down to, "Let's check and make sure." Yet we are increasingly unsure, paralyzed by the same doubt I experienced in my own life. The welter of opposing voices in my head mirrors the bitterly conflicting interests of modern politics and other institutions that thwart any purposeful transformation. They leave us helpless, trundling forward under the momentum of the past, buffeted by forces beyond our control.

Masters of the Universe

Scientific thought is essentially power thought—the sort of thought that is to say whose purpose, conscious or unconscious, is to give power to its possessor.
—Bertrand Russell

At its purest, the purpose of science is to better understand the world Or, we could say, it is to bring new worlds into the human domain of understanding. Science begins as an exploration of the unknown, and later becomes a conquest that subjugates that unknown to human purposes. It is highly significant that the Scientific Revolution coincided so closely in time with the European Age of Exploration. In both we see the same missionary zeal, the same sense of a new world of possibility, the same ideological roots, and the same tragic consequences.

The Age of Exploration led to the Age of Imperialism, both geopolitical and scientific. The urge to discover new lands was never innocent of the power motive. The sense of mission that drove the Europeans to civilize and colonize the world also infuses science. To civilize: to make tame, to bring order to. To colonize: to make subservient, to administer as a source of raw materials. Science colonizes the world for technology, finding ways to put materials to use, "harnessing" the forces of nature. Each new world that science discovered—the microscopic and the celestial, the electromagnetic and the chemical—was first explored and then exploited as a new dominion. Both campaigns of conquest, the scientific and the terrestrial, are expressions of the same aspiration: to make the world ours.

Starting about five hundred years ago, scientists and explorers issued forth from the Old World into a new. One frontier after another succumbed: the heavens, the sea, the poles, the archeological past, Everest, the cell, the genes, outer space, the atom. Concurrent with the expansion of the territory of civilization, the human realm broadened with each scientific conquest and the realm of the mysterious, the wild, shrank. By the end of the 19th century both conquests seemed nearly complete: only a few scattered hunter-gatherer tribes remained in the earth's remotest regions, and only a few recondite phenomena, it seemed, still eluded the onward march of science.

The exhilarating promise of the new worlds sparked an optimism and a zeal that was to last for several centuries. Some vestige of it remains today in persistent hopes that nanotechnology or genetic engineering will bring the same easy riches (or even the Fountain of Youth) once sought in the terrestrial New World. But as I observed in Chapter One, confidence in this promise is wearing thin.

Whether or not its promise will ever be redeemed, perhaps we *have* entered a new world. Certainly the astonishing, nigh-miraculous technologies of air and space travel, instantaneous communications, and information processing would have seemed fantastical to people five hundred years ago. But if we have entered a new world, we have indubitably brought the old world with us, just as the European colonists brought along and perpetuated the violence and injustice they sought to escape. The new realm that science has opened to us is just like the old: it bears just as much uncertainty, just as much want, just as much suffering and just as much savagery, if only in somewhat different form.

This should not be surprising, because the Scientific Revolution was not really anything new. It was not a cultural discontinuity, but rather the crystallization of trends far preceding it. Science is just the culminating articulation, indeed the apotheosis, of trends of objectification going back thousands of years. Science takes the objectification of nature to its extreme, but conversely, a preexisting objectification of nature is necessary to even articulate its basic tenets and methods. It was only in the 17th century that our separation was sufficient for science to take off. The great names of the Scientific Revolution—Galileo, Newton, Descartes, Leibnitz, Bacon—merely gave expression to ideas whose time had come.

Before the 17th century human beings had not even the basis to dream of the Scientific Program to understand everything and the Technological Program to control it. The mysteries were too great and the powers of nature too awesome, our knowledge too scant and our technology too feeble. However, the slow accumulation of technology and empirical science through the Renaissance period gradually eroded nature's forbidding immensity, bringing us to a point where such an assault on its mysteries became conceivable.

The conceptual underpinnings of this assault were formulated by Kepler, Galileo, Bacon, and Descartes in the early part of the 17th century. The key physical insight (discovered by Galileo and formalized by Descartes) seems quite innocuous: A moving body continues to move forever at the same speed and the same direction, unless a force (friction, for example) acts upon it. Before Galileo, people naturally assumed that it takes a constant force to keep something in motion: When the ox stops pulling, the cart stops moving. Galileo said no, without force, nothing new ever really happens. Moving bodies keep moving in the same direction; resting bodies stay at rest. To change anything, force must be applied.

Why was this such a big deal? We live in a world of movement. Before we' had digested the physics of Galileo, Descartes, and Newton, it seemed obvious that in order for there to be movement, there must be a Mover, a being to keep the sun and the moon in motion, to blow the wind and to rain the rain, to grow the plants and animate the animals. With the new laws of motion, no such Mover was necessary. Once set in motion, everything keeps moving by itself. At most, motion can get transferred from one object to another. God was no longer necessary to *animate* the world.

Parallel logic led naturally to the thought, developed by Descartes and others, that maybe animals are machines too, that no *anima*, no spirit, is needed to animate them either. The law of conservation of momentum is thus a direct denial of the ancient religion of animism. Consider a Native American term for God: "the spirit that moves all things". With Galileo and Newton after him, no such spirit is necessary. Nothing is innately animate, but only moves by the application of physical force. Matter is inherently dead.

Descartes, Galileo, and the rest still believed in God, but they removed Him from the world of matter. God became a watchmaker, and creation became a discrete act and not an ongoing process. The universe became, essentially, a machine. The divine, once wholly identified with nature, and gradually abstracted through the age of agriculture and the Machine, was now completely removed from the world of matter.

With God no longer participating in the moving of the world, there is nothing to stop human beings from becoming the world's masters. And the tools of our mastery are the tools of force. There is nothing we cannot alter if we can only apply enough force in the right way. Our power over the universe and each other is limited only by the amount of force at our disposal, and our understanding of where to apply it. Herein lies an intriguing definition of technology: a system of techniques for the application of force.

And how do we know the correct way to apply force? By applying reason to the quantitative, objective description of reality that science provides. Our power, in other words, comes through the faculties of the mind. And what is the domain of the mind? Which aspects of the universe are to be included? Kepler's answer was this: "As the ear is made to perceive sound, and the eye to perceive color, so the mind has been formed to perceive quantities." Galileo heartily concurred. The brain, he believed, is wholly concerned with the apprehension of what he called primary qualities: size, shape, quantity, and motion. Everything else, even and especially sensory experiences such as sounds, odors, and colors, was secondary, outside the province of mind and outside the province of science. After all, we share those experiences with animals, but the abstraction and quantification that Galileo attributed to pure mind is a singularly human trait. By implication, the more fully we devote ourselves to that function, the more separate we are from the animals, the more ascended.

In exiling quality from reality, Galileo banned subjectivity from science and denied the importance of how we experience the world. Science today still strives to remove any dependency on individual subjective experience. Following Galileo, it concerns itself with that which is independent of subjectivity. Lewis Mumford puts it succinctly: "Following Kepler's lead, Galileo constructed a world in which matter alone mattered, in which qualities became 'immaterial' and were turned by inference into superfluous exudations of the mind."[6]

So deeply has the gospel of objectivity taken hold that it pervades our very language, so that when we use words to deconstruct it, we risk unconsciously reinforcing it. Witness the odd phrase above: "in which matter alone mattered." Here, to matter is to be significant, to be effectively real. Matter, turned into verb form, means to be real. Implicit in that very verb is that only matter is real. (And what about weighty matters?) If we tried to posit the opposite sentiment, say, "Spirit matters more than matter," we are actually reinforcing the primacy of matter via the tacit assumptions embodied in the language itself.

Even more subtly, every declarative "is" sentence also reinforces objectivity by making a peremptory claim about an absolute reality independent of anyone's subjective experience. You see, that's just the way it *is*.

If, as modern physics suggests, the observer is inseparable from the observed, then any "is" sentence is at best an approximation and at worst a lie. Such a conclusion inheres already in the abstraction of symbolic language, as described in Chapter Two. In symbolic culture, the alienating conclusions of Kepler, Galileo, and Descartes are already present. These thinkers merely formalized separation as an ideological principle. A long-gathering undercurrent had now risen to the surface and would soon sweep all before it.

Galileo's excision of God from the world of matter mirrored the even more audacious banishment of subjective experiences from the domain of rigorous intellectual exploration. Not only their knowability was questioned, but even their reality. Science is the study of reality; what is not measurable is not a valid subject of science; therefore what is not measurable is not real. A century later, David Hume took up this position with great enthusiasm: "Let us ask, Does it contain any abstract reasoning concerning quantity or number? No. Does it contain any experimental reasoning concerning matter of fact and existence? No. Commit it then to the flames; for it contains nothing but sophistry and illusion."[7]

In defense of these philosophers, it helps to see where they were coming from. The ideology of objectivity doubtless had a salutary effect initially, liberating thought from the stultifying Scholastic tradition that had long sequestered knowledge in the arcane volumes of Aristotle and the Church theologians. The new scientific knowledge, in contrast, was accessible to anyone; scientific experiments were replicable by anyone seeking to see for himself. No faith in dogma was necessary; all knowledge was to be open to first-hand verification. Truth was taken out of the hands of the ecclesiastical hierarchy. The Scientific Revolution sought to free thought, not to bind it.

Ironic indeed, then, is the present state of science, in which once again vast areas of inquiry are off-limits; in which experimental results that contradict orthodoxy are excluded from publication; in which knowledge is restricted to those initiated into the language of its abstruse texts; in which whole fields wallow in fruitless hyperspecialization; in which the public can only await the pronouncements of this new quasi-ecclesiastical hierarchy, holder of the keys to the gates of knowledge. Can we say that we have not replicated the old world within the new? Upon the Scientific Method, which freed thought from the institutionalized, authoritarian superstition of the Middle Ages, we have built yet a new orthodoxy more totalitarian, though more subtle, than the first.

Galileo's assertion that the universe is "written in the language of mathematics" potentially subordinates all its mysteries to human understanding and human control. Accordingly, we attempt to understand the world by (1) gathering data, and (2) manipulating that data according to mathematical models. Nature is thereby rendered tractable, promising a reliable foundation to the Technological Program of control. Mathematically, the ambition of subordinating the universe to numbers took form in Descartes' system of coordinates, which associated every point in space and time with a number. Descartes was also among the first to fully grasp the potential power of this new approach to knowledge, as in this famous passage:

> For by them I perceived it to be possible to arrive at knowledge highly useful in life; and in room of the speculative philosophy usually taught in the schools, to discover a practical, by means of which, knowing the force and action of fire, water, air the stars, the heavens, and all the other bodies that surround us, as distinctly as we know the various crafts of our artisans, we might also apply them in the same way to all the uses to which they are adapted, and thus render ourselves the lords and possessors of nature. And this is a result to be desired, not only in order to the invention of an infinity of arts, by which we might be enabled to enjoy without any trouble the fruits of the earth, and all its comforts, but also and especially for the preservation of health.[8]

Here Descartes articulates quite clearly the relationship between science and technology that was to dominate the next three centuries. Science achieves understanding, upon which basis technology achieves control. If we can understand precisely how something works, then we can conceivably control it with infinite precision. And the purpose of all this, the motivation and the justification, is to dominate nature, eliminate labor ("enjoy without any trouble the fruits of the earth"), ensure comfort, and conquer disease. He doesn't go so far as the techno-utopian ideal of overcoming death itself—such audacity had to wait the twentieth century—but he nonetheless lays out the Technological Program in all its essential details.

While Galileo and Descartes posited the mathematization of the universe, the first promising claim to having actually achieved such a feat had to wait until the defining figure of the Scientific Revolution, Isaac Newton. With his famous equation F=MA (force equals mass times acceleration), Newton put Galileo's discovery into rigorous mathematical form. Force, and only force, causes acceleration, a change in the rate and direction of movement.

Newton also furthered the removal of spirit from the world of matter by uniting heaven and earth. Up through the Middle Ages, Heaven was not an abstract concept, but literally identified with the sky. That's where God lived. The sky, the heavens, was the abode of God during the agricultural phase of humanity. The Greeks put their gods first on Mount Olympus, and then later on an invisible, supernatural Olympus in the sky. The same identity existed in classical China: in Chinese, the word *tian* means both heaven and sky, and the semi-divine emperor was the *tianzi*, the "son of heaven".

Before Newton, the heavenly realm and the earthly realm remained separate. The heavenly realm was the realm of perfection, where heavenly bodies moved in perfect circles (actually ellipses) and along predictable paths. The earthly realm was chaotic, and whatever order there was (tides, day and night, seasons, and so forth) seemed to originate in the heavens. Naturally, then, people associated the heavenly realm of order and mathematical perfection with God. Heavenly bodies were not subject to earthly laws—the moon does not fall out of the sky the way Galileo's weights fell from the leaning tower of Pisa.

Newton's accomplishment was to show that, by understanding gravity as a force, the same equation, F=MA, could be made to describe both realms, the heavenly and the earthly. A single equation replaced the empirical laws of both Galileo and Kepler, which had seemed so entirely different. One equation described both the motions of the planets and the motion of an apple falling from a tree, an astonishing unification. Here was the first candidate for a "theory of everything", still the Holy Grail of physics. Here was the first plausible hope that maybe the whole universe and everything in it really could be understood in the form of mathematics, just as Galileo said.

Interestingly, even as they furthered the reduction of nature to mathematics, Newton's Laws required a new advance in that reduction to even be conceived. The derivation and application of Newton's laws required a novel mathematical technique—calculus—that solves problems by treating time as a succession of infinitesimally brief instants, essentially reducing process to number and becoming to being. Even as mere mathematics, calculus smuggles in a very different mode of conceptualization that perhaps could not have occurred outside the context of increasing objectification of the world. Maybe this is why Archimedes did not invent calculus two millennia beforehand, despite having applied the basic technique to numerous problems in geometry. Similarly, maybe it is an unconscious rejection of this leap in abstraction that renders so many students, even those who were "good at math" in high school, seemingly unable to learn calculus. And you thought you were stupid!

Newton's Universal Law of Gravitation purported to be just that, universal. The mind of man had finally penetrated the deepest secret of the universe. The greatest mystery had been revealed. Newton had discovered the key to the mechanism of God's creation. The human realm of the understood now encompassed the entire cosmos via a single governing equation. All that was now needed was to accumulate data.

No wonder Newton's discovery was so exhilarating and why Newton himself was such a celebrity. Poets spoke of him discovering the key that unlocks the universe. His epitaph, penned by Alexander Pope, reads, "Nature and Nature's laws lay hid in night / God said: 'Let Newton be' and all was light."

It is significant that the canonical founders of modern science were so preoccupied with the sky, an unearthly realm well suited to a mode of inquiry that strove to be independent of human subjectivity. This focus links them back to the priesthood of the ancient builder civilizations with their semi-divine rulers, the sons of heaven, the earthly representatives of the solar god. Even in those days, the court priest-scientists gazed upon the skies for purposes of astrology and calendar-making. Scientists, with their heads in the clouds, are not too concerned with earthly affairs—hence the stereotype of the absent-minded professor. They have also generally been politically innocuous—as long as they "stick to science" and "don't enter politics". The worldly realm is supposed to be separate from the realm of science, which is the non-earthly, the celestial. Metaphorically this holds even more strongly. Science, especially "pure" science which is loftier than applied, is a rarefied plane of pure thought, inaccessible to all but the most highly trained intellects. It is wholly in the realm of the mind. And since intellect or mind is itself a uniquely human realm, pure science represents the loftiest human ascent, and the scientist is the most highly ascended human being. Yes, scientists are the modern priesthood, gazing with their mysterious instruments at invisible worlds to divine the truth. We the uninitiated stand outside their temples awaiting their pronouncements.

The work of Kepler, Galileo, and Newton amounted to a conquest of the heavens, a bringing of celestial phenomena into the human domain of abstract mathematics. The literal "conquest of space" had to wait a few more centuries, but the ambition to do so was inevitable.[9] Space travel was to be the fulfillment of human destiny, holding all the promise of a new world and the final transcendence of the old. Yet when we finally landed on the moon, nothing much happened. Our leaders, channeling a generalized aspiration, had their heads in the clouds, the heavenly realm. But the earthly realm proved not so easy to leave behind. Space exploration was an unprecedented and literal "ascent" of humanity. The original abode of God had been physically breached. We had literally entered the heavens, and we found that we had taken our earthly nature with us into our New World. We had not left biology or the world behind; in fact, space travel required that we take them with us, a bit of earth enclosed in a space capsule.

Neither are our forays into the realms of the mind ever unsullied by worldly matters. The culture of science is no more immune than any other human sphere to pettiness, vanity, politicking, cheating, favoritism, and prejudice. And like space travel, any attempt to divorce a rational society or a rational life from the organic supporting matrix where it belongs requires tremendous effort and incurs tremendous danger. Such a life or society is tenuous, fragile, and short-lived. It cannot exist for long without reconnecting to the wellspring of life.

No more independent of nature are we than an astronaut is independent of the earth. Only a very foolish astronaut would think that he has no more need of the planet: "Hey, I've got food, I've got water, I've got oxygen... I'm fine!" Such is the myopia of the civilization of fire, encapsulated in its own vehicle of exploration, fueled and sustained by the supplies—natural, social, cultural, and spiritual capital—that it has taken along. Our voyage has taken us far, but to what end?

We reached the moon, and it was barren. The bleak moonscape of rocks and dust is a fitting metaphor for the landscape of separation, whether the emotional desolation of the man of reason, or the ugly homogeneity of suburbia. Yet our sojourn—the entire course of separation—is not without purpose. To convey a hint of what that purpose might be, I've selected a few quotes from astronauts describing their experiences as they gazed upon the earth from the vantage point of the most extreme literal separation human beings have ever known:[10]

> From the moon, the Earth is so small and so fragile, and such a precious little spot in that Universe, that you can block it out with your thumb. Then you realize that on that spot, that little blue and white thing, is everything that means anything to you — all of history and music and poetry and art and death and birth and love, tears, joy, games, all of it right there on that little spot that you can cover with your thumb. And you realize from that perspective that you've changed forever, that there is something new there, that the relationship is no longer what it was. — Rusty Schweickart

> When I was the last man to walk on the moon in December 1972, I stood in the blue darkness and looked in awe at the Earth from the lunar surface. What I saw was almost too beautiful to grasp. There was too much logic, too much purpose — it was just too beautiful to have

happened by accident. It doesn't matter how you choose to worship God... God has to exist to have created what I was privileged to see. — Gene Cernan

On the return trip home, gazing through 240,000 miles of space toward the stars and the planet from which I had come, I suddenly experienced the Universe as intelligent, loving, harmonious. — Edgar Mitchell

The first day we all pointed to our own countries. The third or fourth day we were pointing to our continents. By the fifth day we were aware of only one Earth. — Sultan bin Salman al-Saud

It isn't important in which sea or lake you observe a slick of pollution, or in the forests of which country a fire breaks out, or on which continent a hurricane arises. You are standing guard over the whole of our Earth. —Yuri Artyukhin

With all the arguments, pro and con, for going to the moon, no one suggested that we should do it to look at the Earth. But that may in fact have been the most important reason of all. — Joseph P. Allen

Like its most iconic achievement, space travel, science has taken us on flights of intellect to a cold, barren, alien realm, reducing life to a collection of forces and masses. And yet, this new vantage point has revealed a previously unsuspected splendor. Gazing through the lens of accumulated scientific knowledge at a body or a cell, when we really get its complexity and orchestration, its order and its beauty, the perfect mesh of levels and systems, then we know we are in the presence of a miracle. Awe is the only authentic response. Science has brought us to a place where we can walk in living awe of the ongoing miracle that is the world. In analogy to Joseph Allen's thought above, perhaps it is this, and not control, that is the true purpose of science. It is to apprehend new realms of the awesome.

Certainly, the alternative goal of science—to bring all nature into the human realm—has failed. Just as the conquest of the heavens that was to be the consummation of science has proved a mirage, Newton's candidate for the Theory of Everything that would make us masters of the universe soon turned out to be incomplete. So we added new laws for electricity and magnetism, and by the end of the 19th century, a physics combining Maxwell's equations and Newtonian kinetics seemed complete—except for a few pesky anomalies, minor details such as quantized radiation and the invariance of the speed of light. These led to quantum mechanics and relativity, whose unification we are still striving for today. Reading the popular literature, one gets the impression that we are almost there. Soon the remaining mysteries will be revealed. The latest candidate is String Theory—scientists are working on the details right now!

The Quest for Certainty

Despite the elusiveness of a Theory of Everything, all seemed well with the project of perfect understanding and perfect control up through the early 20th century. The new laws of electricity and magnetism shared with Newtonian kinetics the key features of determinism, reductionism, and objectivity. Under these laws, all that is necessary to understand and predict any phenomena is to add up all the forces on all the parts.

In the kinetics of Newton, if we know the initial state of a system (e.g. the initial velocity and angle of a cannonball, plus such details as air resistance if you want to get technical) we also know, with mathematical certainty, its state at any time in the future. The universe of classical science is itself such a system, composed of masses large and small governed by the same deterministic laws. If you can measure the state of anything in the universe accurately enough, you will know its future (and past) with perfect certainty. If you measure the state of everything in the universe, you know the future of the whole universe! Understanding of the whole comes through the measurement of its parts. Knowing the parts, you know the whole. This is the doctrine of reductionism, a key component of what Fritjof Capra called "the Newtonian World-machine".

Perfect deterministic, reductionistic knowledge suggests the possibility of perfect control.[11] Control the initial conditions of any system, and you control the outcome. If the outcome differs from expectations, that can only be due to some unknown variable or inaccuracy in the initial measurement. Either our knowledge is imperfect, or our control insufficiently precise.

The Technological Program depends on the principle of determinism. If nature is inherently unpredictable and mysterious, if it somehow eludes complete description in the form of numbers, if identical setups produce different results each time, then the goal of perfect control is unattainable. The Technological Program also depends on reductionism. An irreducible whole is essentially mysterious, impervious to analysis and only conditionally subject to control. If we can understand something in terms of the interactions of its parts, then we can engineer it by altering or replacing those parts. It is under our control.

Science is the logical extension of the labeling and numbering of the world described in Chapter Two. The original conceit was that by naming and measuring nature, we make it ours. Science fleshed out this primitive intuition by saying, in essence, that if we can name and measure everything, we can apply to everything the tools of mathematics and deductive logic. The final destination of language, the categorization and naming of everything in the universe, and of number, the measurement and quantification of everything in the universe, now manifests in science as the profusion of technical vocabulary that defines every field and in the program of quantification that has virtually usurped the "soft" sciences.

In the centuries after Newton, more and more fields of human inquiry aspired to that magisterial appellation, science. The category of science expanded to include biology, medicine, archeology, anthropology, economics, sociology, and psychology, progressively annexing the territory of the qualitative and the subjective. Chemistry was to be reduced to physics, biology to chemistry, the organism to the cells, the brain to the neurons, economics to individual behavior. The goal of all this was to ground all science upon the certainty of physics, expressed as a system of axioms and therefore borrowing its infallibility from mathematics.

To this day, every field of thought that presumes to the status of a science turns to mathematics for validation, either directly or by implying through hypertrophied technical jargon an exactitude of meaning that admits to the methods of deductive reasoning. We buy into this every time we succumb to the urge to start a paper by "stating our definitions" or "stating our assumptions"—a blatant imitation of the axiomatic method. The same reductionist mentality motivates our pedagogic emphasis on "analysis", which literally means to cut apart. To analyze a situation is to break it down into its constituent pieces. That's what we do collectively too, with the splintering of knowledge into fields, subfields, and specializations and subspecializations within these.

When outright reduction to something more physical was impossible, the aspiring field would imitate physics instead. Thus in the "social sciences" we hear constantly of "laws"—the laws of history, the laws of economics, the laws of human behavior—as well as psychological "tensions", historical "forces", economic "mechanisms", and political "momentum". These justify an engineering approach to human institutions and foster the illusion that they, too, might be understood—and eventually controlled—through the abstract methods of scientific reason. Even in fields where outright physicalism of metaphor is absent, the quest for certainty remains in the reductionist campaign to explain, in parallel with physics, the complex in terms of the simple, the measurable, the quantifiable, the controllable.

Witness economics, which applies the "law" of rational self-interest to its atomic units, individual human beings. Higher-order laws like that of supply and demand are akin to the laws of planetary motion, in that they derive from the lower-order laws. Once the axioms are set, the rest is just mathematics. Of course, success has been notoriously elusive in economics' quest for mathematical certainty, judging by its inability to produce accurate predictions based on "initial conditions". But economists do not therefore conclude that their subject is impervious to mathematical method; quite the opposite, they believe that the mathematization has not gone far *enough*. Better data, perhaps, or more precise characterization of various uncontrolled variables of human behavior, and economics would finally live up to its scientific pretenses.

At the bottom of the entire reductionist program is nothing less than a Theory of Everything, a modern version of Newton's Universal Law of Gravitation that would encompass all known forces. From this would be derived, progressively, all the higher-level laws of chemistry, biology, psychology, sociology, geology, all the way up to cosmology, so that every question would become a question of science, mathematically deducible from physical laws and data. This ambition was articulated very early on in the Scientific Revolution by Leibnitz, who wrote, "If controversies were to arise, there would be no more need of disputation between two philosophers than between two accountants. For it would suffice to take their pencils in their hands, and say to each other, 'Let us calculate.'"[12]

Of course, such a reduction of nature to mathematics is only as powerful and as reliable as the math that lies beneath it. The Scientific Revolutionaries considered math to be a source of infallible knowledge. According to Immanuel Kant's highly influential reasoning, mathematical knowledge consists of "necessary truths" that, like 2+2=4, cannot be any other way, as opposed to the "contingent truths" of empirical observation. (Even here we find an unstated assumption of objectivity, that the laws of physics can be separated from the selves that ponder them.)

Because mathematics is the bedrock upon which the entire edifice of science rests, much energy in the early 20th century went into establishing mathematics on a firm axiomatic foundation. This program hit a brick wall in the 1930's with the work of Gödel, Turing, and Church as described in Chapter Six, but like the rest of the Newtonian World-machine it still exerts its influence in the general unspoken assumption that the most reliable, most valid knowledge is that which can be put in the form of numbers.

The world of control that determinism opens up extends far beyond science and technology. In politics too, and indeed in personal life, control rests upon a similar foundation: the application of force based on precise data about the world. Accordingly, power-hungry politicians and totalitarian governments are obsessed with controlling the flow of information because, in their view, knowledge is power. The same goes for all controlling personalities. They want to be privy to inside information; they want to know your secrets; they hate it when something happens behind their backs.

That knowledge is equivalent to information is a direct consequence of the world-view that arose with Newtonian kinetics. We try to discover all the "forces" bearing on a situation; knowing them, we can control the outcome, provided we have enough force of our own. Whether in physics or politics, force plus information equals control.

Keep that formula in mind next time you read the news. F + I = C.

Actually, this strategy only works in certain limited circumstances. It fails miserably in non-linear systems, in which effects feed back into causes. Tiny errors lead to huge uncertainties and radically unpredictable results. The situation spins out of control. Trapped in a Newtonian mindset, we are helpless to respond except with more control. The current Bush administration is a good example of this, desperately lying, hiding, and manipulating information to control the effects of earlier lies and manipulations. I also think of an addict trying to manage a life that is falling apart, or an adulterous spouse trying to hide the multiplying evidence of his infidelity. Newtonian-based logic says that this should work. It is simply a matter of being thorough, of finding all the possible causes of a breakdown. With enough information and enough force, we should be able to manage any situation.

If only I could "figure it out", then maybe my life too would become manageable. If only I could apply enough force, enough willpower, maybe I could conform it to the image of my desire. Whether collectively or personally, what a fraudulent promise that has proven to be!

Another way to look at it is that trying harder will never work. Yet that is our typical response to failure, both on the individual and collective level. When we break our New Year's resolutions, what do we conclude? We didn't try hard enough, we hadn't enough willpower. Similarly, we act as if more technology, more laws, more vigilance can succeed where previous technology, laws, and vigilance failed. But greater effort from our present state of being only serves to reinforce that state of being. Using more force promotes the mentality of using force. Methods born of separation exacerbate separation. Does anyone today remember that World War One was fervently believed to be the "war to end all wars"? Despite that stupendous failure, equivalent logic lives on in the current "war on terror".

Thanks to science, we have more information about the world than ever before. Thanks to technology, our ability to apply force is likewise unprecedented. Yet despite centuries of progress in the technologies of control, we have made little overall progress in improving ecological, social, and political conditions. To the contrary: our planet veers toward disaster. What is the source of this failure? We have attempted to apply the linear strategy of F + I = C beyond its proper domain. Faced with a complex problem, the engineer or the manager breaks it down into parts, in which each process is a discrete module performing one of a series of specialized functions. Organic interdependency and feedback is fatal to such systems of management and control. The performance of any part of an organic non-linear system cannot be understood or predicted in isolation from the rest, but only in relationship to the rest. Such parts are no longer freely interchangeable, and the methodologies of reductionism are impotent.

The solution for several centuries has been to attempt to make linear what is actually not. Sometimes we do so quite benignly: In mathematics, for example, we use numeric methods to approximate solutions to nonlinear differential equations, which are generally unsolvable analytically. Much more sinister is the attempt to impose linearity on human institutions and human beings, which necessitates the destruction of all that is organic, traditional, and autonomous. All is to be rationally planned out, and each person made into a discrete standardized element of a vast machine. In medicine the consequences of reducing the organic to the linear are equally horrific. These are but two facets of the totalizing "World Under Control" I describe in Chapter Five.

I am not saying that reductionistic science, and the technology based upon it, is powerless. On the contrary, it has generated the entire infrastructure of our society, from bridges and skyscrapers to batteries and microchips, using standard components and according to generalized principles. The reduction of the infinitely diverse natural world into a finite set of standard inputs and processes actually *works*—for applications in which the inevitable remaining differences don't matter. For these practical purposes, each electron, each iron atom, each drop of water, each block of granite, each "qualified human resource" is the same. For these practical purposes, yes, but not for *all* practical purposes—that is the delusion into which our success has led us. Reductionistic science and rationality itself, both based on the abstraction of regularities in nature, have allowed us to build very high indeed, an edifice reaching farther than ever before toward the heavens. Yet paradoxically, we are no closer to Heaven than when we started.

We have achieved mastery of the linear domain, and attempted to expand that domain to cover the universe. Most real-world systems, however, including living organisms, are hopelessly non-linear. From this realization will arise a new approach to engineering and to problem-solving in general that does not start by breaking the problem into pieces. Our whole concept of "design" will evolve as well, away from hierarchical, modular approaches toward those based on self-organization and emergence. In doing so, a certain degree of certainty and control will be lost. Our relationship to nature, to each other, and to the universe will be radically transformed. This shift can only happen as part of a profound transformation of our sense of self, of who we are in relation to the universe. We will have to let go of control and face the fear behind that compulsion. Lewis Mumford identified it this way: "Today this almost pathological fear of what cannot be directly examined and brought under control—external, preferably mechanical, electronic, or chemical control—survives as a scientific equivalent of a much older atavism, fear of the dark".[13] Letting go of control means, then, that the Age of Fire, in which the circle of domesticity defined the human realm as the illuminated, is coming to an end. We will become once more at home in the dark: at home in mystery, in uncertainty, in unreasonableness. Or at least we will no longer fear to venture there.

Reducing Reality

Man must at last wake out of his millenary dream; and in doing so wake to his total solitude, his fundamental isolation. Now does he at last realize that, like a gypsy, he lives at the boundary of an alien world. A world that is deaf to his music, just as indifferent to his hopes as to his suffering or his crimes.
—Jacques Monod

Reductionism does not deny other types of explanation, but holds that it is the reductionistic explanation that is fundamental or primary. "Why did I scratch my nose just now?" The reductionistic explanation is something like, "An excitation of a nerve receptor in the mucosal lining of my nose sent a bioelectrical impulse to a neuron in my brain, which was transmitted to another neuron via a synaptic release of neurotransmitters, and then to another, eventually causing a set of muscle fibers to twitch in a particular sequence that lifted my hand to my nose." Of course, another explanation is "because it itched", but this higher-level phenomenon is nothing more than the sum of various (ultimately biochemically characterizable) states of nerve cells and so on. That is what "itched" really means, right?

And, why did I hug my five-year-old son when he was crying? Again, through a long sequence of completely deterministic chemical and electrical cause-and-effect, sound waves stimulated neurons which triggered a set of neural firing patterns and hormone releases; these in turn stimulated other neurons which caused the various muscular contractions that produce comforting facial expressions, sounds, and hugging. Of course, I could say, "I hugged him because I wanted to comfort him," but like everything else, this "wanting" reduces to a certain state of matter. "Yes, you wanted to hug him, but what that *really* means is a secretion of hormones A, B, and C, neurotransmitters D, E, and F, a particular firing pattern in the reticular formation of the brain, etc. That is what 'wanting to hug' really is."

The absurdity of the above reveals reductionism as an ideology, and not necessarily the way science is really practiced. Even if we fully reduced a hug to an ensemble of elementary particles and forces, that would not *explain* anything at all unless we interpreted various states along the way as higher-level functions like love, comfort, and so forth. Moreover, teleological explanations in science are quite common, as in

"Why are the salmon swimming inland? To get to their spawning grounds." The ideology of reductionism says that such statements are mere code words for "deeper" explanations; for example, that some set of genetic factors generate biochemical "mechanisms" to produce the spawning behavior.

Living up to its name, reductionism attempts to explain the complex in terms of the simple. Just as Newton used one simple formula to explain Kepler's complex empirical laws of planetary motion, reductionism assumes that all complex behavior arises from the summation of a few types of simple interactions. The kinetic theory of gases, for example, is derivable from the statistical properties of lots and lots of small particles (molecules) bumping into each other. The higher-order laws of pressure, volume, and temperature arise from the lower-level, more fundamental laws of Newtonian kinetics. These are the reality underlying the appearance. Having reduced the complex to the simple, we can then follow Leibnitz. Let us calculate.

To avoid an infinite regress catastrophe, the reductionist program must eventually rest on fundamental, irreducible building blocks or "elements". The frequent appearance of that word in introductory textbooks ("Elements of… ") bespeaks our reductionistic assumptions. Start with the basics and build up from there. This approach to pedagogy is not only dull but also ineffective, as anyone who has struggled through a course on organic chemistry or differential equations can attest. A much better way to teach math and science is from an historical context. In an introductory course in abstract algebra, for instance, don't start by listing the axioms that define a group. Start with a key historical problem, say the algebraic insolubility of the quintic, and trace the steps toward increasing abstraction by which it was solved. Imitating history, the axioms are arrived at as the end point, not the starting point, of a field of mathematics. "The basics" acquire context, interest, and motivation.

The idea that all appearances are merely different permutations of a few basic elements goes back to ancient times—the five elements of the Chinese, the four elements of the Greeks, the three doshas of the Hindus—but it is mostly associated with the Greek Atomists.[14] It fosters the belief in the possibility of complete control of nature: all we have to do is master these few building blocks. Whether to a Greek physician reducing the body to the Four Humors, or a modern nano-engineer seeking to build matter atom by atom, reductionism offers the same vision of unlimited power over nature.

Reducing the complex world of our experience to the simplicity of a finite number of elements parallels Chapter Two's reduction of the unique to the generic through language and measure. Uniqueness is an illusion: all objects are merely different permutations of identical, generic protons, neutrons, and electrons (let's keep it simple for now). The reductionist program considers two examples of a single element to be identical. An electron is an electron is an electron. The same goes for a proton, a neutron, and thus everything built from them—all atoms, all matter. Certainty and control both demand this genericness. Once we reduce a thing to its elements, then we have fully characterized it. There are no more variables, no individuality or uniqueness beyond our grasp. Reductionism says that the reduction of infinite reality into a finite set of labels and data is not actually a reduction at all. If taken far enough, label and number can *fully* encompass reality, leaving nothing out.

It would be a disaster for reductionism, then, if each bit of matter in the universe were unique, if each drop of water were different from any other, if each electron had its own personality. Of course, this is precisely how pre-technological people see the world. Animists see a spirit in everything, so that no rock is just a rock, no dandelion is just a dandelion, no drop of water just a drop of water. Each has an individuality that no fineness of labeling can ever capture. Strangely enough, quantum mechanics appears to confirm precisely this. In identical experimental circumstances, two electrons will behave differently, as if each had a different personality. Conventional interpretation deals with this situation probabilistically, insisting that the two are still identical. The possibility is rarely considered that any two bits of matter in the universe are *irreducibly* unique, for this would torpedo the reductionist program and put reality forever beyond complete description and control. Quantum mechanical interpretations such as that of David Bohm, whose "hidden variables" encode an admission that electrons that behave differently *are* different, are therefore anathema to the scientific establishment. In the end, they lead us back to animism.[15] The same goes for the dawning realization that water is not a uniform fluid but that, indeed, each drop of water on the planet is structurally unique. Mainstream science has long ignored this information, because it does not accord with the program of reducing reality to a handful of generic elements, or "building blocks". In this phrase we see again the program's motivation: that we too might use these "building blocks" to construct an improved reality of our choosing.

The power and certainty we get from reducing nature to its elements only applies to the extent that these elements are indeed fundamental. Perhaps that explains the enormous institutional resistance to the research of Louis Kervran, an eminent French chemist who produced, back in the 1960s, compelling evidence for the routine, low-temperature biological transmutation of certain elements.[16] It also drove the (successful) attempt to explain the seemingly arbitrary, messy, complex properties of the chemical elements in terms of just a few simple subatomic particles, followed by the (less successful) attempt to unify various subatomic forces—with their apparently arbitrary values—into a single unified force comprising the "Theory of Everything" referred to above. Today, the ambition to reduce everything to a few basic building blocks is in deep trouble as a burgeoning menagerie of "fundamental" particles, outnumbering even the original 92 chemical elements, is required to account for all the observed interactions of physics. We are told that physicists are closing in on a Unified Field Theory that would reduce these particles, again, to just so many permutations of something even more fundamental. The parallel between this Theory of Everything, just around the corner since the time of Einstein, and technological Utopia is quite striking. It's just a matter of a few more discoveries.

Can we build a better version of reality with finite building blocks? Only if reality itself is finite too. Can we build a tower to heaven? Only if the sky is a finite distance away. Interestingly, the cosmology that is orthodox at the time of this writing posits a finite universe: bounded in the macroscopic direction and discrete in the microscopic. The first limit is the product of Big Bang cosmology; the second of the quantization of time and space. All possible variables in the universe have a finite range of values. The number of resulting permutations is more than astronomical, and for all practical purposes infinite, but it reinforces a conceit that has been with us since the seventeenth century: the entire universe is nothing but a number. There is no aspect of reality that might not someday be brought into the human realm.

At the heart of the reductionist program is a deep assumption about the nature of reality. Contrary to the argument of Chapter Two, that we reduce and impoverish the world through our labeling and numbering of it, reductionism assumes that nothing essential is lost. That means that whatever appears to have vanished—sacredness, beauty, meaning, spirit—must never have really existed to begin with. They are illusions, human projections, not part of cold hard reality. When we take the world apart, they are not there. If they exist, where are they?

Where is the human spirit? It is not, contrary to Descartes, in the pineal gland. It is not in the heart. It is not in the pituitary gland, the liver, the stomach. Take a person apart and it is not there. Reductionism holds that if something exists, we can extract it, isolate it, separate it out. (Notice that here again, religion agrees with science. The soul is distinct from the body and can be separated out.)

Where is beauty? It is in a butterfly, but when we chloroform it, lay it out on the dissecting table, and cut it apart, beauty is gone. Beauty is in a poem, but when we over-analyze the poem to find exactly what is beautiful about it, beauty disappears from that too. Beauty is in a painting, but can we reduce it to quantitative measures of color and proportion, and then apply these to the standardized production of beauty? No. Beauty is a relationship, not an objective property, and the mass-production of generic relationships produces, necessarily, an aesthetic that is equally phony, generic, and cheap.

Where is sacredness? Following the same deep ideology as their scientific brethren, the religious authorities have sought to isolate sacredness as well, limiting it to Bibles, crosses, and churches. The furthest extreme of this separation coincides in its genesis with its scientific counterpart, originating in the sixteenth and seventeenth centuries. The Protestant movement progressively excluded the divine from more and more of the human world. Earlier, the Catholic church had removed divinity from ordinary people; now the Protestant reformers began to remove it from Mother Mary and the saints as well, so that all that was left of our original panentheistic world was a single, isolated mote of divinity embodied by Jesus Christ.

Can anything really be understood in isolation from the rest of the universe? The culture of science that emerged 400 years ago says, "Yes." We explore reality by dis-organizing it: isolating pieces, eliminating variables, shielding outside influences. Thus the entomologist brings dead "specimens" back to the lab; the geologist brings back samples; the physiologist dissects cadavers and the chemist seeks purified substances purged of the chaotic contamination of the world. Such methods have their uses. Indeed they have created a world far different, at least superficially, from what we knew 400 years ago. They are, however, incapable of apprehending anything that exists only in relationship, anything that, when you disassemble the whole and isolate the parts, is no longer there. What are those things that exist only in relationship, that are properties of wholes and not of parts? Here are a few examples: consciousness, spirit, sacredness, life, beauty, selfhood, divinity, love, truth, emotion, purpose, transcendence, will; in short, all that makes us human. Yet when you take the human apart, none of them are there. The epigons of the Scientific Revolution conclude, therefore, that they don't "really" exist. They cannot countenance the possibility that these properties, in Mumford's words, "are not accidental by-products of mass, energy, and motion, but are aboriginal components of the same system."

Think about the phrase I used a few paragraphs back, "cold, hard reality," and notice how naturally it rolls off the tongue even today. Such is the legacy of Galileo, for it reveals which qualities we consider real, and which we do not. Scientifically obsolete long ago, the conception of reality as the summation of a near-infinity of tiny, hard masses lives on in our metaphors and intuitions.

The attempt to understand the universe and thereby bring it into the human realm often proceeds by way of metaphor. We project human creations onto the world at large. We cannot avoid doing this; even by using language, we connect words to things. Not accidentally, the technology that began to dominate Europe as the Scientific Revolution progressed also provided its thinkers with their most potent metaphor: the machine. Like the Newtonian universe, a machine is something we understand by taking apart. Like the Newtonian universe, it too is composed of a finite number of generic, interchangeable parts. Like the Newtonian universe, it too runs along deterministically according to the designs of its maker. No wonder that "the universe is a gigantic machine" was a constant refrain of scientific pioneers from Galileo and Descartes onward.

It gets worse. Can you think of a machine that does not *do* anything? A machine whose only function is to function, and to do so precisely and unvaryingly? All machines are meant to do that, but this machine does *only* that. It is the epitome of eternal, regular, yet pointless movement, of repetitive routine. The machine I speak of is, of course, a clock. Think then of what a clockwork universe connotes. Think of what is implied by the watchmaker conception of God. The universe, and our own lives within it, ticks on and on, pointlessly. (No wonder we, living in a society ruled by the clock, so often feel like we are just marking time.)

Galileo and his contemporaries drew great inspiration from the clever automatons of the day, whose clockwork animated whole scenes. In likening living beings to these clockwork simulacra, they imputed to life the very qualities of the clock: automatic, preprogrammed, mechanical.

Not life at all, but a semblance of life. The society we live in is also an outgrowth of the clockwork conception, wedded to technologies that themselves embody the regularity, precision, and automaticity of the clock. Such a society offers a life that too is but a semblance of a life. Galileo's conception of the real stripped reality of all subjective content, depositing us in a forlorn shell of a world whose empty, repetitive routines can never compensate for our lost belongingness.

Universe-as-machine finds an unlikely ally in theistic religion. A machine is something designed by an intelligence for a purpose. The metaphor of machine is inherently teleological. How ironic, considering that the determinism that motivates that metaphor explicitly denies any such purposiveness. Moreover, as Mumford observes, "By turning man into a 'machine made by the hands of God', he [Descartes] tacitly turned into gods those who were capable of designing and making machines."[17] The mechanical metaphor for life turns human beings into gods. No wonder we so brazenly aspired to the Olympian powers of flight, control of the weather, eternal youth, and so on. Some of these, including the power to immolate the planet in nuclear destruction, we have achieved; others remain forever on the horizon. Hubris, we know, carries an inevitable price, and in our aspiration to conquer the universe and supplant God we have inflated ourselves to the furthest imaginable degree. Let us hope the final price is less total than our degree of hubris portends.

Here is another great irony: While the quantification of the universe and the reduction of reality to mathematics brought all existence into the human realm, it also excluded from reality everything that makes us human. The most significant parts of life became secondary, reduced to force and motion, chemistry and electricity, or their existence denied altogether. I am not saying that some other element infuses human emotion, consciousness, and perception—an immaterial spirit added to the electrochemical-physical mix. No. My issue is with what is primary. Galileo's distinction between primary and secondary qualities is nothing but an ideology. We could equally call motion, force, and extension secondary, and say that these are but how the human qualities are enacted. Matter, we could say, is the mere agent for enacting what is primary: emotions, beliefs, ideas, perceptions, consciousness, dreams, spirit, poetry, art, love, beauty, divinity.

What a monstrous doctrine it is, to declare all these things unreal, and what a monstrous civilization it has produced. How can it be that we have created a religion—science—that explicitly denies the very fabric of human experience? I want you to be amazed and aghast. Implicit in this doctrine is the devaluation of everything that makes us human, and our reduction, therefore, to the status of automatons. It is a doctrine that robs us of life itself. Can you see the audacity of this robbery? We will take your emotions—they are, after all, mere configurations of hormones, blood vessels, and neurons. We will take your perceptions—they are, after all, merely the neurological result of absorption spectra and other properties of matter. We will take your sacredness, your poetry, your art. We will exile you to the cold hard world of force and reason.

If I have exaggerated the psychological effects of the doctrine of reductionism, I have not exaggerated by much. The evidence is all around us. What is it that drives our society? What is it that makes the world go 'round? What provides the units by which we reduce more and more of the world to numbers? The answer, of course, is money. The Galilean presumption that everything real can be described by numbers takes practical form in the world we live in today, in which nearly everything has a value, a price tag, and in which human beings act according to their "rational self-interest". Money is the unit-of-account for determining what, indeed, our rational self-interest is. It is what denominates the "cold hard world of force and reason" into which we are thrust. (And what is the realest kind of money? Cold, hard cash.) The practical person "looks at the numbers" and tries to see whether things "add up".

The world has been turned upside down. Numbers, the ultimate in abstraction, have become more real than our subjective experiences. Try explaining *that* to the Piraha! Yet collectively, we act as if it were true. In economics, for example, an object or activity doesn't even count as a "good" or a "service" unless it has been exchanged for money. Gifts freely given do not count. Reread that last sentence, please. Do not count. Here again, embodied in our language, is the very assumption I am questioning. If something "counts", it is real.

The growing ubiquity of the money economy, and the underlying dynamic that drives its growth, are the subject of Chapter Four. As you read it, notice how the progressive monetization of life over the past few centuries parallels the progressive conquest by science of the former realms of the subjective.

Clearly, the effects of Galileo's ontological exclusion of human subjectivity go far beyond philosophy. It generates (or at least reinforces) an alienation that erodes the heart out of everything our civilization has ever achieved. The banishment of the human being takes concrete form in the outright replacement of human labor by machines, in the mechanization and dehumanization of even those tasks still performed by humans, and finally, in the commoditization of more and more forms of human relationship. Conceptually and practically, science and technology have extended the qualities of the machine further and further into organic life.

Paradoxically, the same principles of mechanism, reductionism, and determinism that promise certainty and control also afflict us with feelings of powerlessness and bewilderment. For when we include ourselves among the Newtonian masses of the universe, then we too are at the mercy of blind, impersonal forces that wholly determine our life's trajectory. In the ideology we inherited from the Scientific Revolution, free will, like all the other secondary qualities, is a mere construct, a statistical approximation, but not fundamentally real.

To recover meaning, sacredness, or free will apparently requires dualism, a separation of self out from the deterministic laws of the universe—an ultimately incoherent solution which alienates us all the more. Yet the alternative is even worse: nihilism, the Existentialist void—philosophies which, not accidentally, emerged at the peak of the Newtonian World-machine's reign in the early 20th century. This worldview so deeply imbues our intuitions and logic that we can barely conceive of a self that is neither dualistically distinct from matter, nor a deterministic automaton whose attributes of mind or soul are mere epiphenomena. Prior to the 20th century, these were the only alternatives science presented us, a bleak choice that remains with us today like a burr in the shoe and will continue to generate existential unease until the day comes when we finally digest the ramifications of 20th century science.

This choice reflects an apparent incompatibility of science and religion. Intuitively rejecting the "deterministic automaton" of science, evangelical friends of mine choose instead to disbelieve vast swaths of science—all the physics, biology, archeology, paleontology, geology, and astronomy that conflicts with the Biblical story of creation. Meanwhile, scientifically-oriented people occupy the equally unenviable position of denying their intuitions of a purpose, significance, and destiny to life. I often detect a wistfulness in self-described atheists, as if they wished there were soul, God, purpose and significance—Wouldn't it be nice!—but that unfortunately, sober reason dictates otherwise. Sometimes they cover up this wistfulness or sense of loss with an aggressive display of self-righteousness along the lines of "I can handle the merciless truth, but you need to comfort yourself with fairy stories." Others are aggressively cynical and reflexively derisive. The emotions, anger and sadness, that underlie these responses arise from the monstrous robbery I describe above. Again, this robbery is not the removal of God from Heaven—it is the removal of divinity from the world. Whether God has been removed to Heaven, as by religion, or extirpated altogether, as by science, matters little.

One purpose of this book is to establish an organic conception of divinity that draws strength from the wonders science has discovered rather than depending on their denial. Related to this is an organic conception of self as an emergent property of complex relationships, not a separate soul merely passing through the world and thus alien to it, not a mote of consciousness merely observing the world from within a prison of flesh, and yet, neither a soulless biological machine programmed by its genes to survive and reproduce. It is to the scientific origins of the modern sense of self that we turn next.

The Ghost in the Machine

Life is but a motion of the limbs.... For what is the heart, but a spring; and the nerves, but so many strings; and the joints, but so many wheels, giving motion to the whole body?

—Thomas Hobbes

Following the Galilean split of the universe into the objective and the subjective, the next step would be to eliminate the latter entirely by quantifying the qualitative. In many fields this has been achieved. Sounds, for instance, we reduce to just so many sine waves, which when added together (as by a synthesizer) can replicate any natural sound. Similarly, we can simulate any visual experience with another dataset of numbers representing the red, blue, and green components of a finite array of pixels. Yes, we have made great strides in converting the world into numbers.

Of course, as citizens of the universe, that which we do to the universe we also do to ourselves. It did not take long for the clockwork paradigm to be applied to life in general, and human beings in particular. If all the world is a machine, then we who are of the world are machines too.

The equation of human beings to machines is a proposition so flagrantly outrageous to common sense that it took centuries of preparation before it could be articulated and accepted. Machines, after all, are built, but human beings grow. Machines only move as directed; human beings move autonomously. Machines are built to standard specifications; each human being is unique. Machines are generally hard; human beings are soft. Machine movements are regular and predictable; human beings' irregular and spontaneous. Machines do not repair themselves; human bodies can.

Yet the mechanistic conclusion became inevitable when Galileo excluded subjectivity from reality and God from the everyday workings of the world. The experiencer of subjective qualities is no longer eligible for participation in the world of matter, and so becomes, at best, a mere onlooker. Outside that onlooker, there is just the mechanical world of matter, including the body. As for those pesky subjective experiences, the solution in the centuries after Galileo was to convert them into just so many measurable inputs and outputs, thereby, according to Galileo's criterion, making them real once more. The Behaviorism of the mid-20th century went so far as to deny the reality of subjective states explicitly. Current neuroscience is not so brazen, but it is programmatically akin. Seeking to characterize subjective states according to measurable patterns of electromagnetic, chemical, and physical activity, neurology brings these states into the realm of science—the quantifiable—and, potentially, into the realm of technology—the controllable. The mind, they say, and not space is the final frontier, whose conquest might enable us to achieve, once and for all, the end of human suffering. By converting suffering from a subjective to an objective state, we might control it through the appropriate projection of force through electrical or pharmaceutical means.

Most researchers acknowledge that the quantification of even the most basic states of pain and pleasure has run up against insuperable subjectivity. This realization has yet to significantly impact the practice of psychiatry, however, which prescribes happy drugs in record numbers on the mechanistic premise that happiness *is*, or is caused by, quantifiable levels of serotonin and other neurotransmitters.

Hidden underneath Galileo's vitiation of subjectivity is a concept of objectivity that seems reasonable to our own warped intuitions, even though it is contrary both to ancient worldviews and modern physics. It is the absolute Cartesian coordinate system of space and time referred to in the first section of this chapter: a matrix in which objects and events have a discrete existence independent of any observer. It is significant that the originator of our present concept of mathematical coordinates, Rene Descartes, also gave us the defining statement of the modern self.

Descartes immortal declaration, "I think, therefore I am," took to its most extreme conclusion Galileo's separation of mind and the experienced world of the senses. For if mind is separate from the world, and being inheres in mind, then being too is independent of the world. To be human is to be separate. Separate from what? From the world we experience; that is, from nature in the most general sense of the word.

In this statement, Descartes took the dualistic division of the universe into two parts, self and other, to its logical extreme, consummating a process that had been developing since the beginning of time. Whereas the primitive self is defined by intimate relationships with people and nature, the distancing effects of symbolic culture, cultivation, and technology led to the emergence of a new self: the freely choosing rational actor. Joseph Campbell writes:

> Along with—and as a consequence of—this loss of essential identity with the organic divine being of a living universe, man has been given, or rather has won for himself, release to an existence of his own, endued with a certain freedom of will. And he has been set thereby in relationship to a deity, apart from himself, who also enjoys free will. The gods of the great Orient, as agents of the cycle, are hardly more than supervisors, personifying and administering the process of a cycle that they neither put in motion nor control.[18]

The stage was now set for Descartes to reduce the self to its absolute minimum: a mote of consciousness observing, but not identified with, the body, the brain, the sensations, and the thoughts. Descartes' "I am" is the thinker, not the thoughts, the feeler, not the feelings; it is the observer and the rest of the universe is the object. Descartes thus took to its nadir the shrinkage of the self that had been going on ever since the origin of technology and symbolic culture, and whose counterpart is a corresponding extreme of alienation from nature, other people, and now, thanks to Descartes, even our own bodies, thoughts, and feelings.

In the final analysis, the dualism articulated by Descartes reduces the entire universe to the status of a mere object, a "bare, depopulated world of matter and motion: a wasteland."[19] It is utterly alien to the Cartesian self, trapped in its automatic prison of the flesh. One reason this belief is so psychologically devastating is that it implies that the rest of the universe would really be no different without us. We are dispensable, separate, unnecessary. Our experience in the anonymous society of the Machine bears this out: it reduces each person to a role, a standardized performer of a function. In Newton's universe, each object is similarly reduced to a *mass* characterized completely by generic properties such as position, velocity, mass, and later, electric charge. Not only is the universe an object in the sense of being external to self, but the self becomes an object as well, one among many, operationally defined according to the above physical properties, interacting—just like any other matter—with the rest of the universe according to Newton's impersonal, deterministic laws. You, my friend, are a mass.

The assumption of objectivity took mathematical form with Newton's formulation of his laws of gravity and motion, which operate against the backdrop of Descartes' absolute coordinate system. The coordinate system is unchanging and eternal. It is the fabric of reality across which we move; it is more fundamental than the objects it contains; it is, moreover, prior to the observer, who is irrelevant to the properties of other objects and the forces acting upon them.[20] The discrete and separate self is thus written into the very bedrock of Science. No matter that the absolute universal coordinate system passed into scientific oblivion in 1905—it is still alive and well in our intuitions about what is rational, objective, and scientific. There is an absolute reality out there, and science is the way we discover what it is.

Viewing the universe and even living creatures as essentially soulless machines, moral compunctions about the treatment of a living, feeling being cease to apply. Max Velmans observes, "According to Descartes, only humans combine *res cogitans* (the stuff of consciousness) with *res extensa* (material stuff). Animals, which he refers to as 'brutes', are nothing more than unconscious machines."[21] Accordingly, Descartes' followers had no compunctions about nailing dogs up to boards and cutting them open to see how the parts worked, understanding their cries of pain as nothing more than the wheezing of bellows and the creaking of wheels. Fontenelle, one of Descartes' contemporaries, described it like this:

> They administered beatings to dogs with perfect indifference, and made fun of those who pitied the creatures as if they had felt pain. They said that the animals were clocks; that the cries they emitted when struck, were only the noise of a little spring which had been touched, but that the whole body was without feeling. They nailed poor animals up on boards by their four paws to vivisect them and see the circulation of the

blood."[22]

See—here's the pump! Here are the bellows!

By this logic, the other objects of the universe, including living ones, do not really matter. They lack something that the self possesses. Morality applies to them no more than it applies to a blender, a clock. Suppose I take a soft plastic toy cat and replace its squeaker with a device that when squeezed made a sound just like a cat in mortal agony. When I stamp on it with my boot, I am not really causing suffering, only the appearance of suffering. I haven't done anything immoral (a little twisted, maybe, but not evil). If animals and indeed the entire universe are similarly insensate, bearing only the illusion of feeling, then the same moral license applies to the whole universe. Such is the implacable conclusion of the Galilean banishment of the subjective from the realm of scientific reality.

The aforementioned cat example is highly relevant to modern life. The entertainment industry calls it a "sound effect". In an artificial world, in which the separate human realm has engulfed all else, what difference is there, really, between a depiction of gruesome death in a video game, and the same in a photograph or newspaper article? What is the difference from the perspective of the audience? The only difference lies in how the words or images are interpreted—as real or as unreal. But we have been trained, as reductionists, to understand things out of context, to remove the specimen from nature to the laboratory. No matter what the source, whether a prison in Baghdad or a video game studio in Silicon Valley, the viewer experiences the same pixels on the screen. We keep young children away from violent movies because watching them would be too traumatic for them—they would think the violence was real. But soon enough, the endemic exposure to violence in our culture desensitizes them to it. The suffering of others takes on an unreality, without which we could never continue to perpetrate it. Yet we cannot blame this unreality wholly on the media. Is it not built into the Galilean conception of science? According to that, the suffering of others *is* unreal, except to the extent it is quantified. Unfortunately, the numbers used to quantify violence—casualty statistics, acres of rainforest cleared, parts per billion of toxic chemicals, homes destroyed, and so on—keep the actual suffering at a safe distance, remote and therefore unreal. When does it become real to us? When it exits the realm of objectivity to become stories and images connected to actual human beings. Understanding this, politicians prevent us as much as possible from seeing actual photos of the devastation of war, knowing that when we get the reality of the suffering, we will call for its cessation. Galileo had it exactly backward. It is the subjective that is real.

With Galileo and Descartes, the distancing from the victim, incipient already in label and number, received its full ideological enunciation. From the world of the primitive animist, in which we are, in Wendell Berry's words, "holy beings living among other holy beings in a world that is itself holy", we have arrived at a world in which we are mechanical beings, living among other mechanical beings, in a world that is itself a gigantic machine.

Descartes himself balked at the final step of this process, reserving for the human being a shred of subjectivity, a soul, a discrete point of awareness. Like all of us, Descartes did not *feel* like a machine! But the same logic that Descartes applied to animals can be applied to human beings as well—and has been, repeatedly, throughout history with devastating consequences.

The logic, articulated first by LeMettrie in the 1748 essay "Man a Machine" and culminating in present-day works such as Daniel Dennett's trenchantly argued *Consciousness Explained*, is merciless. There is no kernel of awareness, no seat of the soul, no "Cartesian Theater" (to use Dennett's term) where the soul (spirit, consciousness) views incoming sensory information. In other words, there is no refuge for subjectivity, whose conversion to the measurable then suffers no limit. The progress we have made in analyzing perceptions such as vision and hearing can be extended indefinitely. Not even consciousness itself—the sanctuary of the Cartesian soul—is exempt. According to recent research, mystical states of unitive experience—oneness with the cosmos or with God—correspond to measurable activity in certain parts of the brain.[23] Perhaps with the right electrical stimulation, someday we'll be able to experience them on demand. There is no state of consciousness that does not arise from or correspond to some state of the brain, which, after all, is composed of matter obeying the same physical laws as any other matter in the universe.

Got that? All your thoughts, all your feelings, even your religious experiences, are nothing but the interaction of the various masses that make you up. Science seems to negate our very souls.

Despite the negation of the soul that seems to follow from mechanism's fullest expression, organized religion does not seriously challenge mechanism. Religion essentially agrees with the orthodox scientific view that the universe operates according to mechanical principles, except for certain special circumstances in which a being external to matter, called God, interferes in the universe's deterministic laws in occurrences called miracles. By separating God out from the universe and consigning Him to the role of clockmaker and occasional miracle-producer, the Church's response to science abetted its despiritualization of life, as should be expected given that both institutions spring from the same fundamental cultural forces. In both the scientific and the religious view, the human being is essentially alone: in the former case because there is no God at all, and in the latter because God has been separated out from the material world in which we live.

Dennett's dismantling of the Cartesian Theater is just the final step in the separation of spirit and matter, because once spirit has been reduced (as by Descartes) into a mote of self-consciousness disconnected from the flesh machine of the body, it becomes entirely irrelevant to the physical world and therefore to science. Disconnected from matter, for all practical purposes it may as well not exist at all.

In the face of Cartesian dualism and the burgeoning explanatory power of science, religion retreated inward, away from its former role of explaining the workings of the world, to concern itself exclusively with "spiritual matters". But no retreat could be far enough to evade the long reach of science, which demolished one by one the remaining mysteries, converting spirit to mind and mind to brain. With psychiatry and neurology elucidating the biological basis of thought and emotion, there is little room indeed for the non-material soul.

Some people try to rescue meaning, significance, and sacredness by citing mysteries that "science can never explain." Ironically, as these mysteries succumb one by one, the universalist claims of reason and the Scientific Method emerge all the stronger. More ironically still, these attempts actually reinforce the core assumption responsible for the desacralization of the world in the first place: in short, that the sacred and the miraculous are to be found outside the mundane workings of the world. Actually, they are to be found within.

This proposition, which I will develop in Chapter Six, circumvents the entire culture war between science and religion while subverting the fundamental assumptions of both. In this war, the forces of science are represented by a group of philosophers sometimes called the "New Humanists." Led by Daniel Dennett, Jared Diamond, Steven Pinker, Marvin Minsky, Richard Dawkins, and Lee Smolin, they uphold the ideology of the Scientific Program, proclaiming the imminent revelation of the last of nature's mysteries: free will, love, consciousness, and religious experience.

These philosophers wage a crusade against their chosen number-one enemy, the forces of religion, which appeal to our intuitive sense of a purpose and significance to life in positing an external god, spirit, or equivalent force that infuses life and the universe with meaning. The philosophical problems of this dualism are well-known: primarily, if an extra-material spirit interacts with matter in any way, how can it still be extra-material? If it is material, then in which constituent of matter does it reside? The main physical forces are (it is claimed) already known and described in equations. And as the New Humanists like to show, the realm of the mysterious, which might otherwise seem to require new forces, is ever shrinking, leaving the theist with only two apparent choices: (1) to relinquish the world of matter entirely to science, rendering spirit a mere ghost in the machine, wholly impotent and wholly inconsequential, or (2) simply refuse to see the evidence of science, disbelieving in (for example) evolution despite enormous evidence.

Accordingly, we have in our society two distinct trends in institutional religion. Corresponding to the first choice, religious belief becomes increasingly irrelevant to life in mainstream society, having little influence over the way we live; concurrently, corresponding to the second choice, large numbers of fundamentalists are dropping out of mainstream society into a polarized world of the saved versus the unsaved. In the first instance, religion has no bearing on material life: regardless of religious affiliation, all watch the same TV programs, root for the same sports teams, buy the same brands, go to the same schools. Because it is inconsequential to the material world, religion can be kept out of the classroom, out of the boardroom, out of the conversation. In the second instance, paralleling their denial of the consensus of the facts of the world, some religious groups withdraw into an insular subculture in which religion once again imbues every aspect of life. Accordingly, they educate their children at home, associate only with others of like religious persuasion, protect their children from the "demonic influences" of such institutions as trick-or-treating, Harry Potter, and Pokémon, abstain from television, rock music, and popular culture, and even form their own communities, sometimes within fortified compounds. Their insularity is another permutation of "removal from this world" that characterizes modern religion.

The apparent diametric opposition between the New Humanists and the religious fundamentalists is a facade that masks a fundamental agreement: that to accept the scientific worldview is to lose meaning, purpose, significance, and sacredness. We are at a loss to conceive of the sacred non-dualistically, in the absence of something external to make it sacred, to infuse it with the spirit that plain matter lacks. One purpose of this book is to offer a conception of spirit that is not dualistic, and therefore a conception of spirituality that does not remove us from the life of this world.

While Daniel Dennett is a resolute champion of scientific orthodoxy and a firm believer in the Scientific Program, his work sets the stage for a return to a wholly enspirited universe. For he has dismantled the dualistic division of reality into two separate aspects, spirit and matter, pointing to the illusory nature of the discrete self—the Cartesian observing point of consciousness. His is not a new insight—Buddhists have been saying essentially the same thing for thousands of years—but his work is significant for its explanation in non-dualistic terms of the properties of consciousness, and for its exposure of lingering dualistic assumptions in science.

We have come full circle. From our animistic beginnings in which matter and spirit were one, we have progressed through millennia of widening separation between them until, eventually, spirit became wholly non-material and therefore non-existent. We were left with matter alone. How is this any different from the animist?

There is in fact a key difference. The difference is not in the attitude toward spirit, though; it is in the attitude toward matter! The respiriting of the world lies not in bringing an extra-material spirit into matter, but in understanding that matter itself possesses the properties formerly attributed to spirit. The whole world is spiritual. It does not contain or possess spirit; it *is* spirit.

This book proposes a conception of self that is not a discrete, separate entity but an emergent property of complex interactions encompassing not just the brain but the entire body and the environment too, both physical and social. To pretend otherwise is to cut ourselves off from most of what we are. Taken to the Cartesian extreme, it cuts us off even from our own bodies—which are no longer really self but just matter—as well as from our feelings. "I think, therefore I am." Am-ness lies in thought alone. Or less, because the logic of Descartes says that the self is not the thoughts but that which is aware of the thoughts, the "mote of self-consciousness" mentioned above. The logical conclusion of the self-other duality is to reduce the self to nothing at all. Could this be yet another source of our society's ambient anxiety? Could this progressive reduction of the self to a non-existent point of awareness in denial of all that we are be a reason why we compulsively add on so much not-self, so many material and social possessions, in a futile attempt to recover lost being-ness?

The Origin of Life

Despite the conjectures of Descartes and LeMettrie, the phenomenon of life seemed for a long time to deny the dispiriting ramifications of Newtonian determinism and reductionism. The incredible order and complexity of life seemed incompatible with the simple deterministic laws of physics: from whence this complexity? The processes of biology are so finely coordinated, so tightly coupled, that it seemed in the 17th century (and still seems to many to this day) that life must be designed and orchestrated by some superior outside intelligence. Furthermore, the growth and spontaneous movement of living beings does not in any obvious way reflect simple, mechanical laws like the conservation of momentum. We do not behave as mere masses subject to forces, but initiate action on our own. Intuition tells us that this animation, this spontaneous movement and growth, is a key feature of life. Thales recognized as much in the 7th century BCE when he said, "The lodestone has life, or soul, as it is able to move iron", as did Aristotle in associating life with movement ("The soul creates movement").[24] As recently as the 19th century, the Vitalists articulated the same intuition, that only some *elan vital* could produce life's spontaneity in a universe of dead matter.

To make the Newtonian worldview work, some explanation was necessary of how purposeful, animated, complex life forms could emerge from dead chemicals.

The original mechanical conception of life—heart as pump, lungs as bellows—was spectacularly unsuccessful in explaining much more than the circulation of the blood (and actually, not even that[25]). It was only in the 19th century, with the work of Mendel and Darwin, that there was any real progress reconciling the simplicity of physics and chemistry with the animation and purposiveness of nature. By the end of the 20th century, the sciences of molecular biology and genetics purported to provide, if not a solution, at least the outline of a solution to all the fundamental mysteries of life, from its origin, to its evolution, to its present behavior. Not coincidentally, the Grand Synthesis of Darwinian evolution and Mendelian genetics solved these problems in a way that reinforced the defining ideologies of our civilization: the discrete and separate self, the program to control the world, the primacy of competition as an agent of progress, and the destiny of humankind to transcend nature through technology.

At present, the dominant understanding of change in nature is Neodarwinism, which Lynn Margulis has summarized as "an attempt to reconcile Mendelian genetics, which says that organisms do not change with time, with Darwinism, which claims they do."[26] In Mendelian genetics, variation within a species comes only from recombination of existing DNA, which denies the possibility of genuinely new traits ever emerging. However, overwhelming evidence from paleontology, embryology, genetics, and other fields make it clear that life indeed evolves over time, and that genuinely new features arise repeatedly across the eons.

Neodarwinism ascribes the source of this change to random mutation, and its direction to natural selection: the competition for the resources to survive and reproduce. Biological evolution is supposed to happen through a gradual accumulation of random point mutations, frameshift errors, accidental deletions and insertions, and other chance alterations to DNA, which are then "tried out" (expressed in actual organisms) in various combinations. Most mutations are either deadly or harmless, but occasionally one of them will create some new characteristic which confers a competitive advantage, enabling the new organism and its descendants to dominate the old in its ecological niche, or to occupy a new niche altogether. Over time, the accumulation of these new characteristics comes to define a new species.[27]

The traditional view of biology is that the self is the "expression of the genes", which comprise the blueprint for morphology and the underlying determinant of behavior. Only in humans, it is thought, does the countervailing determinant of culture (sometimes) override or at least modify genetic "programming". This is an old idea in new garb—in an earlier age it was not "genetic programming" but rather our "bestial nature", "Original Sin", or "the temptations of the flesh". Either way, the conclusion is the same, that we are, with the rise of culture, transcending nature. We are transcending our biology and rising to a new estate, a uniquely human realm. Aside from the human exception, it is thought, all living beings follow their genetic program, acting on the environment according to the genes' instructions (mediated through the machinery of proteins and their products).

To engineer an organism, then, is really just a matter of engineering its genes, whose properties are fundamentally isolable from the environment. The conception of the genes as the blueprint and program for the organism therefore abets the program of control as well as identifying them as the kernel of the biological self.

The blueprint-and-program genetic paradigm dovetails nicely with Cartesian objectivity. The genes are separate from the environment which they try to manipulate, and themselves only affected by the environment accidentally, through random mutation. This view parallels the view of man separate from nature, subject to its random "whims" perhaps, but nonetheless the master, the conscious manipulator of unconscious matter. It also parallels the Cartesian conception of the soul, residing at the body's control center and supervising its actions. Ironically enough, it parallels as well the conventional Christian conception of God, running nature from a separate vantage point *outside* nature. Orthodox genetics fits in very well with all our culture's assumptions about self and world. No wonder there has been such resistance to alternative understandings of the genes that see them as instruments of environmental purpose, and not the kernel of autonomous, discrete biological selfhood.

In Chapter Six we will examine some of these alternatives, from those of Jean-Baptiste Lamarck to present-day iconoclasts like Bruce Lipton. When the blueprint-and-program model of the genes falls, much else will fall with it.

A good way to see the cultural ramifications of modern biology is to start with its account of the origin of life, and its answers to the eternal human questions of "Where did we come from?" "Why are we here?" and, "Where are we going?" Evolutionary biology has given us the outlines of a story of the origin of life—our creation myth. Like all creation myths, it encodes deep-seated cultural values as well as our sense of ourselves as a people. I will introduce it with an eye for the cultural assumptions and biases it implies. "In believing this," I will ask, "what do we necessarily believe about ourselves?" Why are we here? What is the purpose of life? Where are we going? Our culture's answers to these questions arise from, and predispose us toward, the creation story we have chosen.

Richard Dawkins' 1990 book, *The Selfish Gene*, lucidly presents the prevailing view of biogenesis, conceived a hundred years ago by Oparin and Haldane. Like the Neodarwinian theory of evolution, it hinges on two key features: random mutation and natural selection. The story starts with a prebiotic soup of the organic molecules that are the building blocks of life. The crucial event is the appearance, by chance, of a complex molecule with the very special property of catalyzing the formation of a new copy of itself. Admittedly, the probability of this happening randomly is exceedingly low, but it only had to happen once, because as soon as such a molecule existed, it began creating copies of itself. Dawkins calls this molecule a replicator—the precursor to the first gene.

Soon the ocean was full of such replicators. Not all of them were the same though, because random copying errors created a number of variants. Some of them varied in ways that rendered them unable to create copies of themselves. Such species quickly disappeared. Other variants were more stable or more fecund, so they became more common in the "soup". At some point, the chemical building blocks became scarce, and some replicator variants no longer could reproduce so well. Other variants did much better—for instance, those that were able to cannibalize other replicators and use their parts. Conditions changed, and as they changed, some random variants did better than others. At some point, again through random copying errors (mutations), some replicators emerged with new properties: protective protein coats, for instance, and eventually, protein-based machinery to exercise homeostasis within a lipid membrane. Innumerable failures accompanied each evolutionary jump to a higher level of complexity. For each mutation that enjoyed an advantage in survival and replication, there were thousands that were doomed to extinction.

A sponsoring assumption in the "selfish gene" biogenesis story is that there is a clear demarcation between organism and environment. There are the replicators, which are distinct from one another, and there is the substrate. The key event in the origin of life is the appearance of a molecule, presumably a strand of RNA or something like it, that can replicate *itself*. The separate individual is seen as primary. How naturally this fits in with the beliefs of our own culture: that human beings are separate from nature, and that each of us has a distinct, separate existence independent of other human beings. How naturally it accords with the attitude that nature is a collection of resources for us to use to our best advantage. Replicator and substrate, humanity and natural resources, self and environment. The very word "environment" encodes an inside/outside, self/non-self distinction.

So deeply are these assumptions ingrained that only with difficulty can we even recognize them as assumptions and not objective facts about reality. We can hardly conceive of an origin of life that does not start with an original living creature, discrete and separate from its environment, because that is how we conceive an organism, a "being". Yet other cultures recognized a more fluid identity in which the defining unit was the family, the tribe, the village, the forest. The very meaning of "I" is culturally determined. The shaman Martin Prechtel speaks of a culture in which a man beseeching a shaman to cure his ailing wife says not, "My wife is sick" but rather, "My family is sick."[28] The sickness is as much his own as his wife's. Or if a few individuals in the village are sick, he might say, "My village is sick." Even if a Western doctor might judge him a magnificent specimen of bodily health, he would not agree with the statement "I am healthy" because to him, "I" means something different than it does to us. Its boundaries are more fluid. For him to say, "I am healthy but my family, village, forest, or world is sick" would be as absurd as to say, "I am healthy but my liver, kidneys, and heart are sick." Someone immersed in such a culture might not see the appearance of a replicator as the key event in biogenesis at all.

This is not merely an enlightened understanding that no person can be truly healthy if his family, village, or ecosystem is not healthy; it is a broader definition of self that includes family, village, and ecosystem. It is an understanding written into such spiritual teachings as the hermetic principle "As above, so below," the Taoist concept of an internal universe embodying all the relationships that exist externally, and the Buddhist teachings of karma—that anything you do to the world, you do to yourself—and the unreality of the self. It is very difficult for us moderns to understand the non-dualistic teachings of ancient religions, including Christianity, except through the lens of dualism. Thus karma is misunderstood as an outside universe somehow exacting revenge for misdeeds and rewarding good deeds. Animism is misunderstood as all things *possessing* a separate substance, spirit. "As above so below" is misunderstood to imply an independently existing line of demarcation (the self) between two independent realms. Christianity is misunderstood to posit a God fundamentally separate from this life and this earth. These misunderstandings all stem from our concept of self. Similarly, our concept of self predisposes us toward (and then draws further support from) the

Neodarwinian story of biogenesis and evolution, and blinds us to competing tales of potentially greater explanatory power. Blind also are we to evidence for them in nature, which we either ignore, dismiss as an interesting curiosity or exception, or awkwardly attempt to explain away in the terms of competition and survival.[29] Nonetheless, scientifically cogent theories of biogenesis that do not start with a separate-and-discrete replicator do exist. In these, ecosystem and not organism is primary, or life is a fundamental property of the universe. I explore such theories in Chapter Six; predictably, they will have consequences far beyond biology, since they suggest a very different concept of what a "self" is.

Another key feature of Neodarwinian biogenesis and evolution is the absence of purpose or intention. No replicator planned on developing a protein coat or any other feature. Each new mutation happened by chance and survived only because this new configuration was better able to survive and replicate. Evolution is the random, undirected exploration of the space of possible replicating molecules. All the life forms we see today in their bewildering complexity are nothing more than the devices by which these replicators, now genes, survive and reproduce themselves.

Accordingly, the fundamental explanation for the behavior of living beings, including humans, is that genes program such behavior because it confers a competitive advantage on the organism, enabling it to better survive and pass on those genes. So in observing animal behavior, the Darwinian biologist asks, "What competitive advantage does such behavior serve?" and looks for an explanation in those terms. It is assumed that the genes essentially "program" the morphology and behavior of living beings; a mutant gene might program a different behavior, which may or may not aid the chances of reproduction. Those that do, survive; those that don't, go extinct. Analogous thinking underlies the anthropological and paleontological study of human beings: Again the scientist asks, "What competitive advantage does such behavior serve?" or, "How does this adaptation (biological or technological) contribute to survival?" Implicit in Darwin is the assumption that competition and survival are the proper terms by which to understand the living world.

Now, most people I know don't behave like ruthless maximizers of their genetic self-interest. Does that mean that, thank goodness, culture has sufficiently conditioned us into moral, ethical, socially responsible beings? Or is it the other way around? Is it instead that our true nature is love, wholeness, and creativity, and that culture drills into us a survival anxiety that thwarts the expression of our true nature? Do you sometimes truly desire to "do the right thing," to dedicate yourself to serving other people, to doing good work, but feel you cannot "afford to"? This is an important question. Paralleling the broader technological conquest of nature, the contemporary understanding of genetics motivates a war against *human* nature.

Because it is based on *random* mutation, there is no directedness, no purpose to evolution. There is therefore, at bottom, no purpose to human existence either, except for that programmed into us by our genes. To put it in the simplest terms, the purpose of life is to survive—to survive and reproduce. That is, as Dawkins puts it, "the ultimate rationale for our existence."

So that we do not underestimate the ramifications of that statement, let's examine what the "ultimate rationale" for love might be. The genes program a mother to love her offspring, because the behavior that love comprises helps the offspring survive to adulthood and pass on those same genes to the next generation. Our genes code for the production of certain proteins which are the construction materials for the glandular and neurological machinery that is in turn the production site for the hormones and neurotransmitters that we experience as love. Romantic love also contributes to the replication of our genes in the most obvious way, and through similar mechanisms. These confer such a huge advantage in survival-and-replication of our genes, that it is worth having them even if they get "misapplied" sometimes: directed at other people's children, for instance, or at humanity in general or life in general. Forget all sentimental religious or spiritual theories about love—its real essence is simply a way to maximize the survival of your genes.

This is "natural selection"—the struggle to survive. Darwinism sees the world of life as essentially a vast competition for survival, in which the life forms that survive are those that outcompete the others for scarce resources. How like the society of Darwin's time! In the age of Laissez-Faire economics, described as a dog-eat-dog world, the theory of evolution by natural selection must have seemed intuitively obvious. And not only is life a competition, but competition is *good*, the engine of progress—not just biological progress (evolution) but technological and economic progress as well. This view helps motivate the Technological Program, because it enables us to dominate (that is, outcompete) the rest of nature. It is explicit in liberal economic theory, whose models draw out positive consequences of competition. The parallel in economic discourse to Darwinian evolution is nearly exact: the fittest firms survive, the less fit go extinct, and the competitive environment pushes all of them to continually improve through efficiency and innovation.[30] Without competition, what would push firms ever to improve on last year's model? Images of the "dinosaur" firms of the old Soviet Union come to mind.

In fact, the competition that we see as the driving force of life and evolution is very much a projection of our own cultural beliefs. Just as we project our own anxiety onto primitive peoples, so we project the relentless competitiveness of modern human life onto nature. Competition is an important part of nature, of course, but not the prime mover, the defining feature, nor the engine of progress. In Chapter Six I will lay out another paradigm, based on cooperation and symbiotic merger, that invites a very different set of cultural values and which might inform a very different system of economics and technology.

The competition-based Darwinist paradigm not only drew support from its cultural milieu, but reinforced it as well. It seemed to offer a justification for the terrible social inequalities that existed between rich and poor, as well as for the relationship between the various "races of man". The rich and powerful were more "fit" and therefore *deserved* to thrive, while the poor, bless their souls, were less fit and their demise nothing more than the course of nature. Liberals and conservatives agreed on this fundamental premise even as they disagreed on what to do about it. Liberals lamented the plight of these inferior people and races and sought to soften the blow, arguing that man could rise above nature, red in tooth and claw; conservatives, on the other hand, believed this genetic triage to be a good thing. Some even went so far as to oppose any sort of government intervention to bring better conditions to the poor on the grounds that it would weaken the gene pool, allowing the less fit to survive. Such thinking taken to a further extreme led to the movement in the United States, coordinated in Cold Spring Harbor, New York, to purposefully strengthen the gene pool through the forced sterilization of "inferior" people such as mental patients and prisoners. In Germany, of course, this "science of eugenics" helped motivate the Holocaust.

This is one way the principle of reductionism can lead to the program of engineering and control and, quite literally, "reduces" life. For the eugenicists of Cold Spring Harbor and Nazi Germany, here finally was a "scientific" basis for the ultimate engineering project: the improvement of the human race. The rationalistic garb of the Nazi program was not just a cloak for barbarism; it was basic to the whole enterprise of eugenics. Moreover, the distancing of the researcher from the object of study that defines objective scientific inquiry also contributed to the dehumanization of their victims. Naturally, then, bland technocrats such as Adolf Eichmann played key roles in orchestrating the logistics of the Holocaust, which, when reduced to the language of the factory—inputs, outputs, job specializations, and budgets—is cognitively no different from any other engineering problem.

I do not mean to lay the blame for social Darwinism, and certainly not for ethnic cleansing, at the feet of humble, humane Charles Darwin (although hints of these ideas may be found in his writing[31]), but rather to illustrate that scientific theories do not exist in isolation from their social environment. Darwinism was born in an environment conducive to its acceptance, and its concepts, in turn, reinforced and accelerated preexisting trends. Darwinism was both a cause *and an effect* of ideas like "Competition is good," "Competition is the way of nature," and "Competition is the engine of progress." Equally, and on a deeper level, Darwinism was a consequence of the millennia of separation that preceded it, as well as a major impetus accelerating that separation to its present climax. Darwinism constitutes a necessary scientific apparatus for the projection of our defining ideology—the discrete and separate self—onto biology.

The primacy of competition in nature (and human nature) dovetails closely with the ideology of the discrete self living in an objective universe. Life as a competition for resources necessitates that there be competitors, discrete subjects whose conflicting interests drive the constant struggle for survival. It also implies a distinction between life and resources, which translates in our perceptions to a distinction between self and world.

These interdependencies ensure that as long as the Age of Separation holds sway over our thoughts, the Neodarwinian synthesis will reign as orthodoxy no matter what practical difficulties it encounters. Indeed, current attempts to explain the emergence of the first replicator have run into severe, probably insuperable difficulties, which theorists explain away along the lines of, "Something like this *must* have happened, there is no other way." Analogues to these difficulties plague the entire story of evolution-based-on-random-mutation-and-natural-selection. The principle difficulty is what various writers have termed the "bootstrap problem" or "irreducible complexity". I will examine irreducible complexity in detail in Chapter Six, because the nature of the problem points, conspicuously, to the different conception of self that I have hinted at throughout this book. Whether in Darwin's time or today, the standard theory of life's origin and evolution owes its acceptance more to its integration into our overall conception of self and world than to its scientific merits.

Note to the scientific reader: If you still suspect I have an agenda of "Intelligent Design", you haven't been reading very carefully. From my perspective, mechanism and Intelligent Design are opposed only superficially. Both imply that any spirit must come from the outside; they disagree only on whether such an outside spirit, intelligence, or organizing principle exists. The alternative I offer builds on substantial groundwork that I develop in Chapter Six, but here is a preview:

> I offer the reader not a mundane universe in which nothing is sacred because there is no God, nor a split universe in which some things are holy, of God, and others just matter, but rather a universe that is fully sacred, pregnant with meaning, immanent with God, in which order, pattern, and beauty arise spontaneously from the ground up, neither imposed from above by a designer nor projected from within by the observer, and of which God is an inseparable property. The marvelous complexity and beauty of nature is not some consolation prize for science's denial of the sacred, but evidence that the universe is itself sacred.

Such a shift in worldview, of course, carries profound implications for the way we live. Conventionally, technology has been seen as a logical extension of natural evolution, an acceleration of the biological imperative to outcompete other life forms and extract more energy from the environment. The history of technology seems to bear this out, as it has allowed us to usurp enough of the earth's solar income and biomass to support six billion humans today, as compared to a few million in the Stone Age. It is, however, becoming increasingly apparent that our present way of life and relationship to the planet is not sustainable. The Technological Program is failing. Perhaps a new understanding of life's evolution based on symbiosis, interconnectedness, and emergent order will inspire a new mode of technology as well, that does not put us into opposition to the rest of life, that does not separate us from nature, and that does not consign us to the misery, alienation, loneliness, and powerlessness that have resulted from the specialization and mass scale of life that technology has brought.

Alone in the Universe

Out, out, brief candle!
Life's but a walking shadow, a poor player
That struts and frets his hour upon the stage
And then is heard no more; it is a tale
Told by an idiot, full of sound and fury,
Signifying nothing.
—Shakespeare's Macbeth

However distasteful the dispirited world of deterministic forces acting on objective particles, however dispiriting a biology, an economy, and a psychology rooted in the struggle to survive, for several hundred years now it would seem that we have no viable alternative to believe in. Whatever religion teaches or intuition suggests, science has told us: Sorry, the world is just like that. Remember Richard Dawkins: "The universe we observe has precisely the properties we should expect if there is, at bottom, no design, no purpose, no evil and no good, nothing but blind, pitiless indifference."[32]

In other words, it has been the sober view of science that no matter what meaning or purpose we impute to the world, all that is really happening is fundamental particles interacting according to impersonal, objective, deterministic laws. There may be other "higher level" contingent explanations for things, but the most fundamental explanation—the real reason—for anything, even love, boils down to "particle A bumps into particle B."

It was the genius of Darwin to explain how complex life could develop from a foundation of deterministic material laws, collapsing the last stronghold of religion. Thanks to random mutation and natural selection, with the addition of subsequent developments in genetics and biochemistry, no longer was any kind of animating force necessary to explain life. Chance alone was the only "reason" why human beings—or indeed any life—arose on this planet. As Jacques Monod puts it in *Chance and Necessity*, "The universe was not pregnant with life nor the biosphere with man. Our number came up in the Monte Carlo game."[33] We are alone in the universe, in which the only meaning is that which we create.

Elsewhere in the book, Monod quite astutely identifies the source of morality, values, and ethics in animism, the belief that nothing is truly inanimate but that all things are infused with spirit and purpose. Yet science, the fruit of objectivity—a "new and unique source of truth"—utterly confounds animism. The vestiges of animism in modern thought he sees as a pusillanimous refusal to face up to the truth. We accept the power and gifts science has brought us, but our refusal to accept its full philosophical consequences leaves us living a lie. Speaking of scientific objectivity, Monod writes,

> If we accept this message—accept all it contains—then man must at last wake out of his millenary dream; and in doing so wake to his total solitude, his fundamental isolation. Now does he at last realize that, like a gypsy, he lives at the boundary of an alien world. A world that is deaf to his music, just as indifferent to his hopes as to his suffering or his crimes.[34]

Monod's remonstrance is but a restatement of his own unconscious assumptions, assumptions rooted in the Galilean banishment of subjectivity. We live at the boundary of an alien world, yes, but it is we who have defined that boundary into existence. Science at its very foundation *defines* the world as alien. Once we have bought into that definition and accepted it as an unquestionable axiom, extremes of separation, alienation, and despair follow as a matter of course. Fully immersed in the ideology of science, Bertrand Russell, for one, could see no way out:

> Even more purposeless, more void of meaning, is the world which science presents for our belief. Amid such a world, if anywhere, our ideals henceforward must find a home. That man is the product of causes which had no prevision of the end they were achieving; that his origin, his growth, his fears, his loves and his beliefs, are but the outcome of accidental collocations of atoms; that no fire, no heroism, no intensity of thought and feeling, can preserve an individual life beyond the grave; that all the labors of the ages, all the devotion, all the inspiration, all the noonday brightness of human genius, are destined to extinction in the vast death of the solar system, and the whole temper of Man's achievement must inevitably be buried beneath the debris of a universe in ruins—all these things, if not quite beyond dispute, are yet so nearly certain that no philosophy which rejects them can hope to stand. Only within the scaffolding of these truths, only on the firm foundation of unyielding despair, can the soul's habitation henceforth be safely built.[35]

At least Russell was forthright about it. It has been far more common to obfuscate "these truths" and to mute that unyielding despair. Russell's logic is impeccable; his conclusions follow inexorably from the mythology of science that immersed him and still immerses us. The distractions and addictions that mute the despair, the frenetic pace and empty acquisitiveness of modern society, point to an inner vacuity, the end result of the progressive extirpation of subjectivity. While we associate Russell's sentiments with the pinnacle of classical science at the turn of the 20th century, these were merely the finale of a process that began long ago, when label and number first began to strip the world of particularity.

How ironic it is, that the endpoint of this vast campaign to make the world ours was to exclude ourselves from the world entirely. The human realm has expanded to encompass all reality, yet we are alone in the universe.

A key theme that will emerge later in this book is that this aloneness is an illusion, an artifact of our self-definition, a byproduct of our way of relating. In your heart of hearts, you know this. Even if your intellectual opinion is the same as Monod's or Dawkins', your behavior betrays you. Beliefs are not just ideas in our heads, they are not just opinions but reveal themselves as actions. All of us, even the most cynically ruthless, act from time to time as if life were not, in Shakespeare's words, a sound and a fury, signifying nothing.[36] We might profess to believe in the complete indifference of the universe. We might profess to believe that the events of our lives are mostly random, that there is no purpose to our existence, that we are indeed as helpless as a Newtonian mass to alter our fate, except to master a greater force than the forces buffeting us. And the structures of our society conspire to reinforce that belief. But none of us actually believes it. I know because we don't act that way. Monod is right: we are still animists at heart.

What the Knower in our hearts knows is that the people and events of our lives are connected according to an ineffable logic, proceeding as if by divine orchestration toward a destiny that flows from who we are, and that changes according to who we choose to be. Not just some events, but all events are significant; none are random. Yes, I am asserting the "magical thinking" shared by all primitive cultures. Science disagrees. Science says we merely project non-existent patterns and relationships onto a random reality.

Let me revisit a statement I made earlier in this chapter: "By this logic, the other objects of the universe, including living ones, do not really matter. They lack something that the self possesses. Morality applies to them no more than it applies to a blender, a clock."

But maybe morality does apply to blenders and clocks. Maybe they are no more insensate brutes than plants, animals, or people. Science understands the affection we have for inanimate objects to be a jejune projection of human qualities onto things. Have you ever felt sorry for an old car you finally sent to the junkyard? An old baseball mitt? A doll? Our childish minds tend to impute human feelings to them, even though they are, actually, just composed of lifeless, unconscious matter. It is this childishness that Monod rails against—echoed by a chorus of voices calling for higher scientific literacy, a more technocratic society. Let's put the experts in charge. And as individuals, let's grow up. Be reasonable. Be dispassionate. Be rational. Your car is just a hunk of metal. Your baseball mitt is just a piece of leather. Well, this book says otherwise. I say your affections are sourced in valid intuition rather than puerile fantasy.

I align myself with the intuitions of children and primitives who ascribe consciousness and spirit to all things of the universe, animate or not. The Native American term "all my relations" is not limited to living beings; it includes mountains, rocks, waterfalls, lakes, the wind, the soil. All have spirit, perhaps even life. The error would come not in applying morality to inanimate objects, but in applying the same morality to them as we do to living beings. Being good to your car does not mean covering it with a blanket on cold nights. Today we commit an error far worse than that. Cut off from our animistic love of the material world, our treatment of it is devoid of affection, devoid of love, devoid of morality. Cut off from animism, we wreck the world with moral impunity.

I will not offer proof that animism is true. As you will discover if you try to apply it to your love life, the whole quest for certainty can invalidate even the possibility of what it is trying to establish. No, you can never be sure that your car, baseball mitt, doll, or for that matter pet, friend, or lover really has subjectivity. In the end, belief comes down to a choice. Unfortunately, our choice for several hundred years has been colored by an emptying ideology that has now become obsolete. No longer need divine orchestration depend on a divine Orchestrator, imposing His design upon the world. No longer need consciousness or spirit be infused from without, a ghost inhabiting a machine. Drawing on newly emerging paradigms in science, an alternative far more magnificent than that is now available.

With a sad sigh, we turn away from our heart's knowing and live a life under control, applying a personalized version of the Technological Program to ourselves. The nature that we conquer includes human nature. We do so out of fear. We are scared out of living according to our knowledge of a purpose, significance, and sacredness to each act, event, person, place, and thing. In a million ways, culture conditions us to accept all the premises of separation. Deep down we know it isn't true, but *we are afraid that it is*. We are afraid that there is just this, just a bunch of discrete, separate beings in a "blind, pitiless, and indifferent" universe of force and mass.

Impelled by both the fear and the inner knowing, people have tried for several centuries to find a way out of the seemingly unbreakable stranglehold of deterministic science. Often their efforts amount to the naïve bravado of, "Science can never explain everything." But as one mystery after another succumbs to scientific explanation, the bravado sounds increasingly desperate, and the dread grows that maybe, science *can* explain everything. Maybe the Scientific Program can be fulfilled.

Take heart, skeptics, I do not intend in these pages to trot out unexplained mysteries and deduce from them that "see, there is room left for God after all." Such an approach actually subtly reinforces the Cartesian worldview. It posits two realms, one (which encompasses vast swaths of solved mysteries—almost everything in fact) that is material, blind, pitiless and indifferent, and the other that is spiritual but increasingly inconsequential as one mystery after another succumbs to scientific explanation. No, the truth is far more splendid than an indifferent material universe run by a creator god, or a disappearing mote of spirit inside a robot made of flesh.

That is not to deny that there are still unexplained mysteries, just that we need not rely upon them as a source of sacredness and meaning. In fact, the New Humanists' implication—echoing the ideology of the Scientific Program—that most of the mysteries have been solved is laughable. What has happened is that the scientific establishment has excluded vast areas of reality from consideration, simply because the phenomena do not fit into current paradigms. I am reminded of Richard Feynman's reaction to Uri Geller's demonstration of spoon-bending: "I'm smart enough to know I'm stupid." By that he meant that there must be some sleight-of-hand trick involved, some stage magic that an uninitiated observer, even a brilliant physicist, could not detect. Because, it couldn't have been real. To Feynman, that was a possibility not even worth investigating. And so it goes for the vast realm of "anomalies": psi, precognition, remote viewing, past-life memories, and others, which are excluded from the realm of the mysterious on dogmatic grounds. The logic is circular: there must be a mundane explanation (e.g. fraud, delusion, sloppy science, or overlooked physical mechanisms), because we already can account for everything.

Notice here again the trap of language. The word "mundane" simultaneously means "of the world" and "unremarkable". Something mundane is not magical, mysterious, sacred, or amazing. By using the word, we imply that the world itself lacks those qualities. We imply that the world is not an ongoing, living miracle.

Starting with Fritjof Capra's seminal *The Tao of Physics* and Ilya Prigogine's *Order out of Chaos*, various thinkers have testified to a sea change in the fundamental underpinnings of science, a shift away from the Newtonian World-machine. Awareness is growing that 20th-century science has obliterated the Newtonian principles of determinism, reductionism, objectivity, dualism, and mechanism. We are just now beginning to experience the effects of this shift, which implies nothing less than a crumbling of certainty, the death-knell of the Scientific Program, and the end of the regime of separation. No longer need we appeal to bravado or intuition to deny it. Science itself has developed to the point where its own assumptions have become transparently untenable. The same has happened to the illusion of separation generally. All lies tend to grow, even to colonize all of life for their maintenance, but in so doing become all the more fragile and all the more transparent. We have invested deeply in ours, but the game is up. The energy required to maintain the lie of the discrete and separate self far outweighs the benefits. We are beginning to let go.

In Chapter Six I will describe how the second Scientific Revolution that is now under way implies a very different conception of self and world, one that will inevitably undo the regime of control, the ambitions of the Technological Program, the conversion of life into money, and all the other bitter fruits of the Age of Separation.

Whether held by the rational atheist or the religious fundamentalist, the belief that the material world is not in itself sacred has the same devastating consequences. Internally, they are the existential void, the cosmic alienation, the "unyielding despair" upon which we must attempt to build a philosophy of life. And externally, we mirror this internal wasteland with a corresponding campaign of destruction that treats the earth and everything on it as nothing more than inconsequential lumps of matter. What else could we expect, having defined it thus? The only limit to our despoliation of reality is a vestigial compunction, the remnant of our innate love of life. Well-meaning people seek to rationalize it along the lines of, "We should protect the environment because we depend on it."

However, since the Technological Program implies that this dependence is temporary and diminishing, such a rationale is rarely compelling, and indeed contributes to the underlying ideology that values nature only for its practical usefulness to man. What other conclusion could there be, in the absence of the sacred?

How much of humanity's depredation, violence, and ruination arises from the cosmic alienation implicit in a mechanical universe? It is a universe in which, in Jacques Monod's words, "man knows at last that he is alone in the universe's unfeeling immensity, out of which he emerged only by chance. His destiny is nowhere spelled out, nor his duty." He could have gone on, as Russell does in "A Free Man's Worship", to say that destiny and duty, and therefore goodness, morality, and purpose, are for us to create ourselves. Ultimately, then, they are artificial.

The artificiality or culturally constructed nature of meaning is a cornerstone of the Deconstructionism that so influenced 20th-century thinking about literature and art. Deconstructionism attempts to grapple with the nihilistic void I have described, and once served a useful purpose in its critique of the reductionistic quest for objective certainty. But eventually it foundered on its own premises (or lack thereof), degenerating into an irrelevant corpus of sterile, opaque, frivolous, and often silly texts. (Or maybe it's just that I feel stupid whenever I try to read them!) Texts, and texts about texts, and texts about texts about texts... another artificial realm mirroring the one we find ourselves in generally.

This book is not just another version of Postmodernism that says reality only comes into being through our interpretation of it. All of those qualities that elude reductionism—purpose, meaning, order, beauty, sacredness—emerge organically as a function of relatedness, with or without human participation. The Deconstructionists and Postmodernists are the most recent incarnations of a long line of cultural sensitives—Heidegger and Nietzsche, Sartre and Camus, Foucault and Derrida—who sought to come to terms with our separation from the matrix of organic divinity and keep nihilism at bay, but none of them truly rejected the premises of that separation. Nihilism, Existentialism, Deconstructionism are all different expressions of Russell's unyielding despair. The glittering promise of the New Worlds of the Enlightenment hid that despair for a while. In the 20th century, as the failure of that promise grew increasingly undeniable, the despair too reared its ugly head. In the present day, our final escape has been apathy, cynicism, and resignation. Camp and kitsch. Whatever. I don't care, and I don't care that I don't care. Or, devotion to the acknowledgedly unimportant and absurd. Sports teams. Reality TV. Game shows. Soap operas. Trivia. Separated completely from the real world, we content ourselves with its most inane counterfeit.

Except, of course, we are not content. The pleasures our escapes offer can barely assuage the pain of the interior wound even in the best of times. All the more feeble they are revealed to be, when the crises of real life find a way in. It is then that the pathetic illusoriness of the manufactured realm becomes apparent. Exposure to passion, heartbreak, loss, pain, illness, and death bring us back to a realm that, despite the ideology that so completely possesses us, is undeniably, experientially *real.*

When this happens collectively, as it will with the impending convergence of crises, we will look back upon our former preoccupations—the Superbowls and celebrities—with jaws agape. What world were we living in? How could we have cared so much about so little?

In a world in which nothing matters, the most atrocious events are no longer horrifying; the most piteous victims no longer stir our compassion; the most frightening possibilities, like nuclear war and ecological destruction, no longer frighten us. Sometimes we explain it away as "compassion fatigue", but really it is a disconnection from reality. None of it seems real. We sit back, benumbed, watching the world slide slowly toward a precipice as if it were an on-screen enactment. Similarly, we watch the years of our own lives march on, indifferent to the preciousness of each passing moment. Only once in a while an alarm goes off; we panic for a moment with a thought like, "This is real! This is my life! What am I here for?" And then our environment tempts us back into stupor.

This sense of unreality facilitates the unrecognized flipside to nihilism: a corresponding license to unlimited dominion over the universe. There is no purpose to fulfill, no natural role or function, nothing aside from superstition and temporary, practical limitations to prevent us from becoming, indeed, the "lords and possessors" of nature. The will to dominate nature mirrors our self-imposed exile from nature.

The more we dominate, own, and control, the more separate we experience ourselves. The more separate we experience ourselves, the greater the urge to dominate, to own, to control.

We are attempting to take, by force, that which is already ours. Like the affluence of the hunter-gatherer, what we desire most has always been available, without much effort. The compulsion to add more and more to the self arises from our denial of all that we are. How do I know this, that we are more? I know it in the same way you do. Certain moments have shown me. They came unbidden and without effort or contrivance.

Although a semi-conscious shift is under way, we as a species have still not quite experienced the furthest extreme of separation that will mark the Capraian "turning point" toward an Age of Reunion. We are almost there, though. Once so wholly embedded within nature that we could not even conceive of it as something separate from ourselves, we now possess an ideology that allows no other conclusion but our complete alienation. In our age, the consequences of our pretense to dominion over life and world have born full flower: a world in which our ambition to become the "lords" of nature manifests as a totalizing program of control, and in which our ambition to become nature's "possessors" manifests as a totalizing regime of money and property.

These two trends go back thousands of years. Finally now, the reduction of reality into representation described in Chapter Two is approaching its ultimate expression in the conversion of life, time, and the world into money; while the Technological Program of control is culminating in the effort to bring all aspects of human existence into its domain. The next two chapters will explore these, the consummating expressions of the Age of Separation.

CHAPTER IV

Money and Property

The Realm of Me and Mine

The idea of property occurs naturally to the discrete and separate self. Just as Cartesian objectivity divides the world into self and other, property divides it into mine and yours. Just as Galilean materialism insists that only the measurable is real, so economics denominates all value in money units.

The urge to own arises as a natural response to an alienating ideology that severs felt connections and leaves us "alone in the universe." Shorn of connectedness and identity with the matrix of all being, the tiny, isolated self that remains has a voracious need to claim as much as possible of that lost beingness for its own. If all the world, all of life and earth, is no longer me, I can at least compensate by making it mine.

It has been said of infants, "Their wants are their needs." The same is actually true of adults too, except that the want has been so distorted that its object can no longer satisfy the need, but may even intensify it. Such is the case with greed. Greed is not some unfortunate appendage to human nature to be controlled or conquered. It arises from a hunger for identity—for the richness of relationship from which identity is built. Ironically, following the pattern of any addiction, indulging in greed only exacerbates the underlying need, because enclosing more of the world into the domain of *mine* separates us all the more from the connected interbeingness for which we hunger.

Perhaps this realization can temper our judgmentality toward the greedy. The next time you witness greed, see a hungry person instead.

The next time you feel greedy yourself, take a moment to touch the wantingness, the existential incompletion, underneath that greed. The same goes for selfishness generally, that constricted feeling of wanting to manage and control the world outside the self so as to turn it toward the self's benefit. Selfishness in all its forms seeks the benefit and inflation of a self rendered artificially small, a self which is in fact an ideological construct.

As that word *mine* indicates, ownership implies an attachment of things to self. The more we own, the more we are. The constellation of me and mine grows. But no matter how large the discrete and separate self grows, it is still far smaller than the self of the hunter-gatherer. The pre-separation mind is able to affirm, all at once and without contradiction, "I am this body," "I am this tribe," "I am the jungle," "I am the world." No matter how much of the jungle we control, we are smaller than the one who knows, "I am the jungle." No matter how dominant we are socially, we are far less than one who knows, "I am my tribe." And far less secure, too, because all of these appendages to our tiny separate selves are easily sundered from us. We are therefore perpetually and irremediably insecure. We go to great lengths to protect all these accessories of identity, our possessions and money and reputations, and when our house is burglarized, our wallet stolen, or our reputation besmirched, we feel as if our very selves have been violated.

Not only does our acquisitiveness arise out of separation, it reinforces it as well. The notion that a forest, a gene, an idea, an image, a song is a separate *thing* that admits ownership is quite new. Who are we to own a piece of the world, to separate out a part of the sacred universe and make it *mine?* Such hubris, once unknown in the world, has had the unfortunate effect of separating out ourselves as well from the matrix of reality, cutting us off (in experience if not in fact) from each other, from nature, and from spirit. By objectifying the world and everything in it, by making an other of the world, we necessarily objectify ourselves as well in relation to that other. The self becomes a lonely and isolated ego, connected to the world pragmatically but not in essence, afraid of death and thus closed to life. Such a self, cut off from its true nature and separated from the factitious environment created by its own self-definition, will always be insecure and will always try to exert more and more control over this environment.

The extent to which we identify ourselves with our bodies, possessions, and the domain of our control is also the extent to which we are afraid of death. I am speaking here not of the biological terror that drives any animal to struggle with a predator, but to an ambient dread that drives us to pretense and hiding. More than any other crisis, death is the intruder whose mere approach crumbles the fortress of the separate self. A personal brush with death, or even the passing of a loved one, connects us to a reality beyond the constructs of me and mine. Death opens our hearts. Death reminds us, with a clarity that trumps all logic, that only love is real. And what is love, but a melting of the boundaries between self and other? As many poets have understood, love too is a kind of death.

To a person identified with tribe, forest, and planet, the death of the body and all it controls is far less frightening. Another way to describe such a person is that he or she is in love with the world. Love is antidote to fear of death, because it expands one's boundaries beyond what can be lost. Conversely, fear of death blocks love by shutting us in and making us small. And fear of death is built into our ideology—the self-definition implicit in objectivist science.

Money and property simply enforce this self-definition. They are concrete manifestations of the separate self, the self that is afraid of death and closed to love. Money, in its present form, is anti-love. But it is not the root of all evil, just another expression of separation, another piece of the puzzle. Other systems of money are possible that have the opposite effect of our present currency, structurally discouraging the accumulation of me and mine. Curious? Keep reading... I'll get to them in Chapter Seven.

Something like a money system cannot be changed in isolation. Not only does it correspond to our sense of self and our identification with ego; it also flows from the meta-historical process of separation I have described thus far.

Incipient already in fire and stone, label and number, the objectification of the world crystallized into a new phase with the advent of agriculture: domestication of plants and animals, the turning of nature to human purposes. Then the Machine propelled separation to yet a new level: its promise of transcending natural limitations set us above and apart from nature, while machine society's mass scale and division of labor unraveled human communities. Finally, the methods and logic of the machine achieved their apotheosis in science, which elevated the long-emerging ideology of the discrete and separate self to the status of sanctified truth.

The stage was now set for this ideology to play itself out in the material and social realm. When the world becomes a collection of objects (as in symbolic culture), when these objects are subordinated to human use (as in domestication, agriculture), then they inevitably become property, things that may be bought and sold, defined by their utility for human ends. When science and machine technology then totalize the subjugation of nature, the conversion of the world to money and property tends toward totality as well. The propertization and monetization of life discussed in this chapter grows inevitably from the separation that began with agriculture or before, and that reached its conceptual fulfillment with the Newtonian World-machine.

Money is the instrument—not the cause, the *instrument*—by which our separation from nature, spirit, love, beauty, justice, peace, and community approaches its maximum.

Immersed in the logic of money, we actually see this separation as a good thing. If that seems an outrageous statement, consider what is meant by "financial independence" and the closely related goal of financial security. Financial security means having enough money not to be dependent on good luck or good will. Money promises to insulate us from the whims of nature and the vicissitudes of fate, from the physical and social environment. From this perspective, the quest for financial security is but a projection of the Technological Program into personal life. Insulation from the whims of the environment (which is to master it, to bring it under control) is the age-old quest of technology. And its fulfillment (perfect control over nature) also means perfect security, the elimination of risk.

The campaign to make oneself fully free of the whims of fate, of the vicissitudes of nature, and of reliance on one's community can never actually succeed (just as the Technological Program can never succeed in its campaign to fully control nature), but the semblance of success may persist for some time: the all-American upper-middle class suburbanite with a good job (plus the resumé to get another good job if something should happen), good health (plus plenty of insurance should something happen), diversified investments (just in case), and the rest. Such a person is, in a very real sense, not dependent on anybody—not on any specific person, that is. Of course he is dependent on the farmers who grows his food, but not on any particular farmer, not on any individual person. The goodwill of any individual person is unnecessary because he can always "pay someone else to do it." He lives in a world without obligation. He is beholden to no one.

Not only is perfect independence (financial or otherwise) forever beyond our grasp, it is an illusion cloaking an even greater dependency. It is not the dependency that is dangerous though—it is the illusion. It is the illusion that separates us from, and thus allows us to destroy, so much of what we actually depend on. What does it take to pierce that illusion? Usually it takes a crisis: an encounter with death as described above, or another of life's catastrophes such as divorce, bankruptcy, illness, humiliation, or imprisonment. We stave these off as long as possible with our programs of management and control, but eventually one or another finds its way into even the most secure fortress of self. These events transform us. We let go as we discover that the only lasting, dependable security comes from controlling less not more, opening up to life, loosening the rigid boundaries of self, letting other people in, and become tied—that is more dependent, not less—to a community of people and the community of nature.

As above, so below. Each of these personal crises has a collective counterpart that humanity is facing today. In our depletion of natural resources—soil, water, energy—we face bankruptcy; in the breakdown of our communities and the rending of the social fabric we face divorce; in the mounting ecological crisis and the threat of nuclear war, we face death. The conventional response is to try to hold everything together, to maintain the illusion of independence by extending it still further. It is to remedy the failure of control by applying even more control.

The importance we place on independence from the social and material world has deep roots in our basic mythology. In the fundamentally indifferent universe of Newton or the fundamentally competitive world of Darwin, independence from the rest of the world is surely a good thing. By owning more and more of the world we make it safe, make it ours. We gain mastery over the random forces buffeting us, and we maximize the resources available for our own survival benefit.

This chapter explores the ways and means by which money has been the instrument of the destruction of love, truth, beauty, spirit, nature, and community. At the conceptual level, reductionistic science foretold their disappearance several centuries ago, for all exemplify the eliminated Galilean secondary qualities that "when you take it apart are not there." Money, the unit of account for the reduction of life, has brought reductionism into the daily realm. This chapter tells the story of our impoverishment. My goal is not to make you bitter, however palpable my indignation. My goal, rather, is to raise your expectations and inspire within you a sense of lofty possibilities. By identifying what has been lost, and how, we may forge a path to its recovery. I am speaking to your sense of disenfranchisement—whether you are rich or poor, powerful or oppressed. Indeed, the disenfranchisement I speak of may be even more extreme in society's winners, for two reasons. One is that the impoverishing dynamics of money are often more advanced in their lives; the other is that the vacuity of society's rewards is all the more evident for having acquired them. No longer can the pursuit of them obscure the hunger for the lost wealth of connection and being.

I am speaking, in other words, to your sense that a more beautiful world is possible. We need not wallow in grief over the beautiful things that have passed, nor wallow in resentment toward the forces that have taken them from us. It is important, however, to acknowledge and be sad for what is lost, so that we may complete the past and create the future in wholeness. Drawing on the Sanskrit root *sat*, to be sad means to be full. Sad, satisfied, sated. Complete. Ready to live and create from a full experience of reality.

This chapter will put into a vast context much of what troubles us about the world today. Here are some of the questions that for ten years or more have inspired the explorations that resulted in this book. I invite you to hold them in your mind as you read this chapter.

Why are my adult friendships so superficial?

Why are people so busy?

Why are children so highly scheduled?

Why do I see so many fewer children playing outdoors as compared to my childhood?

Why do Americans rarely sing in public?

What happened to all the great storytellers?

What happened to the extended family?

Why have houses and yards gotten so huge?

Why are prices so distorted that it is cheaper to buy a new appliance than to repair an old one?

Why do corporations composed of nice people do awful things?

Why don't people know their neighbors very well anymore?

Why are there "No Trespassing" signs everywhere?

Why are lawsuits and liability concerns so prevalent these days?

Why has society become so idealistic?

Why are we seemingly helpless to slow down the destruction of the ecosystem?

Why is work unpleasant?

Does it have to be frustrating and lonely to be a parent taking care of young children?

Why do people voice the desire to create community, but then show themselves by their actions to be uninterested?

Why are television and video games so addictive?

What bothers me so much about copyrights, trademarks, and intellectual property?

Why don't people do things for themselves anymore?

Alone in a Crowd

It should not be surprising that money is deeply implicated in the dissolution of community, because anonymity and competition are intrinsic to money as we know it. The anonymity of money is a function of its abstraction. The history of money is the history of the gradual abstraction of value from physical objects. Early forms of money possessed intrinsic value, and were distinguished from other objects of intrinsic value by their portability, storability, and universality. Whether camels, bags of grain, or jugs of oil, early media of exchange had an inherent value to nearly every member of the society.

As society specialized and trade flourished, more abstract forms of money developed that depended not on inherent value but on collective belief in their value. Why trade actual bags of grain when you can just trade representations of those bags? Paper money, and to a great extent coinage, depends for its value on collective perceptions rather than practical utility. You can't eat gold.

The next stage of the abstraction of value came with the divorce of money from even the representation of physical objects. With the abandonment of the gold standard in the 20th century, a dollar came to be worth… a dollar. Currency has become a completely abstract representation of value; indeed, the abstraction is so complete that it no longer really represents anything at all. The parallel with language is uncanny. Just as words have lost their mooring in the reality of our senses, "forcing us into increasingly exaggerated elocutions to communicate at all," so also has money become not just a representation of value but *value itself.*

The last thirty years have witnessed the final step of this abstraction: the gradual elimination of physical currency altogether in favor of numbers in a computer.

Just as words increasingly mean nothing at all, money is also nearing a crisis in which, so disconnected from the utilitarian objects it once represented, it becomes nothing more than hunks of metal, pieces of paper, and bits in a computer. Our efforts to stave off this eventuality (of hyperinflation and currency collapse) mirror the logic of the technological fix, postponing the day of reckoning.

Money is abstract not only with regard to objects of utility, but also with regard to people. Anybody's money is the same. While camels or jugs of oil or any tangible object has an individuality connected with its origin, money is completely generic and thus completely anonymous. Nothing in the digits of your savings account statement tells you whom that money came from. One person's money is as good as another's. It is no accident that our society, based increasingly on money, is also increasingly a generic and anonymous society. Money is how the society of the Machine enacts the standardization and depersonalization implicit in its mass scale and division of labor. But more than just a means to implement depersonalization, money also pushes it further.

To see how, let us return to the paradise of financial independence, ignoring for now that the security it promises is but a temporary illusion, and instead look at the results when it is actually achieved. Often, it is when the semblance of independence is achieved that its emptiness becomes most apparent. Simply observe that the financially independent individual, among other equally independent individuals, has no basis for community except for the effort to "be nice" and "make friends". Underneath even the most well-motivated social gathering is the knowledge: We don't really need each other. Contemporary parties, for example, are almost always based on consumption—of food, drink, drugs, sports, or other forms of entertainment. We recognize them as frivolous. This sort of fun really doesn't matter, and neither do the friendships based on fun. Does anybody ever become close by partying together?

Actually, I don't think that joint consumption is even fun. It only passes the time painlessly by covering up a lack, and leaves us feeling all the more empty. The significance of the superficiality of our social leisure becomes apparent when we contrast that sort of "fun" with a very different activity, play. Unlike joint consumption, play is by nature creative. Joint creativity fosters relationships that are anything but superficial. But when our fun, our entertainment, is itself the object of purchase, and is created by distant and anonymous specialists for our *consumption* (movies, sports contests, music), then we become consumers and not producers of fun. We are no longer play-ers.

Play is the production of fun; entertainment is the consumption of fun. When the neighbors watch the Superbowl together they are consumers; when they organize a game of touch football (alas, the parks are empty these days) they are producers. When they watch music videos together they consume; when they play in a band they produce. Only through the latter activity is there the possibility of getting to know each other's strengths and limitations, character and inner resources. In contrast, the typical cocktail party, dinner party, or Superbowl party affords little opportunity to share much of oneself, because there is nothing to do. (And have you noticed how any attempt to share oneself in such settings seems contrived, inappropriate, or awkward?) Besides, real intimacy comes not from telling about yourself—your childhood, your relationships, your health problems, etc.—but from joint creativity, which brings out your true qualities, invites you to show that aspect of yourself needed for the task at hand. Later, when intimacy has developed, telling about oneself may come naturally—or it may not even be necessary.

Have you ever wondered why your childhood friendships were closer, more intimate, more bonded than those of adulthood? At least that's how I remember mine. It wasn't because we had heart-to-heart conversations about our feelings. With our childhood friends we felt a closeness that probably wasn't communicated in words. We did things together and created things together. From an adult's perspective our creativity was nothing but games: our play forts and cardboard box houses and pretend tea parties and imaginary sports teams and teddy bear families were not real. As children, though, these activities were very real to us indeed; we were absolutely in earnest and invested no less a degree of emotion in our make-believe than adults do in theirs.

Yes, the adult world is make-believe too. Roles and costumes, games and pretenses contribute to a vast story. When we become aware of it, we sense the artificiality of it all and feel, perhaps, like a child playing grown-up. The entire edifice of culture and technology is built on stories, composed of symbols, about how the world is. Usually we don't notice; we think it is all "for real". Our stories are mostly unconscious. But the new edifice that will rise from the ruins of the old will be built on very different stories of self and world, and these stories will be consciously told. We will go back to play.

As children the things we did together mattered to us. To us they were real; we cared about them intensely and they evoked our full being. In contrast, most of the things we do together as adults for the sake of fun and friendship do not matter. We recognize them as frivolous, unnecessary, and relegate them to our "spare time". A child does not relegate play to spare time, unless forced to.

I remember the long afternoons of childhood when my friends and I would get totally involved in some project or other, which became for that time the most important thing in the universe. We were completely immersed, in our project and in our group. Our union was greater than our mere sum as individuals; the whole was greater than the sum of the parts. The friendships that satisfy our need for connection are those that make each person more than themselves. That extra dimension belongs to both partners and to neither, akin to the "fifth voice" that emerges in a barbershop quartet out of the harmonics of the four. In many of my adult relationships I feel diminished, not enlarged. I don't feel like I've let go of boundaries to become part of something greater than my self; instead I find myself tightly guarding my boundaries and doling out only that little bit of myself that is safe or likeable or proper. Others do the same. We are reserved. We are restrained.

Our reservedness should not be too surprising, because there is little in our adult friendships that compels us to be together. We can get together and talk, we can get together and eat and talk, we can get together and drink and talk. We can watch a movie or a concert together and be entertained. There are many opportunities for joint consumption but few for joint creativity, or for doing things together about which we care intensely. At most we might go sailing or play sports with friends, and at least we are working together toward a common purpose, but even so we recognize it as a game, a pastime. The reason adult friendships seem so superficial is that they *are* superficial. The reason we can find little to do besides getting together and talking, or getting together to be entertained, is that our society's specialization has left us with little else to do. Thus the teenager's constant refrain: "There's nothing to do." He is right. As we move into adulthood, in place of play we are offered consumption, in place of joint creativity, competition, and in place of playmates, the professional colleague.[1]

The feeling "We don't really need each other" is by no means limited to leisure gatherings. What better description could there be of the loss of community in today's world? We don't really need each other. We don't need to know the person who grows, ships, and processes our food, makes our clothing, builds our house, creates our music, makes or fixes our car; we don't even need to know the person who takes care of our babies while we are at work. We are dependent on the role, but only incidentally on the person fulfilling that role. Whatever it is, we can just pay someone to do it (or pay someone else to do it) as long as we have money. And how do we get money? By performing some other specialized role that, more likely than not, amounts to other people paying us to do something for them. This is what I call the monetized life, in which nearly all aspects of existence have been either converted to commodities or assigned a financial value.

The necessities of life have been given over to specialists, leaving us with nothing meaningful to do (outside our own area of expertise) but to entertain ourselves. Meanwhile, whatever functions of daily living that remain to us are mostly solitary functions: driving places, buying things, paying bills, cooking convenience foods, doing housework. None of these demand the help of neighbors, relatives, or friends. We wish we were closer to our neighbors; we think of ourselves as friendly people who would gladly help them. But there is little to help them with. In our house-boxes, we are self-sufficient. Or rather, we are self-sufficient in relation to the people we know but dependent as never before on total strangers living thousands of miles away.

Times of crisis still can bring us closer to our neighbors. When a health crisis renders us unable to perform the simple functions of daily survival, or a natural disaster or social crisis ruptures the supplies of food, electricity, and transportation that make us dependent on remote strangers but independent of our neighbors, we are glad to help each other out. Reciprocal relationships quickly form. But usually, we don't help out our neighbors very much because there is nothing to help them with.

For the typical surburbanite, what is there to do with friends? We can cook together for fun, but we don't need each other's help in producing food. We don't need each other to create shelter or clothing. We don't need each other to care for us when we are sick. All these functions have been given over to paid specialists who are generally strangers. In an age of mass consumption, we don't need each other to produce entertainment. In an age of paid childcare, we hesitate to ask each other for help with the children. In the age of TV and the Internet, we don't need each other to tell us the news. In fact, not only is there little to do together, there is equally little to talk about. All that is left is the weather, the lawn, celebrities and sports. "Serious" topics are taboo. We can fill up our social gatherings with words, it is true, but we are left feeling empty, sending those words into an aching void that words can never fill.

And so we find in our culture a loneliness and hunger for authenticity that may well be unsurpassed in history. We try to "build community," not realizing that mere intention is not enough when separation is built into the very social and physical infrastructure of our society. To the extent that this infrastructure is intact in our lives, we will never experience community. Community is incompatible with the modern lifestyle of highly specialized work and complete dependence on other specialists outside that work. It is a mistake to think that we can live ultra-specialized lives and somehow add another ingredient called "community" on top of it all. Again, what is there really to share? Not much that matters, to the extent that we are independent of neighbors and dependent on faceless institutions and distant strangers. We can try: go meet the neighbors, organize a potluck, a listserve, a party. Such community can never be real, because the groundwork of life is already anonymity and convenience.

When we pay professionals to grow our food, prepare our food, create our entertainment, make our clothes, build our houses, clean our houses, treat our illnesses, and educate our children, what's left? What's left on which to base community? Real communities are interdependent.

Now we have reached the sinister core of financial independence: it isolates us in a world of strangers. It is strangers whom we pay to perform the functions listed above. It doesn't really matter who grows your food—if they have a problem, you can always pay someone else to do it. This phrase encapsulates much about our modern society. When all functions are standardized and narrowly defined, it does not matter too much who fills them. *We can always pay someone else to do it.* As an individual, it is hard not to feel dispensable, a cog in the machine. We feel dispensable because, in terms of survival, in terms of all the economic functions of life, we *are* dispensable.

If you buy food from the supermarket deli, the people behind the kitchen doors, whom you never meet, are dispensable. If they quit, even if they die, someone else can be hired to fill their role. The same goes for the laborers in Indonesia who make the clothes you buy at the superstore. The same goes for the engineers who design your computer. We rely on their roles, their functions, but as individual humans they are expendable. Maybe you are a nice, friendly person who actually exchanges friendly greetings with the cashier who's worked for five years at the local supermarket, but while you may be dependent on her role, the specific person filling this role is unimportant. It does not really matter if you get along with this person, or even know her name. She could be fired or die and it would make little difference in your life. It would not be much of a loss. Unless you live in a very small town, you probably will never know what happened to her or ever think to ask. All the more so for the vast majority of the people who sustain our material lives. They, unlike cashiers, are utterly faceless to us.

Because the economy depends on our roles, but does not care which individuals fill these roles, we suffer an omnipresent anxiety and insecurity borne of the fact that *the world can get along just fine without us.* We are easily replaced. Of course, for our friends and loved ones—people who know us personally—we are irreplaceable. But with the increasingly fine division of labor and mass scale of modern society, these are fewer and fewer, as more and more social functions enter the monetized realm. Thus we live in fear, anxiety, and insecurity, and justifiably so, because we are easily replaceable in the roles we perform to earn money.

We can get along fine without you. We'll just pay someone else to do it.

The Anonymous Power

The monetized life is a lonely life because it reduces the people in our lives to anonymous occupiers of roles, and also because financial transactions are by their nature generally free of obligation. Once the money has been paid and the goods delivered, the transaction is over. No future relationship is implied. Each party has discharged his or her obligation.

To be financially secure means to have enough money to discharge *all* of our obligations, leaving no need to depend on anyone's favors or gifts. It means we are now free of obligations—in other words, we are independent. But another word for obligations is "ties". In fact, the word "obligate" means to place a tie upon someone (*ob* = on, *ligate* = to tie). The closed money transaction leaves little or no tie between the transactees. Someone enjoying financial security will never have to rely on personal relationships with other people. He won't need anyone to do him any favors—he can *pay* for your services, thank you. Financial security means that you are not dependent on the goodwill of any individual person. If the farmer who grows your food decides he doesn't like you, no matter, you can pay someone else to grow it.

Gift transactions are quite different: they are open-ended and personal. The transaction is incomplete, leaving an obligation—a tie—between the giver and the receiver. Giving a gift creates or affirms a social relationship; it connects the giver and the receiver. Gifts usually imply future gifts, whether reciprocal or to someone else down the line. Moreover, gifts usually don't have a standard value; their value depends on the unique relationship between giver and receiver, and it reinforces that relationship. In fact, that's one of the main reasons we give gifts in the first place: to become more tied to, and thus less independent of, the person receiving. Money is the opposite. Money is suitable for conducting trade among strangers, such as when any one of us buys something at a supermarket, a Wal-Mart, or on line, and it does not make them into anything more than strangers. Money has the same value no matter who gives it to you—it does not imply or necessitate a relationship. In fact, money issues often complicate or destroy relationships.

Economics and common sense associate money with *self-interest.* It should not be surprising, then, that the gift transactions that so sharply contrast with money transactions also generate a different sense of self, as well as different perceptions of what is in the interest of that self. In gift-based societies, the ties created by the gift usually extend beyond two people to involve the whole tribe or village. Lewis Hyde, in his classic *The Gift*, observes that gifts move in circles, and as these circles spread outward, the sense of self spreads as well to include the entire gift-giving network. "The ego's firmness has its virtues, but at some point we seek the slow dilation... in which the ego enjoys a widening give-and-take with the world and is finally abandoned in ripeness."[2] The contrary process, in which commodity exchange replaces gift-giving networks, corresponds to a narrowing of self into a lonely, alienated ego.

When the relationships of gift-giving in a community are replaced by monetary transactions, the fabric of the community unravels. The tight-knit communities of primitive societies were held together by intricate customs of gift-giving; indeed, the renegade ex-financier Bernard Lietaer cites the still-surviving gift-giving customs of Japan as a reason why that society has resisted some of the community-destroying effects of the modern economy. Although it is perhaps an unwarranted conclusion to say that communities arise out of gift-giving (and not also the other way around), Lietaer's definition of a community is an elegant one: "a group of people who honor each other's gifts, who can trust that their gifts will be reciprocated some day, in some way." In the absence of such trust, the interests of the self are the accumulation and control of resources that defines self-interest in a money economy.

The gift mentality of hunter-gatherers extended beyond kith and kin to include their entire environment, as did their trust that their needs would be provided. Whether by human or natural agency, the gift circles back to enrich the giver. Thus it was that the hunter-gatherer was uninterested in accumulating property. In the realm of the gift, accumulation is senseless.

All this changed with the transition to agriculture. Agriculture, which requires an input of labor to reconfigure nature into a more productive species mix, fosters a mentality of taking, not merely receiving gifts freely offered. This psychological transition was a gradual one: we still speak of the "gifts of the land", but is it a true gift when we must manipulate the giver? Moreover, the rhythm of agriculture includes a phase of accumulation and storage, which are essential to the security of the farmer. The wealthiest farmer is the one with the greatest store of grain. The greatest lord in an agricultural society is he who controls the greatest productive assets. The long, slow demise of gift mentality goes hand in hand with the shift in thinking that arose with civilization. Money is a manifestation of this shift, but not its deepest cause. It only reinforces the scarcity-based accumulative thinking that inevitably accompanies the separation of self.

The assumption of scarcity is at the very root of economics, in which exchange happens when one person has a "need" or a "want" that is difficult or impossible to fulfill oneself, but that another person can fulfill more easily. Hunter-gatherers had virtually no such needs, and subsistence farmers have very few. We can see economic growth, then, as reflecting an escalation of neediness, an intensification of the state of being in want. Today, inundated in an unprecedented deluge of material conveniences and luxuries, we are nonetheless desperately in want. Think of that phrase, to be in want. I want, I want, I want. To be constantly in want is the very definition of poverty, no matter how large one's house or bank account. By that measure, ours is perhaps the poorest society the world has ever known.

The breakdown of community under the assault of the money

economy is well-documented wherever money has taken the place of traditional reciprocity. Helena Norberg-Hodge gives an especially clear account of this process in her book, *Ancient Futures*. She describes the impact of monetization in the Himalayan region of Ladakh, where generations-old customs of labor exchange among neighbors allowed each farmer to harvest his crop in time without the need to hire help. After labor became a commodity and people grew dependent upon money to survive, this custom faded away. Farmers had to use hired laborers, drawing them further into the money economy and making them dependent upon the sale of their crops (and thus susceptible to fluctuations in global commodities markets).

In sharp contrast to the monetized world of financial security, which inexorably separates everyone from everyone else, a gift economy is an economy of obligation and dependence. Financial security is not true independence, but merely dependence on strangers, who will only do the things necessary for your survival if you pay them. Would you rather be dependent on strangers, or on people you know? Well, that probably depends on how you treat the people you know. Thus the monetized life removes some of the incentives for people to adhere to social and ethical norms. Dissolution of community is *built in* to our system of money. The monetization of life dissolves communities, and the dissolution of community necessitates the further monetization of life.

The distancing and anonymizing effects of money make it unlikely that the people we depend on economically are our friends. Meanwhile, our social circles, our friends, usually perform specialized functions that are not directly related to our lives at all. Work and socializing are separate. In fact it is often considered in bad taste, or a threat to the friendship, to become economically involved with someone even when it is possible. Consider as an example a hypothetical group of friends: a software engineer, a professor at a university, a podiatrist, a lawyer, a real estate agent, an insurance agent, and an artist. None of these people depend on each other to meet any other need but socializing. The lawyer need not buy art from the artist. The real estate agent can easily go to any podiatrist. The engineer's children probably won't be educated by that particular professor. The group of them get together every other weekend and have dinner. Sometimes they'll watch a football game or go on a picnic. A typical group of friends. They get along just fine, and there really is no reason for them not to get along, simply because there is no opportunity for any conflicts of interest to arise.

Conflicts of interest only arise when, say, the podiatrist has a legal problem and chooses his friend the lawyer, when the artist wants to buy a house and chooses the real estate agent, when the professor buys insurance from the insurance agent. All of a sudden things become touchy and awkward. When it comes to money, people start to wonder, "Am I being taken advantage of?" The artist feels like the real estate agent is treating her "like a client," trying to sell her a house. The podiatrist is secretly outraged by the high fee her lawyer friend is charging. The professor wonders whether perhaps he is overinsured. It's so much simpler to keep friends and business separate. Indeed many people make it a matter of principle never to mix the two. They keep the relationship between themselves and their lawyer, doctor, and insurance agent all purely "professional." It would seem that the secret to preserving a friendship is to remain completely independent of one another, at least in the economic realm.

Whereas in the past we were intimately connected to the people we materially depended on, today our economic relationships are increasingly separate from our social relationships, to the extreme where money issues often ruin friendships. When we no longer have economic ties to our friends, we must pay someone instead. Pay whom? If we are "rational", whomever offers the cheapest price for the same service. Whenever any product or service becomes a standardized commodity, price becomes the sole basis of differentiation. Competition is the flip side of depersonalization. We are set into competition with each other, not necessarily because this is human nature, but because the ubiquitous pressure of money removes any other basis for choice.

Money opens the door to pressures that are distinctly unfriendly. It is not very nice to maximize your rational self-interest. Yet that is what our present money system compels us to do. Yes, better to keep money out of the friendship. The only problem is that when all of life has been converted to money, we must then keep all of life out of the friendship too, leaving us with superficialities. This separation of the spiritual needs of friendship and the material needs of economic relationships reflects the Cartesian separation of spirit and matter. Friendship without material interdependency is usually just as anemic, just as superficial as spirituality divorced from the real world.

I suppose I do enjoy some of my friendships in which all we do is talk, but niceness, partying, eating and talking are nothing on which to base a community. In real communities, people depend on each other.

They must find ways to get along. They must learn to accommodate each other's faults. Acceptance is crucial to each person's survival. This stands in sharp contrast to most so-called "on-line communities," where exiting the community is as simple as hitting the delete key.

In former times, mutual dependency and the need for joint creativity cemented friendships and gave ample opportunity to become, in common cause, part of something greater than ourselves. In small farming communities, for example, neighbors and relatives would help each other with the harvest. They would help each other with the many projects that were too big for one family, like building a barn or a house. They would care for each other's animals during times of illness, mourning, weddings, and the like. They also depended upon each other—people they knew personally—for entertainment, music, stories, and other forms of culture, of which they were not mere consumers but coproducers. Sometimes their very survival depended on cooperation.

That is why in the past, it was a very serious matter to be shunned by the community. If the local doctor, the local grocer, the local weaver, the local blacksmith refused to serve you, you couldn't just "pay someone else to do it." Moving to another part of the country was a big deal, because you had to become part of a new community. Nowadays moving to another part of the country necessitates very little change in lifestyle—you can procure all the necessities (and luxuries) of life in the same franchise stores anywhere you go. All you need is money. You need not get along with anyone or even, thanks to this anonymous power, know anyone by name.

There is another reason, besides mutual material dependency, why ostracism was one of the worst punishments possible in ancient societies. It wasn't so much that we depended on our social relations for survival: hunter-gatherers and primitive farmers can generally survive pretty well on their own, though life is harder without the sharing and reciprocity of a group. Rather, one's sense of identity, the answer to the question "Who am I?" was derived from our station in the social web. Primarily, we were defined by our kinship ties; secondarily, we were defined by our skills, experiences, and vocation. Who am I? I am John's brother, Jimi's dad, Laura's uncle, Cathy's cousin; Dana's brother in law. In a traditional society, each person's recognized kin network extended through hundreds of familiar faces, united by customs of reciprocity and by stories of grandfathers and great-grandmothers, fading back into the time of legend and myth. To be expelled from the village or the tribe was to be shorn of one's identity, a fate perhaps even worse than death.

In America today not only the community but even the extended family has been shattered as a viable social unit. The nuclear family is the rule; as often as not the father and mother live separately. Grandparents, uncles, and cousins are occasional visitors, and many adults go years without seeing their cousins, or even their brothers and sisters. In our highly mobile age, where it is nothing to move across the continent to take a new job, our relatives are often scattered across the country. Outside the immediate family, school, and workplace, the faces we see daily are the faces of strangers. And of the faces known to us, really only the family is at all intimate, at all familiar with the stories of our lives.

The result is that we lack the means to establish a strong self-identity. No one knows our story. Human beings have always defined themselves in great part through their relationships with others, building a common story defining each of its actors. Now these stories have been splintered into tiny four-person units (I am exaggerating a little bit) that are below the threshold for robust self-definition. Unlike in the traditional small village or tribe, where *everyone* knew your story, and where, knowing everyone else's story, you had a context to create a solid story of self, today we interact day-in and day-out with outsiders. We maintain our private lives, and know little of the lives of our coworkers, our customers, our colleagues, our students, our teachers, our neighbors, or anyone else outside the home.

In tandem with the ascent of money-based relationships, the realm of the private has expanded at an accelerating rate through modern history. Societies always had some realm of the private, a certain scale of intimacy in which some functions, such as lovemaking, childbirth, and defecation, were carried out in seclusion, and in which other functions were completely public. Today we encase our entire lives, almost, in the private boxes of our homes, which have, not coincidentally, doubled in size since the 1950s—a literal, physical manifestation of the burgeoning private realm.

In addition to an eroded sense of identity, the physical and social isolation that occurs in the boxes of modern society contributes to a near-universal loneliness and boredom. The lonely housewife isolated in her suburban cage is an emblem for our disconnected lives, insulated by the walls of our specialized roles as well as by the infrastructure of suburbia, interfacing with other lives through the impersonal medium of money.

The fragmentation of society that has followed from the economics of the Machine presents a tremendous business opportunity. When tribe and village, clan and extended family have been shattered, the resulting emotional void creates a demand for substitute relationships. The weak sense of self-identity that springs from our isolation leaves us extremely vulnerable to consumerism, which seeks to define who we are by what we own. We are tempted to define ourselves through our sneakers— "Be like Mike," the Nike ads used to say—and through our cars, our houses, our watches, our clothing, our sports teams. Even more insidiously, we seek out new stories to replace the missing stories of kin and community in which we once embedded ourselves. While some vestiges of "family stories" remain, the storytelling function has by and large been usurped once again by remote professionals: the TV and movie producers on the one hand, and the news media and educational establishment on the other. These institutions provide us with new stories to answer the question, "Who am I?" The entertainment world feeds us stories about total strangers. TV dramas and soap operas bestow the illusion of being intimately familiar with people's lives. Such shows tap into the inborn identity-building function of the psyche, but that function is truncated when those intimately viewed lives off in TV-land never feed back into our own.

Meanwhile, the larger stories of Where did we come from? and Why are we here? have been professionalized as well. Instead of myths and legends, we have history and the news that exploit, for political and economic ends, our need for a Story of the People by which to identify ourselves. In the worst case, people in their desire for a story by which to define themselves will adopt one of racial or national chauvinism, such as have underlain much of the horrific history of the last century.

People who are firmly ensconced in a local, kinship-based community are less susceptible to consumerism and fascism alike, because both base their appeal on a need for self-identity. Therefore, to introduce consumerism to a previously isolated culture it is first necessary to destroy its sense of identity. Here's how: Disrupt its networks of reciprocity by introducing consumer items from the outside. Erode its self-esteem with glamorous images of the West. Demean its mythologies through missionary work and scientific education. Dismantle its traditional ways of transmitting local knowledge by introducing schooling with outside curricula. Destroy its language by providing that schooling in English or another national or world language. Truncate its ties to the land by importing cheap food to make local agriculture uneconomic. Then you will have created a people hungry for the right sneaker.

The transition from a society based on gifts, through the stages of barter, commodity currencies, precious metals, and the present-day financial system, has taken thousands of years in some places, and has been introduced quite suddenly in others. It continues to this day: More and more of our human abilities, skills, relationships, and culture are becoming the subject of property and therefore of money. We are nearing the culmination of a vast historical process: the conversion into financial capital of a variety of other forms of wealth that were never before the subject of purchase, sale, and ownership, that were never before associated with money, but were instead held in common, by a community or a society—the commonwealth. As they become monetized, the money-associated qualities of anonymity, scarcity, and alienation encroach ever-further on those remaining areas of human beingness where the dynamics of the gift still hold.

These other forms of wealth are sometimes referred to as social capital. I have found it illuminating to distinguish them further into social capital, cultural capital, spiritual capital, and natural capital. The distinctions among these four kinds of non-money capital are somewhat artificial, but have been useful to me in perceiving just how deeply the conversion of life into money has reached. I introduce the new terms because they help access dimensions of our impoverishment that often go unrecognized.

The story behind the conversion of each of these kinds of wealth, and the effects that conversion has had, sheds light onto the nature of property, the nature of money, and the fundamental understanding of self and world that goes along with it. The dissolution of community around the world, the filling of our neighborhoods with strangers, the loneliness and anonymity of modern society, the demise of the extended family, the plunder of the planetary ecosystem, the shortening of children's attention spans... all spring from our money system. And money, the great anonymizing power, has even deeper roots in our sense of self. The long transition from gifts to money, from giving to keeping, is written into our very self-definition. Together, our self-definition and its monetary manifestation constitute a pattern that is rapidly propelling us toward social and environmental calamity.

Social Capital

Social capital is the totality of human relationships that sustain life and make it rich. On the physical level social capital comprises the life-sustaining relationships whereby we provide each other food, shelter, and clothing, as well as care for the young, old, and infirm. Less tangibly, it includes such wealth as community, friendships, fun, teaching, and a sense of belonging. Together these constitute a cultural inheritance, a treasure passed on from generation to generation in the form of learned skills, customs, and human connections. Among them are the "ties" discussed in the previous section.

There are many ways by which to convert social capital into money. Let's start with the most primal, ancient, and ubiquitous example: food, or more precisely, the relationships by which we feed each other, including food production, processing, and preparation.

Starting even before the Industrial Revolution, the proportion of the population engaged in primary food production has steadily dropped, reaching 1-2% in present-day America. The standard interpretation of this statistic is that technology, economy of scale, and so forth have freed us from the drudgery of food production, so that most of us can now "choose not to be farmers."A side-effect has been that not only do we choose not to produce our own food, but we have forgotten how. What was once a ubiquitous skill rarely involving money is now something we pay distant specialists to do for us.

The same thing has happened to food processing, as huge factories and global distribution have taken over what was once done at the household or local level. Completing the transformation from social capital to financial capital, today even the final stage of food production—food preparation and cooking—is disappearing from our generalized repertoire of skills and passing into the hands of distant specialists.[3] From half to two-thirds of all meals in America are now prepared outside the home, either in the form of restaurant meals or as ready-to-eat take-out from supermarket delis.[4] Even meals cooked at home are often prepared from ingredients that are already highly processed; when we do cook, we perform only the final stages of cookery, taking advantage of pre-made mixes, sauces, canned soups, and so forth. How many people make their own pie crusts any more, or fry their own french fries, or bake their own bread, or can their own vegetables, or make their own soup stock? These skills have gradually become industrialized and therefore lost to the average household. We have essentially sold away the ability to cook food, converting this form of social capital into financial capital—money.

What has been lost? More important than the skills and know-how are the relationships that once revolved around food. To give food to another is an intimate act, a primal expression of nurture that creates a powerful bond. The oldest way to befriend an animal or a stranger is to offer her a meal. Do you see something monstrous, something obscene, about the routine purchase and sale of an intimate act? Converted into a mere service, the act of feeding another being loses its potency, and a primary generator of relationships is shut down.

Many other primitive, physical means of caring for other humans and building intimate bonds have suffered the same fate. The skills and relationships that once revolved around clothing, shelter, and medicine have been dismantled, converted into an array of services to be carried out by remote machines and their anonymous functionaries. Removed from the domain of self-sufficiency or personal reciprocity, these functions become the objects of purchase and sale—generic commodities rather than personal relationships. As they are converted into money, money dynamics increasingly rule our lives and we lose the substantive basis of our relationships. The cost of living rises, our dependency on money grows, our connections to local people and local ecosystems wither away, and we live increasingly in a monetized realm.

It is hard to imagine, but there was a time when food was rarely purchased for money, when the typical peasant was not part of the money economy at all. In some parts of the world this was true up through the latter part of the 20th century; in some remote corners it may still hold true today: to have no money, yet to enjoy a sufficiency of food, clothing, shelter, medicine, and entertainment. Such a society represents a tremendous business opportunity, because social capital can be mined and exported just as mineral wealth can. Here's how it works: Any region in which the non-monetary social relationships still predominate will have a very low cost of living. Prevailing wages will be low, as will the cost of food, shelter, and clothing. In a process called "development", we destroy existing networks of non-monetary reciprocity by introducing consumer products that are unavailable locally and can only be purchased with money. Meanwhile, we establish factories or other facilities to remove people from the non-monetary economy to earn a wage. The former is known as "opening the market" and the latter as "investment"; resistance to this project is called "protectionism", which the developed countries (those whose social capital is nearly depleted) seek to overcome by fair means or foul. The process is quite similar to what is necessary to make people "hungry for the right sneaker". It happened in the West a long time ago, for as Kirkpatrick Sale explains, it is a prerequisite as well as a consequence of industrialization:

> It was the task of industrial society to destroy all of that [reciprocity]. All that "community" implies—self-sufficiency, mutual aid, morality in the marketplace, stubborn tradition, regulation by custom, organic knowledge instead of mechanistic science—had to be steadily and systematically disrupted and displaced. All of the practices that kept the individual from being a consumer had to be done away with so that the cogs and wheels of an unfettered machine called "the economy" could operate without interference, influenced merely by invisible hands and inevitable balances...[5]

Want a good business idea? Simply find something—anything—that people still do for themselves or each other. Then convince them that it is too difficult, tedious, demeaning, or dangerous to do themselves. (If necessary, make it inconvenient or illegal.) Finally, sell it to them instead. In other words, take something away and sell it back again. Many of the hottest growth niches in the economy fall into this category: after-school child care, lawn care, housecleaning, takeout food. There are companies that will put up your Christmas decorations for you. There are even—are you sitting down?—professional gift-buyers who, for a fee, will go out and purchase the Christmas gifts that you are too busy to buy. When these sectors are saturated, when their associated social capital has been fully converted into money, what will come next? What else do people still do for themselves?

Usually, of course, there is no conscious plot to make the tasks of life inconvenient or illegal. The accelerating pace of modern life alone is enough to drive us to pursue the convenience and efficiency that purveyors of services and technology have to offer. They are merely taking advantage of opportunities that our system makes available.

From this perspective, technology is a mechanism for the transfer of functions and skills away from the general population into the hands of paid specialists. Technology in the form of the "labor-saving device" represents not only (and not necessarily) the reduction of labor but rather its transfer, from ourselves and people we know, to the anonymous functionaries of the Machine. A vacuum cleaner embodies the combined efforts of innumerable engineers, miners, petroleum workers, assembly line workers, truck drivers, and so on, all helping you clean your floor. They are all strangers to you, of course, but they are also far more efficient. All the objects of technology we use similarly embody vast economies of scale and exquisitely fine division of labor. Rather than a community, millions of strangers help us live. Our relationship with them is that we are consumers of their products, and they of ours.

Anonymity, technology, and specialization are closely interlinked. Technology, starting with agriculture, provided the food surpluses that enabled division of labor to occur in the first place, and its transfer of functions away from the average individual did not stop with food production. The same dynamics that initiated the ascent of humanity are still in place; therefore the transfer of functions to paid specialists—the conversion of social capital into financial capital—has little likelihood of slowing down until all social capital has been consumed—a point we are nearing today.

As she is stripped of her traditions, self-sufficiency, customs and relationships, the individual becomes less a human being and more a pure consumer. Remember, contrary to Descartes, it is our relationships that create our identity. The sell-off of social capital thus represents a sell-off of our very being. Surely, to be a consumer is to be less than human. But the sell-off of social capital leaves us nothing but that, consumers, except within the specialized niche of our vocations.

In the discourse of the business world, and often in politics too, that is what you are called, a consumer. One who consumes. The very ubiquity of the word inures us to its full significance, so repeat it to yourself a few times. I am a consumer. I am a consumer. I consume. When all other relationships are gone, that is what is left.

The word "consume" brings us back to the origins of separation in fire, a metaphor for our civilization's consumptive linearity. Indeed, the conversion of human bonds into money resembles an oxidative reaction in chemistry, which replaces existing internal chemical bonds with new, lower-energy bonds to something from the outside, an oxygen atom. Money is like the thermal energy liberated thereby, no longer bound to any individual and thus free to perform work. Incinerating relationships and converting them into services, the original bonds among individuals are severed. A purchased "service" is as generic as the money that buys it.

Social relationships embody a kind of wealth in the form of bonds or ties, which can in principle be liquidated, turned into a commodity, and sold back as a service. Almost no relationship is exempt from this colonization of the social space: not parenting, not sex, not friendship, not trust.

Let's start with an example of this colonization that also exemplifies the inauthenticity of commoditized relationships: the "hospitality industry". This is the name by which the hotel industry calls itself, but what a perversion of the original meaning of the word hospitality! Hospitality bespeaks a generous, welcoming attitude of sharing one's home. How would you feel if you shared the home of a friend for a few days and were then presented with a bill, with 500% markups for each telephone call, bottle of water, and item of food you consumed? Consider also the perversion of the word "guest" in this context. In a saner context we hear words like, "I wouldn't dream of asking you to pay for dinner; you are my guest." Of course, such relationships depend on community, custom, and other (non-monetary) ties. The hotel industry represents an aspect of the conversion of these ties to money, the liquidation of this form of social capital. Now, when we want someone to treat us as a guest, we must pay them.

Nor has friendship itself escaped the ravages of professionalization. While it is a truism that money cannot buy friends, it can buy many of the functions that friendship once served. One of the most insidious transformations of our social capital is the commercialization, in the form of celebrity news, TV dramas and the like, of intimate involvement in other people's lives and stories. As described earlier in this chapter, these replace the caring relationships and intimate knowledge of people around us with voyeuristic glimpses into the lives of distant or fictitious personages, temporarily assuaging the hunger for intimacy and community. Of course, this feeling of involvement is an illusion; it is a one-way involvement that can therefore never be truly nourishing. It leaves the real need unmet, and ultimately even stronger—a perfect recipe for addiction.

A further example of the professionalization of friendship is to be found in the proliferating professions of life coach, grief counselor, psychologist, spiritual adviser, and so forth. Wise advice and a steadying hand, a person to turn to and a shoulder to cry on—these too are now for sale. The rapid growth of these "services" can mean only one thing: again, that something people once did for themselves and each other has been taken away from them and sold back. Cut off from community and alienated from our own intuitive wisdom, we find ourselves increasingly dependent on professional advice.

Another thread in the mosaic of the monetized society is the replacement of such social functions as reputation, word-of-mouth, credibility, and trust with standardized, objective substitutes. Since the professionals we pay are unknown to us and outside our shrunken social networks, we rely on various kinds of certification and licensing to assure us that these professionals are competent and responsible, protection we need in the absence of personal connections. We don't know anything about these people, except for that tiny sliver of their lives that remains public. We know much less about the people around us than ever before. We are likely to know more about the "private lives"—matters of sex, family, and health—of our celebrities than of our next-door neighbors. And while word-of-mouth retains some power to enforce responsible professional behavior, especially in small towns, in the huge anonymous cities and suburbs, where much of the population is new to the area, we know nothing about our architects, doctors, contractors, and other professionals except what expert opinion, embodied in licensing exams and the system that confers credentials, tells us. In essence, exams and credentials represent the conversion of a social function—reputation—into money. A similar dynamic is at work in the laws to enforce contracts—a replacement for trust—and the penal code to enforce responsible behavior—a replacement for community-based social pressure. These are only necessary in a mass-scale, anonymous, monetized society.

One thing people have done for themselves for millions of years is to take care of their own children, a sacred responsibility that would only be entrusted to the most intimate friends and relatives. According to my "good business idea", parenting is thus a great business opportunity. Why do it yourself when specialists wielding economies of scale can do it so much more efficiently? Indeed, child care has been another huge growth industry in recent decades; once a matter of the greatest intimacy, it is now within the realm of professional services. Someone is paid to have a relationship with your child. One student described her job in a day care center like this: "Some of the parents would drop them off at eight in the morning and pick them up at eight o'clock at night. We feed them lunch and dinner, change their diapers, read them stories, comfort their little hurts, potty-train them… we're the ones raising these kids, not the parents." Admittedly, this is an extreme example, but the widespread trend is toward parenting—many aspects of it at least—becoming a paid function. It is after all more economically efficient for specialists to care for many toddlers all at once in a dedicated facility than for each set of parents to do it themselves.

Before I move on, please understand that we must not blame the parents. They aren't lazy or uncaring, but merely victims of a monetization of life that has spread the need for efficiency into life's every corner. The monetization of life leaves life dependent on money, as the old networks of reciprocity evaporate. Many mothers have little choice but to join the (paid) workforce. Add to this the physical and social isolation of suburbia, especially the fragmentation of the extended family, and the day-to-day life of the stay-at-home mom can become lonely and exhausting. I know because I was a stay-at-home dad for several years.

The later stages of child-rearing are also the province of strangers. We pay the schools, schoolteachers, and a whole educational apparatus to do it. As the loss of cooking skills exemplifies, a capacity unused will eventually atrophy. Increasingly unsure of our own parenting abilities, we turn to experts and professionals; indeed if we refuse to, all manner of legal and bureaucratic mechanisms conspire to coerce us. For example, what is the proper balance between freedom and limits for my child? Should he wander the greenbelt by my house unsupervised? Wisdom in such matters has been transferred from its original matrix of family, community, custom and tradition into the hands of professionals and politicians. A huge array of books, counselors, social workers, psychologists, pediatricians, teachers, and legal regulations provides that guidance today. They are our professional child-rearers.

At the other end of life, care for the elderly, as well as for invalids, has also changed from an unpaid function of the extended family and community into a professional service. Don't blame the families though, because this phenomenon interrelates with many other trends in modern society: the geographical scattering that has followed the dissolution of community, the monetization of life that pressures both parents in a household to work, the medicalization of old age and the professionalization of medicine, as well as the general decline in health among old people that renders them dependent for decades.

One day my six-year-old son Matthew asked me, "Which would you choose, fame, money, or love?" I told him money of course... just kidding. I explained that love is never for sale, and how sometimes rich people are worried that people only like them for their money and not for themselves. Then a chill passed over me, when I thought about what love is in action. One aspect of love enacted is simply to take care of someone, to help them on a nurturing physical level—the way, for example, my massage therapist and herbalist help me. They soothe my aches and heal my hurts. And I pay them for it. Don't get me wrong—neither of them are especially mercenary people, but the society in which we live gives us no choice, or makes it look as if we have no choice, but to charge money for healing, loving, caring, nurturing. As I have already mentioned, other primal expressions of love, to feed, clothe, and shelter another being, have also become professional services. Yes my friend, love itself is for sale, if not the actual emotion at least many aspects of its expression.

When you choose a preschool for your child, will you consider whether it provides a "loving atmosphere"? Would you pay more for one that does? Should love be part of the job description? What about companies who advertise that they "care about the customer"? Their employees answer the phone full of enthusiasm to talk to you. The cashier at the supermarket wishes you a nice day. We recognize all these expressions of caring as phony, because we know that someone cannot be paid to care. At most we can muster the semblance of caring. Similarly, commodified services are never as authentic or as nourishing as the personal relationships they replace.

The result is a near-universal loneliness and an inauthenticity so pervasive we hardly notice it. What can we expect when the people who perform crucial life-giving functions only do it because they are paid to? In many cultures, to prepare food for another was considered an intimate act. To take care of someone on a physical level is intimate; it has been the basis of family, friendship, and community from time immemorial. Now that these functions have been sold off, the relationships they once fostered are left hollow. No wonder that divorce has risen wherever the modern economy has replaced traditional "home economics", leaving the family with a domain reduced, in Emile Durkheim's words, to "emotional release and the sharing of affections."[6]

We live in a world that literally does not care. Your telephone company, despite their recorded messages, does not actually care about you, the customer. The clerk at the supermarket does not actually care whether you "have a nice day." How could he, not even knowing you? The programmer of the automatic checkout machine does not actually harbor feelings of gratitude toward you, whatever the message on screen.

Your waiter is not actually interested in how you folks are doing today. These are all part of the ubiquitous matrix of lies I mentioned in Chapter Two, and they reveal so clearly what exactly it is that we have converted into money. We have sold off authentic human relationship.

No discussion of the monetization of human relationship would be complete without mentioning sex, whose commodification is so widely condemned as to be almost a cliché. Prostitution is only the tip of the iceberg; equally significant is the use of sexual associations to sell product, as if sex itself were for sale. Pornography, too, is a reduction of that most sacred of human interactions into a commodity. Reduction is the operative word, because sexual feelings and sexual relationships are potentially a key to transcending the prison of the discrete and separate self, as well as the means to create new life. I can't think of anything more sacred than that! Yet, the images of advertising and pornography suggest that no, sex is but a matter of tits-and-ass, getting hard, getting laid, getting off.

The metaphors of "having" and "getting" so commonly applied to sex bespeak the degree to which it has been commodified. Why do we usually not speak of "giving sex" or "sharing sex"? Even in the absence of any outright financial transaction, quasi-economic concepts of loss and gain infuse our culture's thinking about sex. It is unavoidable, written into our self-definition as separate, discrete beings. Yet precisely because its deepest spiritual function is to melt the boundaries that enforce this separation, sexual love, more than any other relationship, is diminished and debased by its commoditization. For the same reason, sex has an enormous subversive potential. The sharing of self it involves explodes the very basis for the world of separation in which we live, and its associated pleasure hints at the ecstasy awaiting us when we throw off separation's shackles. Perhaps that is why repressive political regimes typically exhibit great hostility toward sexual licentiousness—a form of repression that George Orwell identified as a key feature of totalitarianism. Our own society takes a different, more insidious approach to defusing sex's explosive revolutionary potential, attempting to excise its transcendental core. The husk that remains is, depending on the context, an inconsequential pleasure, a biological function, rank animality, obscene temptation, or a frightening taboo. None of these honors the sacred dimension of sex, which ancient Taoist and Tantric practice saw as nothing less than a gateway to the transcendence of cosmic polarities. Potentially a touchstone for reconnecting with the true unity behind our all-consuming play of individuation, potentially a secret window through the veil of our illusory separateness, sacred sexuality has been reduced to a fuck.

Of course, whatever we pretend to make of it, sex is far more than this. Its soul-shattering and life-creating potential remain despite the cultural pretense that it is a casual commodity. The result of this delusion? Heartbreak, emotional wounding, guilt, rape, abortion, and a feeling of betrayal stemming from the inner knowledge that we have "bought in" to something infinitely inferior to what life can offer. Hence the near-universal acknowledgment of casual sex as spiritually vacuous and emotionally unsatisfying. The same could be said, though, of any relationship that has been depersonalized.

Thankfully, our culture's attempts to tame sexuality have never been completely successful, and sex remains a potent force capable of smashing apart the most secure fortress of self, the most ordered life, the most tightly controlled personality. In college my male teammates and I spoke of sex in the coarsest imaginable language of having and getting, but this was mostly a pretense. Our youthful forays into sexual love were no less shattering. Alone with our girlfriends it was a different matter entirely. Our outward casualness could not insulate us from the wrenching, liberating, shattering power of sexual love to open a door to the soul. I wonder if any of my girlfriends from that time will read this? If so, I want you to know that even if I then seemed a hopeless cad, your love turned deep invisible keys in my soul. Your heartbreak was not in vain. What you gave me, I needed for my future opening.

Typically, it is we men who explore the farthest reaches of separation. Like Theseus in the Labyrinth, we would probably wander in it forever were it not for the lifeline Woman provides. Of course, these male and female principles exist in all of us; in each man and woman there is both the yin and the yang. The feminine principle in all of us—intuitive rather than logical, organic rather than analytic—brings us back from our journey of separation toward wholeness.

Accordingly, the present extreme of separation to which our society has "ascended" is a yin-yang imbalance, reflected in patriarchal religion and patriarchal society. As the ancient Chinese understood, the extreme of yang is also the birth of yin. Only in the convergence of crises to which separation has brought us will the Age of Reunion begin. On a less cheerful note, in Chinese medicine there is also a condition called "the collapse of yang", which happens when maintaining a prolonged imbalance so depletes the body's resources that nothing is left to sustain the process of rebalancing. Then all intervention is useless. It is too late. The patient dies.

For civilization, the collapse of yang would mean that when we finally see the source of our crises, we will be too weak to do anything about it. Today we have not yet reached that point. Sufficient social and natural resources still exist to create a beautiful world for all of us. Yet we continue to deplete social, natural, cultural, and spiritual capital at an accelerating rate. How long until we so exhaust it that, when we wake up to the urgency of our condition, we find we lack the strength to create the beautiful world we can see clearly at last?

Cultural Capital

"How is it possible to own the stars?"
"To whom do they belong?" the businessman retorted peevishly.
"I don't know. To nobody."
"Then they belong to me, because I was the first person to think of it."
—The Little Prince

Cultural capital refers to the cumulative products of the human mind, including language, art, stories, music, and ideas. Only recently have they been considered valid objects of property rights, and only recently have they been produced by specialists for the consumption of the masses.

In hunter-gatherer days, while some people probably had exceptional talents in art or music, there was no separate category of "artist" or "musician" because everybody was both. To sing and play an instrument, to dance, to draw, to create beautiful objects from the materials available in the environment, is as natural a part of being human as it is to walk, talk, and play. The atrophy of these functions began with the advent of the Neolithic division of labor, and accelerated as the division of labor intensified with the Industrial Revolution, which also created the means to replace them with mass-produced substitutes.

Let's look at the example of music. Before the invention of the phonograph, practically any social gathering involved the playing of instruments and the singing of songs. People sang all the time. There was no one (except perhaps the profoundly tone-deaf) who "couldn't sing." If in 1880 you went to a dinner party at one of the large Victorian houses still standing in central Pennsylvania, chances are that after dinner, everyone would adjourn to the piano parlor to sing. People sang as they worked and sang as they played, they sang in chain gangs and around campfires, they sang alone and they sang at virtually any gathering, they sang traditional songs from the Old Country and they sang the folk songs of North America. But with the advent of the phonograph, then the radio, and finally the explosion of electronic devices that make recorded music ubiquitous today, the human capacity to sing began a long, gradual decline among the general population. Why? Because it was no longer necessary. When we buy recorded music, we are essentially paying someone to sing for us, paying for a function that was once just part of being human. Remember the eternal business idea of finding something people do for themselves and then convincing them to pay for it instead? That is exactly what happened with recorded music. It attracted people with its novelty and the fact that they could listen to the very best singers and musicians in the entire country, whereas before they had to content themselves with the best singer in the family or the town. In comparison, their own voices didn't sound so good anymore, and they became convinced of their own inferiority.

Nonetheless, the decline of singing took a very long time. As recently as the 1940s, my father remembers his whole neighborhood getting together every week in the summer in suburban St. Louis to have a picnic and sing the old songs. Such neighborhood get-togethers have become rare these days with the generalized breakdown of community; even when they do occur there isn't much singing going on. In colleges a half-century ago, students would sing the college songs at parties and football games, or someone would bring a guitar to the campus lawn and people would gather around and sing. Today they listen to recorded music (at high volume, insulating themselves from the world behind a wall of sound).[7] By the 1970s the only people still singing were children. They were the only people for whom song was integrated into daily life. I remember sometimes on the elementary school bus we would sing all the way to school, the whole busload of us. That behavior disappeared by high school, however; we had grown out of it. We had grown out of singing just for the fun of it (and not as a performance). How sad. Today singing on the bus is probably against the rules.

Some vestiges of our inborn musicality remain: musically-inclined friends who get together for jam sessions and the like. Such activities are a powerful reclamation of our cultural, social, and spiritual capital. By and large, though, music has become a paid function, a commodity.

Something very similar has happened to storytelling. Television has replaced traditional tales as well as family stories and community stories. The old-fashioned storyteller, someone who can spin a good yarn, is a rarity these days; rare too are the types of venues and occasions where her stories might be heard. Instead, through our consumption of television and movies, we pay remote specialists to produce our stories for us. Significantly, the producers of these stories now own them, an unprecedented development. For most of human history, no one imagined that you could own a story. Stories were simply not conceivable objects of property, but constituted in each culture a vast commonwealth. Today, corporations such as Disney mine that commonwealth, wall off parts of it for themselves, and convert it into money.

A related erosion of cultural capital is the hollowing out and commoditization of our holiday customs and religious traditions. Starting with Christmas, one by one our holidays have been reduced to the buying of things; each is a fountain of profit. For Valentine's Day we buy chocolate and flowers, for Easter we buy candy, for July Fourth we buy firecrackers, for Halloween we buy costumes (last year my children were the only ones I saw with homemade costumes). For Christmas, in addition to gifts of all descriptions, we buy wreaths, decorations, and cookies. And for all holidays we buy greeting cards. A new trend in the restaurant business is Thanksgiving dinner with all the trappings—why cook it at home when it is so much more efficient to have restaurant chefs do it? Remember again the quintessential business idea: Find something that people still do for themselves, and sell it to them instead.

The commoditization of our cultural capital has received increased attention recently due to the ongoing controversies over intellectual property and the preservation of the commons in cyberspace. Before discussing these less tangible objects of property, let's examine what property actually means. Functionally, property is merely a social agreement that a subject (individual or corporate) has certain exclusive rights to use a thing in a certain way. These rights vary depending on their object and the society upon which they depend. For example, land ownership confers the right to prohibit trespassers in America, but not in Scandinavia.[8] Trademarks confer the exclusive right to use a word or words for specific commercial purposes. For example, we can say that Wal-Mart owns the word "Always" even though I can still use that word in this book. Wal-Mart's exclusive right is to use it in a specific commercial context.

Not too long ago, I asked a Penn State class how many of them download copyrighted materials from the Internet. The show of hands was unanimous. "According to the legal definition," I said, "you are all thieves. But stealing has a moral as well as a legal component. So legal definitions aside, how many of you *feel* like a thief, the way you would if you shoplifted a CD?[9]" This time, not a single hand was raised.

I continued, "So no one feels like a thief. The record industry says that the reason you don't feel like a thief is that you are morally or ethically deficient. Maybe you are ignorant; maybe you are just plain *bad*."

But perhaps there is another explanation, an explanation more trusting of the moral instincts of that vast majority of young adults who download MP3 files. Maybe the reason my class did not feel like thieves lies not in deficient ethics, but with the concept of intellectual property itself.

We understand the purpose of copyrights and patents to be to protect the interests of creators—artists, musicians, and inventors. But interestingly enough, the framers of the United States Constitution cited a rather different reason for authorizing Congress to establish copyrights and patents. Their reason was, "To promote the Progress of Science and useful Arts, by securing for limited Times to Authors and Inventors the exclusive Right to their respective Writings and Discoveries." In other words, they deemed it socially beneficial for creators to be able to profit from their works. But did they think that a person who comes up with an idea should "own" it? Here are the words of Thomas Jefferson:

> "If nature has made any one thing less susceptible than all others of exclusive property, it is the action of the thinking power called an idea, which an individual may exclusively possess as long as he keeps it to himself; but the moment it is divulged, it forces itself into the possession of every one, and the receiver cannot dispossess himself of it. Its peculiar character, too, is that no one possesses the less, because every other possesses the whole of it. He who receives an idea from me, receives instruction himself without lessening mine; as he who lights his taper at mine, receives light without darkening me. That ideas should freely spread from one to another over the globe, for the moral and mutual instruction of man, and improvement of his condition, seems to have been peculiarly and benevolently designed by nature, when she made them, like fire, expansible over all space, without lessening their density in any point, and like the air in which we breathe, move, and have our physical being, incapable of confinement or exclusive appropriation. Inventions then cannot, in nature, be a subject of property."[10]

Reflecting the Founders' reservations about owning ideas, the Constitution prescribes that patents and copyrights are to be valid only for a limited time. The earliest legislation enacted limited terms indeed: copyrights lasted just 14 years and could be renewed for 14 more. Today, copyrights last the entire life of the author plus an additional 70 years, or 95 years for corporate copyrights![11]

Artists and inventors were not considered to own their ideas at all, but simply to enjoy an exclusive but highly circumscribed right to profit from them. It is this distinction that justifies the institution of the lending library. But today this distinction has crumbled: it is not only illegal to sell copies of a movie; it is illegal to copy it at all. In a former time, the students' actions would have been neither illegal, nor considered unethical. If books were invented in today's legal climate, the lending library would surely not exist.

The reasons for setting time limits on patents and copyrights are both practical and moral. Practically speaking, when someone has exclusive rights over a creation for too long, the effect is to stymie innovation, not promote it. That is because art, music, and technology build upon themselves. Their history is one of constant borrowing and self-referencing. Art and music draw from the cultural milieu that surrounds them, which is itself composed, in part, of the art and music already in currency. Today, because of intellectual property law, many types of artistic expression are essentially illegal. For example the digital age enables us to re-edit music and movies, to weave new material into them, to manipulate them in countless ways—that is, to use them as raw material for continued creativity. It is technologically feasible for any PC owner, but legally impossible for anyone without vast amounts of capital to purchase rights.

More generally, when culture is private property, then artistic creation, which draws on the culture, must trespass or be severely limited. When the propertization of the cultural space, the enclosure of the cultural commons, becomes total, it will become impossible to create art at all without asking permission. Already something like this is happening in the film industry, as Lawrence Lessig explains in *The Future of Ideas*:[12]

Wait! In an extremely ironic turn of events, just as I was about to quote Lessig I saw a notice that said, "No part of this excerpt may be reproduced or reprinted without permission in writing from the publisher." Not wishing to go through the hassle of obtaining such permission, I'll paraphrase instead. Lessig quotes filmaker Davis Guggenheim that in making a film it is necessary to procure rights to all the images it contains. A poster in the background, a Coke^{tm} can, a piece of furniture (because it was created by a designer), a building (whose image might belong to the architect) might all require copyright clearance. Soon, there will be nothing left to film, except with permission, as the propertization of images proceeds toward totality.[13] Increasingly, says Guggenheim, the lawyers determine the content of a film. These issues generate actual court cases. Lessig gives several examples: *Twelve Monkeys*, suspended when an artist claimed that a chair in the movie resembled a piece of furniture he'd designed. *Batman Forever*, threatened when an architect demanded money for the image of a courtyard the Batmobile drove through. *The Devil's Advocate*, because of some sculptures that appeared in the background.

There, I think I did that without breaking any laws or encroaching on Mr. Lessig's property rights to those sequences of words he now owns. It wasn't easy though, because the clearest articulation was the one Mr. Lessig employed himself. Similarly, a filmmaker can try to use only images from the public domain, but because modern life happens in a proprietary realm, certain sentiments are difficult to express. McDonald's, for example, is a potent cultural symbol; there is no good substitute. The filmmaker must settle for a reduced stock of images, just as I had to settle for a reduced stock of phrases in paraphrasing Lawrence Lessig.

A similar situation pertains to technological innovation, as the free use of scientific discoveries is increasingly limited by legal ownership over their applications. In the realm of biomedical science, the traditional free exchange of information, strains of organisms, and so forth is crumbling because genetically engineered microorganisms can now be patented; that is, made into property.[14] In the past, scientific progress was based on cooperation more than competition: People shared new results in journals and through informal exchange, protected by their university salaries from the compulsion to profit financially from their work. Now that careers often depend on corporate research funding and now that researchers and universities have a pecuniary interest in patenting results, a new era of secrecy has dawned. It originates from the profit potential inherent in the patentability of new pharmaceuticals and genetically modified organisms. Contrary to the Founders' intentions, intellectual property in this case stymies scientific development, because science progresses by building on the results of previous science. When the incentive is to conceal instead of to share, this is no longer possible. What else could we expect from the conversion of scientific knowledge into property, which is nothing else than the enclosure of part of the commons into a private holding?

The moral case against property rights over ideas rests on a similar basis. Ideas do not just appear out of thin air. They coalesce from the raw materials of the cultural commonwealth, and depend on the ambient culture for their resonance and relevance. Mickey Mouse, for example, is appealing mostly because (A) he is a small, humble creature—a mouse, and (B) he looks cute, an effect achieved by drawing him with large eyes, large ears, and a small body relative to his head. Mickey's appeal, therefore, and his tremendous commercial success, draw upon factors—cultural raw materials and transcultural features of human perception—that are part of the commonwealth.

Recently there was a court case between Exxon and Kellogg over the cartoon tigers used to market their brands: the Exxon gas tiger and Tony the Tiger. The two look quite similar (and similar as well to "Tigger" of Winnie the Pooh fame, and to "Hobbes" from the comic strip Calvin & Hobbes). One of these two corporations claimed that the other infringed on its trademark rights, in effect asserting ownership over a certain rendering of an animal—the tiger—that has very deep cultural resonance. The idea and image of a tiger, with all its associations of strength, power, and beauty, is a kind of cultural capital. Since there are really only a limited number of ways to draw a tiger and have it still look like a tiger and carry these associations, the litigants were in effect asserting private ownership over an item of cultural capital that was once public, unowned.

Whenever someone gets an "idea for a story," is it really original? Perhaps all stories are only variations on a handful of archetypal plots. And certainly all stories draw from the storyteller's experience with real people and real events. Certainly she may put them together in a unique way, but can she in all modesty lay claim to ownership of that story? Reading academic exegeses of literary works it is impossible to avoid the feeling that the original authors could not possibly have intended all that. Such intent is unnecessary if they are but channels, and not creators, of their stories. I agree with Lewis Hyde that any creative work comes from a source greater than ourselves. Through us, the Muse delivers great archetypes and universal themes. To claim ownership of them is to subordinate a greater to a lesser. The true artist is humbled before his work.

I say this as an author myself trying to earn a living from my books. I have expressed some ideas in a new way, and applied some ancient ideas to certain aspects of modern life, but I do not presume to own the ideas themselves, nor the words used to express them? How dare I presume? In one form or another, these ideas have inhabited human minds for thousands of years; in each era they take on a new appearance. It would be hubris to enclose part of our common human heritage as my own personal property. That is one reason I'm putting the entire content of this book on line. And while I'm at it, I hereby give everyone permission to "reproduce, reprint, store in a retrieval system, or transmit by any means, electronic, mechanical, photocopying, recording, or otherwise" this book, as long as it isn't for commercial purposes. I do not give you permission to repackage and sell these words—any copy you make, you may only give away for free.

The current rush to own melodies, words, images, algorithms, and even the code of life itself is a new enclosure movement, much akin to the privatization of the village commons by the English Enclosure Acts of the 17th and 18th centuries. Words that were once public and available for all to use are now the property of corporations and other institutions, unavailable to others for certain uses. "I'm lovin' it" is the property of the McDonalds Corporation (it has the exclusive right to a certain use of that phrase). "Make every drop count" is the property of Coca-cola. "Ideas for life" is the property of Panasonic. "Always" is the property of Wal-Mart. "Making life better" is the property of Penn State. You can still use these words for most purposes, but not as a slogan for your organization. Penn State has taken them. Nike has taken "Just do it." Donald Trump has taken "You're fired!" These are but a few of the tens of thousands of common phrases that have been expropriated from the public language.

Even the words "Love thy neighbor," which appear in both the Torah and the New Testament and lie at the foundation of moral philosophy, are now subject to ownership. In 2001 a Michigan jewelry merchant using "Love your neighbor" sued a Florida charity named "Love thy neighbor" for trademark infringement, claiming that the similar name confused customers and resulted in lost profits.[15] The defendant said he was flabbergasted that it is possible to register rights to an expression that "has been around for 5,700 years." Legalisms aside, does the plaintiff really have a right to this phrase? Did he create it and invent the concept? Or has he merely cordoned off a part of the cultural commonwealth and claimed it for his own use and profit, solely on the justification of having gotten there first?

What of music? Do we just invent new songs out of nothing? Do we really invent new melodies and tell new stories? Or are these things cobbled together out of the myriad works of creators past: free-floating ideas plucked out of the noosphere and arranged to appeal to a given audience? To a medieval minstrel, to claim ownership over a song and demand others not play it would have seemed a brazen conceit; to a tribal storyteller such an attitude may have seemed well nigh blasphemous. Stories and songs were sacred gifts from the gods.

A similar argument pertains to technological inventions—they too are born of a complex creative matrix, inspired by ideas ambient in the culture. Hence Lewis Mumford defined the patent as "a device that enables one man to claim special financial rewards for being the last link in the complicated social process that produced the invention."[16]

To claim ownership over what is not rightfully yours is theft. To stake out a permanent claim on the commons of our cultural heritage is to steal from us all. Seen this way, students downloading music and movies are simply resisting our era's new land grab on the intellectual commons. They sense what Thomas Jefferson articulated so well: no person has a moral right to own an idea.

It is tempting to say that the riches of our cultural heritage should remain "public property", but to use this term reinforces a dangerous underlying assumption that the word "property" can be made to apply to them at all. The cultural commons, like the village commons before it, was once the world outside the sphere of human ownership. Today, there is no such sphere. The separate human realm of the owned has expanded to cover all.

I am advocating a revolution in human beingness that goes much deeper than a mere Marxist shift of property from individual to common ownership. Property itself will become an outmoded concept.

A great hubris is implicit in the very concept of ownership. Ownership subordinates a thing to human being, makes what was at large and wild into a possession, something of ourselves. Wendell Berry has said, "He (God) is the wildest being in existence. The presence of His spirit in us is our wildness, our oneness with the wilderness of Creation. That is why subduing the things of nature to human purposes is so dangerous and why it so often results in evil, in separation and desecration."[17]

Of course, the things of nature, culture, and spirit that we attempt to subdue by making them into property are not subdued in reality, but only in our perception. Only in our perception are they reduced to possessions, something no longer larger than ourselves but smaller, no longer the mystery of the unknown but the catalogue of the owned. For what is it for a thing to be mine? Do I change its essence thereby, by imagining I own it? All of what we call "intellectual property"—patents, phrases, text, images, sounds—are pieces of the cultural universe that we separate out and make private. What is it to make something private, to make property of something? What really changes? Does the song know when its royalty rights have been transferred? Does the story know when the copyright has expired? What has changed is really only our collective perception of it. Property is, after all, a social convention, an agreement about someone's exclusive right to use a thing in specified ways. However, we seem to have forgotten this. We seem to think that property *belongs* to us in some essential way, that it is *of* us. We seem to think that our property is part of ourselves, and that by owning it we therefore make ourselves more, larger, greater.

Significantly, it is force that backs up the social agreement defining property. If I defy that agreement by trespassing on your land, you can threaten or apply physical force (through police proxies, for instance) to maintain your socially-defined exclusive rights. Property encodes power relationships among human beings. Less commonly recognized is that property embodies a relationship to the world at large no less fraught with force and control.

Changing our concept of ownership and abandoning the conversion of the universe into property involves a fundamentally different conception of ourselves in relation to the world. It involves a letting go, a relaxing of boundaries, a trusting in what was once so quaintly known as "Providence" instead of reining in the whole world, bringing it under control, *making it ours*. In the propertization of everything we again find an echo of the technological program of perfect control and the scientific program of perfect understanding upon which it is based. Our socioeconomic system and our way of life is inseparable from our beliefs about reality: our ontology, our cosmology, and our self-definition. It is all about making it ours, conquering nature, conquering the mystery. To conquer means to make ours, and that means to own. To assign property rights to all reality is indeed to become the lords and masters of the universe.

The world of property is precisely the "separate human realm" whose emergence we have traced back to fire, stone, language, and number. It is a world labeled, numbered, and subordinated to human ownership; it is a world whose value we define in terms of money—a purely human abstraction, and a proxy for the interest of the separate self. As more of the world enters the money economy, the separate human realm grows and the wild shrinks. All becomes fuel for the fire that defines the circle of domesticity, blazing so high now that nowhere can we avoid its heat.

The realm of the owned continues to expand. Corvette owns a certain shade of red, UPS a certain shade of brown. These are parts of the electromagnetic spectrum. Did these companies create these colors, or just enclose them, wall them off and call them theirs? Harley-Davidson has done the same thing to the sound of its revving motorcycle engine. To the extent that music, images, and text can be digitized, intellectual property comes down to owning numbers, a natural next step after the conversion of the world to numbers described in Chapter Two. Thus we have extended ownership to the fundamental stuff of reality: electromagnetic waves, numbers, DNA, sound waves. That these are considered "intellectual property" again bespeaks our hubris, that we presume dominion over something far prior to human beings. We have merely taken what was there already, the substrate of reality. Whether words or land, at the beginning of the chain of purchase and sale, someone must have simply appropriated it. As P.J. Proudhon proclaimed in 1840, "Property is theft."

Our possession of this world we have made ours is a grand larceny. And its victim is the commonwealth: the land, the genome, mother culture. We could say that property is what has been stolen from us all, or from Nature, or from God. In any event, our progressive ownership of the world naturally and inevitably accompanies our progressive estrangement from the world, so that in the end we languish in the prison of me and mine which, no matter how great our possessions, is far narrower and dingier than the unbounded Wild from whence we came.

Natural Capital

The same people who brought the sewer pipes turn another small piece of the Pine Barrens into money every few weeks. Probably when there's nothing left, and the last of the watershed is poisoned, the people who are responsible for killing the beaver will know the helpless sense of irreversible loss Rick and I had felt.

-- Tom Brown, Jr.

In no realm is the campaign to "make it all ours" so concrete as in the conversion of natural capital to financial capital. Natural capital refers to the earth itself: the earth's minerals, land, soil, oceans, fresh water, genomes, and biota; everything, that is, that was not created by human beings. I would like to say "natural resources", but in this locution lies the very assumption we are examining. To think of the planet as consisting of "resources" already implies it is ours, defines the world in terms of its usefulness to us, and sets us separate and apart in relation to it.

The most commonly recognized form of natural capital is fossil fuel, comprising coal, oil, and natural gas, which we are depleting at an accelerating rate. These are Thom Hartman's "last days of ancient sunlight" upon which industrial society depends. However, while the energy crisis certainly does contribute to the convergence of crises that will undo the Age of Separation, it is only a tiny part of the picture. Other forms of natural capital are probably even more significant. First is the earth's capacity to absorb the toxic effluvia of our conversion of wealth to money: the radioactive waste and the pharmaceutical residues, the agricultural runoff and the incinerator sludge, the sulfur dioxide and the ozone, the CFCs and the PCBs. Once localized or subclinical in their effects, these poisons now threaten entire bioregions and ecosystems. Wait, did I say "threaten"? That isn't the right word anymore, because many of them are already dead, and many more are dying. As the damage intensifies, the consequences will be harder to avoid. Indeed, they are already upon us, in the rise of birth defects, autoimmune diseases, and cancer. But that is nothing compared to the possible disruption of climactic systems and the mechanisms of atmospheric homeostasis that depend on a healthy ecology.

Another urgent problem is the loss of fertile soil and clean water. Deserts have been swallowing fertile land ever since linear methods were first applied to agriculture. Ancient Sumer, the cradle of civilization, was once known as the Fertile Crescent. Now it is the desert of Iraq. North Africa, once the breadbasket of the Roman Empire, is now wholly given to the great Sahara. The same fate awaits the United States, where topsoil losses of two billion tons per year are seventeen times higher than the rate of soil formation,[18] and especially the drier West, where dependency on irrigation exposes vast tracts to salinization. Meanwhile, chemical agriculture contributes to an alarming depletion of soil minerals, as USDA statistics on fruits and vegetables—which have as little as ten or twenty percent the mineral content of two generations ago—demonstrate.[19] As for water, water tables are dropping on every continent. Great rivers such as the Yellow and Colorado routinely run dry, and the Nile and Ganges barely make it to the sea. The Aral Sea has shrunk to half its former size, killing all fish due to higher salinity.[20] Meanwhile, according to the World Medical Association, over half the world population lacks access to potable water,[21] while even in America tap water is routinely treated with chlorine to destroy waterborne pathogens. Increasingly, it is also polluted with various carcinogens, mutagens, and endocrine disruptors. Of course, like any other destruction of the commonwealth, this presents a great business opportunity. The fastest-growing beverage category is now bottled water, now a commodity like everything else, while huge corporations vie for control of newly privatized water utilities around the world.

Sometimes people object to these doom-and-gloom scenarios by pointing out examples like the Great Lakes where water quality has actually improved. Yes, there have been temporary, localized improvements. Too often, though, the pollution has merely been exported. Factories in China, producing the same heavy industrial products once made in Cleveland and Milwaukee, spew the same pollutants as before into different rivers. Air quality may have improved in Los Angeles, but the worsening air in Bangkok, Manila, and Shanghai more than offsets that improvement. Some pollutants are now banned, and some practices abandoned, but others more insidious have taken their place. Nonetheless, mine is not actually a doom-and-gloom viewpoint. I am an optimist. My optimism does not depend, however, on ignoring the gravity of our situation. It will take an extraordinary transformation in human beingness to create the beautiful world my heart tells me is possible.

I'll touch upon just one other category of disappearing natural capital: biodiversity. The totality of the planet's plants, animals, bacteria, and fungi contain an enormous reservoir of genetic material that, once lost, can never be recovered. Typically, environmentalists cite two reasons for preserving this irreplaceable form of wealth. First, diverse ecosystems are more robust and better able to exercise their function in the homeostasis of the biosphere—to manufacture oxygen, for example. Second, the innumerable species presently going extinct might contain valuable medicinal compounds, genetic material, or other components useful to human beings. These are good reasons, but they understate the case. Science is learning that much, if not most, genetic material is never expressed. Once considered junk, this DNA is potentially a reservoir of adaptability that may become essential to the planet's survival and/or transformation. Of course, the idea that genes are waiting, quiescent within the genome for the time for their expression violates basic principles of Darwinian evolution (as there is no selection mechanism). I will develop this line of reasoning more fully in later chapters.

When people become aware of the catastrophic destruction of natural capital, they either go into denial ("It couldn't be that bad" or, "I can't do anything about it"), or they quite rightly become extremely alarmed and think, "We must mend our ways." However, the origin of "our ways" is much deeper than we think. We can dream of tighter regulations or more reverent attitudes toward the earth, more responsible governments and better technology, but the tragic fact is that our civilization is constitutionally incapable of reversing the annihilation of natural capital, or even slowing it down. That destruction flows inevitably from money system and the sense of self that underlies it.

Let me repeat that: Our civilization is constitutionally incapable of reversing the annihilation of natural capital, or even slowing it down. Get used to that. When we really understand that, the project of reconceiving civilization itself will gain powerful impetus.

The subjugation of all the earth's land and everything on and under it starts with a conceptual separation, an objectification of the world that facilitates its conversion first into "resources", then into property, and finally into money. Older history textbooks speak of "How the West was won" while newer, more progressive editions might give lip service to the idea that the North American continent was stolen from its original inhabitants, the Native Americans. From the indigenous perspective, though, the true crime is much greater than that. The crime of the Europeans went far beyond murdering the Native Americans for "their" land, which perhaps would not have been unthinkable to the indigenous mind—after all, territorial disputes are not unknown among hunter-gatherers. The crime, the sin, the sacrilege, was to presume to take the land not from humans, but from something much greater: from Nature, God, the spirit that moves all things.

America was not stolen from the Indians, because the Indians never owned it. The land was not property, and the crime was far worse than mere theft. While pre-agricultural peoples often have a tribal territory, they would be appalled at the idea that land could be owned. Is not the earth a being greater than any human, or even any group of humans? How can a greater *belong* to a lesser? To presume to own a piece of the earth, to say it is *mine*, is from the indigenous perspective a sacrilege so audacious as to be unthinkable. To reduce the earth to property and eventually to money is indeed to make a greater into a lesser, to turn the sacred into the profane, the divine into the human, the infinite into the quantified. I can think of no better definition of sacrilege than that.

If the mass murder of the world's indigenous peoples constitutes a "crime against humanity," then the objectification of the world constitutes a crime against Nature, God, and spirit. In fact, the former crime flows naturally and inexorably out of the latter. How much easier it is to kill someone or something that we see as an Other. The original source of both crimes is nothing other than separation itself, a process as old as time that received successive boosts from fire, agriculture, the Machine, and science. But it was agriculture that most powerfully accelerated the conversion of land into thing.

It is easy to see how the concept of land ownership arose with agriculture. Agriculture involves applying labor to a piece of land: tilling the soil, planting seeds, pulling up weeds, watering the plants, fertilizing the soil, and so forth.[22] A farmer has a right to the fruits of his labor, and would certainly object to someone coming by and "gathering" the abundant edible crops he had grown there. However, there is still a huge conceptual leap from owning the crops to owning the land itself. In some societies the only "owner" of land was the king. As the king was considered a semi-divine being, this was tantamount to saying that land was beyond the domain of human ownership. Early ownership of land was probably more akin to a generally-recognized right of stewardship, ensuring people had an incentive to work the land and could justly benefit from their labors, just as early intellectual property rights meant to encourage creativity by bestowing the temporary right to profit from an idea, but *not* ownership of the idea itself.

Whether in the case of land or intellectual property, the transformation from a right-to-benefit into outright ownership was a gradual one. Let's keep in mind that this was a conceptual transformation (the land doesn't admit to being owned), a human projection onto reality. Land ownership (and indeed all forms of ownership) says more about our perception of the world than about the nature of the thing owned. The transition from the early days when ownership of land was as unthinkable as ownership of the sky, sun, and moon, to the present day when nearly every square foot of the earth is subject to ownership of one sort or another, is really just the story of our changing view of ourselves in relation to the universe.

The ending of serfdom in late medieval Europe is a case in point. Before feudalism gave way to a money economy and the commons migrated into private hands, land was generally not a fungible asset. Lewis Hyde writes,

> Whereas before a man could fish in any stream and hunt in any forest, now he found there were individuals who claimed to be the owners of these commons. The basis of land tenure had shifted. The medieval serf had been almost the opposite of a property owner: the land had owned *him*. He could not move freely from place to place, and yet he had inalienable rights to the piece of land to which he was attached. Now men claimed to own the land and offered it rent it out at a fee. While a serf could not be removed from his land, a tenant could be evicted not only through failure to pay the rent but merely at the whim of the landlord.[23]

Complementing the reduction of land to just another thing, in the same period Martin Luther and his ilk absolved worldly rulers of any obligation to rule by Christian principles, accelerating the separation of reality into two realms, worldly and divine. We became lords and masters of the real world, banishing God to a materially inconsequential realm of the otherworldly. Land, like the rest of the world, was no longer sacred. Of course, the peasants resisted their dispossession from the commons, fomenting the bloody struggle known in Germany as the Peasants' War. It is a struggle reenacted time and again around the globe, whenever people resist the incursion of property rights into yet another sphere of human relationship. As Hyde puts it, "the Peasants' War was the same war that the American Indians had to fight with the Europeans, a war against the marketing of formerly inalienable properties." Relationships, land, water… what's next? The air? The sky? Here is a corollary to the eternal business idea of the previous section: Find something so fundamental to life that no one could imagine it being separated off as property. Then by some means deprive people of it and sell it back to them.

As recently as 17th-century England and frontier America, there was much land that was not owned at all—the commons. Today we have reached the opposite extreme from the days of hunter-gatherers: it is hard for people to conceive of land *not* being owned. Surely someone must own it? It can't just *be there* unowned, can it? I like to stump my students by asking them for examples of unowned land. National parks, state forests and the like don't count, because they are owned by the government, not unowned. Basically, the only unowned land, the only commons still remaining in present-day America is the street. The street is public space, maintained and regulated by municipal governments, but not the subject of any deed. Other than the street, and perhaps Antarctica, all the world's land surface has been made into mine, yours, his, theirs, ours.

When I say that as hunter-gatherers we were embedded in nature, part of nature, or not separate from nature, it does not mean only that we "lived more naturally". It means we did not see "nature" as a separate thing from ourselves. The dualism between man and nature did not exist. Tom Brown Jr. speaks of a boyhood experience lying beneath the stars: "We lay for an hour looking up into the black, star-filled sky until at some point, although I never closed my eyes, I was no longer lying in a field. I had become part of a pattern that the stars and the breeze and the grass and the insects were all part of. There was no awareness of this until I heard the first deer coming through the grass. Then I was suddenly aware that I had been lying there without thoughts or sensations other than just *being.*"[24]

Aside from being conceptually embedded in nature, hunter-gatherers also had a practical reason for having a very limited concept of property: They were largely nomadic and could not carry many possessions with them. Accumulation of possessions was impossible. Agriculture was the practical and conceptual foundation of the age of property. Practically, it permitted a sedentary lifestyle and the accumulation of possessions. Agriculture also gave rise to greater concentrations of population, social hierarchies, and specialization, all of which further propelled the development of property rights and money. On the conceptual level, the fact that a farmer applies work to the land rather than simply taking what the land has to offer contributes to the feeling that the land is his. More generally, agriculture and the entire corpus of technology that developed in agricultural civilizations sought to control or improve upon nature, thus turning nature into an other, the object of manipulation, and thus the object of ownership.

Once the dualistic man/nature conception had been established, it is quite understandable how people would feel free to convert that of nature into that of man. That is precisely the process that the term "natural resources" implies. Resources: things for us to use. Or to conserve and use later, or not to use at all. In any event, things for us.

The millennia-long conversion of natural capital into financial capital is this process in action. Environmentalists have told us for many years that we are "living off our capital", creating an illusion of wealth through the unsustainable consumption of our natural capital. The root of the problem is that our economic system does not recognize the things of nature as wealth. It is written into the dynamics of business that natural capital be converted as rapidly and efficiently as possible into financial capital, regardless of the ethics, morals, or good intentions of business-people. (Government regulation can at best slow down this conversion by offering obstacles to its efficiency.)

One of the gravest errors activists make is to demonize their opponents; for example, to assume that businesspeople and "the rich" are more greedy, less conscious, or less spiritually evolved than themselves. But greed is a result, and not a cause, of our economic system. Imagine that Acme corporation owns a mine but the CEO, being a nice guy, refuses to operate it because to do so would deplete the precious groundwater of the region. A mine that would otherwise produce $12 million a year lies abandoned. Acme's profits, being $12 million lower than they could be, are reflected in the total value of the company, which is perhaps $200 million lower—$1 billion instead of $1.2 billion. When the shareholders find out about the mine, they put pressure on management to exploit it, perhaps even suing the CEO for failing to act in the financial interest of the shareholders. If he resists, tremendous pressure will be brought to bear. Competitors who have no such scruples will undersell Acme's mine products, and the company may attract the attention of a corporate raider who will leverage the $200 million equity that the mine represents into a $1 billion loan from an investment bank, buy the company, exploit the mine (perhaps selling it for $200 million), pay back the loan with $50 million interest to the bank, and keep $150 million for himself.

The same fate befalls a company that spends extra money on pollution control, higher wages, or any other socially or environmentally responsible behavior. The reason is that the costs of such behavior are external to the company's balance sheets. Such costs are called "externalities" in economics —the financial counterpart to technology's making an *other* out of nature.

Until natural capital is converted into a commodity denominated in dollars, it is invisible to our economic statistics and balance sheets. It is hard to even articulate the value of nature otherwise: hence the profusion of environmentalist arguments based on cost-benefit considerations.

Why should we save the rainforests? Because of all the medicines that might be produced from the undiscovered plant species there? Because of the economic value of their contribution as a carbon sink? The economic value of their pollinating species? In essence, these arguments try to persuade us to protect the environment because the long-term cost to the economy of environmental destruction far exceeds the economic cost of preservation. Well-meaning as they are, such arguments actually exacerbate the root problem, which is the basic Benthamite assumption that goodness can be quantified, that the way to make life better is to maximize financial returns, and, even more deeply, that nature can be made ours, and, yet more deeply, the illusion of our separateness. Such arguments grant the disastrous premise that nature is indeed a thing, best disposed of according to the financial consequences.

Cost-benefit arguments for environmental protection have the further disadvantage that they are usually ineffective even as a short-term tactic. I am inspired in this regard by Gandhi's exhortation to "appeal to their reason and conscience," and by Edwin O. Wilson's invocation of a universal "biophilia"—a love of living beings—innate to each one of us, however deeply buried. In the long run and probably even in the short run, it may be more effective to appeal to people's sense of beauty and their desire to do the right thing. "Let's save the environment because otherwise it will cost too much" is an appeal to a baser instinct. It disrespects its audience by assuming that greed is their strongest motivation. (It is especially counterproductive when facing people who stand to gain financially from consuming natural capital.) It is also on some level dishonest: I do not know any environmentalist motivated by the long-term economic savings of environmental protection. Let us instead appeal to what is highest in other people: their sense of rightness, beauty, and justice; their desire to be a good person; their longing to enact their innate love for our beautiful planet. The greed behind the plundering of the planet, and the insecurity and anxiety behind the greed, is after all a product of our money system as well as an inevitable effect of our separation from self, spirit, nature, and each other. It is not our true essence.

The conversion of all forms of wealth into money violates our sense of beauty, rightness, and purpose. It has made the world uglier. In the realm of art, epithets like "commercial art" or "a sellout" are not compliments. Nor would most people maintain that a lumber yard is more beautiful than a forest, or that the beauty of a well-constructed highway exceeds that of the landscapes destroyed by the quarries, mines and so forth that provided its raw materials.

On a personal level, how often do issues of practicality—which more often than not involve money—seem to interfere with our longing to live a beautiful life? Usually, rational economic interests seem to directly contradict Joseph Campbell's urging, "Follow your bliss." I cannot count the number of times when I didn't do something in the most beautiful way because I "could not afford to."

Just as the conversion of the world to money makes less of the world, so does the conversion of life to money (Time is money!) make less of life. Adam Smith's economic man, making choices according to his rational self-interest, affirms the presumption that the value of all things can be tabulated in terms of money. All things admit of a value, a quantity, a measure. Here is the translation (and reduction) of the immeasurability of the Wild into the numbers of human abstraction. And thus the wild, nature, is brought under control. Here is another parallel with the program of science, which seeks, through its equations, to reduce nature to human terms.

The disregard that modern human economies—and the science of economics—have for the environment is inseparable from our fundamental understanding of self and world. As Herman Daly says, the materialistic, mechanistic worldview implies that "the natural world is just a pile of instrumental accidental stuff to be used up on the arbitrary projects of one purposeless species."[25] If there is no real purpose beyond personal survival, comfort, and pleasure, if the order and beauty of nature is mere accident, and life a seething scum on an insignificant ball of rock hurtling through space, and all existence "a sound and a fury, signifying nothing," then what does it matter, really, the fate of the earth?

At this point it is customary to trot out God as an imposer of meaning, purpose, and sacredness onto our world of dead matter. Such a response, unfortunately, leaves matter no less dead, life no less mundane, and beauty no less arbitrary. When God is separate from creation and spirit is separate from matter, we naturally construe teachings such as "be fruitful and multiply" as divine license to pillage and wreck in converting natural wealth into personal, financial wealth. Too often, the theistic treatment of our planet is no less destructive than the atheistic, and even more.

A non-dualistic religion sees things differently. Wendell Berry, arguing from scripture, puts it eloquently: "Creation is not in any sense independent of the Creator, the result of a primal creative act long over and done with, but is the continuous, constant participation of all creatures in the being of God."[26] Instead of granting us divine license to plunder Creation, non-dualistic religion would see the destruction of nature as, to quote Berry, a "most horrid blasphemy. It is flinging God's gifts into His face, as if they were of no worth beyond that assigned to them by our destruction of them."[27] By converting natural capital into financial capital, we are assigning a monetary value to the work of God—excusable, perhaps, if matter is separate, "mundane", unspiritual, but not if nature is the sacred, ongoing expression of divine Creativity. In Psalms 24:1 it is written, "The earth is the Lord's and the fullness thereof: the world and they that dwell therein." Seen in this light, the conversion of the earth into assets, resources, and property that started with agriculture is nothing more than an attempt at robbery, a sacrilege of the highest order. Only the illusion of our separateness gives us the temerity.

Sadly, because our separateness is indeed only an illusion, we cannot escape the damaged and diminished world that our despoliation has rendered. We can, however, start to reverse the uglification of the world by using our creative gifts, including our technology, toward the divine purpose for which those gifts are meant: to create beauty. If the things of the world are put here for our use, as some interpret the Bible to say, then what other use could it be besides as a participation in and an extension of God's ongoing work of creation? A fallen branch can be made a flute. Every time we cut down a tree or dig a new quarry, the question we should ask ourselves is, "Does this augment the divinity of Creation by making it more beautiful? Or does it detract from the divinity of Creation by making it more ugly?" And the answer must apply to the whole of Creation, not just a part of it; not just to the sleek new automobile, but to the mine pits, polluted air, and ravaged landscapes that go hand in hand with present industrial processes.

What shall we make of this beautiful world into which we have been born?

Spiritual Capital

It is not only the physical and cultural "wild" that has been converted and sold off as property. We have done the same to the wild within ourselves: our imagination, creativity, attention span, playfulness, and spontaneity. I call these things spiritual "capital" because they are indeed productive assets, generators of wealth. In describing their cooptation and conversion into financial capital, I describe how the ascent of humanity has in fact eroded our humanity, made us lesser beings.

It is a process that has been going on for some time, at least since the origin of agriculture. Interestingly, there are allusions to this degeneration of human capacity in several ancient spiritual traditions, particularly Taoism. While perhaps better understood as metaphors for the submerged capacities of childhood, Taoism abounds with references to ancient humans who possessed wisdom and abilities far exceeding those of people today. Similar references exist in ancient stories of the Golden Age in Greek and nearly every other mythology.

In discussing social capital I talked of technology as a means by which activities that people once did for themselves are turned over to paid specialists. We pay others to do for us what we once did for ourselves. Perhaps the most profound aspect of this shift is in a realm that few of us have ever considered as a form of capital: the capacity to imagine and play.

I was fortunate enough to experience childhood just before the age of video games, Nintendo, Gameboy, and the like, and was protected by idealistic parents from at least some of the damage of television. I remember spending hours in my room with my stuffed animals, rock collection, and other toys, weaving elaborate stories about them. To each I would assign a personality, each would become a character in an imaginary world existing either solely in my own head or among siblings and playmates. Today, this creative function of imagining worlds and the characters in them has been taken over by faraway adults in television studios and software development companies who provide children with the ready-made worlds and personalities of TV shows and video games. A capacity of the human mind has been stripped from the individual for profit's sake.

And what shoddy substitutes these commercially motivated worlds are for the spontaneous creations of the child's mind! Most of them are worlds of dichotomous good and evil where problems are solved by violence, that are devoid of nuance and detail, and that are disconnected from the rest of the world. Worse yet, they are (unlike the mind of a child) finite, limited by the medium of the story. They therefore constrain the freedom of the child's mind to develop and play out various elements of the unconscious.

If childhood play is practice for life, then our television raises children to be passive consumers of it. As for video games, they condition us to the mindless acquisition of meaningless rewards (points), to the destruction of generic enemies, and to accepting choices defined by remote others, the programmers of our lives.

The kind of adult that results from a childhood bereft of the opportunity for spontaneous self-directed world-making is someone who will continue to be vulnerable to stories created by others. Not only will he always be in the market for entertainment, but he will be easily manipulable by politicians and advertisers seeking to profit from the acceptance of a certain story. Such adults will be compliant subjects rather than active citizens, for they will have had no practice in creating a world for themselves. They will be content with meaningless choices contained in a box. They will be passive, determined by the forces around them much as Newtonian massive bodies move according to the deterministic forces of physics. The economics implicit in our Newtonian world-view has conspired to create human beings whose lack of autonomy and free will confirms the mechanistic conception of dead matter moved by impersonal forces. Our cosmology has been projected onto ourselves.

Meaningless choices contained in a box. Can you think of a more apt description of a video game or television?

While we speak of children "playing" with their video games, these are not so much the objects of play as they are *substitutes* for play, and they are merely one small aspect of the disappearance of play from our culture. Play is properly a creative activity: Give a child some blocks and they become trucks, a city, a forest, a zoo. Put a few children together and they create worlds of the imagination that incorporate whatever materials are available, both physical and cultural. The play worlds that children create, that together constitute the Kingdom of Childhood, are practice for the creative work of the empowered adult fashioning a good life and contributing to a beautiful world. We are meant to be creative beings, not just to live out the lives that are handed to us. As Joseph Chilton Pearce puts it, "As a child, reality is whatever one makes of it."[28] This is potentially true for adults as well, but unlikely if we have never experienced it as children.

Unfortunately, the types of toys and activities we provide our children give them little liberty to construct their own kingdoms. Childhood has become a process of breaking the human spirit so that we are indeed content to live the lives that are offered us by the modern economy; or, if not content, unable to imagine anything else; or, if able to imagine, disbelieving of our ability to create a different life. Modern life therefore leaves us numb, defeated, or hopeless. How has this happened?

First, in the last fifty years the number and variety of toys available to children has proliferated into piles and piles of plastic junk, making it less necessary for them to apply their imagination and creativity to turn ordinary things of the world into toys. Joseph Chilton Pearce observes, "When today's toddler sees her mother making cookies and wants to take part, she need not resort to jar top, stick, and mud, like some primitive. She probably has a complete miniature kitchen, scale-model perfect with battery operated appliances."[29] Imagination is less necessary.

Even worse, today's toys are not passive objects of play that can only come to life in the child's mind; they are animated by electronics. The "toy" takes charge and the child assumes a passive role. In most video games the child does not create the story, but instead moves through a story that has been created for her. (Ah, how like modern adult life!) If there is room for exploration, as there is in some virtual worlds, it is still—unlike the real world—finite and *programmed.* The world is finite and its limits set by someone else. The only mysteries are those that have been fabricated and doled out.

I remember as a boy spending hours and hours with my globe, tracing the mountain ranges with my finger, connecting the oceans, making up stories about the different countries, comparing sizes and latitudes. Today's toy globes come complete with computer chips and voice recordings that turn the child into a passive absorber of "information". Advertising and packaging for the profusion of computer-enabled devices on the market today trumpet them as "educational", as if education were the acquisition of facts. It would seem that for children today, a toy cannot be fun if it doesn't "do" anything. A toy that just sits there and depends on the child to do something to it is boring.

But what has actually happened is that the child's imagination has withered. Children who never had toys that do the play for them are rarely bored. Boredom is actually a kind of a withdrawal symptom from the addiction to the ever-intensifying sensual stimulation provided by today's electronic toys, games, and media.[30] The TV show is over, I'm bored. Play some Gameboy. I'm sick of that now, I'm bored. Eat some cookies. I'm full now, I'm bored. Moving from one stimulus to another, the mind develops a higher and higher tolerance to stimulation. And so the dosage must increase: video games that are even more exciting, television that is even faster-paced, more dramatic. Thus, movies and television programming have over the last few decades become increasingly fast-paced, the editing faster, the scenes shorter, the special effects more dramatic. Old movies are kind of boring, aren't they?

Boredom is the beginning of the mind's healing process. Like any withdrawal symptom, it is painful, but very soon latent, undeveloped abilities of the human mind start to manifest. Without diverging into a long discussion of the benefits of meditation, silence, and solitude in nature, suffice it to say that children and adults alike benefit from situations that provide the raw materials of creativity and models of creativity, but leave the process of creativity up to the individual.

This is also how people learn. The ability to learn constitutes yet another form of spiritual capital that has become grist for the money machine.

Boredom is a defining characteristic of modern society for another reason: It represents a hunger for authentic experience. The consumer economy takes advantage of this hunger by selling us artificial experiences, of greater and greater variety and intensity. Television and video games are attractive because they (temporarily) assuage our hunger for experience; yet at the same time they are a primary contributor to the separation from reality. Because they are more lurid, more dramatic, and louder than real life, real life becomes boring in comparison. The child becomes so adapted to the extreme stimuli of electronic entertainment that he loses sensitivity to more subtle sensations. Also, because the barrage of stimuli in electronic entertainment is so rapid, the child's attention span shrinks to the point where he is incapable of patient, sustained observation. In order to command viewer attention, programmers put a "technical event"—a zoom, frame shift, new object—into television content every few seconds, engaging the neurological orientation response. Also called the "startle effect", it is subject to habituation and thus requires increasing intensity over time—effects more violent, louder, more dramatic, more shocking. No wonder reality is so boring. Compare today's bored, jaded teenager with the Piraha described in Chapter Two, who are enthralled watching a boat slowly appear from afar and disappear over the horizon. Today's commercial entertainment embodies the sell-off of our very wonderment at the slow, subtle processes of nature. What is a chick breaking out of a shell compared to cartoon dramas in which the fate of the universe is at stake every half-hour?

I will never forget a trip I took with my wife and children to the aquarium in Baltimore, Maryland. The place thronged with parents dragging their hyperactive children from one exhibit to another, but most of the children simply weren't interested in looking at the fish. They'd look for a few seconds and then run off, pulling on their parents' clothing. Even the sharks failed to hold their attention for very long. Then at one tank I heard a chorus of voices, children's and adults', saying something about "Finding Nemo". Here, finally, was a spark of interest from the children—but it came from a movie. Apparently, one of the fish was identical to a character in a popular children's movie that takes place in a coral reef. The fish in this story were rendered in life-like 3D animation, and you can bet the events of that movie were far more exciting, far more stimulating, than anything you could see in a fish tank. No wonder the children were bored. Reality is boring.

My aquarium experience illustrates a profound phenomenon: the conversion of experience—that is, life—into a consumer product. This is the ultimate conversion of spiritual to financial capital, because it means that *being* itself, the experience of living, costs money. Paid specialists, usually strangers working for enormous organizations, create the life we experience. Of course it has not yet come entirely to that extreme, but consider: whereas once children would go out into forest, lake, and field to experience nature, now the experience is packaged and sold to them in the form of an aquarium or nature TV program. By this process, nature is transformed from part of life into a spectacle even as we rely on others to provide our experiences. In yet another realm of life we have become consumers.

The conversion of life into something that is consumed means that we pass through it instead of creating it for ourselves. Early training for such a life is provided by the activities we offer children, which increasingly amount to little more than to go somewhere and look at whatever is provided and listen to whatever is said. Or, we put them through the motions of some prescribed, programmed sequences of actions in imitation of the creative process, converting learning into the following of instructions. A couple years ago I was a driver for a school field trip to a llama farm. The children were not allowed to touch the animals, their feed, or anything else: all they did was look at the llamas and listen to the employee tell them information. None of them asked any questions, for they were not interested. In fact, it took quite an effort on the part of the teachers to keep order and restrain the children's natural curiosity and desire to explore. They might as well have been watching a program about llamas on television, though I suppose they at least got to smell them.

Our town has a children's parade every year, but the children are not in charge of the parade. They merely walk through it, carrying cardboard animals on poles or dressed in costumes created by someone else. Reread the last two sentences. Do they metaphorically describe adult life too?

I have witnessed "crafts" that children are trotted through at Sunday school and day care, which consist of children doing as they are told, step by step, sometimes without even understanding what they are doing, and rarely taking any pride in their finished work. Indeed it would be too messy, too chaotic, to let the children loose with the raw materials of creativity. Of course, creativity that must be constrained to a controlled area and result in a predetermined product is not creativity at all; it is labor. No wonder the children are so lackadaisical, so unmotivated. They are preparing for an adulthood of following instructions. They are preparing for an adulthood of work, of labor, of being functionaries of the Machine. Your creativity will be constrained to a controlled area and result in a predetermined product. Is that your job description too?

One of the most pernicious manifestations of the control and, therefore, elimination of play has come through organized sports, which started with Little League and extend today into all sports and all age groups. The sandlot, the playground, the neighborhood, and the vacant lot, where children were once free to hammer out their own rules, choose their own teams, and resolve their own disputes, have been replaced by yet another supervised realm where play means going through a prescribed set of motions and where the social interaction is mediated and guided by authority. In a way, the children do not play the game at all: the game plays the children, who are just accessories, placeholders fulfilling functions determined by rules created and enforced by adults. Here play has lost its essential creative nature. I do not know what to call it, but it really is not play any more at all. It is, however, an excellent conditioning method for producing people who look to authority for instructions on how to "play the game". Have you ever felt like you are, still, one of these "placeholders" fulfilling functions determined and enforced by some unchallengeable authority?

Then, of course, there is school, which plays a crucial role in the loss of all forms of capital: social, cultural, and spiritual. Here again, children progress through a more-or-less preset sequence of steps (the curriculum), their natural desire to explore and create confined to specified times, places, and subjects. They read about the world without experiencing it. At school, knowledge comes from the absorption of information, facts, and data provided by authority. We don't develop confidence in our ability to learn for ourselves through first-hand observation.

The restrictions we place on children arise out of two related concerns: safety and practicality, both of which boil down to some version of control. It is not as safe to let your children roam the neighborhood or the forest as it is to keep them at home. Prefabricated, programmed "experiences" are safer than real experiences in the world, which is beyond the human realm of predictability and control. Similarly, the educational objective of gaining the skills and credentials necessary for a secure position as a paid specialist also attempts to avoid the inherent uncertainty of life. It *makes* nature provide instead of trusting nature to provide. It is the old distinction between the agriculturalist working to coax food from the land, and the hunter-gatherer accepting nature's gifts. In this case, "trusting nature" refers to trusting that the natural fecundity of the child as a creative being will result in survival and even abundance. But there is more, because creativity *is* risky, as is unfettered exploration of the world. It is safer to keep Junior at home. But why has safety and security seemingly become our society's highest priority? Just as "homeland security" can and is being used to justify any repressive measure, so also can child safety justify any limitation on children's freedom to create, explore, and direct their own lives. At bottom, the emphasis on safety is a manifestation of survival anxiety, and the belief that the purpose of life is to survive. From it springs our preoccupation with safety as well as the technological program to control reality.

How do we keep our children safe? By confining them to a controlled environment where every possible danger has been eliminated. But this essentially takes away the possibility of real experiences, those that haven't been set up and planned out for them. An experience that is programmed, laid out, all its parameters known by another, is somehow phony, like a public relations pseudo-event. It would seem that we are bent on eliminating risk from life and particularly from childhood. What is risk? It comes from the unknown. Testing the boundaries of our world, which are by definition unknown until we explore them, is inherently a risky activity. Since this is how we learn who we are in relation to the world, the regime of safety, confinement, and supervision in effect prevents children from discovering who they are; it keeps them, that is, from self-realization.

Our controlling of children reflects in two ways the technological program to control nature. First, it implements upon our children the program of security through control, which stems from the survival anxiety implicit in our scientific paradigms and underlying our social structures. Second, and more striking, is this: Our children *are* nature; they represent the very thing we are trying to bring under control. Their spontaneity, creativity, and playfulness, their unruly nature, *is* the wild that we seek to conquer or, to use less inflammatory language, that we seek to mold into the "responsible", "mature" domesticated adult, someone whose behavior rarely sacrifices the rational self-interest of safety, comfort, and security (embodied to a large degree in money) for the creative risks of the unknown. In precise parallel, we use science to subordinate the unknown universe to human understanding, and we use technology to domesticate the world. The motivations for doing so are identical to those we try to foster in the mature adult: safety, security, and predictability.

The subjugation of children to a safe, controlled, programmed semblance of life does not end with high school graduation. By the time we reach adulthood we have become so conditioned to be consumers of a life prepared for us by anonymous others, and so helpless and fearful of creating our own, that we remain forever dependent on fabricated experiences. Another word for experiences fabricated by others is entertainment. In the absence of these, having lost or never developed the capacity for autonomous creativity, we experience the discomfort we call boredom.

In my earlier discussion of anxiety theory, I related boredom to Stephen Buhner's "interior wound" of separation from nature, a hole in the heart so painful that we constantly crave distraction, entertainment, something to take us away from the pain. At the same time, we try to fill in the hole by acquiring more and more possessions, whether tangible or intangible: a futile attempt to fill up the void inside by adding more to the outside. In the context of the loss of spiritual capital, this hole in the heart is nothing less than life itself, our own life, the life we could create for ourselves but that has been sold off to the demands of technological society.

The globe of my childhood was already a step away from real life, the open-ended infinite that is nature, the real world. Already, what I was exploring was a manufactured representation of the world. At least, though, my experience within the confines of that representation was unprogrammed. Through my imaginings, I still *played* with that globe. It did not play me, turn me into its operator, the button-pusher that moves the game from start to finish.

Like a child moving through his lessons at school, making his way through the prescribed body of information that constitutes an "education," like a bunch of confused six-year-olds trying to follow the instructions of their T-ball coach, so do we move through modern life. We go through the steps of a life prepared for us by others, a life that is not our own. As the final step in the conversion of the whole world into money, we are selling off our very lives.

What is important here is not so much money but control. The destruction of the human spirit, accomplished mostly during childhood and maintained to the grave, is another aspect of the taming of the wild, the conquest of nature, the fulfillment of the technological and scientific program at the level of the individual. Life, in other words, has been brought under control, or so we would persuade ourselves. Why else the emphasis on safety, security, and practicality? As observed, these boil down to an emphasis on survival, which, in our specialized society, is a function of money. Again, we are selling off our very lives, purchasing experiences instead of creating them, being run by our machines, our schedules, our clocks and our calendars. Time is money.

There are other aspects to the conversion of spiritual capital into money, the diminishment of the human spirit for the sake of profit. Each human being is born as a magnificent, creative, spontaneous spirit, an enormous spiritual being capable of incredible feats of learning. To reduce that spirit to something willing to occupy one of the narrow, meaningless roles in society as we know it, to make that spirit accept the imitation of life today's world offers (and upon which our economy depends), is an enormous enterprise and a shameful crime. We do not really understand what is happening when, as children and teenagers, vast swaths of our spirit are sold off to the demands of the monetized world. We know only that we have been robbed. Occasionally we meet someone who, by some accident of fate, has survived with a full complement of human abilities intact and we are amazed; we then label that person a genius and dismiss her almost as something not human, and certainly a being in a different category from ourselves. Instead, I urge you to see such people as an indication of your own potential, and a promise of a future in which each person's unique brilliance contributes to the co-creation of a beautiful world. For while such potential can be suppressed, even indefinitely, its spark can never go out. It is in you and in me. Once you know it's there, you will have a burning desire to rekindle it. That is the desire that will change the world.

Time, Money, and The Good

Social, cultural, natural, and spiritual capital are tightly intertwined. When we privatize cultural capital such as songs and images, we also destroy the social capital of the human relationships that create them, and erode the spiritual capital of individual creative capacity. Moreover, as music and images originate in nature's sounds, vibrational ratios, and electromagnetic spectra, they may be considered natural capital as well.

To these four forms of non-money capital could be added many others. Aesthetic capital, for instance, would consist of the unspoiled views, natural landscapes, and unbroken quiet that has nearly disappeared from most of the country. We rarely assign a financial value to such pleasures as a sky full of stars, or see it as a robbery when the lights of civilization blot out all but the thirty or forty brightest. Airlines do not understand what they are taking from the public when their jet trails mar every sky from New York to Los Angeles. Power companies don't appreciate the tragedy of power lines across nearly every panorama. Most children these days have never experienced the stillness of nature. The beeps of construction vehicles backing up and the roar of motors are audible in every park and forest left standing in my own corner of suburbia.

Add to all this the general uglification of the landscape that accompanies superhighways, industrial facilities, superstores, and strip malls, and we have lost a form of wealth so basic that we hardly ever realize it is missing: to open our eyes and look upon beauty. Increasingly, there is no longer anywhere to escape the separate human realm that we have created.

Paradoxically, even as it drives the conversion of beauty into ugliness, money is increasingly required to purchase beauty: a corner of the world insulated from the ugliness without. Houses with a beautiful view command a premium on the real estate market. We pay top dollar for vacation getaways to remote locations. Unique, handcrafted items are more expensive than generic mass-produced ones. Aesthetic wealth, once free to all, is increasingly the exclusive province of the rich. Down the street the bulldozers crush shrubs and trees to put up a new supercenter. Over the hill, a new highway roars. Pavement, squares, and noise proliferate around me. Maybe when I get rich I can buy a house out in the country away from it all.

A related form of wealth is a kind of mental silence whose commoditized form is called mindspace. Our attention is for sale, to advertisers and anybody else with a product or idea to sell. The result is a cluttering of the mental environment with an unremitting stream of commercial messages: on billboards, on buses, on invoices, before movies, at sports events. When I was a child, I don't remember seeing corporate logos at the ballpark or on the program of our local arts festival. Every patch of empty space and every second of empty time is ripe for conversion into money. I once read that advertisers can buy time on a special radio station that broadcasts commercials into the Lincoln Tunnel for the seven minutes or so of empty time when cars are out of reach of normal broadcasters. Today there is even talk of new sound technologies that will deliver location-specific messages to pedestrians as they walk past each store.

A variant of social capital might be termed "civic capital". Civic capital is the generalized, culturally-transmitted participation in politics and government, including both the confidence and the political skills to participate effectively in democratic society. In his book *Bowling Alone*, political scientist Robert Putnam chronicles a generation-long decline in civic participation that has reduced democratic citizenship to a mere matter of voting—more "meaningless choices in a box." Few people these days participate meaningfully in local government as the community networks that once exercised political clout have evaporated along with other dimensions of community. Increasingly (though recently there are signs that this trend is reversing) we are content to choose among the options presented us from on high.

Finally, we are losing what we might call physiological or health capital: the biological resources and abilities of the human body. It is evident that many of our human capacities have atrophied, to be supplemented with technology. For example, not only do most people not know how to procure food and shelter in nature, but many Americans over the age of fifty cannot survive without a variety of pharmaceutical medications. Nor are many women able to give birth to a child in the absence of technology. Soon they may need it even to conceive: infertility is a growing problem worldwide, and recent studies have confirmed an annual decrease of around 2% in sperm counts.[31]

The aids and comforts of technology have become props to survival on the biological level. Admittedly, most of the dependence is an illusion, a function of ignorance: in the case of pharmaceuticals, an ignorance of natural medicine and the maintenance of the body; in the case of childbirth, an institutional distrust of the woman's body and a taught fear that compels women to seek expert assistance. While the dependence on technology is not yet total, we are slowly but surely losing our abilities to walk, sleep, squat, sit cross-legged, run, defecate (why else so many laxative commercials on TV), and even to breath. Technology accomplishes for us the functions these abilities used to serve.

In most cases, this dependency has not (yet) been integrated into our genes and is in principle reversible. Nonetheless, we have indeed sold, or at least mortgaged, much of our physical health to the demands of the money economy. My health is an asset, a resource that I can convert into money, for example by working overtime in an office or working in hazardous conditions. Remember the eternal business idea of taking something away from someone and then selling it back to them? Technological society, through its conveniences and demands, has taken away our health and now sells it back to us, for money, through the machinations of medicine, supplements, fitness centers, and so forth, all of which allow us to cope in today's world. The semblance of health they bestow is usually enough to get by in a technological society, where we rarely have to endure cold temperatures for a long time, or climb trees, or walk for ten miles, but it is a far lower level of health than a human living in nature enjoys.

If current trends persist, the dependence on medical technology is likely to grow more and more acute, until we reach the science-fiction world where medical devices and other technology are incorporated into the human body from birth, or even before. Already there is talk about "designer babies" whose genes are artificially engineered or selected for specific traits. Already there are implants of computer chips and various timed-release drugs. And while the talk is of someday creating superhumans with vastly augmented mental and physical capacities, most of the applications so far are simply to help unhealthy people to cope, to get by.

We are experiencing today an epidemic of mysterious new diseases, mostly malfunctions of the immune system. "Science is making progress toward a cure," we are told, but another way to look at it is that a (perhaps life-long) course of drug therapy, gene therapy, or whatever will effectively incorporate technology into bodies that are not viable without it. Since the diseases of the 21st century are, in the opinion of most alternative practitioners, caused by technology (environmental toxins, sedentary lifestyles, and industrial food processing, to name a few), our use of technological medicine indeed amounts to the buying back of what technology has taken away from us to begin with. In other words, we have converted our physiological or genetic capital into financial capital. We have sold away our health only to buy back an inferior version of it.

Earlier I wrote, "What else do people still do for themselves?" What else could be converted into a "service"? What about body parts and functions? I am thinking here of the kidney industry, which induces poor people around the world to sell their own kidneys for transplants in the West. Another growth industry is the surrogate womb business, outsourced increasingly to Third World countries.[32] From conception to gestation to birth, from nursing to day care to school, from little league to summer camp to electronic entertainment, the process of making a human being becomes, in the extreme, a series of services. Not too far, really, from the nightmares of Aldous Huxley.

When we become aware of the magnitude of our destitution, we naturally desire to reclaim for ourselves some of that lost wealth. However, to reclaim our lost health and connectedness, stories and imagination, capacities and relationships is no trivial matter. Many of the old relationship-sustaining structures are broken, leaving us alone and helpless. We know not where to begin. Moreover, the monetization of life engenders a relentless insecurity and anxiety that bludgeons us into submission. When all our survival skills have been sold off we are left dependent on money—a supreme irony, given that people sacrifice life for money precisely for the stated purpose of achieving "financial security." Money, once incidental to survival (as in the case of the subsistence farmer) now becomes synonymous with the means to survive. A natural corollary of the conversion of the physical, social, and cultural world to property is that we must pay to live. Could this, and not a sinful human nature, be the reason for the anxiety and greed that so pervades our culture?

Dependency on money, coupled with the ceaselessly intensifying competition it implies, means that it is not only difficult but also *irrational* to reclaim life for ourselves. The remote specialists and professionals who discharge our life functions do so much more efficiently than we can. To cook from scratch, garden, fix our own cars, play our own music and make our own clothes are economically inefficient activities that set us back in life's great competition. In economics, the very word "rational" means the maximization of one's financial interests—a very significant assumption. How much is your time worth? How many dollars per hour? Go ahead and calculate how much money you net by changing your own oil. (Include the "cost" of your own labor.) It is more rational to let the specialist do it. It is more rational to let the specialists at the supermarket bakery bake your son's birthday cake (the cake costs only $7 and it would take you more than an hour). It is more rational to let the specialists with their vast economies of scale cook all your food. It is more rational to let dedicated specialists clean your house and take care of your children. Time, after all, is money.

The emptiness of this way of thinking is self-evident, for it reduces life to money (just as the phrase "Time is money" implies). Yet the reduction of life to money is exactly what is happening whenever technology makes our lives more convenient. The reduction of life to money is exactly what is happening through the vast conversion of all forms of social, spiritual, cultural, and natural capital into financial capital.

The equation "Time is money" brings us to the heart of the matter. Having turned the world and everything in it into property, we apply the same equation to time itself, subjecting our every moment to the calculus of economic justification. In this saying is encoded nothing less than the complete monetization of human life. There is something monstrous about the very idea of a wage, the sale of one's time that is the sale of one's life. You sell the very hours of your life. Such a concept indeed sparked tremendous resistance during the early Industrial Revolution, before which, Kirkpatrick Sale writes, "Time was a medium, not a commodity." But as the routine, mechanical actions of the assembly line worker and industrial machine operator replaced the skills of the independent craftsperson, skills became increasingly superfluous until workers had naught to offer but their time. They resisted heroically this monetization of life itself, even when the alternative was utter destitution, and the question of how to instill and enforce "labor discipline" became a leading topic among intellectuals of the day.

Unfortunately, self-evident as its emptiness may be, the reduction of life to money is written on a deep level into the assumptions of economics and the criteria for public policy. Its main conceptual justification originated in the philosophy of Jeremy Bentham, who thought that the duty of government should be to maximize total happiness. True to the

Galilean tradition, he decided that the way to proceed would be to quantify goodness, or as he called it, "utility". Then it would be a simple matter to calculate the total utility resulting from Policy A, compare it to the total for Policy B, and choose the one generating the most "utils" of happiness.

Now you'd think that to quantify happiness or goodness is a notion so absurd as to require no comment. I wish that were true! In fact, we have actually attempted to implement Bentham's suggestion, though in a disguised form. Instead of the "util", we denominate the good in dollars.

Economics asserts that we are rational actor seeking to maximize a self-interest or personal good measured in dollars. The good of the nation, too, finds expression in dollars whenever we assume that economic growth is a positive good. Economic growth is defined in terms of gross domestic product—the total value of all "goods" and services *measured in dollars.* Another significant quantification of goodness is embodied in cost-benefit analysis, which assumes that all relevant costs and benefits can be assigned a monetary value. In so doing, it is often necessary to assign a value even to human life. Is it worth it to spend a billion dollars on safety equipment that will save ten lives? How much money is a human life worth? How much would you demand in exchange for yours? The monstrous implications of this way of thinking—a *literal* reduction of life to money—are clear. Ethicists tie themselves into knots over this issue, but no satisfactory solution is possible as long as we continue to quantify the good.

Yet some economists seem to think that this quantification has not gone far enough. Exemplified by Gary Becker, they proclaim that an economic calculus governs all human interactions, from crime to marriage to the pursuit of education. Becker, who won the Nobel prize for "having extended the domain of microeconomic analysis to a wide range of human behavior and interaction, including nonmarket behavior,"[33] advocates extending market mechanisms into these remaining nonmarket realms.[34] Ah, if only the monetization of life could be completed, then life would be perfectly rational, perfectly efficient.

The use of the word "goods" to denote the salable products of human activity reveals some very deep assumptions. First, harking back to agriculture, it suggests that goodness comes from human manipulation of nature and not from nature itself. A good is something produced, extracted from its original place in the ground, the water, or the forest and then subjected to other forms of processing. Unimproved, nature is not good. Oil in the ground or a forest left standing are invisible to economic accounting: they are not yet "goods", just as unpaid mothering is not yet a "service". A second implication is that if something is good, it can be assigned a price—commoditized, bought, and sold. Got that? Our definition of a "good" is that it is exchanged for money. Money = Good. That some good things are not yet commodities merely means that the conversion of social, cultural, spiritual, and natural capital is not yet complete—good news for anyone but an economist. From the economist's point of view, the equation of economic growth with more and more "goods" adds a note of moral imperative to the quest for economic growth. More and more goodness, more and more happiness. The ascent of humanity. "To cry enough or to call a limit was treason. Happiness and expanding production were one."[35]

The Economics of Other

The conversion of life into money means that there is ever more of the latter and less of the former. In our economic calculus, however, this is seen unambiguously as an increase in wealth, which is a quantitative concept denominated by money. Anything without a monetary exchange value is invisible, outside economic logic. These "externalities" are the counterpart of Galileo's excluded subjective properties (only that which can be measured, denominated, counts). They are also counterpart to the "other" that technology makes of the world. Herein lies the fundamental difference between modern economies and natural ecologies.

In nature there is no waste (except for heat radiated out into space); as Paul Hawken puts it, "Waste is food." Natural processes are therefore cyclical. What comes from the earth eventually returns to the earth in a form usable to other living beings. There is no linear buildup of waste, no linear drawdown of essential resources. Industry, on the other hand, is linear in that it starts with resources and ends with waste—economically valueless, even biologically hazardous substances that must be disposed of. Since "resources" such as the social, cultural, natural, and spiritual capital described in this chapter begin outside the domain of money, their commoditization and depletion makes us by definition richer, adding to gross domestic product. Meanwhile, because these resources are not endlessly recycled, their depletion accompanies a corresponding growth of material, social, and spiritual waste: slag heaps and slums, toxic waste dumps and toxic bodies, dying lakes and wrecked cultures, degraded ecosystems and broken families.

The linear character of the modern economy is obviously unsustainable, because both resources and the earth's capacity to absorb waste are finite. The modern economy therefore represents an outright denial of humanity's participation in nature, and embodies a belief that the laws of nature do not apply to us.

Classical economics denies the finiteness of resources by saying their depletion will cause prices to rise, stimulating innovation in the search for replacements. In other words, when we deplete oil to the point where it costs $500 a barrel, huge incentives will encourage development of alternative energy sources. When we deplete topsoil to the point where soil-grown food becomes prohibitively expensive, we will find other ways to grow or synthesize food. When we destroy the ozone layer, innovation will be brought to bear to invent the Lifeskin® personal security suit and the Ecodome® ecological containment enclosure. The unstated assumption is that our ability to engineer and control the universe is infinite—as the price of a depleted resource rises toward infinity, so does the incentive (and by implication the capacity) to innovate. You should by now recognize the telltale signature of the technological fix, and the Technological Program that carries us farther and farther from nature. It is written into the assumptions of economics. If the oceans are depleted of fish, no matter, we'll just farm them. If the soil becomes unusable, why, we'll just make new soil. If the earth becomes uninhabitable, why, we'll just construct a new earth.

The assurances of classical economics and the ideology of control are beginning to wear a little thin, though, because as described in Chapter One, the world seems to be spinning out of control. Our problems proliferate faster than we can manage them. Life goes on only with increasingly frantic efforts to keep everything together for now, while vast problems are sequestered away for later. We do this collectively, in the form of temporary containment of toxic and radioactive waste (science will surely find a permanent solution before it starts leaking in forty years), as well as individually, when we ignore huge contradictions in life and pave over festering wounds the way chemical companies might pave over buried toxic waste. But as one popular title says, "Feelings buried alive never die." This is just as true of our collective garbage as it is of our personal garbage.

If the certainty of the Technological Program and the promise of economic logic are misguided, then the depletion of natural capital is simply a drawdown of capital reserves, not the creation of new wealth, and the accumulation of waste is simply a bill to be paid, and not a disposal into a limitless outside. How could we think otherwise? Only if we conceive of ourselves as existing apart from nature, so that there is indeed a place to throw things "out". How else could we countenance the production of persistent bio-accumulating poisons such as PCB's, mercury, and dioxin? A culture that knew nature as sacred wouldn't dare. A culture whose sense of itself included plants, animals, forests, and land wouldn't dare. That we indeed dare is simply a product of our own self-misconception. Only if we see ourselves as fundamentally separate from nature is it reasonable to think that the poison won't eventually affect us.

The same logic of externalization applies on the level of the individual or corporation. Profits accrue to those who most successfully externalize their costs. This is Step Two of the eternal formula for business success. Step One was to take something away from people (e.g. social capital) and sell it back again. Step Two is to make someone else pay at least some of the costs, while you get the profits.

Do the printers of unwanted junk mail have to pay the costs of disposing it in landfills? Do the makers or users of pesticides have to pay the costs of cleaning the groundwater they eventually contaminate? Do the makers or users of nitrogen fertilizer have to pay the costs of eutrophication (algae blooms that deoxygenate water and kill the fish)? Would the mounds of plastic junk we buy every Christmas still be so cheap if they incorporated the medical costs of toxic petroleum byproducts? When cancer rates rise 500% near an incinerator, refinery, or paper mill, does the manufacturer pay the medical costs? Paul Hawken writes,

> Gasoline is cheap in the United States because its price does not reflect the cost of smog, acid rain, and their subsequent effects on health and the environment. Likewise, American food is the cheapest in the world, but the price does not reflect the fact that we have depleted the soil, reducing average topsoil from a depth of twenty-one to six inches over the past hundred years, contaminated our groundwater (farmers do not drink from wells in Iowa), and poisoned wildlife through the use of pesticides.[36]

In some cases, such as a polluting factory, it is the neighbors who pay the costs. Build the smokestacks higher, and it is fishermen thousands of miles away who pay the costs. But only part of those costs, because acid rain does more than kill lakes; it also contributes to massive tree die-offs decades later. The costs of ecosystem disruption are untraceable to a single source, highly distributed, impossible to predict accurately in advance, and often paid only by future generations.

As long ago as 1920, economist Arthur C. Pigou realized that for markets to promote the general welfare, producers must pay the full costs of production; means must be found to "internalize" external costs like those listed above. His solution was governmental imposition of "Pigovian" taxes or subsidies to make products reflect their true costs. Forty years later, Ronald Coase demonstrated that such government intervention is unnecessary—if transaction costs are zero (a favorite assumption of economists in their mathematical games) and if property rights are "properly assigned".[37] Although Coase is often cited to justify a libertarian, non-interventionist policy based on property rights, his work actually implies that this is impractical because of the problem of transaction costs. When costs are highly distributed, untraceable to a single source, and hard to calculate, the transaction costs in allocating them are prohibitive. Even more disturbing is the second condition that property rights be "properly" assigned. Coase's logic essentially demands that *everything* become the subject of property and be assigned a monetary value. Or in other words, that everything have a price. Yet when residents of the Rocky Mountains were polled on how much money it would take for them to accept air pollution in their area, they said that *no amount of money* would be adequate compensation for their loss of fresh air and clear views. No amount of money. According to economics, this is a profoundly irrational response. But I bet that you'd answer the same if asked how much money you'd take in exchange for your vision, your right leg, or your child. Indeed, when such things are assigned a monetary value the consequences tend to be monstrous (as in the "kidney industry"). What else could we expect when the infinitely precious is reduced to a finite sum? A fundamental assumption of economics is thus shown to be nonsense, and worse, anti-life. So come on, how much do you want for your health? For your friendships? For your self-respect? For your hours? For your life?

"Every man has his price." Is that saying really true? Or is it just a symptom of how broken we are to the enslavement of money?

The business model of profiting at the expense of others fits in quite well with traditional biology, in which organisms compete for resources and excrete "wastes" into the "environment". Money economics similarly views human beings as separate subjects competing for resources, seeking to maximize self-interest—an exact parallel to the Darwinian view of biology. "Human beings are basically selfish"—does that seem like a truism to you? Actually, such behavior is built in to the structure of money. Consequently, the more that money transactions replace human relationships, the more life becomes a struggle among competing "others". Thomas Carlyle describes the inevitable result in *Gospel of Mammonism*, "We call it a Society and go about professing openly the totalest separation, isolation. Our life is not a mutual helpfulness; but rather, cloaked under due laws-of-war, named "fair competition" and so forth, it is a mutual hostility."[38]

Economics is just another facet of the dualism of self and other, a distorting lens that warps our entire understanding of the universe. But now we are beginning to understand that nature is not like that, and that we are not separate from nature. Traditionally, we have seen organisms as using resources and excreting waste into the environment, which by some lucky chance has other organisms that have evolved to recycle the waste through the system. What does the individual rock-weathering bacterium care that its wastes eventually provide calcium carbonate for some sea creature? No positive reinforcement reaches it quickly enough to affect natural selection. Is it by some lucky chance that so far no organism except for man has created waste that is unusable and deadly to the rest of life?

The totalizing trend of money and its conceptual equivalence to utility or goodness is responsible for the lunacy of current economic accounting, in which phenomena like cancer, toxic waste leaks, divorce, imprisonment, and so forth contribute to GDP—the total value of all "goods" and "services". As long as the damage caused to people, cultures, and ecosystems is not denominated in money, it is in the realm of other, off the balance sheet. The same goes for business accounting, in which costs can only be externalized when the payer is, again, off the balance sheet—an other. To the extent that we identify with our communities, we cannot export costs to them. Both social pressures and our own conscience will stop us. To the extent that we identify with nature, neither can we see as profit anything that diminishes the overall wholeness and beauty of the earth.

Yet as anyone who has ever tried to do business with a conscience knows, powerful forces seem to conspire to enforce a ruthless approach to business. Do-gooders get forced out of business by those who more efficiently externalize their costs. Everyday experience confirms competition as a fact of life. Why? Because, due to the monetization of all the forms of capital that allowed us to survive without it, money has become essential to life. And money, in turn, has built-in characteristics which reinforce all the processes of separation, alienation, competition, and ongoing conversion of life into money.

Interest and Self-Interest

In the empires of usury the sentimentality of the man with the soft heart calls to us because it speaks of what has been lost.
—Lewis Hyde

Our system of money and property contributes in many ways to the process of separation: of people from nature, from spirit, and from community. For one, the monetized life depends on distant, impersonal institutions and the anonymous specialists that compose them, rendering us less tied to our neighbors. Secondly, the nature of money transactions is closed, in contrast to open-ended gift-giving which creates obligations—literal ties that bind a community. Thirdly, by its very nature as an abstract representation of value, money creates the illusion that utility, the good, is something that can be counted and quantified. Fourth, the concept of property makes the world into a collection of discrete things which can be separated, sold off, and owned. To own something is to separate it out from the commons and subordinate it to oneself. Accumulation of property, and particularly money, represents an annexation of the wild into the domain of self, creating a perception of security that is not really there, separating the self even more from the rest of the world, and reinforcing the illusion that pieces of the world can be separated out and made mine.

Yet the original purpose of money is merely to facilitate exchange. On the face of it, exchange should bring people closer together, not separate them. In Chapter Seven I will describe a money system that will do precisely that: undo separation, build community instead of breaking it down, bring us closer to nature instead of distancing us. Such money systems already exist in embryonic form, and to understand what characteristics they need to have, it helps to understand what characteristics they must not have.

There are two central characteristics of present-day money that drive the conversion of social, cultural, spiritual, and natural capital into financial capital, and which are also bound up with our self-perception as separate beings in a universe full of discrete objects. These two, deeply interrelated, characteristics are *scarcity* and *interest.*

Scarcity and interest are products of the way modern-day money is created: it is lent into existence by banks.[39] While this has been partially true for several hundred years, until 1971 there was (at least in theory) a commodity backing—gold—for the U.S. dollar, and therefore for other currencies pegged to the dollar. But since the dismantling of the Bretton-Woods system in 1971, currency has not been based on gold or any other commodity. The amount of new money banks can create by making loans is limited only by their own reserves (and ratio requirements and the discount interest rate). The total level of these reserves economy-wide is determined by the Federal Reserve (or outside the U.S., the central bank) through the purchase and sale of government securities.

Bernard Lietaer comments, "For a bank-debt-based fiat currency system to function at all, scarcity must be artificially and systematically introduced and maintained."[40] When a bank extends a loan, the borrower must pay it back with interest, which he competes with everyone else to procure from the limited amount of still-to-be-created money. Governments and their central banks must exercise careful control—through interest rates, margin reserve requirements, and, most important in the present era, purchase or sale of government securities on the open market—over the rate at which this new money is created. Theirs is a difficult balancing act between tightness, which creates more scarcity, intensifies competition, and leads to bankruptcies, layoffs, concentration of wealth, and economic recession, and looseness, which creates less scarcity, higher inflation, and increased economic activity, but at the risk of runaway inflation and complete currency collapse. In order to prevent the latter eventuality, money must be kept scarce, consigning its users to perpetual competition and perpetual insecurity.

The bank-debt fiat currency system is not the deepest source of scarcity and insecurity, however, because both are built in to the phenomenon of interest, which itself has deeper roots in our self-conception. The bank-debt currency system we have today is founded upon interest. That's the motivation for banks to create money in the first place. Creating money is only a side effect, irrelevant to the commercial bank, of their main purpose of earning a profit. Another side effect is the necessity of perpetual economic growth and, consequently, the conversion of all common wealth into private monetary wealth described in previous sections.

Let's trace how interest leads to scarcity, competition, and the necessity of perpetual growth. Since nearly all money in the economy is being lent out at interest through one mechanism or another (deposits, loans, etc.), it follows either (1) that some of these loans must end up in default, or (2) that the supply of money must grow. If I am to pay back a loan with interest, I must obtain that extra amount beyond the principal from somewhere else. If the money supply is not growing, then a percentage of wealth-holders corresponding to the prevailing interest rate must go bankrupt. In other words, if there are one thousand dollars in the world, and they are lent out at ten percent interest to ten people, then one must go bankrupt to supply the other nine with the money to pay back their loans after one year. That is how interest sets us in competition.

In the real world, of course, the money supply is not static, it grows. But that does not alter the underlying dynamic of scarcity and competition. At any given moment, we collectively owe more money than exists right now. Where will the new money come from? In today's fractional reserve banking system, new money does not come from mining more gold and minting more coins. It appears every time a bank or other institution makes a loan. To whom will a bank lend money? Preferably to someone with "good credit", which quantifies a judgment of one's ability to compete for money and therefore to pay back a loan with interest. In today's system, money does not exist without debt, debt does not exist without interest, and interest drives us to earn more and more money. Some of us can take the money from others, but collectively we must create new goods and services. That is ultimately what our employers pay us to do. Either as entrepreneurs or employees, lenders or borrowers, we participate in the conversion of social and natural capital into financial capital. Interest generates an irresistible organic pressure that endlessly expands the monetized realm.

When new money is loaned into existence system-wide in amounts exceeding the ability of the economy to create new goods and services, the result is inflation. More money chases fewer goods. The currency will lose its value, an eventuality distasteful to those who hold lots of it (creditors, the rich and powerful). It would seem good for the rest of us, we who were once known as the "debtor class", but unfortunately inflation is subject to powerful positive-feedback mechanisms that cause it to "run away" and collapse the currency. To prevent this, non-inflationary economic growth—an increase in the production of goods and services—is structurally necessary for today's money system to exist. That is what drives the relentless conversion of life into money I have described in this chapter.

The need for growth, the built-in scarcity of modern money, the phenomenon of interest, and the pervasive, continual competitive basis of the modern economy are all related. Wherever money has landed, traditional gift economies have deteriorated as competition replaced sharing as the basis of economic interaction.

And for what purpose, this artificially induced competitiveness, this omnipresent scarcity, anxiety, and insecurity of modern life? We are, after all, living in a world of material plenty. As perhaps never before in human history, we have the capacity to easily fulfill all the physical wants of every human being on the planet. Ironically, our system of money, the very symbol of wealth, creates scarcity in the midst of this plenty. To what end? The purpose of money, after all, is first and foremost to facilitate exchange. Bernard Lietaer writes, "The current money system obliges us to incur debt collectively, and to compete with others in the community, just to obtain the means to perform exchanges between us."[41] In classical economics, the competition inherent in scarce money is thought to be a good thing, for it induces efficiency. But in a world of plenty, is efficiency really the highest good? No, especially when efficiency equates to the swiftness with which common wealth is converted to private capital.

The question to ask, then, is whether another kind of money system might serve the need of facilitating exchange (and certain other needs) without inducing the relentless incineration of social, cultural, natural, and spiritual capital that fuels economic growth today. For it would be wrong to pin the ultimate blame on money *per se*. Money as we know it developed in the context of our entire civilization; it is not only a cause but also an effect of our general economic system as well as our worldview, our cosmology, and our self-definition. Money as we know it embodies and reifies our deep cultural assumption that the world is amenable to cut-and-control: division into discrete pieces, objects, that may be labeled and transferred. Money scarcity grows from and reinforces the idea, implicit in the technological management of the world, that we must manipulate and improve upon nature in order to obtain the means of survival; it is linked to survival anxiety.

The phenomenon of interest boils down to the belief that "money costs money". Interest is the price we pay or extract for the use of money, which in the present age of specialization equates to survival. Interest, therefore, encodes the belief that the means of survival are precious, rare, scarce, and therefore the objects of competition. Lending money with interest amounts to, "I will help you survive, but only if you pay me." In a world of plenty, if you give someone food today, would you ask them to pay you back in even greater quantity tomorrow? It would be illogical and unnecessary in a world where survival, not to mention abundance, were not linked to labor, scarcity, and anxiety. Thus interest not only creates a mentality of scarcity, it also naturally arises from a mentality of scarcity.

The mentality of scarcity, which impels us to keep and hoard, is contrary to the mentality of the gift, which relaxes the boundaries of self and ties communities together. "[Modern man] lives in the spirit of usury, which is the spirit of boundaries and divisions."[42] A cardinal feature of an authentic gift is that we give it unconditionally. We may expect to be gifted in return, whether by the recipient or another member of the community, but we do not impose conditions on a true gift, or it is not really a gift. Clearly, lending money at interest is utterly contrary to the spirit of the gift.

Lewis Hyde identifies another universal characteristic of a gift, which is that it naturally increases as it circulates within a community, and that this increase must not be kept for oneself, but allowed to circulate with the gift. Interest amounts to keeping the increase on the gift for oneself, thereby withholding it from circulation in the community, weakening community for the benefit of the individual. It is no accident that many societies prohibited usury among themselves but allowed it in transactions with outsiders, who could not be trusted to recirculate a true gift back into the community. Hence the prohibition in Deuteronomy 23:20: "Unto a stranger you may lend upon usury, but unto thy brother thou shalt not lend upon usury."

The ramifications of this injunction when combined with Jesus' teaching that all men are brothers are obvious: interest is forbidden entirely. This was the position of the Catholic Church throughout the Middle Ages, and is still the rule in Islam today. However, starting with the merger of Church and state and accelerating with the rise of mercantilism in the late Middle Ages, pressure mounted to resolve the fundamental tension between Christian teaching and the requirements of commerce. The solution provided by Martin Luther and John Calvin was to separate moral and civil law, maintaining that the ways of Christ are not the ways of the world. Thus spirit became further separated from matter, and religion retreated another step toward worldly irrelevancy.

Interest also violates yet a third feature of gift networks, that the gift flows toward he who needs it most. Hyde explains,

> The gift moves toward the empty place. As it turns in its circle it turns toward him who has been empty-handed the longest, and if someone appears elsewhere whose need is greater it leaves its old channel and moves toward him. Our generosity may leave us empty, but our emptiness then pulls gently at the whole until the thing in motion returns to replenish us. Social nature abhors a vacuum.[43]

Interest, on the other hand, directs the flow away from the one with the greatest need, and toward the one with the greatest wealth already.

The imperative of perpetual growth implicit in interest is what drives the relentless conversion of life, world, and spirit into money. And yet because money is identified with Benthamite "utility"—that is, the good—this entire process is considered rational in traditional (neoclassical) economic theory. Quite simply, whenever anything is monetized, the world's "goodness" level rises. Accordingly, such things as toxic waste, cancer, divorce, weaponry, and so forth count as "goods and services" and contribute to GDP, which is conventionally accepted as a measure of a nation's well-being. The same assumption appears in the euphemism "goods" to describe the products of industry. The very definition of a "good" is anything exchanged for money.

In terms of conventional economics, it may actually be in an individual's rational self-interest to engage in activities that render the earth uninhabitable. This is potentially true even on the collective level: given the exponential nature of future cash flow discounting, it may be more in our "rational self-interest" to liquidate all natural capital right now—cash in the earth—than to preserve it for future generations. After all, the net present value of an eternal annual cash flow of one trillion dollars is only some twenty trillion dollars (at a 5% discount rate).[44] Economically speaking, it would be more rational to destroy the planet in ten years while generating income of $100 trillion, than to settle for a sustainable level of $3 trillion a year.

If this seems like an outlandish fantasy, consider that it is exactly what we are doing today! According to the parameters we have established, we are making the insane but rational choice to incinerate our natural, social, cultural, and spiritual capital for financial profit. Amazingly, this end was foreseen thousands of years ago by the originator of the story of King Midas, whose touch turned everything to gold. Delighted at first with his gift, soon he had turned all his food, flowers, even his loved ones into cold, hard metal. Just like King Midas, we too are converting natural beauty, human relationships, and the basis of our very survival into money. Yet despite this ancient warning, we continue to behave as if we could eat our money: I once read of an East Asian minister who said his country's forests would be more valuable clearcut and the money put in the bank to earn interest. Apparently, the effects of destroying the planet are of little concern to economists. William Nordhaus of Yale proclaims, "Agriculture, the part of the economy that is sensitive to climate change, accounts for just three percent of national output. That means there is no way to get a very large effect on the US economy." Oxford economist Wilfred Beckerman echoes him: "Even if net output of agriculture fell by 50 percent by the end of the next century, this is only a 1.5 per cent cut in GNP."[45]

Must we, like King Midas, find ourselves marooned in a cold, comfortless, ugly, inhospitable world before we realize we cannot eat our money?

Because it builds exponentially, interest feeds a linearity that puts humankind outside of nature, which is bound by cycles. Subtly but inexorably, it drives the assumption that human beings exist apart from natural law. As well, interest drives a relentless anxiety by demanding always more, more, more, propelling the endless conversion of all wealth into financial capital. Part of this anxiety is encoded in the very word, "interest", which implies that self-interest too is bound up in ever-lasting increase.

Interest is a necessary counterpart to the mentality of externalization. Like interest, externalization involves a denial of nature's cyclicity by treating it as an infinite reservoir of resources and an infinite dumping ground for waste. Interest is also akin to fire, the foundation of modern technology. To keep it going requires the addition of ever more fuel, until the whole world is consumed, leaving but a pile of dollars or ash.

Money is a most peculiar kind of property, for unlike physical inventories of goods, "rust doth not corrode nor moths corrupt" it. Cash does not depreciate in value; on the contrary, in its modern, abstracted form of bits in a bank's computer, it grows in value as it earns interest. Thus it appears to violate a fundamental natural law: impermanence. Money does not require maintenance like a plot of farmland to maintain its productivity. It does not require constant rotation of stock like a store of grain to keep it fresh. No accident, then, was money's early and enduring association with gold, the metal most famously impervious to oxidation. Money perpetuates the fundamental illusion of independence from nature; financial wealth endures without constant interaction with the environment. Other forms of wealth are bothersome, because they require a continuing relationship with other people and the environment. But not money, which is now wholly abstract from physical commodities and thus abstract as well from natural laws of decay and change. Money as we know it is thus an integral component of the discrete and separate self.

It is a curious fact that most people are extremely unwilling to share their money. Even among relatives, sharing money is bound by strong taboos: I know countless poor families whose brothers, cousins, or uncles' families are very wealthy. And how many friendships have disintegrated, how many family members have shunned each other for years, over issues of money? Money, it seems, is inextricably wrapped up in the very essence of selfishness—a clue to its deep association with self. Hence the intense sense of violation we feel upon getting "ripped off" (as if a part of our bodies were being removed) when from another perspective all that has happened is pieces of paper changing hands or bits turning on and off in a bank computer.

We do not usually share our money because we see it almost as part of our selves and the foundation of our biological security. Money is self. Meanwhile, conditioned by science and the origins of separation underlying it, we see other people as essentially just that, "other". Mixing these two realms invites confusion and conflict. The problem is, the more of life we convert to money, the more territory falls into one of these dichotomous realms, mine or yours, and the less common ground there is to share life and develop unguarded relationships. The conversion of life to money reduces everything to an economic transaction, leaving us the loneliest people ever to inhabit the planet. The propertization of the whole world means that everything is either mine, or someone else's. No longer is anything *in common.*

The violation we feel at being ripped off is much akin to the violation an indigenous hunter-gatherer must feel at witnessing the destruction of nature. When "I" is defined not as a discrete individual but through a web of relationships with people, earth, animals, and plants, then any harm to them violates ourselves as well. Even we moderns sometimes feel an echo of this violation when we see the bulldozers knocking down the trees to build a new shopping center. That is because our separation from the trees is illusory. The buried connectedness can be resisted through ideology, narcotized through distractions, or intimidated through the invocation of survival anxiety, but it can never die because it is germane to who we really are. The love of life that Edwin Wilson has named biophilia, and our natural empathy toward other human beings, is ultimately irrepressible because we *are* life and life is us.

The regime of separation has deadened us to the self-violation inherent in the wrecking of the planet and the degradation of its inhabitants. In an attempt to compensate for our lost sense of beingness, we transfer it to possessions and particularly to money, setting the stage for disaster. How? Because money (bearing interest) is an outright lie, encoding a false promise of imperishability and eternal growth. Identified with self, money and its associated "assets" suggest that *if we stay in control of it*, the self might be maintained forever, impervious to the rest of the cycle that follows growth: decay, death, and rebirth.

Obviously, there is a problem when something that does not decay but only grows, forever, exponentially, is linked to commodities which do not share this property. The only possible result is that these other commodities—the social, cultural, natural, and spiritual capital of this chapter—will eventually be exhausted in the frenetic, hopeless attempt to redeem the ultimately fraudulent promise inherent in money with interest.

The (interrelated) characteristics of scarcity and interest are not accidental features of our system that, if only someone had made a wiser choice, could be different. They are implicit in our Newtonian-Cartesian cosmology in which, by definition, more for me is less for you. As this cosmology rapidly becomes obsolete, hope is emerging for a transition to a new money system embodying a very different conception of self and world. Without such a transition, there is little hope that the current conversion of social, cultural, natural, and spiritual capital into money will ever abate.

The Crisis of Capital

The logic of interest is the logic of an addict. It assumes that always and forever, we will find some way to feed an endlessly growing need to consume. We will always be able to find some new resource, some new form of capital to convert into money. Like the craving of an addict, interest demands that more and more of life be directed toward feeding it. I'm not just speaking metaphorically here, as anyone knows who bears heavy student loans or credit card debt. More and more of life is directed toward feeding it. Interest enforces the perpetual sacrifice William Wordsworth understood as a key feature of Machine civilization. The higher that interest rates are, the more powerful the exigency that speeds the conversion of social and natural capital into money.

The longer any addiction is maintained, the greater the depletion of life, and the more extreme the measures required to perpetuate it. Just as the addict cashes in life insurance policies, borrows money from friends, and eventually converts every physical resource and social resource into money, so also does our civilization seek out every possible source of unexploited social, natural, cultural, and spiritual capital. However good our intentions to preserve it, interest generates an unstoppable force that assails it from all directions, always seeking a way in.

Marxism describes this process and its inevitability in terms of the imperative of (financial) capital to constantly find new domains of exploitation in order to delay the crisis of falling profits, destructive competition, and concentration of ownership. The endless growth demanded by our system of money and property, the endless conversion of life to money, conceals a latent contradiction that has been known since the inception of the Industrial Revolution. This is the crisis of overproduction, described explicitly by Karl Marx, but also frankly discussed by industrialists and their intellectuals to the present day.

Simply put, under the present system of money and property, it is generally to the individual producer's advantage to produce as much as possible to take advantage of economies of scale—the hallmark of industry. Yet when every producer does this the result is overproduction and hence falling profits, falling wages, bankruptcies, concentration of ownership, unemployment, and finally, to complete the vicious circle, falling demand. Marx foresaw it all culminating in a revolution, when the misery entailed in the above trends became unbearable to enough people. Faced with this possibility, capitalistic society is driven by an organic imperative to postpone—hopefully indefinitely—the frightening denouement predicted by Marx. One way to do this is to limit production (for example by raising interest rates or through a command economy); another is to incinerate overproduction through warfare; a third is to find "new markets" through technology or imperialism. The first cannot work because, most fundamentally, it is inimical to our money system and its demand for endless growth, and thus sparks the very deflationary crisis it means to avoid. The second option works just fine, but requires an endless intensification that came to an end with the advent of thermonuclear weapons. The continual state of war that exists today is insufficient to absorb an exponential increase in production.

That leaves us with the third option. The imminent and inevitable revolution that Marx predicted has not come about, principally because technology has constantly created fresh new markets. Each time, a new industry starts off with a vibrant pandemonium of myriad competing firms, providing new jobs, new wealth, new opportunities for entrepreneurship. Eventually, firms fail or merge, profit levels fall, and layoffs set in, but by this time some new sector of the economy has emerged to take up the slack. Marx's analysis thus appears to be quite accurate in describing the evolution of a particular industry, but inapplicable to the economy as a whole as long as there are always new technologies, new industries, and new markets.

Is there a limit to the ability of technology to open up new domains for capital investment? If not, then Marx's crisis of capitalism will never arrive. But if we understand technology as a means to convert non-monetary forms of capital into financial capital, then there will come a point when these other forms of capital are exhausted. In other words, what has been going on all along is not the creation of new wealth, but the conversion of existing (non-monetary) wealth into monetary wealth to give "capitalism" a lease on life. I have argued that "new" sectors of the economy usually involve the transfer of things people once did for themselves into the hands of specialists. What will happen when there is no more to be transferred?

The situation is closely analogous to the standard Marxist explanation of colonialism and imperialism. In Marxist theory, when a country has developed to a certain point, the mounting crisis of lower and lower returns on investment and falling profits generates enormous pressure to find new areas where wages can go still lower, where raw materials can be extracted still more cheaply, where there is new demand for the overcapacity of industrial products, and where prices have not been driven near the break-even point by extended competitive pressure. In other words, the crisis is exported to a colony and temporarily relieved until that colony too has no more to offer in terms of lower wages, cheap resources, etc.; i.e. when all its social, cultural, natural, and spiritual wealth has been taken from it. Eventually the developed nations would run out of colonies; then, the Marxists said, finally we will have our Revolution.

Unrecognized by the classical Marxist thinkers was another kind of colonization that has happened simultaneously with the geographical: the colonization of non-physical territories. The nature of financial capital, which is written into our current system of money, which is itself a projection of the way we understand the universe, constitutionally requires constant growth. Maintaining the growth machine requires constantly finding new sources of demand. These can be found in other countries for existing products; alternatively, new demand can be created at home by finding "unmet needs" (as they are called in standard economics). But what is an unmet need? Usually it is some aspect of our humanity that already exists—it is not really new—but that has heretofore existed outside the money economy. In other words, it is a form of wealth that has not yet been converted to money.

William Greider, Daniel Korten, and many others argue persuasively that the crisis of capitalism is already imminent. Greider in particular, writing in 1998, gives a compelling account of all the elements of the Marxian crisis: overcapacity of production, falling profits, concentration of ownership, and the downward spiral of wages on a global scale.[46] Eight years later, the deflationary depression he warned against has not come to pass—and not because any government took his prescriptions. But perhaps it is not that Greider was incorrect, but only that he underestimated the non-geographical domains that have not yet been fully colonized.

The Socialist solution to the Marxian crisis fails because it doesn't get at the root of the problem, which is not the private ownership of property but rather the concept of property to begin with. And the concept of property, as we have seen, itself depends on our definition of ourselves and the way we understand the world. Greed, competition, anxiety, and scarcity are built into our cosmology and our science, and will not go away until and unless the other pieces of the pattern change as well. Specifically, they will not go away when we see the world, nature, language, thought and idea as things that can be cordoned off and owned as objects.

The characteristics of capital upon which Marx based his theory stem from the characteristics of the money system that currently dominates, which features a "scarce" interest-bearing currency created by banks. Very different economic dynamics would result from a money system with different characteristics, and a different money system will spring forth naturally from a different understanding of and relationship to the world. A very different form of capitalism will then emerge, to which Marx's dynamics do not apply, which fosters cooperation over competition, sharing over exploitation, community over separation; in other words, which values and develops all the non-monetary forms of wealth described herein. It will be a system that makes us richer in life, rather than poorer in life but richer in money. And significantly, it will be a system that is not *designed* in the usual sense but which is allowed to grow. This is the true revolution: not a superficial overthrow of whatever powers happen to be, but a radically new understanding of self and world. Marx criticized the bourgeois revolutions of America and France for merely replacing one set of owners with another. Yet isn't his own revolution similarly shallow, leaving untouched the concept of property, the duality of labor and leisure, the ideology of growth, and the assumption of human domination over nature?

Why is such a deep revolution necessary? The history of technology, at least back to agriculture and probably back to the pre-human technologies of fire and stone, is closely linked to an increasing objectification of the world. We conceptually separate ourselves from the environment in order to manipulate it; equally, our successful manipulation of the environment spurs our conceptual separation from it. The concept of property naturally flows out of such an objectification, and the kind of money we have today arises naturally out of such property, which is mine and not yours, which can be accumulated and measured. It is foolish to think that any system other than the capitalism we are familiar with could arise from such a prior foundation.

A revolution that leaves our sense of self and world intact cannot bring other than temporary, superficial change. Only a much deeper revolution, a reconceiving of who we are, can reverse the crises of our age. Fortunately, to use the language of Marx, this deepest of all possible revolutions is *inevitable*, and it is inevitable for precisely the reasons Marx foresaw. The conversion of all other capital into money is unsustainable.

Someday it will run out. As it does, our impoverishment will deepen. Misery and desperation will overcome whatever measures can be invented to suppress or narcotize them. When at last the futility of controlling reality becomes apparent, when at last the burden of maintaining an artificial self separate from nature becomes too heavy to bear any longer, when at last we realize that our wealth has bankrupted us of life, then a million tiny revolutions will converge into a vast planetary shift, a rapid phase-transition into a new mode of being.

It will happen—must happen—perhaps sooner than we think. Indeed it is already happening. Our social, natural, cultural and spiritual capital is almost exhausted. Their depletion is generating crises in all realms of modern life, crises which are seemingly unconnected except that they all arise from the monetization of life or, underneath that, from our fundamental confusion as to who we are, our separation from nature, ourselves, and each other. This is the link that connects such disparate phenomena as peak oil,[47] the autoimmune disease epidemic, global warming, forest death, fishery depletion, the crisis in education, and the impending food crisis. Both monetization and separation are nearing their maxima, their greatest possible extremes. The former is in the completion of the conversion of common wealth into private wealth that I have described in this chapter; the latter is in the complete sense of isolation and alienation implicit in the world of Darwin and Descartes: the naked material self in a world forged by chance and determinism, where purpose, meaning, and God are, by the nature of reality, nothing more than self-delusory figments of the imagination.

Paradoxically, it is in the fulfillment of these extremes (each of which is a cause and an aspect of the other) that their opposites are born. Yang, having reached its extreme, gives birth to Yin. The depletion of social capital launches the revolution that will reclaim it. The agony of separation births the surrender that opens us to a larger version of the self, to nature and to life. But the extremum must be reached.

As any environmental scientist knows, it is certain that things will get much much worse for the bulk of humanity before they get any better. Certain forces must play themselves out. The momentous rise in spiritual, humanitarian, and ecological awareness will not save us, not because it is too late (though it is), but because the course of separation has not yet reached its finale.

Like an alcoholic whose resources of goodwill, money, pawnable assets, friends, and credibility are almost exhausted, our way of life is on the verge of collapse. We continue to scramble, applying new technological fixes at greater and greater cost to alleviate the problems caused by the last fix. The addict will keep on using until life becomes completely unmanageable. Ecological awareness, localism, green design, herbalism, community currencies, ecology-based economics are all like the drunk's moments of clarity on the way down. They will not so much save us as serve as the seeds for a new way of living and being that we will adopt after the collapse. Indeed they will all come naturally, as a matter of course—if there is anything left at all.

CHAPTER V

The World under Control

The Total Depravity of Man

Starting with the earliest technologies of fire, stoneshaping, and symbolic culture, the ascent of humanity is a history of ever-greater control over nature and human nature. Its culmination would be to make that control complete. Although most people no longer think it possible to win a complete victory over disease, suffering, and death (that is, a complete victory over nature), the Technological Program lives on in unspoken assumptions and attitudes, the dream of a technological Utopia. Life will get better and better, safer and safer, more and more convenient, efficient, modern, clean, automated, secure.

Underlying this belief is the assumption that we are on the right track and indeed that we have come a long way already. It assumes that life in the raw is, as Thomas Hobbes so famously put it, "solitary, poor, nasty, brutish and short." The tenacity with which we cling to this prejudice reveals just how much depends on it—nothing less than the whole ideology of progress. Progress (in the sense we know it) is only meaningful if we are ascending from a lower to a higher state—from the natural realm to a separate human realm. Extrapolating backward, we must believe that original nature and original human nature are bad. Hence, the goal of control: technology to control nature, culture to control human nature.

This chapter will outline the all-encompassing consequences of the program of bringing life and the world under control. While collectively we seek to dominate and subdue nature through science and technology, the goal of control casts a shadow into our personal lives as well. We feel we need to control ourselves, our bodies, our emotions, our impulses, our appetites. Why? Because in a personalized version of the Hobbesian view of nature, our natural selves are bad too. The psychological "technologies" we use to achieve self-control are legion, internalized from earliest childhood: guilt, willpower, discipline, shame, motivations, threats, rewards. We must govern and control ourselves, just as we must govern and control nature through technology to make it good—an ordered garden as opposed to an inhospitable wilderness. Indeed, as the uncivilized human *is* in a state of nature, the two statements of inherent badness (of wild nature and uncontrolled human) are identical.

In religion this assumption is embodied in the concept of Original Sin and, more broadly, the idea found across institutionalized religions that spirituality consists of a struggle at self-improvement. We strive to raise ourselves above the temptations of the flesh into the realm of the spirit, to overcome our bestial inclinations, to exercise restraint and self-control. We "hold ourselves" to a code of morality.[1] Whether it is Hinduism, Christianity, or any other world religion, it is a basic doctrine that we should try hard to be nice. The same applies, perhaps even more strongly, to ethical and moral systems that are not explicitly religious—another indication that the seeming cultural divide in our society between the secular and the religious is mostly a matter of appearances.

The idea of Original Sin is central to most Christian churches today, although it was once hotly disputed by such Church fathers as Pelagius and fudged by later figures such as Thomas Aquinas. The founders of Protestantism, Martin Luther and John Calvin, argued for the "total depravity of man," the innate sinfulness of human beings.[2] For them, trying hard to be nice wasn't good enough! This is a crucial doctrine because it is the foundation of the entire dogma of Christ as the supernatural Redeemer, an agent of divinity outside ourselves, our one and only salvation.[3] Also, as Abraham Maslow observes,

> any doctrine of the innate depravity of man or any maligning of his animal nature very easily leads to some extra-human interpretation of goodness, saintliness, virtue, self-sacrifice, altruism, etc. If they can't be explained from within human nature—and explained they must be—then they must be explained from outside of human nature. The worse man is, the poorer a thing he is conceived to be, the more necessary

becomes a god.[4]

It is not just religion that depends on the innate depravity of the human species. The doctrine extends beyond theology to atheistic psychology, most famously in the writings of Sigmund Freud, but more recently in works of various Darwinian sociobiologists and cognitive psychologists such as Richard Dawkins and Stephen Pinker.[5]

Here is a revealing irony: although Protestant fundamentalists vilify Darwinism and seek to keep it out of the public schools, their view of the innate sinfulness of man is in complete agreement with the Neodarwinian account of life's origins and evolution. Each dovetails beautifully with the other. This should come as no surprise, since both Darwinism and Original Sin arise out of the same deep cultural conceptions of self and the universe.[6]

The nature implicit in Darwinism, including human nature, is not very nice. It is a nature in which competition is the rule, in which the deepest purpose of life, the deepest motivation of behavior, is to survive and reproduce, and in which cooperation is an occasional, coincidental product of an alignment of interests. Cooperation is okay, but it is even better to trick other organisms into helping you while you refrain from expending any energy helping them. As Richard Dawkins writes, "Natural selection favours genes which control their survival machines in such a way that they make the best use of their environment. This includes making the best use of other survival machines, both of the same and different species."[7]

The definition of the self implicit in the dominant theories of biogenesis and evolution is that each of us is a discrete, separate being struggling against other such beings to survive and reproduce. In this view, from the unicellular stage to the present, successful organisms are those that are better able to look after their own interests at the expense of their rivals, a rival being defined as anything competing for the same resources. Any organism programmed by its genes to enact behaviors that diminish its chances of surviving and reproducing—for example sharing resources when there may not be enough for itself—is less likely to reproduce, and those genes will die out of the gene pool. In other words, we are programmed—life is programmed—to profit at the expense of other beings. Can you think of a better definition of "not nice" than that? Selfishness being our nature, of course we need laws, morals, and self-control to rein in that selfishness and become civil beings.

Thus, mainstream science and mainstream religion agree that we are by nature bad; that therefore, just as we must control Nature, we must control, regulate, improve upon, and yes, dominate ourselves in order that we may be good. Let us add to this agreement economics, which holds as a central axiom that people are driven to maximize self-interest. If science, religion, and economics all agree, the doctrine of innate badness must have deep roots indeed. It is no accident that it is orthodoxy in both the scientific and religious realms. The ideology of our civilization—progress, ascent—depends on it. It is therefore built into many of our reflexive assumptions of what is true. It is also built into our money system, thus generating the very same behavior that we mistake as fundamental.

Because it is imbued so deeply into our mythology, our ideology, our culture, and our economy, the doctrine of Original Sin is actually *correct*—correct insofar as we are immersed in our culture. It is correct given our ideological infrastructure and the motivations built in to our cultural institutions.

Hobbes wrote *Leviathan* long before Darwin ever conceived of evolution via natural selection and long before the theory of the selfish gene was ever invented. The "(human) nature is bad" idea goes back way before Darwin, before Hobbes, before Luther and Calvin. It is implicit in dualism itself, which finds its origins in the so-called Neolithic Revolution if not before, even if its full articulation was not to come until the Scientific Revolution.

Dualism is the idea that the universe is divided into two parts, which go by the names of matter and spirit, God and creation, human and nature or, most fundamentally, self and other. These two parts are by no means symmetrical: self is more important, other less. In religion, soul is self, body is other. The soul is important, while the flesh is at best irrelevant and at worst an impediment to the life of the spirit. Outside religion, the same dualism, following Descartes, manifests as mind = self, body = other. Either way, we identify with our minds and not our bodies, not other life forms, not the world at large. Even if we try hard to cultivate compassion, a narrow identity with an illusory separate self is built in to our deepest worldview. (In fact, that we feel we need to "try" to be compassionate is symptomatic of that worldview.) No wonder we are so out of touch with our bodies and suffer such chronically poor health. No wonder we treat our bodies and our material planet so cavalierly. No wonder we visit such violence on these (unimportant) others—other religions, other nations, other races, other species.

On a collective level, dualism manifests as a distinction between man and nature, again dividing the universe into two parts, one of which is Self, and therefore important, and the other of which is Other, and thus only important to the extent that it affects Self. Dualism motivates well-meaning arguments for "conservation" of natural "resources," locutions which imply the subordination of Nature to man and define the things of the world in terms of their usefulness to ourselves. Rainforests are to be preserved because—who knows?—we might derive important medicines from them some day. And just imagine the economic losses from topsoil erosion! Such arguments are counterproductive because they end up reinforcing the very mindset that is at the root of the problem to begin with: that the world is an Other, here for our use.

The logical conclusion of dualism—that Other is important only to the extent it affects Self—is a hidden abscess constantly leaking poison into the body of our civilization. It is a universal acid that erodes away the core of any system of morals, ethics, and responsibility, which readily succumb to a succession of pragmatic "Why should I's?"

Excepting pragmatic reasons, why should I care about anything outside my self? The usual religious answer, "Because God says to" or "Because God will punish me otherwise" gets us nowhere, because that morality too comes down to pragmatism—avoiding divine retribution for our sins. It is really no different. Must we be scared into being good? Must we exercise self-control, knowing that we are foregoing our natural self-interest that we could maximize by being ruthless to the world? Is eternal struggle the only alternative to depravity?

Why should I, as an individual, do anything for anyone else? Why shouldn't I pollute to my heart's content? Why shouldn't I steal your wallet provided I know I can get away with it? Here is a dialog I recently had with my class:

Me: Why shouldn't I just pollute as much as I want to?

Class: You might poison your own environment.

Me: Yes, you're right—I'd better make sure only to pollute my neighbors' yards, not my own.

Class: If you get caught you'll suffer consequences that outweigh the benefits of polluting.

Me: Okay, I will only pollute to the extent that the benefits outweigh the risks. I'll do it in secret, while maintaining the appearance of being a nice guy. I'll get an unscrupulous business to pay me cash to dump toxic waste into a sinkhole at night. No one will ever know. Or, let's say I'm a corporation and I want to pollute to maintain higher profits. I'll fund biased studies and create PR that says pollution isn't so bad after all. Why shouldn't I?

Class: But if everyone did that you would be doomed.

Me: True. That's why I support laws and morals that prevent others from doing it. But why shouldn't I do it myself if I can get away with it? (Why shouldn't I manipulate "other survival machines, both of the same and different species"?)

In other words, what does the rest of the world matter? The discrete and isolated self of Descartes implies a discrete and separate universe outside that self. As long as I can insulate myself from the world out there—possible in principle precisely because it is separate—I can do anything I want to. The only limits are pragmatic ones. On a case-by-case basis, I can judge whether each action maximizes my benefit. Should I steal that man's wallet that he left on the table when he went to the bathroom? Let me calculate the costs and benefits. The potential payoff is $200, the risk of getting caught in this setting is only about 2%, the fine and loss of face is worth maybe $7,000… calculating, I arrive at an expected payoff of $56 (that is, $200x98%-$7000x2%). So I steal it. Raise the fine to $11,000, and I won't. Right?

It seems absurd, but this is the kind of logic behind a legal system based on deterrence. This system seeks, through the imposition of penalties, to convert behavior that would otherwise be in an individual's rational self-interest into irrational behavior. Without a penalty of some sort, stealing would be in our rational self-interest.

Let us pursue this line of inquiry a little further. Everyone knows that we don't actually perform an explicit calculation every time the opportunity to profit at another's expense presents itself. The decision has been automated through the moral training of our childhood. Our parents, teachers, and other authorities train us to not be selfish by rewarding "nice" behavior and punishing "selfish" behavior. They provide an incentive for niceness that makes it no longer irrational but actually in line with selfishness. Eventually we internalize these incentives in the form of guilt, conscience, and habit. Our natural selves are selfish, ruthless, and depraved, requiring a long period of training to subdue nature and foster morals, ethics, and decent behavior.

Here we see an inextricable link between dualism and control. Bereft of an organic indwelling spirit, the material world that science presents us—indifferent, purposeless, and ruthlessly competitive—cries out for the control we call technology. Ruled by bestial or sinful drives, the physical body living in that world similarly calls out for the control we impose through moral training, education, and culture.

The program of control, in turn, demands the reduction of the world described in "The Origins of Separation"; that is, the finitization of the infinite. In science we seek to control variables, in engineering to account for every conceivable force with our equations. Our efforts at control fail when the infinite leaks back in—an uncontrolled variable in science, an unforeseen circumstance in life, "human error" in a factory. Always, the response to such accidents is to extend control to them too, to make the system failsafe. Yet after thousands of years, no matter how hard we try, infinity keeps creeping back in.

The Winners and the Losers

Under the sway of dualism, we have essentially sought to divide the world into two parts, one infinite and the other finite, and then to live wholly in the latter which, because it is finite, is amenable to control. We are like the frog who jumped into a well and, unable to see anything else or remember the vast world beyond, declared himself suzerain of all the universe. Our lordship over nature is at heart an egregious self-deception, because its first step is to attempt nature's precipitous reduction, which is equally a reduction of life, a reduction of experience, a reduction of feeling, and a reduction of being: a true Faustian exchange of the infinite for the finite.

This reduction comes in many guises and goes by many names. It is the domestication of the wild; it is the measuring and quantification of nature; it is the conversion of cultural, natural, social, and spiritual wealth into money. Because it is a reduction of life, violence is its inevitable accompaniment (I can think of no better definition of violence than the reduction of life); hence the rising crescendo of violence that has bled our civilization for thousands of years and approaches its feverish apogee as we conclude the present wholesale destruction of entire species, oceans, ecosystems, languages, cultures, and peoples.

From the weeding of a strawberry bed to the coercion of a child to the elimination of enemies in the name of national security, the cultivation and control of the world inherently requires violence. Violence is a built-in feature of our worldview; it is implicit in our conception of ourselves as separate beings in a universe of discrete others competing for survival. Moreover, the objectification of other beings, species, people, and the earth itself enables and legitimizes violence toward them. Violence seems not to be violence when it is only a weed, only an animal, only a savage, only an enemy, only a thing. Dehumanization of the victim is a well-known enabling device for torture and genocide, but dehumanization—turning human into object—is just a special case of our off-separation from the rest of the world. To the extent that this is an artificial, illusory separation, the root cause of violence becomes clear. It is simply the result of an ignorance of the very deepest kind—that we know not who we are.

Thus it was that the great avatars of peace in human history counseled not more self-control, not an intensification of the effort to be nice, but rather a surrender into our true selves, which are not the separate, discrete selves of present-day science and religion. Buddha was suggesting no less when, asked "What are you?" by people awestruck at his radiance, answered simply, "I am awake." This is what a human being really is. And Jesus too was saying no less when he spoke of God's love—not for what we might or could be, but for what we truly are. Moses on the mountaintop asked the divine source, "Who shall I say sent me?" The answer: "I AM". I am what? Everything and nothing. When you take us apart, that special part of us we call self (soul, spirit, mind, consciousness) is not there. Thus we are nothing. When you separate us from any other part of the universe, we are less. Thus we are everything.

Because our demarcation of self and other is a false one, the violence we commit upon the other is actually committed upon ourselves. Here again we find a warning in some of our most venerable spiritual teachings. The doctrine of karma states that the effects of our actions are inescapable, that what we do to others we do to ourselves. Yet, characteristically, our religious institutions twist it to mean, "Be good or you will be punished." The Golden Rule works the same way. Since its original meaning, "As you do unto others, so you do unto yourself," is incoherent nonsense to the dualistic mind, we have perverted it into a rule, a standard of behavior to strive toward. Originally, both the doctrine of karma and the Golden Rule were mere statements of fact based on a different conception of self.

The statement "Love thy neighbor as thyself" (*Leviticus* 19:18, *Matthew* 22:39, and elsewhere) falls victim to the same dualistic misinterpretation.

Instead of a rule, we might construe it as a simple statement of fact: As you love your neighbor, so *do* you love yourself. Self and neighbor are not actually separate. Jesus was not going around uttering simple moral platitudes. However, as he was speaking to people immersed in the myth of the separate self, it is no wonder that his teachings were immediately misinterpreted and written down in their current form. The prescriptive and proscriptive forms of spiritual teachings—the do's and don'ts—coincide with the institutional interests of the political powers that coopt all religious movements from the moment of their founders' deaths, if not before.

The way the dualistic mind grapples with the idea of karma or the Golden Rule is to assume the existence of an external Judge, a cosmic referee who weighs one's actions and metes out the corresponding rewards and punishments. When you and your neighbor are fundamentally separate, then there is no intrinsic reason why what you do unto your neighbor should necessarily come back to affect yourself—especially if you take sufficient precautions. An omniscient God is needed to *bring* those consequences to you. The idea of God as a separate power external to nature who enforces morality is therefore a crutch for the mind lost in the myth of separateness, a means of understanding the consequences of our treatment of Others. As such the idea of a supernal God might have a salutary effect in the short term, but runs the risk of reinforcing the illusion of separation. That is what has happened in most contemporary Christian churches, which treat humanity's trashing of the earth as a secular matter, not one of the utmost spiritual urgency.

From the dualistic perspective, what I've heard some Christians say, that there can be no morality without a belief in God, is actually true. "If God does not exist, everything is permitted."[8] Otherwise, what is to stop me from doing ill to my neighbor, as long as I take enough precautions? In theory, I can get away with it. We can go ahead and clearcut the forests. Maybe the consequences won't come back to haunt us, and if they do, we can manage them. This leads again to a world under control. We can get away with treating the world as a resource, a thing, an other, as long as we can manage the pragmatic consequences of that—and there is no reason to think we can't. We can do unto the world whatever we like with impunity, as long as we are clever enough. This all changes with the understanding that self and world, man and nature, are not truly separate. For then there is no escape; then, any effort at control can only postpone the inevitable. The idea of a separate, omniscient, judging, rewarding and punishing God mediates this understanding to the dualistic mind. Unfortunately, the dualistic conception of God invariably leads us to think in terms of pleasing God, obeying God's rules, seeking to gain divine rewards and avoid divine punishments. It leads us, in other words, right back to the regime of control.

By creating an artificial cleavage between self and other, the dualism that divides the world into two parts severs us from part of ourselves. It leaves us partial beings. To fill up the incompleteness we add more and more to ourselves, more property, more material and spiritual baggage, more ego, more self-importance: an expanding territory of self that seeks to subordinate the whole world by bringing it under control. However, by seeking to own as much of the world as possible, we only exacerbate the alienation from the rest of the universe, whose infinity dwarfs us into insignificance no matter how much we acquire. The interior wound—the loss of our inner connection to nature—is never healed by the accretion of more and more self on the outside.

The root of the world crisis is not inherent selfishness or greed that must be overcome. The solution is not to make do with less, sacrifice our best interests, or impose limits on ourselves. These solutions spring from a mentality of scarcity, and it is precisely the mentality of scarcity (which is implicit in the quantization and propertization of the world that makes the infinite finite) that motivates us to hoard and accumulate, possess and own, keep and guard, fence and control. Consider the opposite mentality: "The good things of the world are abundant. There is plenty for everyone. My needs are abundantly provided. I can have my heart's desires." Someone living in such a mindset has no need to own, hoard, or control, because after all, there is plenty. Indigenous hunter-gatherer societies amazed their initial visitors with their guileless, open-handed generosity. As Columbus wrote of the Arawak (before murdering and enslaving them), "They are so ingenuous and free with all they have, that no one would believe it who has not seen it... Of anything they possess, if it be asked of them, they never say no; on the contrary, they invite you to share it and show as much love as if their hearts went with it..."[9] Was an intense acculturation process applied to Arawak children in order to override their inherently greedy, selfish natures and impose the desire to share? I doubt it. Intuitively, we describe their behavior as childish, innocent, or guileless—adjectives that suggest that theirs is the original state of the unspoiled human being. Fully immersed in the reality and the mentality of abundance, sharing came naturally. When "mine" is not a concept, sharing is easy.

Characteristically, some modern anthropologists try to explain away primitive generosity and gift relationships in the terms of economics: costs and benefits to the separate self. Marcel Mauss set the tone in the early 20th century when he tried to analyze gifts as a competitive means to put other under obligation[10], an idea echoed by modern theorists such as Richard Posner.[11] Both see gift-giving as a means to generate prestige and as a kind of insurance policy in the absence of accumulative mechanisms in primitive societies—if I give you a gift you "owe me one". While such motives undoubtedly exist, to see them as primary is to project our own cultural biases. Such ideological contortions are necessary to preserve the assumptions upon which they rest—our civilization's understanding of self and world. From the perspective of the discrete and separate self, altruism simply does not make sense unless some transactional model can reduce it to selfishness in disguise.

Derrick Jensen describes Ruth Benedict's attempt to categorize cultures into good (life-affirming, non-violent, egalitarian, gentle, friendly, and free from cruelty, harsh punishment, exploitation, jealousy, humiliation, and depression) and bad (violent, aggressive, cruel, warlike, competitive, and hierarchical), and to discover what rule or factor distinguished them.[12] Her conclusion: "The social forms and institutions of nonaggressive cultures positively reinforce acts that benefit the group as a whole while negatively reinforcing acts (and eliminating goals) that harm some members of the group. The social forms of aggressive cultures, on the other hand, reward actions that emphasize individual gain, even or especially when that gain harms others in the community." So it is not necessary that "mine" not be a concept, only that social forms and institutions discourage the accumulation of mine. Universal mechanisms of sharing, for instance, obviously make the hoarding of possessions irrational.

But before we conclude that we must change our "social forms and institutions," it would perhaps behoove us to dig a little deeper and ask, "From what basis do these social forms and institutions arise?" While hunter-gatherers indeed had intricate mechanisms to maintain an egalitarian society, these social forms arose naturally out of their sense of self, which was still not so rigidly defined as discrete and separate. Possibly, the objective, outward rules and taboos—Benedict's "forms and institutions"—only became necessary as the separation of self progressed through the Paleolithic. In any event, we cannot hope to produce, *ex nihilo*, new social forms of positive and negative reinforcement in the absence of the proper substructure: a broader, more open sense of the self.

We have become confused. Rational self-interest has become the dupe of our culture's perceptions, so that it is neither rational nor in our interest. Our selfish behavior is only superficially so; actually it conflicts with our true best interests. Chief among such behavior is that which ruthlessly maximizes the perceived benefits of the skin-encapsulated ego. Limiting our destructiveness is *not* a matter of reining in our natural selfish impulses; it is a matter of understanding who we really are. When we do not know who we are, of course our selfishness cannot benefit our true selves. Hence, the endemic misery in our society among its winners and losers alike.

It would be one thing if, indeed, the world were essentially a competitive arena destined to have winners and losers. We would then be justified in making every effort to be among the winners. The sad truth, though, is that in our society, the winners are among the biggest losers of all. Now, that is a bold statement indeed. Reading it, you may suspect I am deficient in knowledge or compassion. What are the petty troubles of the rich compared to the horrendous suffering of our culture's victims? What's a little angst or depression next to starvation, destitution, murder, genocide, tyranny, torture, the smashing of cultures, the looting of ecosystems? Surely I must be oblivious to the true magnitude of the horror.

The litany of our culture's victims is nearly endless—the indigenous peoples, the poor, the ecosystems that have been sacrificed in the interests of wealth and power—but we could hardly blame the exploiters if the alternative were to be themselves among the victims. Who can blame someone for being good to themselves? If the world is in essence "lunch or become lunch" (as I once saw ecology defined), then we cannot blame someone for striving to be in the former category. In such a world, an appropriate ethical system would have the winners be as nice as possible to the losers, offering safety nets to the poor, remediation to the environment, limits on how big a winner you can be. This, in essence, is political liberalism, which does not question any fundamental assumptions. In addition, saintly individuals and their imitators might cast themselves among the losers *on purpose* (even though they could be a "winner" if they so deigned), thereby demonstrating just how nice they are, refusing to take more than their share, nobly sacrificing their chance to enjoy the rewards of privilege. Of course, unless you actually *are* a saint, this self-sacrificial mentality eventually generates resentment at those who decide to enjoy the fruits of being a winner, a resentment often apparent in social and environmental activists. But all of this assumes that the winners really are winners. And that is a deception! Our winners have successfully maximized their "rational self-interest" only to find the promise of secure happiness betrayed.

Do not waste your energy being angry at the rich and powerful. As the Bolsheviks unwittingly demonstrated, nothing much changes even if the rich and powerful are overthrown. Moreover, that anger is in fact counterproductive. Often, the hidden message of activist rhetoric is, "Do not be too good to yourself," or "You are bad for being good to yourself." No wonder so many people are turned off. Those who rely on guilt or shame to persuade us to limit our participation in the destruction of the planet and its people are, in a very subtle way, perpetuating some of the deep axioms that drive the destruction in the first place. They are resorting to a form of control, control over an iniquitous human nature. In a subtle way, they reenact and reinforce the same war of conquest that has left the planet in tatters.

Another hidden assumption is that the good life, whether we unabashedly pursue it or nobly sacrifice it, is actually a good life. It is not. We are chasing a mirage. We have been tricked, duped into the aggrandizement of a narrow self that ultimately doesn't even exist.

The clearest indication of the fraud comes when the program of control (temporarily) succeeds. Even when all the bases are covered, every eventuality anticipated, even at the very pinnacle of health, wealth, and power—even then an enervating ennui creeps into life, starting with the empty crevices, the moments of boredom, and ultimately spreading to engulf the entirety of existence. Initially deniable, perhaps, by intensifying the success, the power, the stimulation, it comes back stronger and stronger until it holds every waking moment in merciless thrall. This is not a new phenomenon, as Lewis Mumford observes:

> An Egyptian story… reveals the emptiness of a Pharaoh's life, in which every desire was too easily satisfied, and time hung with unbearable heaviness on his hands. Desperate, he appealed to his counselors for some relief from his boredom; and one of them put forth a classic suggestion: that he fill a boat with thinly veiled, almost naked girls, who would paddle over the water and sing songs for him. For the hour, the Pharaoh's dreadful tedium, to his great delight, was overcome."[13]

And in Ecclesiastes we read,

> … I gathered me also silver and gold, and the treasure of kings and of

> the provinces; I gathered me men-singers and women-singers, and the delights of the sons of men, musical instruments, and that of all sorts... And whatsoever mine eyes desired I kept not from them; I withheld not my heart from any joy; for my heart rejoiced because of all my labor; and this was my portion from all my labor. Then I looked on all the works that my hands had wrought, and on the labor that I had labored to do; and, behold, all was vanity and a striving after wind, and there was no profit under the sun."

The depredations of our culture in the name of security, ease, and control have created untold misery in this world, *and for no good reason.* Not only is life at the pinnacle of success infamously devoid of intimacy, community, authenticity, and meaning, but even the security and control to which these have been sacrificed is a sham. For the sake of security we have shut off real living, and in return have received not even security but only a temporary semblance of it.

Imagine how fatuous our attempts at security will seem, when one day the multiple crises engendered by the regime of objectification and control converge to engulf us. How pathetic, how futile are our locks and gates, insurance and investments, résumés and expertise, when the intrusion of a personal calamity puts the lie to our illusion of control. All the more for the collective, global calamities that are already written into our future.

To reverse the tide of destruction that has engulfed our living planet, our society, our communities, and our psyches, it will not be enough to try harder to be nice. Religions have enjoined us to that attempt for thousands of years. It hasn't worked. Yet, religion and conventional control-based morality concludes that we must try harder yet—we must do even more of what hasn't worked. If that were the only solution, it would be bad news indeed, and the despair of so many activists would be justified. Even if trying harder succeeded, we would be consigned to an eternity of *trying*, of struggle. Is peace, sustainability, and goodness, individually or for the world, only possible through unending struggle against ourselves?

This book proposes another way: a shift of consciousness that will expand our sense of self and thereby change our selfishness into a force for healing. The problem is not selfishness, it is that we misconceive the self. We need not expend superhuman efforts to build a tower to the sky. The sky is all around us already.

Life, Death, and Struggle

From the standpoint of the separate self, the ultimate victory of the Technological Program would be to triumph over death itself. In one form or another, this goal drives all of our efforts to dominate, accumulate, and control. It exists in dilute form in the pursuit of security; it is written into our ideology of competition and survival of the fittest. Yet, despite propaganda to the contrary, millennia of intensifying control have done nothing to hold death at bay. The predominant causes of death have changed, but not its inevitability (and not nearly as much as we would like to think the human lifespan).[14] As always, we humans are born, we live, and we die.

Failing to overcome death, we instead seek to deny it by hiding it away and pretending it won't happen. In polite company we deny it with euphemisms. Most children's stories of the last fifty years simply never mention it (in contrast to the fairy tales of yore).[15] In our funerary practices we deny death by applying cosmetics and embalming, and by preserving corpses for centuries in lead-lined coffins. We hide away our sick and dying in hospital intensive care units as we go through frantic medical rituals to preserve just a few more hours of life at whatever cost. We buy packaged food that obscures the fact that an animal or plant had to die. In our religion we deny death through various conceptions of an afterlife.[16] In our lives, we pursue goals and ambitions that simply would not make sense if we had integrated the fact of our eventual dying. And in our scientific fantasies, we imagine that someday we can transcend death through nanotechnology, genetic engineering, cyborg technology, or computers.

So successful are we at hiding death that many people grow up having never seen a dead person, surely an unprecedented phenomenon on planet earth. Thanks to our pusillanimous storybooks, polite euphemisms, and hiding of corpses, from the perspective of the child it is as if death did not exist at all. It is an impression strangely consistent with the ubiquitous violence of television cartoons, in which characters survive a ten-ton safe or a bomb falling on them with no permanent ill effects. It is also strangely consistent with the graphic and deadly violence of adult television and movies, in which the killing of legions of bad guys is shorn of nearly everything that makes death death.

The deep reason why our culture finds it necessary to deny death is that death puts the lie to the agenda of our narrowly-conceived selves. Contemplation and integrated awareness of death reveals the unreality, or the conditional reality, of the discrete and separate self. Because the self that we define as our bodies, our names, our knowledge, our possessions, our self-image, and our stories—Alan Watts' "skin-encapsulated egos"—did not exist before we were born and will cease to exist when we die, that self is unreal, impermanent. Along with it, the dualism of self and environment is unreal as well. It is no accident that many spiritual traditions have practices consisting of the contemplation of death. That is also why a close brush with death is so transforming: we stop worrying about the trivial concerns of life and walk in the knowledge, as one near-death survivor put it, that "only love is real." When death exposes the impermanence and conditional reality of the self as we know it, all of the behaviors based on its aggrandizement no longer make sense. Society as we know it is based upon these behaviors; hence society's need to euphemize, hide, and deny death.

Because it accompanies the denial of our true selves, our denial of death is equally a denial of life, and separation from death is separation from life. Death punctures the illusion of the self (as we conceive it) and on a psychological level rends the fabric of the modern world; hence the cultural necessity of denial. When will we cease denying death? Probably only when we no longer can, when it invades our lives so forcefully that the old illusions cannot stand. As the social, ecological, and physiological underpinnings of health accelerate their decay, this invasion is gathering in force and ubiquity. On the personal level, it comes as the realization that the lives we had been living simply do not make sense and cannot go on. On the collective level this realization is precisely that of the "unsustainability" of modern society that the environmentalists keep telling us about. Just as a person often has to become very, very sick or experience a close brush with death in order to wake up to life, so it may also be necessary for the same thing to happen on the planetary level before we as a species wake up to the fraudulence of our dualistic conception of ourselves as separate from nature. When our collective survival is imminently, dramatically, and undeniably in danger, and only then, our present collective behaviors and relationships to the rest of life will cease to make sense.

Cultures less enslaved to the myth of separateness feared death less, seeing it not as an ending but a return. Our fear of death is a product of the same linear thinking that underlies interest-money and all the other expressions of the Age of Fire. It would not occur to someone immersed in the cyclic worldview that underlies the economy of the gift. By virtue of being born, we are all sojourners in the play of separation, but few before us have wandered in so deeply as to forget the wholeness from whence we came. Almost. We have not really forgotten—can never forget—what we are; it remains as a stirring in the heart, enduring forever until the world under control, the world of the separate self, cannot hold. Our interbeingness, banished from waking experience, manifests as a longing, an empathy for other living beings and even inanimate objects.

Characteristically, psychologists consider it a mild pathology to project human attributes onto inanimate objects; even animals and plants we invest with a lesser degree of being than ourselves. (Not *thinking*, they—per Descartes—have less "amness".) Far from mental illness, I see these tugs of identification with the other to be glimpses of an underlying truth; it is only because they deny our presumed separateness that we dismiss them as irrational. Of course, according to the same premise, falling in love is irrational too. What reason is there to care for anyone else? Why should I strive for another's benefit? What's in it for me? Love (and any form of compassion) crumbles the walls of separation. However we may explain it away according to some biological imperative, anyone who has been in love knows there is more to it than a procreative urge or an unconscious bargain for mutual aid. The ecstasy and bliss that suffuses lovers in full flush is an outcropping of the underlying joy inherent in our true state of union. Love, breaking down the barriers between me and thee, connects us to that state. It is our lifeline to sanity, to the wholeness of what we are. (That is also a pretty good definition of Christ, an identity the early Church fathers recognized in their more lucid moments.) Love saves us from our limited ego-selves, breaches the fortress of our separation, rescues us from the inevitable Hell that our addiction to control creates.

As many poets and mystics have perceived, love and death are intimately related. Both involve a dissolution of the ordinary boundaries of self. On the individual level, this dissolution is the felt realization: "You and I are not really separate." On the collective level, it is the realization that human beings are not separate from nature. Another way to view the shift of consciousness that this book heralds is that we will fall back in love with the world.

The realization of the unity of human and nature tells us that "What we do to the world, we do to ourselves"—a concept that I'll refer to as "ecological karma." This goes beyond interconnectedness or the pragmatic realization that we "depend upon nature." The human/nature dualism implies that we can avoid the consequences of our depredations and manage their costs and benefits. We can choose this forest to clear-cut and that mountain to stripmine, because according to our rational calculations they are less important than other ones; they are not "critical ecosystems." We don't really need them. Their value is conditional. We feel we can preserve some and destroy others, as long as we calculate the effects carefully enough, the costs and benefits. In contrast, the understanding of ecological karma tells us that we can never manage or avoid the effects of our actions; that any marring of the health and beauty of the planet diminishes our own health and the beauty of our own lives as well, *inescapably*. It tells us as well that the laws of nature make no exception for human beings.

The new conception of self, personal and collective, that will be born from the collapse of the present one will bring with it a very different kind of technology, economy, medicine, and way of life that seek to create beauty and that model themselves on the processes of nature. It is not so much a "return to nature" as a returning of nature to us, a recovering of nature, a reidentification with nature. Of course, nature has never left us; everything we are is in some sense natural. It is only because we have convinced ourselves otherwise that life has drifted so far out of balance. Our present efforts to deny death, to conquer death, and to artificially prolong life by mechanical means at any cost only increase our fear of death and reinforce our delusions of separateness. They pretend that nature might excuse human beings from her laws, that we might be exempt from the cycles of birth, death, and decay. And fear of death, in turn, really equates to a fear of life, which is about growth, change, and transformation, a continual dying of the old and birth of the new, season to season and moment to moment.

To deny death is to deny life. Think of the godlike ideals to which we aspire, the eternal youth, beauty, leisure, and supernatural power of the Olympian gods. Eternal youth is a kind of eternal death, a stasis more akin to an embalmed Egyptian mummy than to a living being (Egyptian embalming, after all, was directed at preserving the eternal *life* of Pharaoh). "To despise the fact of aging is not only to despise life but to betray a pitiful ignorance of the nature of life.... Youth is not a state to be preserved but a state to be transcended."[17]

The regime of control that denies death holds us apart from life too, both by preventing us, for safety's sake, from living it fully, *but also by cutting us off from even the possibility of fully living* by making an other of the world. Control, in essence, is a struggle against the world. Medical interventions triumph over nature when they interrupt and deny the natural process of death, as does housecleaning when it reverses the migration of dirt—which is quite literally the world—into our houses.

The perfect mastery of nature embodied in the aforementioned Olympian ideal is life-denying even in its more practical incarnation as the secure, ordered life of the modern adult. Thomas Hanna puts it like this:

> One of the myths of aging is that we cannot do all the things that we once could. But the actual fact about aging is that we cease to do all the things that we once did. As our search for a vocation settles into a fixed "job," as our search for a mate settles into marriage, as our many expectations settle into a finite number of fulfillments, as our aspirations settle into steady certitudes, and as our broad range of potential movements settles into a narrow band of habitual movements, we will inevitably find ourselves looking in fewer directions and moving in fewer directions. As the possibilities in our life are sorted through, discarded, and finally edited down to a daily routine of actualities, our living functions become limited and specialized.
>
> The accepted personal goal of "adulthood" is, it seems, to settle down and obtain security, to obtain a fixed pattern of life that allows us to escape from the insecurity of freedom and the incertitude of new aspirations. To the degree that individual human beings are seduced into the accepted belief that personal fulfillment means a settled, secure, and circumscribed mode of life, so, to that degree, do the functions of one's living body adapt, becoming simpler, more straightforward, and rigid. [18]

The fully insured, 100% guaranteed life that our social institutions suggest, channeled through the media, the schools, and often our parents, that is mapped out in advance to minimize uncertainty, is no life at all. What's more, despite all our efforts to mitigate its uncertainty, life is fundamentally uncertain and out of control. Everyone senses on some level that our management of life is little more than a pretense by which we delude ourselves into believing in the permanence and stability of that which is neither. In this gigantic game of let's pretend, we feign belief in the normalcy of modern life, but no one really believes in it; hence, the universal sense of phoniness, emptiness, and void underlying the adult world. Do you ever get the feeling that you are a child playing grown-up? When we act the roles of responsible, right-thinking, civilized adults, we are just pretending, only we have lost ourselves in the pretense. Take heart: it cannot last forever. All of us will one day experience real life smashing through the seemingly secure structures of our lives and selves. When this happens, people think, "How could this be happening?" or "This couldn't be," or "This is something that I thought only happened to other people."

Even worse than the disintegration of the orderly, stable, permanent-seeming life "under control" is for it to smoothly proceed until time and youth are exhausted. This is the true calamity, and not only because of the agony of regret at finding, in old age, that we have not truly lived. From the very beginning, there is the feeling of living a life that someone else has planned out, a life that is not truly our own. My college students sometimes report feeling trapped into living the lives their parents and society at large expect them to live; they are aware of the theoretical possibility of changing their majors to something "impractical" or of dropping out of school and traveling the world, of taking some time off to find themselves, but they cannot find the audacity to take such a step. "My parents would kill me," they say: a very revealing statement for the survival anxiety it encodes. Of course their parents are powerless to stop them, but by now they have internalized the voice of authority, so that what is unacceptable to their parents has become unacceptable to themselves. They are thus fully socialized, fully domesticated, except for a submerged longing, a caged and fenced wildness that can and will explode out in the form of violence—longing denied—whenever self-control lapses or when exceptional outside conditions offer temporary release.

Aversion to death and aversion to uncertainty in life arise from the same basic worldview. Indeed, if uncertainty is the demise of the familiar, then death is the greatest uncertainty any of us have faced since we left the womb. Both involve letting go of previously fixed categories by which we define who we are. A new job, a new relationship, a new life. We can adopt stringent safeguards to avert such happenings, clinging desperately to the familiar—life and self—but only at an increasing cost.

To maintain any system in a state removed from the wild requires a constant expenditure of effort. In *The Yoga of Eating* I use the metaphor of a parking lot:

> Not just the body but all natural things, when left undisturbed, move naturally toward beauty and wholeness. If you don't keep repaving it every few years, an ugly parking lot will crack, grass will come up, and

> after 100 years or so you'll probably have a beautiful forest. Your body is the same way. Stop "paving it over" with artificial ways of being, stop trying to be other than what you are, and it will move towards its natural state of health and beauty. It happens sooner than you think. Why else is rest so healing? Have you ever noticed how beautiful a sleeping person looks?[19]

The opposite of control is to just let things go, let nature take its course. In suburbia, we maintain control by applying dandelion killer, mowing lawns, weeding flower beds, and pruning shrubs so that our lawns remain neat and orderly. Cease any of these activities and the land steadily reverts to its wild state. To maintain the manicured lawn considered beautiful requires effort, which is why a person with an unkempt lawn is considered lazy. A similar effort is required to maintain a clean house by literally keeping the world out. If you want to hold your body away from its natural odor, you must bathe frequently and apply various perfumes and antiperspirants. If you let yourself sleep as late as you want, you miss work. If you don't keep track of time you cannot meet your scheduled obligations.

In none of these areas does a civilized person relinquish control. Culture tells us that to do so would lead to disaster, a rapid unraveling of the fabric of our secure, comfortable modern lives. This parallels the ramifications of Original Sin, which necessitates self-control so that our wanton, depraved natures won't wreak havoc upon the world and each other. The world must be kept under control. Modern life, predicated on the ever-increasing control of the world that defines technology, demands an ever-increasing effort to maintain control—increasingly fine scheduling of time, less rest, more multitasking, more reliance on technological assistants such as cell phones and planners to make us more efficient. The successful person has it all under control. She manages her time and juggles her responsibilities. She has it all together.

The more powerful the control, the greater the effort required to maintain it. Think of the iron discipline of the anorexic, the incessant hand-washing of the germ-phobic, the daily scrubbing and polishing of the clean freak. As is well known, the anorexic or the clean freak is compensating in a safely controllable area of life for a loss of autonomy somewhere else. We ordinary denizens of civilized society are little different. Underlying our superficial structures of control is the same loss of autonomy: a semi-conscious feeling that our lives are not our own. It is a dread, a foreboding, a sense that the most important thing in life—and indeed life itself—has been betrayed. In the pursuit of security we have traded the infinite—the limitless possibilities of life—for the finite: the predictable and safe.

This is the same reduction of life inherent in symbolic culture and domestication, the labeling and numbering of the world, the standardization of the Machine, and the reductionist program of the Scientific Revolution. The end result could be none other than the vigorous application of the Technological Program to all aspects of existence.

The ascent of control is a response to the reduction of self to a separate ego, alone and afraid. Taking many guises, it is a fear of death that motivates the effort to control the world and expand the domain of me and mine. When we begin to view life as other than a zero-sum game where my gain is your loss, then the struggle against the world loses motivation and instead of trying to control life, we let it in. The logic is quite simple: when the other is not an other, then to be good to another is to be good to yourself. A corollary is that the world is not fundamentally hostile. The dirt won't hurt you. A child, not yet acculturated, is not afraid to get dirty. To civilize children, to enlist them in the world of control that we live in, it is therefore necessary to disconnect them from the vastness of their total being and make them afraid. What needs to happen to domesticate our inborn wilderness? How has my wholeness been damaged? Maybe if we can answer that, we'll know how to get it back.

Yes and No

Economics, Darwinian biology, and dualistic religion all agree that the only hope for a livable world is if we all try really hard to be nice. Such a view sees civilization as a fundamental good, for it overlays the bestial inclinations of nature with conditioned behaviors that run counter to nature—counter to its win-at-all-costs, eat-or-be-eaten truths. Since the newborn child is wholly natural, wholly uncultured, education and child-rearing aim to destroy, or at least to suppress, the child's original nature in favor of civilized morals, values, and behavior. Hence the enormous emphasis placed on obedience and discipline.

The alternative to "trying harder" is play, which is spontaneous, improvisational, and *easy* (play only begins when we are at ease). If we try to play, we do not play. Fear and anxiety, compulsion and coercion, obedience and discipline are inimical to play. Many childhood activities that today go by the name of play, such as gymnastics classes and Little League, are actually closer to "skills development" or obedience training. But sad to say, our society begins to deny play at an earlier age than that.

Until she learns to crawl, a baby's freedom to play is restricted only by her physical capacities. She plays constantly, exploring different ways of moving her body and making sounds. Few parents try to restrict and instruct a child younger than eight months old, and there is little reason ever to use the word no. But when she becomes mobile, the baby or toddler can (as all we parents know) get into trouble: make messes, refuse to cooperate, and even endanger herself. Maybe she wants to take out all the pots and pans and bang on them for half an hour. Maybe she doesn't want to get into her car seat. Maybe she isn't hungry right now and would rather throw her food onto the floor. Maybe she wants to climb up the slide backwards. What if some other child hits her on the way down? She could fall and hurt herself.

Situations like these elicit that defining word of early childhood—"no"—which acquires its force from the violence (physical or verbal) that accompanies it, or the implicit threat of such violence. In each case we feel the necessity of asserting control. We cannot let her bang away on the pots and pans because of the horrible racket and the mess, plus it is just so… *disorderly*. She has to get into her car seat, *right now*, because we have to be somewhere at ten o'clock and it is unsafe and illegal not to ride in a "child restraint device" (as I like to call it—"Matthew, get into your child restraint device!"). Throwing her food onto the floor makes such a mess, and besides dinner time is now, not later, and we have other things to do. Immersed in modern life, we are at the mercy of its schedules, we are busy. As for the playground slide, well, maybe if we think about it rationally it isn't so dangerous, but we must teach her to respect rules, after all.

On one level, we seek to corral our children's autonomy and creativity for the same reasons we corral our own: it isn't practical, realistic, or safe. Subtly or not-so-subtly, we try to steer them toward practicality, safety, good manners and good morals. Today crude methods employing outright fear of physical pain have fallen out of favor. Instead we manipulate our children through selective praise and disapproval. We contrive to make them feel good about doing what we approve of, and guilty or ashamed about doing what we disapprove of. Sometimes a mere inflection of the voice is enough, or a subtle comparison to a sibling or friend.

The program of control is both subtle and pervasive. It is nearly always unconscious. If you are a parent, listen to yourself as you speak to your children. Notice when, through word or tone, you create a subtext of "You are bad" or "You are (conditionally) good." If you have parents, listen for that in their communication to you. And finally, no matter who you are, listen for it in your communications with yourself. I have found it pervades my inner dialog. I am good because of this, bad because of that. I can't do X because I'd be bad if I did. I must do Y or I won't be good. You won't necessarily use the actual words "bad" and "good"; instead you might employ substitutes like lazy, indulgent, selfish, greedy, or wrong for the former, and cool, nice, deserving, worthy, or right for the latter. Underneath these words, feelings of guilt, self-rejection, desire for approval, anxiety, and shame shepherd us through a life under control.

The sophistication of this program becomes apparent when we realize that many of the child's behaviors that we seek to control are not actually that dangerous. The justification, "We must teach her to respect rules" takes us to the root of the issue. Overriding the immediate concerns of safety and convenience is the "principle" that she must learn to obey her parents. She must submit to discipline—at first external, then internalized as an adult. Control is actually a goal more important than the safety, practicality, and morality that justifies it. Consider the two cardinal sins of childhood: disobedience and lying. More than nearly any other infraction, disobedience and lying are profoundly disturbing to the typical parent. They provoke irrational and often unbearable frustration, helplessness, and rage, because they reveal that the child is out of control; they amount to an assertion, in action, of the child's autonomy. As such they are sure to incite the harshest punishment!

I remember one time years ago, my son Jimi was riding his trike around our cul-de-sac when I saw a car pulling out of a driveway. "Jimi," I shouted, "a car wants to pull out, move to the side of the road." When he didn't respond, I shouted again, a little louder, "Jimi, get out of the way, there's a car coming!" Still he didn't respond. Fortunately, the driver had seen him and driven around him and out of the cul-de-sac (he had never actually been in much danger), but by then I'd completely lost my temper. I walked up to him and screamed in his face about how he must always do as his father says because what if the car didn't see him? Jimi of course began crying and it was quite a while before I, feeling miserably guilty the whole time, could console him.

After years of reflection on this and similar incidents, I eventually realized that the danger of the car was just an excuse. The real reason I "lost it" was because I felt threatened. I wanted to frighten him into submission. I was furious that he didn't do as I said, and I wanted to scare him into obeying me next time. I was also embarrassed because the neighbor probably disapproved of my lax supervision of his activities. I felt like a bad parent. Jimi's disobedience triggered a deep unease that actually had little to do with any real danger.

I have since learned that to gain a child's trust and obedience, it is far more effective to kneel down in front of him, take his hands, look at him at eye level, and tell him in calm but certain words why it is important to listen to his father. It is so much more effective to trust that a child naturally desires the guidance and protection of the parent, and to speak to that desire. It is not necessary to subdue his spirit. It is not necessary to fight human nature. Can you believe that? Can you believe that my children almost always obey me, and I never threaten or punish them? We don't have to live in a Newtonian world where only force can change the course of matter.

And can you see that my children rarely have any cause to lie to me? There is nothing to evade. Think back on your own adolescence. Did you routinely deceive your parents in order to gain freedom and avoid punishment? Did you have a secret life? How long (if ever) did it take to restore intimacy and trust? Our alienation from our children is yet another example of how control isolates us from the world.

The fashionable locution, "You need to..." is a telling manifestation of the domination of children's spirits. I heard one mother, after her six-year-old daughter had ignored two or three requests to go inside and get ready for bed, switch to a tone of angry, insistent menace: "Maggie, you NEED to go inside right now!" As George Orwell observed, to dominate a person's actions is not enough; his thoughts and feelings must be under control as well. I am not blaming the mother, for she is just a channel for the climate of the culture. But think of the tyranny of dictating to someone what she needs. Any marriage counselor will tell you that it is disrespectful and unproductive to tell another person how they *should* be feeling, but we do it to children all the time through praise and disapproval, reward, threat, and shame.

The subtle forms of control I have described seem gentler than the whippings and beatings of yesteryear, but in essence they are no different. They are merely a different means to access a child's greatest fear.

Far more than pain, what a child (or any young mammal) fears the most is rejection or abandonment by the parent. That is why children have been known to willingly kiss the hand that strikes them, or even to ask for punishment. From this perspective, a putdown is the same as a beating. Both invoke the primal fear of abandonment. In fact, an occasional spanking is probably less damaging than a prolonged campaign of control that makes the child feel never good enough for complete acceptance. Never good enough. When approval is conditional on performance, then no degree of perfection can ever suffice to put the child at ease. The same is true when he grows up and internalizes parental approval as self-approval. Conditional approval means you are perpetually on probation. Your natural self is bad, so you have to try hard to be good. Yet no amount of effort can build a tower to Heaven. No matter what heights we achieve we always fall infinitely short.

The human spirit is so strong that only a threat to survival can subdue it. The primal fear of abandonment is instrumental in breaking a child's spirit. The "training" or acculturation of children taps into survival anxiety at the deepest level. Remember my students who say, "My parents would kill me." That is a code-phrase for fear of abandonment.

Survival anxiety is also what motivates the parents. We typically rationalize this breaking of a child's spirit by saying it is for his "own good", either in terms of physical safety or in terms of winning a secure place in society. What would happen, after all, if we raised a child to do what he wants to do instead of what he has to do? A child who always put play before work? A child who never compromised his dignity? Such a child could not occupy the usual roles society offers, would never submit to the humiliating routines of school nor the routine humiliation of life in the Machine. Derrick Jensen puts it this way:

> I've since come to understand the reason school lasts thirteen years. It takes that long to sufficiently break a child's will. It is not easy to disconnect children's wills, to disconnect them from their own experiences of the world in preparation for the lives of painful employment they will have to endure. Less time wouldn't do it, and in fact, those who are especially slow go to college. For the exceedingly obstinate child there is graduate school.[20]

The tyranny is far more subtle than Jensen describes, because it is actually not those children whose will is especially strong who go on to college and graduate school for some additional will-breaking. Quite the opposite. College and graduate school are a kind of reward allowed only to those who, through good grades, have demonstrated that their wills are sufficiently broken to qualify for elite positions in society. Certainly there are a lucky few who simply love schoolwork, but for most school is a chore, a discipline through which we demonstrate our willingness to do as we are told. Those who cannot bring themselves to fully comply with the instructions, whose attention wanders, who clown around in class, who would rather play outside than do homework, and whose aversion fuels a spirit strong enough to resist the institutional, cultural, and parental mechanisms that enforce compliance, will not get good grades in school. Instead they will earn labels such as stupid, lazy, bad or, increasingly, medicalized versions of these such as ADD (attention deficit disorder), ADHD (attention deficit hyperactivity disorder), or my favorite, ODD (oppositional defiant disorder). Then, where the veiled threats, phony incentives, and other forms of scholastic manipulation have failed, we can resort to pharmaceutical methods of control to rein in the recalcitrant child.

Society will do just about anything it takes to compel the child to take "no" for an answer. This is a vast undertaking, because it is anathema to the human spirit of exploration and creativity. That is why we see toddlers repeating the word to themselves over and over, trying to come to grips with it, trying to reconcile its life-denying force with the creative potency within them. Try this sometime: go to one of those little playgrounds for young children, the ones full of young mothers with their toddlers. Close your eyes and listen. One word will stick out like a sore thumb; again and again you'll hear it, pronounced in menacing tones. "No Jeremy!" "No Ashley!" "No Courtney!" No, no, no. Think of the impact on a small child: the parents are gods in his eyes, enormous omnipotent figures nearly coextensive, especially in the young toddler, with the universe itself. The gods tell us, "No", with menace and the threat of alienation. The universe is not friendly. We are not free.

Internalized at a young age, the relentless refrain of "no" echoes throughout our childhood and adolescence until, by the time we reach maturity, it has silently imbued itself into our fundamental perceptions. The result is that we come to doubt the validity of our own creativity; we are constantly looking over our shoulders and wondering if it is okay, wondering if it is acceptable to venture into new territory. And eventually we become accustomed to this state of being, comfortable with a world in which everything is either expressly permitted or explicitly forbidden, where there is no uncertainty, no ambiguity, no open territory. Our entire lives, in other words, have come to be defined by this innocent-looking two-letter word. That, I imagine, was the intuition underneath the radical Sixties message: "The Establishment says no, the Yippies say yes!"[21] Yes to what? The declaration needs no explanation because its meaning is intuitively obvious, a universal Yes that needs no object.

In bogus compensation for cutting us off from most of life with the word "no," society offers us limited, harmless, and ultimately phony arenas for escapism and indulgence, where we can enjoy a narrow range of trivial freedoms. The model for this is the playpen: a highly circumscribed section of the world where every variable is under control, where we can't knock over a lamp, dirty the carpet, or run amok, and where it is absolutely safe. The exuberance and abandon of real living happens in safe, controlled fragments of life: on vacations, in bars, at parties, at amusement parks, on guided tours, through shopping catalogs, channel surfing, the Internet, and the limitless universe of consumer brands. These are the playpens of adulthood.

Within these playpens our society seeks now to encompass the whole of life. Here we find the counterpart of "no": a conditional, narrow yes defined by our own fears and by the prohibitions of society.

Above I wrote that the world of play still survives intact through the first eight months of life. But just now I glanced at a news article about the latest trend in "baby workouts," where "the class uses music and props to keep the babies focused and helps improve their eye tracking and coordination."[22] In the hands of a good teacher such a class could be wholly playful, but the very concept reveals the general migration of physical movement—originally one of the greatest joys of being alive—into the realm of work: exercise or a work-out, something we have to do for fitness, to control our weight or shape the body. Here is yet another extension of the scheduled, directed, programmed life into unprecedented territory. "Play's End," as Joseph Chilton Pearce sorrowfully puts it, comes at a very early age these days. It has to. The spirit must be broken early to submit to a world under control.

The Pressure to Break Free

When we clamp down and control life, we generate a pressure that must find an outlet. The release is usually guilty and often in secret, and diverts our creative life force—our divine nature—into activities that are generally trivial, meaningless, marginal to our main occupation, and sometimes downright destructive.

To most of the roles society offers, I say, "You are made for more than that." We inhabit, in the words of Ivan Illich, "a world into which nobody fits who has not been crushed and molded by sixteen years of formal education."[23] The very idea of having to be at a job "on time" was appalling to early industrial laborers, who also refused the numbing repetitiveness of industrial work until the specter of starvation compelled them. What truly self-respecting person would spend a life marketing soda pop or chewing gum unless they were somehow broken by repeated threats to survival? To participate in our society's depredations is an indignity. A corporate executive recently confessed to me that his job consisted of lying to the customer; another that his job consisted of frightening customers into accepting digital security products that they really didn't need. An elite lawyer described his job as, "I take money from one rich son-of-a-bitch and give it to another rich son-of-a-bitch." Part of my job at Penn State is to pass judgment on students by issuing them a grade. To be sure, there are many people fortunate enough to have interesting jobs, creative jobs, perhaps even meaningful jobs, but even if you love your work, what do you have to put up with in order to do it? Indignity is hard to avoid when our whole economy revolves around the creation and fulfillment of phony needs. It is hard to avoid when our institutions depend on standardization of roles. It is hard to avoid when the profitability and even survival of a company conflicts with human and community values. We all know this, yet we feel compelled to participate. To participate in any form of injustice, humiliation, or degradation, of people or the environment, is itself degrading. What does it take to get a divine human spirit to participate in this? It requires its breaking, so that we dare not say no.

To cut corners, to do something just well enough for the grade, the customer, or the boss, to do anything with the feeling that it really doesn't matter, to care about your work mostly because you are paid to care about it—all of this is degrading. It is an indignity to make anything less than an art out of your work. That is what we are made for—nothing less. I will never forget the day long ago when I realized that I really didn't care about anything I did at my job. We were having a meeting about some new computer audio capability, and everyone seemed quite interested in it. A lively discussion was going on about how to integrate it into our software. I looked around the room and thought, "Hold on, you mean you people really care about this? I thought we were all just pretending to care." At that moment I realized that I only cared about it because I was paid to care, a realization which fueled a growing sense of panic in the weeks to come. "There's got to be more than this," I thought. "What about caring about something for real? What about devoting my full energy to work that I love? Don't I get to have that this lifetime?"

Very few of society's usual positions can accommodate the enormous creative life-force of an unbroken human being. To keep the world under control demands that we bottle up this creative force and expend as much of it as possible in harmless ways—harmless to the status quo, that is, though not to the individual. All the addicts and alcoholics I have known—all of them!—are blessed (or cursed) with what seems to be an exceptional creative energy that is burned up in their addiction. Other people channel it into obsessions and compulsions, hobbies, nervous tics, excessive exercise, overeating, work, and the like, contributing to a more dilute version of the addict's sense of life betrayed.

When we submit to lesser lives, we cannot avoid a sense of self-betrayal: that we are complicit in the plunder of our most precious possession. The roles society offers do not befit the divine beings that we are. It is not merely that a career as a retail clerk is beneath my dignity; it is beneath *anyone's* dignity. No one is meant to do such work for very long. It can be fun for a while, for the few days or weeks it takes to fully master it and learn what there is to learn. One of the best jobs I ever had was washing dishes in the university cafeteria. It was fun figuring out the series of hand movements that maximized my efficiency, learning how the kitchen worked, having food fights with the other dishwashers, and spraying my friends with the dishwashing hose when they came to bus their trays. Everyone should do jobs like this for a while as I did: a couple of hours a day, a few days a week. It is only when, through poverty and despair, we become enslaved to such work that it becomes degrading.

When a divine spirit mortgages her very purpose in life, which is joyful participation in the creation of a beautiful universe, she will compensate by being good to herself in whatever ways are available. She may thus appear selfish or greedy, but all she is really doing is searching for something that is missing. Of course, no amount of food, material possessions, physical beauty, power over other people, or money can fulfill her need as a divine being to express her creative potential and experience intimate connection to the rest of life.

The sense of entitlement that drives selfishness and greed thus arises from an authentic source: that we have been robbed of our birthright. Something *is* missing, it's just that we are looking in the wrong place for it. An ancient Sufi tale describes the sage-fool, Mullah Nasruddin, groping around under the streetlight. A passer-by asks, "Mullah, what are you looking for?"

"Alas," replies the hapless Nasruddin, "I have lost my house key."

"Well, when was the last time you saw it?"

"I think I lost it over there in the shadow of those trees."

"Well, why are you looking for it here then?"

"Sir, can you not see how dark it is there? I am looking here under the streetlight where I can see better."

What we are missing is nothing less than the key to our homecoming, a reunion with the divine essence of our being. Unfortunately, we dare not look in the frightening shadows where the key actually lies, preferring instead to find that missing something in a safe zone. So conditioned are we to survival anxiety that we dare not leave its domain. Notice how all the things we pursue in lieu of our life purpose—money, beauty, career, power, prestige, possessions—are all linked to survival: the command of people and resources. We have been rendered afraid.

Now for the good news: A human soul can never actually be broken, but will continue to resist a wrong life through whatever means it can. The soul is like the wild that eventually, through roots, rot, and weather, will bring down the strongest house. The soul will engineer situations to bypass our best efforts at control. First there will be a growing sense of unease and discontent. Getting up for work becomes a chore and we find ourselves looking forward to the end of the day, the weekend, or vacation when the day or week has hardly even begun. We try to dispose of our work responsibilities with as little effort as possible; we cut corners and become lazy, doing the minimum required. To stay at the job now requires willpower: to get up in the morning, to at least put on a show of productivity. We control ourselves with alarm clocks, coffee (to enforce attention), and external motivations such as money, promotion, and so forth. We also motivate control through fear—What would happen if I quit? Usually at this point an opportunity arises elsewhere, and we may or may not have the courage to quit the old job and take a leap into the unknown. Failing this, the stage of self-sabotage begins: after a few close calls, we engineer a situation that enables us to quit or that gets us fired. Alternatively, the soul invites in a disaster from another realm of life. Often there is a convergence of crises in health, career, and marriage, and the betrayal of life purpose extends to these other areas as well. To stay the course of a life wrongly lived requires control; without it one quickly drifts away from anything unpleasant.

Just as we cannot permanently protect a house from natural processes of decay, none of the means of controlling or diverting our creative life-force works forever. Each of them eventually becomes intolerable. The most potent of these diversions are addictions such as alcohol, heroin, cocaine, pornography, and gambling. These can consume an enormous amount of suppressed life-force, but they exact a rising cost on body, life, and mind. By consuming frustrated life-force, they make the wrong life tolerable for a while. The same is true of the petty addictions and distractions common to most people. Usually, because our entry into the wrong life originated in the painful breaking of our spirit, the objects of addiction are also means for the temporary avoidance or amelioration of pain. They are a means of control, a means of coping with life-as-it-is. To break free of the forms and structures of our culture means confronting the materials from which this prison is built: it means facing the fear and the pain. It is no accident that two of the most potent addictive substances, heroin and alcohol, have physiological pain-killing effects; what is not generally recognized is that all addictions do the same. Only, they do not really kill pain; they merely postpone it, keep it under control, keep it temporarily unfelt.

Please, then, do not condemn addicts as weak-willed, immoral, or indulgent. They have merely received more than their share of pain. Like all of us, they are suffering under one of the many permutations of the myth of control. Genetic predisposition and the vagaries of circumstance merely determine the form this control of pain takes.

The recovering alcoholic Jack Erdmann writes, "When I started, the alcohol told me there was a way out, that the pain could be killed."[24] In effect, alcohol, drugs, gambling, television, and other addictions large and small are simply technologies, technologies for controlling pain and accessories in making life-as-it-is manageable. But then the trap is sprung: "Then it told me to kill the pain at all costs." Inexorably, the life-as-it-is that demands management becomes wholly the product of the technology of control. As Erdmann puts it, "This is the secret of alcohol. The alcohol creates the symptoms you think it's treating."

It's the technological fix again. Technology, the means to manipulate nature, is driven by the urge for comfort and security; that is, the avoidance of pain and the insurance of survival. Yet the very experience of life as a struggle for survival is a product of technological culture, which seeks to control reality rather than simply accept it. Remember agriculture, which replaced the Edenic existence of hunting and gathering with a life of toil, in which a harvest tomorrow requires labor today. As soon as that happened and the population grew to exceed the carrying capacity of the unimproved land, we became addicted to technology. We could no longer live without it. And as invention after invention came along to ease this burden of labor, to mitigate the omnipresent threat of famine and the travails of civilized life, the underlying anxiety and suffering only grew until it reached the crescendo of the early 21st century. "The alcohol creates the symptoms you think it's treating."

Encompassing both alcohol and technology, control creates the conditions that necessitate control. Control is a trap, a lie, a vicious circle, a one-way train ride. And as Erdmann says, "The last station is Hell."

Looking at the stripmined mountains of West Virginia, the thousands of miles of bulldozed Siberian forests, the bleached coral reefs, the vast parking lots of suburbia, the disintegrating health of Americans, the generation of child cancer victims around the world, the despair of the poor and the ennui of the rich, is there any doubt where we are headed? Is there any doubt that we are creating Hell on earth?

Just as the alcoholic treats the agonizing wreck of his life with another drink, so do we believe that we can fix the mess we have made of the world with more of the same: nanotechnology, perhaps, trumpeted as the final solution to industrial pollution.

The real danger lies when the program of control succeeds in clamping down on our creative life-energy. When internalized coercive mechanisms and external addictions overpower the soul in its struggle to break free, the soul can still play its one remaining trump card. If an escalating series of crises is not enough to dislodge us from the wrong life, then death is the natural choice. Sometimes the rut is just too deep, the labyrinth of self too bewildering, for a person ever to emerge. (I am not speaking here of deliberate suicide, which is usually just another means of avoiding the pain, a futile last-ditch attempt at control.) When hope fails that we may transcend the diminished selves of the world under control, then our true selves engineer the final escape from the illusion of separation.

Now apply this idea metaphorically to the collective level. Are you alarmed? You should be, because the implication is that our civilization—or even our species—may very well choose (albeit unconsciously) to die. If the convergence of crises is not enough to dislodge us from our delusions of separation from nature, then the collective soul of the human race will unleash a catastrophe such as nuclear war. That is what will happen if we hold on too long.

Molding Minds

School is the cheapest police.
—Horace Mann

But who can unlearn all the facts that I've learned
As I sat in their chairs and my synapses burned
And the torture of chalk dust collects on my tongue
Thoughts follow my vision and dance in the sun
All my vasoconstrictors they come slowly undone
Can't this wait till I'm old? Can't I live while I'm young?
—Phish

No discussion of control would be complete without some attention to modern schooling, the linchpin of the program to subjugate the inner wilderness—human nature—just as we use technology to dominate the outer.

Why do we go to school? The universal response is, "To learn", and here we encounter the primary contradiction of modern schooling that leads us to all the others. For if the purpose of school is that children may learn, then school is quite evidently not working.

The massive National Adult Literacy Survey undertaken by the Department of Education in 1992 is especially revealing, so I will quote John Gatto's summary in full:

> 1. Forty-two million [out of 190 million adult] Americans over the age of sixteen can't read. Some of this group can write their names on Social Security cards and fill in height, weight, and birth spaces on application forms.
>
> 2. Fifty million can recognize printed words on a fourth- and fifth-grade

level. They cannot write simple messages or letters.

3. Fifty-five to sixty million are limited to sixth-, seventh-, and eighth-grade reading. A majority of this group could not figure out the price per ounce of peanut butter in a 20-ounce jar costing $1.99 when told they could round the answer off to a whole number.

4. Thirty million have ninth- and tenth-grade reading proficiency. This group (and all preceding) cannot understand a simplified written explanation of the procedures used by attorneys and judges in selecting juries.

5. About 3.5 percent of the 26,000-member sample demonstrated literacy skills adequate to do traditional college study, a level 30 percent of all U.S. high school students reached in 1940, and which 30 percent of secondary students in other developed countries can reach today.

Well, maybe it can't be helped, maybe most people are just dumb. In that case school takes on a benevolent, even a noble aspect, using the sciences of psychology and pedagogy to bring at least a glimmering of knowledge and literacy to the benighted masses, who if left to their own devices and the well-meaning ignorance of their parents would surely be unfit for life in technological society. After all, unlike two hundred years ago, we have to read now, do math… use *computers*. Sure it has some problems, but in general school brings each person up to the level of intellectual competency that their genes and environment allow. Right?

Wrong. Over the last 60 years, while real spending on education nearly quadrupled, illiteracy has quadrupled along with it. Few blacks went to school before 1940, yet in that year black literacy was 80%; today it hovers around 60%. White illiteracy meanwhile rose from 4% to 17%. [25]

Before forced schooling, in 1840, literacy rates in New England approached 100 percent, and the popular best-sellers of the time included books the likes of Herman Melville, James Fenimore Cooper, and Ralph Waldo Emerson—all bought by a population consisting mostly of small farmers. Have you ever sat yourself down for a relaxing afternoon with a little leisurely reading—a thirty-page long Emerson essay loaded with Greek mythological allusions, complex sentences with multiple nested appositive phrases and dependent clauses, intricate logic, and a vocabulary that would challenge most graduate students today? I find I can only access early 19th century literature with a great effort at concentration. My attention wanders, I lose the train of the argument, and soon find myself passing my eyeballs over the page, uncomprehendingly.

John Taylor Gatto observes: "In 1882, fifth graders read these authors in their *Appleton School Reader*: William Shakespeare, Henry Thoreau,

George Washington, Sir Walter Scott, Mark Twain, Benjamin Franklin, Oliver Wendell Holmes, John Bunyan, Daniel Webster, Samuel Johnson, Lewis Carroll, Thomas Jefferson, Ralph Waldo Emerson, and others like them." Today, no more than twenty percent of my *college* students understand the Thomas Jefferson passage quoted in Chapter Four of this book. As for fifth graders, their texts consist mostly of short declarative sentences using simplified vocabulary. A recent trend is the reissue of children's classics such as *Little Women* or *The Wind in the Willows* in special dumbed-down editions.[26] Few children these days have the reading ability to handle *Treasure Island,* and Shakespeare's *Julius Caesar*, once a staple of high school English, is off the curriculum because children no longer have the attention spans to understand it.[27]

Modern schooling is a failure—or is it? That depends on the real purpose of schooling, an enormous topic that I cannot possible do justice to in these pages. Fortunately I don't have to, thanks to John Taylor Gatto's magnificent opus *The Underground History of American Education*, a work of prodigious scholarship, unflinching honesty, seasoned insight, and towering indignation. I will share some of his insights (and his indignation), since they illustrate so well the mentality of control, the logic of the Newtonian World-machine, the liquidation of spiritual capital, and ultimately our culture's fundamental attitudes towards nature and human nature.

Just beneath the superficial justifications for mass forced schooling lies the first level of its true motivation: to create a population suitable for the demands of the industrial economy. School as we know it, like other applications of the technologies and mentalities of mass production, got its start in the early 19th century in the great coal powers of the period: Prussia, England, France, and then the United States.

Early industry faced a problem. Mine and factory work was dull, repetitive, arduous, and dangerous while offering wages barely high enough to sustain life. Office work—the work of clerks, scriveners, and accountants before computers—was equally dull and dehumanizing, if less dangerous. Factory discipline was alien to the independent, self-directed farmers and artisans that made up pre-industrial society, and the question of how to instill labor discipline was discussed at length by the intellectuals of the day. One solution was outright force: driving peasants off the land through enclosure, using militias to enforce strike prohibitions, and extreme economy exigency. However, the inhumanity of this solution offended the conscience. Besides, it was potentially very explosive, as a series of insurrections, revolutions, and bloody labor strikes throughout Europe and North America attested. Wouldn't it be better to somehow condition people from childhood to accept, and even to desire, work that was partial, trivial, mechanical, dull, repetitive, and unchallenging to thought or creativity?

Is this description already reminding you of school? Where learning arises not from curiosity but from authority's agenda; where achievement is adjudged by external standards; where human beings, like so many objects, are numbered, "class"ified, and "graded"; where knowledge is reduced to answers, right and wrong; where children are confined to a classroom or desk except when authority allows them "recess" or a pass; where problems are solved by following Teacher's instructions; where free speech and free assembly are suspended—where, indeed, there are no freedoms at all but only privileges; where bells condition us to follow a regular external schedule; where fraternization is surreptitious (as my teacher once said, "You are not here to socialize!"); where none outside the hierarchical structure of authority have the power to make or change rules; where we must accept the tasks given us; where work is arbitrary and meaningless except for what external reward it brings; where resistance is proved futile in the face of a near-omniscient, omnipotent central authority... what better preparation for adult confinement to offices and factories could there be? What better preparation for accepting unquestioningly the lives given us? Where else can students "learn to think of themselves as employees competing for the favors of management"?[28]

Not only does school prepare us to submit to the trivialized, demeaning, dull, and unfulfilling jobs that dominate our economy to the present time, not only does it prepare us to be modern producers, it equally prepares us to be modern consumers. Consider Gatto's description:

> Schools train individuals to respond as a mass. Boys and girls are drilled in being bored, frightened, envious, emotionally needy, generally incomplete. A successful mass production economy requires such a clientele. A small business, small farm economy like that of the Amish requires individual competence, thoughtfulness, compassion, and universal participation; our own requires a managed mass of leveled, spiritless, anxious, familyless, friendless, godless, and obedient people who believe the difference between "Cheers" and "Seinfeld" is worth arguing about.

The consumer model is written into the very foundations of the modern classroom. Gatto writes: "Schools build national wealth by tearing down personal sovereignty, morality, and family life." This is precisely the social and spiritual capital whose conversion into money was discussed in Chapter Four. It is not just that the broken and stupefied child is unable to stand up for himself in the workplace or to resist his role as a standardized cog in the vast automaton of industrial society; it is that relationships themselves, and all the previously non-monetized functions and exchanges associated with them, have been objectified, depersonalized, and commoditized. When the autonomous relationships (social and spiritual) that define our humanity are stripped away, we naturally becomes consumers of them. When self-directed learning through reading is replaced by programmed teacher instruction—the dishing out of a curriculum—we become consumers and not producers of knowledge, which is reduced to measurable "information". Thus we instill in our children not only obedience, but also intellectual dependency, the reliance on authority for truth. What is the difference between getting truth from books and getting truth from teacher? Reading books as part of a personal search for knowledge does not make one a mere consumer, because the search is self-directed and the information subject to independent, uncoerced selection and judgment. In school quite the opposite holds: the truth—the right answers—has already been pre-selected and pre-judged by the authorities, and the students are to accept it. Or they are coerced into accepting it, at least to the extent that exams, grades, detentions, "permanent records" and so on are effective instruments of reward and punishment.

In other words, school is an instrument of alienation. It alienates children from their families, not only by removing them physically but by replacing and professionalizing a traditionally important sphere of interaction: education. It alienates children from communities, segregating them by age, inducing competition among them, isolating them from adult life, and feeding them a curriculum determined by distant experts. It alienates children from nature and the outdoors, of course, simply by keeping them inside all day—surely an unprecedented condition of childhood until the last century. It alienates children from real experience by substituting for it games, simulations, and lessons, in which everything they do is, after all, only in a classroom, without real consequences, and terminating as soon as the bell rings for the next class. But most importantly, school alienates children from themselves: their own natural curiosity, inner motivation, self-reliance, and self-confidence. As Ivan Illich puts it, "Rich and poor alike depend on schools and hospitals which guide their lives, form their worldview, and define for them what it legitimate and what is not. Both view doctoring oneself as irresponsible, learning on one's own as unreliable, and community organization, when not paid for by those in authority, as a form of aggression or subversion. For both groups the reliance on institutional treatment renders independent accomplishment suspect."[29]

William Torrey Harris, U.S. Commissioner of Education from 1889 to 1906, wrote in the last year of his tenure:

> Ninety-nine [students] out of a hundred are automata, careful to walk in prescribed paths, careful to follow the prescribed custom. This is not an accident but the result of substantial education, which, scientifically defined, is the subsumption of the individual.
>
> The great purpose of school can be realized better in dark, airless, ugly places.... It is to master the physical self, to transcend the beauty of nature. School should develop the power to withdraw from the external world.[30]

Gatto comments, "Nearly a hundred years ago, this schoolman thought self-alienation was the secret to industrial society. Surely he was right." This alienation is nothing else than the separation that is the theme of this book, implicit in all technology and culminating in the pinnacle of modern science, technology, and the Machine.

These features of schooling were designed into it from the very beginning, as stated very explicitly by such guiding organizations as Rockefeller's General Education Board:

> In our dreams... people yield themselves with perfect docility to our molding hands.... We shall not try to make these people or any of their children into philosophers or men of learning or men of science. We have not to raise up from among them authors, educators, poets, or men of letters. We shall not search for embryo great artists, painters, musicians, nor lawyers, doctors, preachers, politicians, statesmen, of whom we have ample supply. The task we set before ourselves is very simple... we will organize children... and teach them to do in a perfect way the things their fathers and mothers are doing in an imperfect way.[31]

"A mass-production economy can neither be created nor sustained without a leveled population, one conditioned to mass habits, mass tastes, mass enthusiasms, predictable mass behavior."[32] The modern institution of school helps *create* the very "human nature" that is *assumed* in liberal economic theory, whose behavior is predictable according to deterministic laws just as are the masses of classical physics.

On a deeper level, the goal of modern education is the perfection of Lewis Mumford's megamachine—the great automaton composed of human parts—that itself provided the original model for the factory, and in which each person is reduced like a machine component to a standardized function. Just as physical machines produced unprecedented wealth and power over the environment, so it was also supposed that the sacrifices made of individual wholeness and self-determination would find compensation in the glorious onward march of science, the eventual conquest of nature, the fulfillment, in other words, of the Technological Program that would take us beyond labor, beyond suffering, beyond death, beyond planet earth across the Final Frontier of space.

The subordination of the individual to the needs of system is a key component of the ideology of "scientific management" associated with Frederick Taylor but tracing its roots back at least to Francis Bacon. Bacon believed that with the Scientific Method humanity had arrived at just that, a "method" that could be mechanically applied to achieve unlimited progress in science. No longer would individual genius be required, just the competent and correct application of method. As Taylor put it, "In the past, man has been first. In the future the system must be first." Gatto comments, "It was not sufficient to have physical movements standardized, the standardized worker 'must be happy in his work,' too, therefore his thought processes also must be standardized."[33] If you aren't happy in your work, that must imply a fault in your production process (socialization, education, training). Fortunately, that can be adjusted with pharmaceutical technology. Indeed, the term "well-adjusted" implies the molding of the human being, a standardizing to the needs of system. The 1933 World's Fair slogan comes to mind: "Science Finds. Industry Applies. Man Conforms." School is simply part of the process of conforming man to machine, the engineering of human nature.

While correctly and compellingly identifying the true historical objectives of schooling as comprising a monstrous violation of the human spirit, Gatto sometimes leaves the impression that it was contingent on a few historical accidents and could easily have been otherwise. If only Humbolt had won the debate with Baron Vom Stein in early 19th-century Prussia, if only the Massachusetts legislature had swung by a mere 36 votes to reject Horace Mann, then the crime of mass compulsory schooling might never have happened. In fact, it was bound to happen, bound by vast historical processes that carried Vom Stein, Mann, Dewey, the Carnegie Foundation, the Rockefeller Board, William Rainey Harper and the rest to victory. Even the behind-the-scenes manipulators—Rockefeller, Carnegie, Ford, and Morgan—who sought the creation of a docile proletariat and orderly society were merely enacting roles dictated by the very processes that brought them to power.

Yes, school is an agent in the dissolution of family and community, in the conversion of citizens into subjects and creators into consumers, in the breaking of children to the demands of institutional life; school is an agent in all this, but these processes extend far beyond the institution of schooling to embed it and guide it. It was inevitable that one way or another, we would apply the same essential technologies to children as we did to nature and everything else—and for the same basic reasons. How to "subordinate the individual to the needs of the system"? The individual must be labeled, quantified, measured, graded, and standardized. That is the only way the methods of management can be applied.

In taking children away from the matrix of family, nature, and social apprenticeship, mass schooling is essentially an enormous experiment in *social engineering*, the fruition of thousands of years of Utopianism going back to Plato in which institutional training of the young was always a crucial component. Up through the Owenite and socialist experiments of the 19th and 20th centuries, children were removed, at least in part, from their families. Sorry, but the family is obsolete; henceforward we are going to raise children *scientifically*. Surely trained experts can do better than ignorant parents. Surely science and reason can improve upon primitive, biological, emotion-driven families. The scientific laws of psychology and child behavior will replace the old irrational customs, and, unclouded by parental subjectivity, we will raise children for modern society. You, the modern parent, can do your best to learn about scientific parenting, but in most areas you'll have to yield to the experts.

The terminus of this trend is nothing other than Huxley's *Brave New World*, in which the factory method is applied to child-rearing from birth and before. All people are graded, from Alpha-plus to Delta-minus—sound familiar?—and each given the stimuli and resources appropriate to their grade.

Like all technology, the social engineering agenda of schooling involves a separation from nature. In this case it is the removal of children from their original biological and social habitat of family and community. The separation from the family, totalized in *Brave New World*, is a necessary, inevitable product of the attempt to engineer society according to the same methods and logic as we engineer the material world. In both there is a replacement of the personal, subjective, and traditional with the abstract, formal, and general. We have not yet reached Huxley's extreme, but a trend in that direction is visible wherever the Technological Program is pursued. When I was a child we listened with horror to stories of the Soviet Union, where *the state was replacing the parent*, replacing the very family, with mandatory "scientific" child care, youth indoctrination, and so forth. But today the same thing is happening everywhere, if not directly at the hands of the state, then with its literal license, or else at the hands of other institutions operating by the usual principles of scientific management. Whether by chance or design, today's system of infant and child care, school, organized sports, counseling, and television conspires to replace the parent and community. The same functions of socialization, education, and identity-building are being provided, but now by institutions and their functionaries who may not really care about your child at all, except that they are *paid* to. Moreover, there is a fundamental conflict between the social engineer's goal of adjusting the child to fit the needs of the system, and the spiritual goal of personal fulfillment. The socialization is socialization to the machine. The identity built is the identity of a consumer.

The agenda of social engineering explains the emphasis that psychology (the "science" of the mind) has always received in pedagogy (the "science" of teaching) ever since Horace Mann advocated phrenology as the key to a successful classroom. As in other realms of humanity's "ascent", we follow Galileo's prescription of applying measurement to learning in hopes of turning it into a science. We can then deploy the whole gamut of technologies based on standardization, efficiency, management, and control. The object of education—the child—becomes the object of technology. School is an aspect of a vast enterprise: the engineering of the human being, the human mind, the human psyche, the human soul. An audacious ambition indeed: not the accidental result of an historical blunder, nor the plot of an evil conspiracy, but implicit in the original audacity of technology. On the deepest level, the purpose and motivation of education is to apply the Technological Program to the ultimate frontier: society and the human being. As technology in general seeks to improve on nature, educational technology seeks to improve on human nature.

The Great Indoors

The white man builds a shelter, and it becomes his prison.... he separates himself from the earth and refuses to budge. Therefore he is always sick.
-- Tom Brown, Jr.

It is significant that school almost always happens indoors, in William Torrey Harris's "dark, airless, ugly places", because the indoors is perhaps the most concrete manifestation of the separate human realm we have created. The engineering of human nature happens in a physical setting from which nature has been removed as well.

Harking back to that most primeval model of separation—cell homeostasis—we create another level of self-other distinction with our buildings. Using bricks and mortar instead of lipid membranes, vacuum cleaners and air conditioning instead of ion pumps, we maintain our buildings in a state far removed from thermodynamic equilibrium by exporting entropy into the environment.

This is more than an adventitious metaphor. The indoors is the epitome of control, a place in which any undomesticated life forms (ants, mice, etc.) are unwelcome, where the processes of nature (such as decay and soil buildup) are arrested, a place where even the climate is controlled. Buildings, like cells, separate the universe into two parts, indoors and outdoors, one of which is regulated and associated with self, and the other wild and associated with other. For the individual human this self-other dichotomy is most pronounced in that unique building called the home: hence the intense feeling of violation when it is burglarized, and the near-existential unease some people feel when they discover an infestation of ants or mice.

In many ways the suburban yard is an extension of the indoors, in that the only life forms in a well-manicured lawn are those expressly permitted to be there, just as the landscaped environment is an extension of the public or corporate indoors. The culmination of this would be the final elimination of the wild altogether: a climate-controlled, genetically-engineered world where no life exists that is not planned and directed. It would be, in effect, the conversion of the entire planet into a Great Indoors, a vision embodied in the bubble-domed cities of futurists since the 1950s.

The domestication of physical space runs parallel to the conversion of the outdoor, public life into the indoor, private life. As communities have unraveled and public life evaporated, the realm of the private has grown correspondingly. One indication of this is that the average size of the suburban home has more than doubled since 1949 (even as average family size fell by one-third)[34]. The realm of the private is the realm of the separate self, and that is the realm of the indoors. Thus, as more and more of life is privatized, functions that were once public and outdoors have migrated into the private home. Among them: theater—having evolved from fully participatory enactments of myth and legend, to small-scale public performances and vaudeville, to the private in-home viewing of mass-produced culture like television and film; music—again having evolved from universal participation to public performance of increasing scale to the private listening experience that reaches its extreme with the earphone; eating—again formerly a community activity; and finally and most significantly, play.

A vast change has overtaken suburbia in the past two generations. The archetypal suburb was first and foremost because that's where people moved to raise families. Lawns and parks and lots of other families with children defined the suburb as a children's paradise. In the cultural mythos of the American Dream, childhood proceeds along the lines of the Little Rascals or Dennis the Menace or the Berenstain Bears: long days playing outside with other children, building clubhouses and forts, jumping rope and playing hopscotch, catching frogs and turtles, biking all over the place… pickup games of baseball and tag… tea parties with the other girls… sledding and snowball fights. Children were seldom at home. They were at a neighbor's house, or over at the playground, or the vacant lot, or down by the pond. It didn't matter as long as they were back for dinner. Until recently, play was outdoors, public, and free of charge.

Where are the children now? This is the question I asked myself one winter Saturday as I walked through the empty suburban streets and past the deserted playgrounds of my home town. Finally I saw a tiny figure dressed in a pink snowsuit, a little girl standing at the edge of her yard, waist deep in the snow. She dipped her mitten into the snow for a taste. Four hundred families in this neighborhood, most of them with children, and only a single five-year-old outdoors on a Saturday afternoon. And I cannot imagine her staying there very long, alone in the snow, the stillness broken only by the passing cars and that odd-looking lone pedestrian. Her life happens indoors.

A major theme of this chapter is how important it is for children to be free to explore their limits and experience the consequences of their mistakes. In large part, this means letting them be out in the world, that is, outdoors, in an environment not wholly under control. I would love to tell everyone, "Just let your kids be free," but unfortunately the matter is not so simple. Embedded as we are in modern society, much of the control imposed upon children is structural in origin and beyond the power of parents to easily alter. Furthermore, control on all levels has engendered the paradox I emphasize throughout this chapter; namely, it has made the world much less safe. In other words, I do not think all you parents out there are control freaks, nor am I going to advocate letting your children out right now to play in traffic.

Wait, let me contradict myself. I do think you—nearly all of us—are control freaks, simply by virtue of our membership in this culture. I also think that it is perfectly fine for children to play in traffic. But not traffic as it is in the world under control (I'll explain that in a moment). While parents do tend to overestimate the dangers "out there," it *is* dangerous—more dangerous than it was a generation ago. By getting the world under control, we have made it more dangerous, a truism we can apply as much to the war on disease and the war on terrorism as we can to that archetype of ordered, controlled modern life, suburbia.

Here are two ways in which control has made suburbia more dangerous for children. First, by moving life into the privacy of indoors, our neighbors have become strangers and we no longer feel at ease knowing that "someone will look out for them." Of course, the anonymity of the contemporary suburb owes itself also to nationwide patterns of mobility, which obliterate the cohesiveness of communities, and to television, which replaces local stories about so-and-so's granddad and the neighbor's uncle down the street with stories beamed in from remote locations about people and events that have nothing to do with the community. Whereas in the past we talked about people we knew, today we talk about characters from TV shows and professional sports teams. Television, originally trumpeted as a means to open us up to a wider world, has had the perverse effect of cutting us off from the people around us. Television is an indoor activity, inherently private and isolating. And when there are no other kids playing outside, and the town is full of strangers, what else is there to do?

The other defining technology of twentieth-century America, the automobile, has also exacerbated separation and brought life indoors. It is ironic indeed that the car, which in theory should bring people closer together by reducing travel time, has had the opposite effect. Consider that without the automobile, suburbs as we know them would not exist. Without the automobile, every house needs to be within walking distance to various shops, post offices, schools, and a train station. It is no accident that older towns and cities, built before the Age of the Automobile, are much denser than the sprawling suburbs-without-a-city that comprise newer metropolitan areas.

In depopulating the outdoors we have conceded it to cars, creating space inhospitable or downright dangerous to people, especially children, driving them even more indoors. The usual vicious circle of technology. The outdoors becomes the road, not a real place at all but merely a distance to be traversed between destinations. The apotheosis of this trend, the superhighway, goes so far as to prohibit pedestrians—that is, people—altogether. By bringing the world under control, we have made huge sections of it off limits, a development presaged by the reduction of the infinite described in "The Origins of Technology".

Whether for children or adults, the migration of life indoors is so extensive that some people feel uneasy to be outside at all. They perceive the outdoors to be not entirely comfortable, safe, or secure—it is the undomesticated realm that is not under complete control. Often this perception manifests physically as intolerance to even modest levels of heat and cold, pollen, biting insects, and so on—another example of how technology generates dependency on technology. The less we utilize and challenge our capacities, the more those capacities wither. Could this be another type of physiological capital we are converting into money? (Not to say life indoors *causes* pollen allergies, though I doubt that many rabbits or deer suffer from such sensitivities. Similarly, animals and pre-technological human beings are far more tolerant of temperature extremes than are civilized humans.)

No matter. When life happens in a box, there is hardly a need to go outdoors at all. In the winter some of my neighbors literally go weeks without spending more than two minutes at a time outdoors—the time it takes to get from parking lot to building. The same is increasingly true in summertime as well, when life takes us from air-conditioned home to air-conditioned car to air-conditioned store or office.

Spend some time in less developed countries and you'll be amazed at how much more time people spend outdoors, and thus how life there is correspondingly much more public. When houses are small, close together, and not very soundproof, as is the case in many parts of the world, and when significant life functions happen outside anyway, then our *selves* are more public too—we are better known and our stories are familiar to all around us. Under such circumstances the illusion of the discrete and separate self is less persuasive. When our traumas and our triumphs happen in an insulated box, it is easy to feel we are alone in the universe. But when we are not so socially isolated—when the whole village can hear us argue or make love, see that a child is ill, watch the baby's first steps—then we see our words, emotions, actions and states of being ripple out into the community and bounce back to affect ourselves. It becomes so much more obvious that we are not discrete subjects alone in an objective universe but rather intimately connected with the world that environs us.

New trends in urban design toward denser development thus have a significance far beyond their ecological and community-building benefits. Just as the isolated suburban box corresponds to an isolated self inside a fortress of security and faux-independence, a community of shared public spaces corresponds to a shared self, for whom the private realm of me and mine is limited. I am not advocating the abolition of that realm, neither in its external manifestation as a private space, nor its interior manifestation as the ego. It's just that today, that realm has expanded like a tumor to consume nearly the whole of life.

By bringing too much of life into the realm of the private, the separate self, the indoors, we end up with less life, not more: less security, less independence, less safety, and an ever-increasing need for more and more control. A recent innovation in traffic engineering gives us a glimpse of another way. Known as "second-generation" traffic engineering, it borrows from Dutch *woonerf* design principles that blur the boundary between street and sidewalk.[35] In many ways it is actually the antithesis of "engineering", which seeks the optimum control of vehicular and pedestrian traffic through the "triple E" paradigm of traffic control: engineering, enforcement, and education. Instead, it foregoes control, dispensing with all traffic lights, stop signs, crosswalks, and lane markings, and inserting trees and other objects right in the middle of roads and intersections. Children can play in the street again!

> "What the early woonerf principles realized," says [urban designer] Hamilton-Baillie, "was that there was a two-way interaction between people and traffic. It was a vicious or, rather, a virtuous circle: The busier the streets are, the safer they become. So once you drive people off the

> street, they become less safe."
>
> Contrast this approach with that of the United Kingdom and the United States, where education campaigns from the 1960s onward were based on maintaining a clear separation between the highway and the rest of the public realm. Children were trained to modify their behavior and, under pain of death, to stay out of the street. "But as soon as you emphasize separation of functions, you have a more dangerous environment," says Hamilton-Baillie. "Because then the driver sees that he or she has priority. And the child who forgets for a moment and chases a ball across the street is a child in the wrong place."

What a beautiful illustration of how control creates the symptoms we think it is addressing! But it also suggests a further point—to relinquish control does not mean to disengage from the world or to be careless. The chaos of the *woonerf* street works because it forces us to be more engaged and more alert, not less. Similarly, the animal or primitive human was far more aware and observant of her environment than we are today, to the point where indigenous people's powers of observation seem well-nigh magical. To relinquish control is not to surrender life; it is to surrender *to* life, to engage life more fully, to let it in, to open to its unreduced repletion.

Life under Contract

The tyranny of yes and no, the migration of life to a wholly domesticated realm, and the dwindling of the uncertain and undefined in favor of the expressly permitted and explicitly forbidden takes another form in laws, rules, and regulations. Here the program of control manifests as a spreading legalism that infiltrates all aspects of modern life.

In the ecological sphere, the shrinking of the wild is domestication, confinement, and the Great Indoors. In the personal sphere, it is the conquest of the child's spontaneity and creativity. In the social sphere, it is the codification of previously informal agreements, ethics, and mores. As legal scholar Paul Campos puts it, "Anyone who compares the legal domains of our society to those of the premodern state immediately becomes aware of a tremendous and ever-increasing contraction of formally unregulated social space."[36]

Our dependence on authority to delimit the realm of the permissible projects onto the collective level as "the urge to regulate—to medicalize, juridify, and police every act of labor or play."[37] Thanks to our schooling, we are accustomed to being told what is allowed, accustomed to authority determining right and wrong, and accustomed to coercive enforcement of these distinctions. Implicit in a world under control is that if something goes wrong, it must be someone's fault. Someone must bear legal responsibility. Someone must be punished, and I should be compensated. And so, children in authoritarian institutions (such as schools) resolve disputes by telling on each other—going to teacher, the source of authority. In religion the same tendency manifests in the idea of God as a referee who will reward the just, punish the wicked, and make everything fair in the end. In law, this attitude motivates the burgeoning realm of the regulated, the "ever-increasing contraction of formally unregulated social space," and our endemic reliance on lawsuits to resolve disputes.

The rule of law has become a necessity as other forms of social coordination have broken down with the dissolution of community. In olden times, informal social pressures were far more potent than they are today because people depended on their neighbors to meet economic and social needs. The goodwill of the community was extremely important and the consequences of social shunning severe. Today in America there are few real communities to speak of, only neighborhoods, and the opinion of the neighbors matters little. As long as I obey all the laws, who cares what the neighbors think? I am safe from anything they might do. I don't need to buy things from them, I don't need them to watch out for my children, I don't use the things they make, I don't depend on them for recreation. The conversion of social capital into money means that just as with all the rest of life's necessities, we rely on remote strangers for dispute resolution. The local and personal has been replaced by the formal and remote.

Suppose my son breaks my neighbor's window with a baseball. My neighbor asks me to pay for it. In the past, if I refused, the opinion of the whole neighborhood would turn against me. If I consistently behaved like that, I would find myself unable to get credit at the local grocer's, unable to receive emergency service from the local doctor, unable to get babysitting help when something happened. People depended on the favors and non-monetary reciprocation that we call neighborliness. In most places in America today, few of those mechanisms of social pressure still operate, and my neighbor has no choice but to sue me, resorting to the outright coercive force of a distant and impersonal authority.

Impersonality is built in to the very concept of law, which is supposed to be objective, rational, and impartial. Disputes are to be resolved not according to who knows whom, not according to one's reputation in the community, not according to whom the judge likes, not according to emotional judgments of right and wrong, and not according to popular sentiment, but rather by the impartial application of legal reasoning. That disputes in general can and should be resolved according to logic and principle is a fundamental assumption springing from the Newtonian-Cartesian worldview and modeling itself after axiomatic mathematics. In formal mathematics we start with basic axioms—the equivalent of a constitution—and reason from those, adding new axioms (legislation) as necessary where the old ones fail to prove or disprove a proposition (resolve a new type of dispute). At the zenith of the Age of Reason in the early 20th century, David Hilbert enunciated the final goal of mathematics: a complete set of axioms from which all mathematical truth could be deduced. Though Hilbert's program crumbled with the work of Gödel and Turing in the 1930s, in the realm of law we seem to think that any failure to rationally resolve the disputes of our day can and should be addressed with yet more laws. The end result would be a Hilbertian complete set of principles by which the legality of *any* act could be determined through unassailable logic. Then the ambiguity of human interaction would be gone. We would have in human relationship the same certainty that the physicists have sought for centuries in their quest for a complete universal law, a "theory of everything," in which everything would be of well-defined legal standing.

Every act would either be legal or illegal. Already we are so thoroughly conditioned to a world under control that I suspect many people would not see this as an alarming statement. It might even be hard to conceive of any other status, just as it is hard to conceive of a piece of land just existing, without being owned. "Someone has to own it—it can't just *be there*!" We find it easy to understand, "Everything is legal unless explicitly prohibited," or "Everything is illegal unless explicitly permitted," but not that some things simply have no legal status.

In any realm, the regime of control demands the elimination of uncontrolled variables, the domestication of the wild. In the realm of law, the wild is the unregulated social space and the variables represent uncertainty. Law represents the social reification of the Technological Program, whose goal is, as Campos puts it, "the final elimination of risk itself." He goes on to interpret "the urge to regulate—to medicalize, juridify, and police every act of labor or play" as a "contemporary by-product of the need to deal with the loss of any broadly held belief regarding... the point of human existence." Intuitively, Campos has come close to a central theme of this book, which is that the compulsion to control the world arises out of an alienation which has shorn life of any meaning except the maximization of comfort, safety, and pleasure. Except it is more than a need to "deal with the loss of a belief" that generates the urge to control; control rather is the inevitable, logical conclusion of our beliefs themselves.

Here's an amusing example of the legalistic program to eliminate uncertainty. Some years ago, I read that the student code at a certain college was amended to require that each new step in erotic foreplay be preceded by the explicit consent of both parties, as in "May I kiss you?" "Yes." "May I caress your breast?" "Yes," and so forth. No doubt this rule was written to clarify certain situations in which sexual consent was ambiguous, as in, "Just because I kissed him doesn't mean I gave him permission to grope me!" The solution: define consent with even more rules. Forget the absurdity of the whole thing for a moment and consider the fundamental assumption that a sufficiently minute set of rules can resolve all ambiguity. It's the technological program again! One can imagine a case where the meaning of "explicit consent" comes into question. Is a sigh of passion enough? Does reciprocation of a caress need prior consent? What about "uh-huh" instead of yes? Where does a sigh end and "uh-huh" begin? No doubt we could create even finer distinctions to resolve these ambiguities, but there would be still more—ambiguities created by the very attempt to define them away.

The parallel with mathematics is uncanny. Gödel demolished Hilbert's Program by proving that given any consistent axiomatic system of sufficient complexity, there will be true sentences in the formal language of that system that are unprovable from the axioms.[38] The true sentence (or its negation) could be added as a new axiom, but no matter how many are added there will always be yet other true statements unprovable from the expanded set of axioms.[39] Similarly, no matter how fine the distinctions of law, there will always be situations that are not logically resolvable from the law. We can add new axioms, new laws, finer distinctions, but still there would be logically irresolvable situations. Steeped as we are in the mindset of the Technological Program, the solution is nonetheless to always add on more laws, more regulations, in hopes of finally encompassing every possible situation within a legal framework: everything either expressly forbidden or explicitly permitted, with no gray areas. The result is, as Campos puts it, "the 100-page appellate court opinion, the 200-page, 500-footnote law review article, the 1,000-page statute, the 16,000-page set of administrative regulations." And yet, as the letter of the law swells in exactitude and scope, the spirit of the law withers away and its power to control human behavior diminishes.

The extension of law into every corner of life is visible in the increasing pervasiveness of the contract. We enter into contracts all the time without even noticing it. Have you ever noticed fine print saying, "By purchasing this product, you agree…" At Penn State, by enrolling in the university the student is legally consenting to abide by the regulations of the voluminous student handbook, and by enrolling in my class is agreeing to the syllabus, which is accorded the status of a contract between the student and the university.

A contract is a legal agreement, one that is "real" in the eyes of the law. Here is the culmination of the confusion between label and object that representational language invites: the reality of an agreement depends on being written down, while consent devolves into a signature. What "counts" in an agreement? Not the unspoken understanding, nor its social context, but only what is in "black and white"—words on a page. Of course, the seeming objectivity of a law or contract is an illusion, since the words' interpretation and enforcement indeed still depend on these, more human, factors.

Expanding legalism rises out of the dissolution of community; equally, it contributes to it. By relying on an outside authority to adjudicate and impose a solution, we are relieved of the necessity of working things out among ourselves. "You won't hear from me again—you'll hear from my lawyer!" These were the last words a friend of mine heard from a business partner of many years. Another variation: "Tell it to the judge."

Another consequence of social governance by lawsuit and threat of lawsuit is the further reduction of life into money that I discussed in Chapter Four. Damages are calculated in dollars, and it is through dollars that we redress matters of betrayal of trust, negligence, pain and suffering, and all the other subjects of lawsuits. Generally speaking when people bring a lawsuit they are seeking money. What else can you sue for? Remorse? Sympathy? Admission of wrongdoing? When someone is financially "compensated" for a wrongful death or dismemberment, the underlying message is that money is the equal of life or limb.

The technological program to make life safe, the conversion of life into money, and the pervasion of the law converge in the realm of liability insurance. The very word insurance hints at the assumption that life can indeed be made secure, that the unsure can be made sure. One of the industry's own terms for itself is, after all, "risk management". The consequences of this assumption are far-reaching indeed. Have you ever wondered why all the fun playground equipment—really high swings, jungle gyms, and slides—has disappeared? Liability. Why doesn't anyone allow skateboarding in their public areas? Liability. When the justification of safety is baldly absurd, liability serves in its place.

For the last twenty or thirty years, deterrence has been the mainstay of our penal code and prison system. The assumption behind deterrence, implicit in the very definition of the word, is that without the penalties people would commit the associated crimes. The law stands between my innate depravity and your life, liberty, and property. Interestingly, this is what Thomas Hobbes was saying in that oft-quoted passage from *Leviathan.* It is instructive to read the whole paragraph:

> Whatsoever therefore is consequent to a time of war, where every man is enemy to every man, the same consequent to the time wherein men live without other security than what their own strength and their own invention shall furnish them withal. In such condition there is no place for industry, because the fruit thereof is uncertain: and consequently no culture of the earth; no navigation, nor use of the commodities that may be imported by sea; no commodious building; no instruments of moving and removing such things as require much force; no knowledge of the face of the earth; no account of time; no arts; no letters; no society; and which is worst of all, continual fear, and danger of violent death; and the life of man, solitary, poor, nasty, brutish, and short.

Hobbes was not talking so much about a pre-technological state, but rather the human condition in the absence of government. Without government, he argued, there could be no commerce, no buildings, no creative achievement, because what would stop someone else from just seizing the fruits of another's labor? We would always live in fear of robbery and would therefore be unwilling to do anything productive. Hobbes is talking about the *ungoverned* state of being.

Because it is based on the Hobbesian-Calvinistic precept of the innate depravity of man, law rests on a coercive basis that seeks to make socially undesirable behavior no longer in a person's rational self-interest. Law is more than codified social agreements, as in "We all agree to stop at red lights." It also includes explicit penalties for breaking those agreements.

This is true not just of criminal law, with its threats to life and liberty, but of civil law too insofar as it seeks the *enforcement* of contracts. Think about that word. Law comes down to the application of force.

But what else could we expect in a civilization so deeply invested in Newtonian principles? Force, after all, is the only way to influence a mass—or, ultimately, change the behavior of a person—in a Newtonian universe.

Inescapably, law based on Hobbesian assumptions and coercive mechanisms disrespects the citizen by assuming that coercion is even needed. Consider a trivial example: a sign that says, "No littering, $300 fine" versus a sign that says, "Keep America beautiful, please don't litter." The first almost implies that if it weren't for the fine, we would want to litter. The fine appears as the reason for not littering. The second comes across as a reminder based on the assumption that the reader certainly does want to keep America beautiful. Or consider an agreement sealed by a handshake. Isn't it a disavowal of trust to insist as well on a financial penalty for breaking it—especially a penalty to be enforced by an outside authority?

Is it really deterrence that prevented you from stealing this book from the store? Is your true nature that of wanton acquisition? Or is that kind of behavior the artificial product of the discrete and isolated self, and the painful response to the survival anxiety implemented through our economic and social institutions? Is that really us? Are you really a murderer, if not for threat of the electric chair? If not, then the laws are insulting.

The implicit insult of a deterrence-based or coercion-based legal system is really no different from the other means by which our culture corrals the human spirit and denies its innate goodness, divinity, and creativity. Really, the essential presumption of deterrence is, to use playground language, "I'm gonna make you do it." How do you "make" someone do something? By having power over them. And the ultimate power over someone is the power to threaten their survival. The threat to survival implicit in the policeman's gun is little different from that in a parent's shaming. The physical restraint of a jail is little different from that of a parent's superior physical strength. The fact that I am not actually afraid a policeman will shoot me is beside the point. The fear was internalized and integrated into concepts of propriety, practicality, and prudence long ago. For most people, the gun is just a little reminder of that, a token of an omnipresent threat to survival. It reminds us of the fear that is "gonna make us do it."

To reform the law along non-coercive lines would be impossible without reforming all the other structures of our society that arise from our conception of self and world. The Hobbesian-Calvinist assumption of our inherent selfishness that underlies our legal system and logically necessitates deterrence is actually true—given the conception of self we have today. However, as our current misconception of the self crumbles, a new system of law will arise alongside a new system of economics, technology, and education. Its forerunners are already being tried out in various communities throughout the world. They do not require a return to a society without specialization of labor, villages and tribes where everyone knows each other, but they do draw upon these social forms for inspiration. Such a system of law fosters community rather than short-circuiting it, respects ambiguity and the developmental function of conflict, and most importantly, assumes the goodness, and not the depravity, of all human beings.

The War on Germs

The same assumptions that shape our legal system also define our medical system. We have seen how the overarching paradigm of the discrete and separate self generates a Technological Program to control nature. In the case of law and education, the control is over human nature; in the case of medicine, it is over the body and its biological environment. In either case, the goal of this control is to achieve security and avoid suffering.

The conception of ourselves as separate subjects in a world of discrete others implies that the interests of the self are fundamentally at odds with the interests of others. More for me is less for you; more for you is less for me. This fundamental ontology manifests in medicine as the germ theory of disease, and until we see ourselves in a different way, any other medical paradigm will remain marginal.

The doomed program to eliminate all suffering by controlling the world takes a specific form in the continuing medical obsession with germs, or more precisely, pathogens—that which "generates" a "pathology"; in other words, that which causes a disease. Bacteria and viruses are the prime culprits, to which we might add various fungi, prions, genetic mutations, and chemicals. To maintain the integrity of the separate self in a hostile world that would love to devour us for its own benefit, we exercise various forms of protection and control: immunization against viral diseases, antibiotics (ponder on the meaning of *that* word!) to destroy "invading" bacteria, quarantines to protect us from contagious individuals, pasteurization to protect foods and beverages from germs, other medicines to help us "fight off" disease.

Apparently this war on germs has been a great success, as all of the feared epidemic killers of the 19th century have been conquered. Actually, the ideology of the germ has greatly exaggerated the role of the two great weapons of modern medicine, the vaccine and the antibiotic, in the demise of infectious disease. According to Ivan Illich, "The combined death rate from scarlet fever, diphtheria, whooping cough, and measles among children up to fifteen shows that nearly 90 percent of the total decline in mortality between 1860 and 1965 had occurred before the introduction of antibiotics and widespread immunization."[40] Adult diseases such as cholera, tuberculosis, and typhoid fever show a similar pattern, which can be attributed more to an improvement in living conditions than to medical advances. Nonetheless, medical propaganda would have us believe that the noble cause of "modern medicine" is to extend it to those backward parts of the globe still in thrall to such diseases. Meanwhile in developed countries, we believe we can be even healthier by extending our vigilance with anti-bacterial soaps, flu vaccines, frequent medical check-ups and earlier screenings.

The campaign of extermination against germs has severe unanticipated consequences for our health. Killed off along with pathogenic microbes are much of our beneficial intestinal flora, which interact with and modulate the immune system in complex ways that recent research is only starting to reveal. Moreover, our native bacteria protect us by monopolizing the intestinal surface to deny competitors a foothold; they even secrete bacteriocins and other chemicals that inhibit the growth of harmful species. As in nature, when the internal ecosystem is destroyed, opportunistic species proliferate, such as pathogenic yeasts and bacteria. We identify the "cause" of candidasis as a species of yeast, but the real cause is a systemic disruption of body ecology. The same is true of our forests, where widespread tree death has also been blamed on various fungi. But why are the trees susceptible as never before? The reason is the same as in the body: systemic toxicity and the disruption of ecosystems.

When antibiotics were first discovered it was thought that bacterial disease was conquered forever. But like most "wars to end all war," the battle against the bacteria has taken one unexpected turn after another. Bacteria have developed resistance to antibiotics with an alacrity far exceeding any expectation—and challenging, indeed, widespread scientific assumptions about the mechanisms of bacterial evolution. The response to the declining effectiveness of antibiotics is—you guessed it!—more antibiotics. If technology seems to have caused a temporary decline in well-being, obviously the answer is more of it: more powerful antibiotics delivered in more potent ways. News articles speak of the "arms race" against the bacteria, in which heroic scientists race to expand the "arsenal" of new antibiotics before the bacteria develop resistance to the old ones.[41] Meanwhile, just as with other examples of the technological fix, each new intensification of the technology exacts a heavier price: side effects such as candidasis. The solution? More control, of course: fungicides to kill the candida! And then other drugs to counteract the side effects of the primary drugs. Somehow we assume that someday a Final Solution will be developed that will once and for all resolve the original problem, and the problems caused by its solution, and the problems caused by *their* solution... and we will all live happily ever after.

Indeed there are scientists working today on just such a final solution: a class of antibiotics that are impervious to all known means of microbial resistance, an endeavor that Stephen Harrod Buhner calls "perhaps the most dangerous actions now occurring on earth."[42] The consequences of such a final solution would make candidasis look like a walk in the park. Consider, for instance, that without bacteria all life on earth would probably perish. Imagine the consequences if herbivores lost the bacteria that allow them to digest cellulose, or if there were no bacteria to fix nitrogen in soil. After all, detectable residues of present-day antibiotics show up regularly and pervasively in groundwater, soil, and the tissues of living creatures. I hope none of these doomsday scenarios come to pass, but one thing is certain: the intensification of the war on the Other can only have one result. Anything we do to the world, we do to ourselves.

The reason is that we are not, in fact, discrete beings separate from the rest of life. The war on germs is but one aspect of our attempt to pretend otherwise, a pretense which, despite our increasingly strenuous efforts, is on the verge of collapse.

Humanity has been compared to a cancer on the planet. A cancer is a tissue that has forgotten its proper function and continues to consume the body's resources even when its headlong growth threatens to kill the very body upon which its own survival depends. Can you think of a more accurate characterization of humanity's role on planet earth? Is it any wonder, then, that what we have done to the world is manifesting physically in our bodies?

On the individual level as well, a part of ourselves exceeds its proper function and commandeers the resources of our entire organism. That part is the survival-based ego, which consumes the years and the vigor that should have gone into fulfilling our life purpose. Instead life is devoted to the pursuit of the meaningless, trivial, and impermanent: all those things that give us a false sense of security, temporary comfort, and conditional self-acceptance. In other words, the psychic organ of self-aggrandizement whose job it is to create a reasonable level of biological safety and comfort has grown into a massive, all-consuming tumor. I think many cancers are the somatization of this self-betrayal.

But how might we interpret disease—and treat it—if not as an invasion? After all, viruses, bacteria, and cancer cells certainly seem to kill in very direct and obvious ways, and sometimes, at least, the allopathic treatments really do work. We "kill the bug" and the patient gets better. It is so tempting to think, "If those cancer cells weren't there, I'd be fine now." "If that candida were gone, I'd be fine now." "If I could kill those HIV viruses I'd be well." The pathogens seem obviously to be the source of the trouble—get rid of them and the disease is gone.

This way of thinking arises from fundamental tenets of Newtonian and Darwinian science, which envisions a generic universe devoid of purpose. Pathogens invade simply because they *can*; there is no purpose other than that. Our interests and theirs are fundamentally at odds. The universe is just like that—innumerable discrete beings competing for resources.

What is it about one body that makes it susceptible to cancer, another to candida, and another to AIDS? Why do only half of study subjects catch colds when their nostril linings are swabbed with virulent rhinovirus cultures? For that matter, why does one section of forest get overgrown with poison ivy, another with jewelweed, another with mustards? Why do the locusts swarm in certain times and places to defoliate every plant in sight? Thinking reductionistically, we generally ascribe these events to chance. I remember one cancer patient, just diagnosed, who desponded, "It looks like I just won the cancer lottery." Statistically speaking, a certain proportion of people get each disease. Will you be one of the unlucky ones? Statistically speaking, plant seeds disperse in a random way according to the variables of wind and other vector paths. Why did a thistle grow here and a burdock there, and not vice-versa? It was random, an accident, a chance.

A scientific paradigm shift is rapidly rendering this idea obsolete. Chapter Six describes how cooperation, symbiosis, and environmental purpose—alongside competition—are the defining forces of nature. Accordingly, we might consider the possibility that diseases too have their purpose, whether on a physical, genetic, or spiritual level.

There is an alternative to the randomness of the locust plague, the burdock and thistle, the cancer lottery. What if the thistle seed sprouted here and the burdock there because each had its own uniquely perfect contribution to make to the soil chemistry at that spot? What if the soil, knowing this, "invited" the seed to sprout and the seedling to grow by providing just the chemistry conducive to each plant? What if every being on earth has a contribution to make to the collective welfare of the whole? Are disease organisms an exception? Perhaps "pathogenic" viruses and bacteria actually have a positive effect on human beings, either individually or collectively.

The paradigm of random "infection" plays into the program of control. When we see germs as predators who seek to steal "resources" from us for their own biological interest (survival and reproduction), then a rational response is to deny them those resources, to hide from the predators or fight them off—the fight-or-flight response. There is no necessary reason why one person rather than another should be the "victim" of the flu. If, on the contrary, there is some reason specific to my own body why the flu has infected me and not you, then the program of control doesn't make sense anymore. If my body is fertile ground for it, eventually the virus will get in and grow. Maybe the body goes through cycles of sickness and health, or of cleansing, rebuilding, and maintaining, that actually necessitate a head cold now and then, or that make it nearly unavoidable. Maybe diseases like colds and the flu serve some kind of eliminative function: toxins being discharged along with the mucus; poisons incinerated in the heat of fever; bacteria invited in as scavengers to help eliminate the byproducts. Perhaps the body needs to go through some intensive housecleaning now and then.

I remember reading one time in a medical advice column that it is a "myth" that people catch cold from getting wet and chilled. "Viruses cause colds" the expert stated, attributing the "myth" to selective memory—we remember the times when a soaking was followed by a cold, and fabricate a causal relationship. Contrarian that I am, I immediately knew that getting a chill does in fact cause colds. Probably what actually happens is that the chill creates an internal climate that is ideal for the virus, which proliferates and eventually causes symptoms through which the body eliminates the excess "cold and damp energy" (to use the nomenclature of Traditional Chinese Medicine (TCM)), perhaps via mucus discharge. The cold and damp energy being gone, the virus no longer does so well and is rapidly eliminated by the immune system. We might look upon the virus then as an *external organ of homeostasis*.

What happens if you scrupulously quarantine yourself from any contact with the cold virus? Well, the cold and damp internal climate persists indefinitely until one eventually gets in—or until something worse happens. As evidence for this conjecture, consider the simple fact that most people feel *better* after getting over a cold or flu than they did before they got sick. You would think they'd be weakened, but no, they often feel revitalized. Traditional Chinese Medicine lends further support to this theory in the adage: "If you don't get little illnesses, someday you'll get a big one." If somehow the housecleaning functions of colds and flu are not allowed to operate, the toxins or energy imbalances build up and result in major illness. Perhaps it would be different in an absolutely pristine environment, but the toxicity of modern lifestyles and environment means that today, occasional illness is a sign of health.

Provided that the symptoms aren't life-threatening, I let them fully express in my body. Cold and flu medications thwart the very purpose for which our bodies invite in these germs. It is much better to support the process and give your body the resources it needs to see it through safely. What those resources are depends on the nature of the illness—and a qualified healer might help you figure out what they are—but invariably rest and quiet are required. Convalescing also gives us a chance to clean out "psychic toxins" along with physical ones: the stress, worries and busy-ness of everyday life.

If disease organisms have a mutually beneficial symbiosis with the human species, then the war on germs is grievously misguided. One of the main conceptual tools I develop in the next chapter is to see symbionts not as separate beings with whom we mutually depend, but as parts of self. The war on germs is therefore another aspect of our culture's self-other confusion, the confusion over who we are. It is not surprising that the great epidemics of our time are the autoimmune diseases. At its most basic, the immune system distinguishes between self and other. Our collective confusion over who we are manifests in the body as immune system disorders. Sometimes the causal link is quite specific, for example the link between the increasing universality of childhood vaccinations and the rise in various immune disorders afflicting children.[43] Best known is the link with autism, which can be understood as an auto-immune attack on myelin in the brain.[44] In my personal experience the most striking is the enormous increase in the prevalence of allergies in children. When I was a child in the 1970s, we had one or two children in our class who were allergic to something or other, but it was fairly unusual. There were many fewer vaccines given then, especially in the first two years. Today, you hardly dare give a child anything to eat without first asking the parents what her allergies are.

What about deadly killers such as cholera, typhus, smallpox, and plague? It is not easy to see *them* as symbiotic friends of humankind, at least on the individual level. Nonetheless, these diseases might have a beneficial function on an ecological, genetic, or spiritual level. In any event, deadly acute epidemic diseases are mostly a thing of the past.[45] Much public health energy is expended preparing for the next epidemic, as authorities raise the specter of terrorist-propogated smallpox, a new virulent influenza epidemic, something exotic like ebola, or a new disease like SARS. In focusing on these, we are fighting the last war with the weapons of the last war. These are the sorts of enemies against which the technologies of control (vaccine, antibiotics, quarantine, etc.) are effective.

The new diseases of the modern era are of a different sort. Cancer, arthritis, chronic fatigue syndrome, Alzheimer's disease, Crohn's Disease, multiple sclerosis, diabetes, AIDS, and so on defy the medicine of control, which has made almost no progress in curing them despite research outlays dwarfing those of the vaccine/antibiotic era. Significantly, most if not all of these new disease involve a dysfunction of the immune system. They reflect the same self-other confusion that defines our relationship to the world.

Helpless against the diseases of the 21st century, we instead take ever-more extreme measures against the microbial world. One manifestation of the war on germs is the proliferation of anti-bacterial soaps, latex gloves for all food service workers, and strap-on mouthguards that appeared (and in some cases were legally mandated) during the Asian SARS "epidemic". These devices constitute a physical barrier between self and world, concretizing the psychological distance that separates us from each other and from nature. Sometimes I have nightmare visions of a future where the very idea of breathing each other's unfiltered air is repulsive and illegal, where everyone wears a gas mask and all human contact is mediated through latex or computer terminals.

Equally frightening is the current hysteria about avian flu. Since it is supposedly spread by wild birds to poultry flocks, some officials are implementing new controls prohibiting chickens from ranging free. Yet it is the caged indoor hens with their debilitated immune systems that are the most susceptible, and it is factory methods of poultry production that actually spread the disease.[46] A related proposal in the United States is the National Animal Identification System (NAIS), which would mandate that a digital tag be embedded under the skin of all livestock. Everywhere, the technologies of control are the same: separation, confinement, and the numbering of all things.

As long as their ideological underpinnings remain intact, none of these trends will abate. Already medical microchips are available that can be embedded under human skin to monitor various physiological states. In Asia during SARS, body temperature readings were taken as condition for entering certain public facilities. Potential epidemics offer a rationale for the quarantine of populations and control over their movements—an internal passport system justified on medical pretexts. All such measures make perfect sense from the mindset of separation and control.

Our phobia of germs is a specific form of a more general aversion that goes by the name of cleanliness—the quintessential manifestation of the urge to control. Cleanliness is next to Godliness, it is said; but that is true only if God is remote from this earth. What is dirt but, quite literally, the world? To maintain absolutely clean bodies and clean houses is to separate ourselves from the world. There are few *faux pas* as serious as showing up filthy and smelly to a social occasion. To be one with the dirt is to be uncivilized, to be less ascended from the earthy world, to be more of the body and less of the mind or spirit. Similar considerations explain why adults feel so uncomfortable when they are sticky. As any parent knows, young children don't mind being sticky at all. But to be sticky literally means that the world (the Other) adheres to your self. Stickiness therefore violates our sense of control by threatening the physical separation of self and world.

Please don't misunderstand: I'm not suggesting that you give up housecleaning and personal hygiene. The separate self has no absolute reality, but it has a conditional reality that is useful within its proper domain. The problem comes when we seek to endlessly expand that domain and make it permanent. I do not advocate the abandonment of the separate self. Instead, it will become more fluid and playful.

A final and very telling aspect of the War on Germs is the insistence on the absolute sterility of the food supply. The pasteurization of milk, beer, and other products has more to do with uniformity and shelf life than with safety. Safety is more a rationalization than a reason.[47] Reinforcing the economic motivations of uniformity, shelf life, and standardization is the more fundamental motivation of control. As long as the milk (for example) is rendered perfectly sterile and kept that way from dairy to processing plant to supermarket, there is no chance of infection—it is perfectly safe and risk-free. Health through control.

The contrasting, ecological paradigm draws on a much broader conception of health, extending to the cows, the land they graze on, and the integrity of the farmers and processors. Raw milk is actually quite safe, but only if it comes from healthy cows—an impossibility under current (industrial) dairying practices. Cows can only be healthy if the soil is healthy, if they graze and live out in the open, if they are not subjected to artificial hormones and antibiotics to boost yields, and in the long run if they are not bred to over-produce milk. Since all these requirements conflict with minimizing cost-per-hundredweight, they are incompatible with the subjection of milk to the commodity markets. All of these aspects of control, from cost control to product sterilization, reinforce each other.

As is the case with so many other technological fixes, the result of food sterilization is to make it less healthful, not more. Not only is pasteurized food depleted of enzymes and vitamins, but the living bacteria in raw dairy and unpasteurized sauerkraut actually contribute to the intestinal flora that help nourish the body and ward off illness. Sterile food is a target for opportunistic contamination by harmful salmonella, E. coli, and other bacteria, which need not compete with the original benign bacteria in unpasteurized milk and other foods. Control, once initiated, must be continually maintained.

Traditional culinary cultures took the opposite approach by making wide use of fermentation, and not the controlled type where a single microorganism is used to inoculate a sterilized medium. Traditional fermentation literally invited the world in, relying on ambient microorganisms as well as complex cultures of hundreds of species of yeasts and bacteria that were symbiotic not only with each other, but with the human beings they coevolved with over generations. The results were somewhat unpredictable: each batch of cave-ripened cheese or naturally fermented wine was unique. Natural fermentation is therefore incompatible with the demands of industrial production and mass marketing, which requires product uniformity and long shelf life.

The latest method of sterilization is irradiation, used extensively on spices, meat, and other foods. Basically, the food is exposed to radioactive nuclear waste at a level sufficient to kill anything alive in it. Amazingly, the food remains relatively impervious to new contamination after the radiation has been administered. How could we think that it can nourish human life if it cannot nourish bacterial life? Only from the mindset of separation.

The current obsession over food hygiene—and the whole campaign of extermination against bacteria—starts to look ridiculous when you realize that the human digestive system is really not so different from that of a dog or pig. The more control we exercise over the world, the more over-sensitive to the world we become. Despite the fact that they regularly eat off the floor and drink out of toilet bowls, dogs don't seem to get sick any oftener than we do, nor have I noticed any appreciable improvement in health since food service workers starting wearing gloves. On the contrary: a lack of regular challenges to the immune system on the one hand primes it to become more sensitive (maybe I'm missing something, it thinks) while on the other hand depriving it of the exercise it may need to take on a real crisis. The immune system becomes at once more sensitive but less able to deal with real challenges. The parallel with the coddled, over-protected modern human is obvious.

The War on Suffering

The war on germs is just one expression of a medical system based on control. Control, in turn, arises from our sense of self, that we are separate beings in an alien and indifferent universe. Not being part of any purpose beyond ourselves, naturally we seek to maximize the security, comfort, pleasure, and other interests of those selves. Technology is one form that ambition takes.

If pain, like the rest of life's events, has no larger purpose or meaning, then why not avoid it? The logic parallels that regarding colds and flus. If we are discrete and separate beings in a fundamentally competitive world, then any confluence of interests must be accidental, any larger purpose must be our own projection, and our well-being must come from gaining as much control as possible over an at-best indifferent but often hostile universe. Remember the logical conclusion of the Technological Program: the elimination of all suffering. That this is a feasible goal is fundamental to the assumption that the world is in essence controllable, and medicine is a key technology in implementing that control.

Despite its colossal disappointments, the myth of the imminent perfection of modern medicine holds that its technological marvels are on the verge of dramatic new improvements on the human body: gene therapy to reverse aging, nanotechnology to cure cancer. "In such a society," wrote Ivan Illich, "people come to believe that in health care, as in all other fields of endeavor, technology can be used to change the human condition according to almost any design."[48] The can-do attitude of the engineer improving on nature motivates such medical fads as the indiscriminate tonsillectomies of the 1930s and 1940s, the routine removal of wisdom teeth in the last two decades, the technological hijacking of the birth process, and, at this writing, pronouncements by certain authorities that at least half the population would benefit from cholesterol-reducing drugs. Just as technology in general says nature could use some improvement, medical technology assumes the same of the human body. Hence we remove organs that we think are unnecessary, such as the uteruses of post-menopausal women, and alter body chemistry with hormone replacement therapy. And we think that yet better health will come from the refinement of control to the cellular and molecular level, with nanotechnology and genetic engineering.

The dogma of control blinds us to the fact of sustained, generalized decline in health despite, or perhaps because of, these interventions. Few doctors realize that the rate of serious birth complications is far lower in Holland, where home births predominate, than in America's high-tech obstetrical wards. They similarly cannot see the horrendous side-effects and marginal benefit of cholesterol-lowering statins. The trend of routine medicalization continues headlong: who knows what "medicines" they'll be adding to the public water supply (in addition to those already there such as fluoride)?

We can already see the final destination of this trend: the medicalization of all life, the conversion of all people into patients. Writing almost thirty years ago, Illich observed that "in some industrial societies social labeling has been medicalized to the point where all deviance has to have a medical label."[49] As children we were horrified at the Soviet practice of locking up dissidents in mental hospitals under the assumption that anyone who could object to the socialist Utopia must be crazy. Today we are witnessing an analogous phenomenon on a vastly wider scale. The tens of millions of people who find themselves unhappy in our modern technological paradise, this world under control, are diagnosed with some psychiatric disorder and medicated with mind-altering drugs. The same thing happens to children who resist the breaking of their spirits: they are diagnosed with "oppositional defiant disorder" or "attention deficit disorder." I look upon these "disorders" as signs of health, not illness. A sane, strong-willed child will resist the mindless routines, the busywork, and the hours cooped up inside a classroom; she will steal moments of play at any opportunity.[50] I also think adult depression can be a sign of health. When we have been frightened away from our creative purpose into a life not really worth living, the soul rebels by withdrawing from that life. This is the paralysis of depression. "I would rather not participate in life at all," says the soul, enforcing its resolve by shutting down the mind, body, and spirit. We no longer feel motivated to live. For a while, sheer willpower and the habit of routine keeps us going, but eventually the soul's call to withdraw becomes undeniable and we sink into the throes of clinical depression, chronic fatigue syndrome, and the like.

Another pathologized symptom of health is anxiety, the feeling that "something isn't right around here, and it needs urgent attention." What is that something? Again, afraid to look in the shadows for the key, we direct it onto trivial fears. However, the underlying emotion is a valid response to the facts of the world. Something, indeed, is terribly wrong around here, and it does need urgent attention. On the physical side, anxiety has a counterpart in newly emergent diseases such as multiple chemical sensitivity and fibromyalgia. (A physical counterpart of ADD is vision problems or dyslexia: you can make me sit at a desk but you can't make me see straight!)

The medicalization of these psychiatric conditions is predicated on the assumption that life and the world as we know it is just fine. Like the dissidents in socialist Utopia, if living this good life renders you depressed, anxious, or unable to concentrate then, well, you've got an illness my friend. The problem isn't the world, and it isn't the life you have chosen; it is the chemistry of your brain. And *that* can be adjusted.

One consequence of the killing of pain, whether physical or psychological, is that it makes a painful world more tolerable. "Any society in which the intensity of discomforts and pains inflicted renders them culturally 'insufferable' could not but come to an end."[51] If it were not for these pharmaceutical methods of controlling the psychological pain of living the modern life, society as we know it would quickly crumble. The same holds on the individual level: just as the medication of society allows it to exist at a much higher level of pain than it otherwise could, so does the suppression of symptoms in the individual allow life to go on as normal. People on SSRIs such as Prozac have told me that the medicine enables them to cope with life. And it does. It allows life as usual to proceed.

The reason that conventional psychiatry—whether pharmaceutical or psychoanalytic—is powerless to substantially help the vast majority of patients is that it does not, and cannot, recognize the wrongness of the world we live in. The psychiatrists have bought in along with the rest of us. Psychiatry operates on the assumption that we should be happy with it. The same assumption of the rightness, or at least the unalterability, of the world given to us underlies the quest to "cope with stress." That life is inherently stressful is not questioned. Psychiatric treatment is infamously impotent to address serious mental conditions because the psychiatrists, as fully enculturated elite members of society, are constitutionally unable to call into question the cultural assumptions in which they are so deeply invested. Their investment blinds them to the underlying rightness and fundamental sanity of a patient's reaction to a world gone wrong. Conventional treatment (particularly pharmaceutical treatment) actually exacerbates the illness by affirming, "Yes, the life society proffers is fine; the problem is with you." I have witnessed dramatic healing simply by affirming to someone, "You are right, this isn't how life is supposed to be"—a realization that empowers change. Not that psychiatry ignores the need for change altogether; it is just usually unable to carry the change deeply enough. In essence, it tells us that we need to adjust to the world and seeks to make us normal again, functioning members of society.

No. We are meant for more. "Everyone can sense the emptiness, the void underneath the forms and structures of modern society." Any psychiatry that seeks to adjust us to such a society is itself insane.

You would be crazy to busily polish the silverware as your house burned down. Similarly, it is insane to live a normal life in today's world. Maybe if millions of children were not malnourished, maybe if torture were not commonplace around the globe, maybe if species and entire ecosystems were not dying, maybe if genocide were forever gone, maybe if all the injustices I've chronicled in these pages did not exist... maybe then the "normal" life presented us would be sane. Maybe then it would make sense to absorb ourselves in professional sports, soap operas, the stock market, material acquisition, and the lives of celebrities. Given the reality of the world today though, the only life that makes sense to live is an extraordinary life.

The teenagers in their idealism and their defiance, the depressed in their rejection of the lives offered them, the anxiety-ridden in their sense that something is not right... all are quite sane. Any psychiatry that doesn't recognize this is doomed from the start. It tells us the problem is not the world, it is ourselves. It merely adds to the chorus of voices telling us, "All is well, all is normal—who are you to think any differently?" That's the same message we get from the media that immerse us, suggesting with their subject matter that we can afford to care about the trivial and the superficial: the sports, celebrities, and so forth. As well, the whole mania for "entertainment" suggests that our world is sound enough that we can afford constant distraction from it. "Things will be fine. Don't worry." I imagine myself on board the Titanic. "Hey guys, we're pretty far north, don't you think we should slow down? Hey guys, isn't that an iceberg up ahead?"

"Charles, relax! Have a drink. Come listen to the band. Everything is fine—see, no one else is worried."

Not only do we live today in a fraudulent, life-denying society into which nobody fits, but the incompatibility of that society with human fulfillment only grows with each passing year. Along with it grows the need for medicating expanding swaths of the population. We have seen this happening with the increasing ubiquity of SSRIs and similar drugs across every age group. The world grows more painful physically as an increasingly toxic environment gives rise to new diseases, as commerce and industry corrupt the food supply, and as the tempo and pressure of life-as-usual quickens. All of these factors accelerate the conversion of citizens into patients.

The medicalization of society contributes to "life under control" in another way by the authority it invests in doctors. A schoolteacher friend of mine told me that school policy only allows his third-graders to visit the bathroom three times each day, the only exception being—you guessed it—children with a note from their doctor. When we give authority to judge health and illness to a doctor—someone external to the person suffering or not suffering—we are attempting to objectify pain, the primary indicator of health. While explicit attempts to quantify pain have failed, the ostensible objectivity of the doctor's note (and its current incarnation in the reams of paperwork that our medical system generates) allows human health to be converted into data. Then it can be treated according to the standardized, anonymous, objective methods of industry. Unfortunately, the objectification of illness and professionalization of medicine have also separated ourselves from our own bodies, rendered us helpless to heal ourselves, and made pain and illness all the more frightening.

The killing of pain is oddly reminiscent of the mentality of agriculture. Just as the farmer pulls out the bad plants (the weeds) to maintain a field of only good plants (crops), so do we discriminate between good and bad feelings. Life under control means eliminating anything that might cause a painful sensation or feeling, an objective tantamount, we think, to eliminating suffering itself.

The confusion between pain and suffering is fundamental. Suffering comes not from pain *per se*, but from resistance to pain and, more generally, from resistance to life. It is no wonder that technological society, predicated on the elimination of pain and the control of nature, has generated suffering unprecedented on planet earth. Technology, which resists the ordinary processes of nature, is mirrored in our psychology as a resistance to the ordinary processes of life. Our internalized dictator—the voice of culture—judges and filters our every feeling and emotion, clinging to the good and weeding out the bad. Attachment and aversion: precisely what Buddha identified as the origin of suffering. The internalized dictator that seeks to maintain and extend control over our every thought and emotion is nothing less than an Orwellian Big Brother, who is always watching. The Thought Police are always on patrol in this ultimate tyranny of the finite part of ourselves over the infinite.

Illich writes, "[Good health] means to be able to feel alive in pleasure *and* in pain; it means to cherish *but also to risk survival*" [emphasis added].[52] The first step to freedom is simply to allow yourself to fully feel whatever there is to feel. This, and not mind control, is where the true benefit of meditation lies. This is also what is meant by accepting God's gifts. If God is good, then every moment is by definition a gift, and to deny it is to separate oneself from God—the Jewish and Christian mystics' explanation of suffering.

The technotopian promise that "the pain can be killed" has led us to believe that pain need not be felt, and this belief generates a resistance to pain that exacerbates the suffering all the more. The promise is a lie, because pain is unavoidable. To be human is to be born into pain. Loved ones pass away. Good things come to an end. Our bodies get old, sick, and injured, and someday we die. Let us not pretend that technology will someday eliminate such occurrences, nor pretend that such occurrences are not painful. Suffering only comes when we do not allow ourselves to feel the pain, when we resist it. That effort is quite logical when we buy in to the big lie of the world under control: that pain need not be felt. When we operate under this delusion, we inevitably become resentful of our pain, and therefore prone to all the abuses of victim mentality and entitlement.

Our deluded conflation of pain and suffering means that the medical program to reduce and eventually eliminate suffering is doomed. What's more, because suffering comes from resistance to pain and not from pain itself, the focus of medicine should not even be on the elimination of pain—yet this is what reductionistic, symptom-oriented medicine naturally tends toward. While such painkillers as morphine are valuable and have their place in medicine, the role of a true healer is not to make life more tolerable, which is essentially what the suppression of symptoms does. No, the role of the healer is just as Illich implies, to help the patient feel more alive in pleasure and in pain. We go to the doctor because it hurts, and our expectation is that he will give us something to make it stop hurting. That is an error. Pain is our friend, never to be sought out, of course, but neither to be resisted. Let it hurt. When we feel what there is to feel, we cease maintaining separation between ourselves and our sensations and therefore come into greater wholeness, greater health. The full experience of pain opens the door to health: whether we are on the sickbed, or facing the psychological pain of life's transitions, or the subdued agony of life in a world gone wrong.

When we allow the full experience of pain, the window to health that opens may indeed utilize the skills of a surgeon, homeopath, herbalist, or other healer. I am not enjoining the reader to forgo all medical care! Please. But bodily recovery is not the only benefit, which is fortunate because the result of healing might not always be what we hope for. There can be healing in death as well as in recovery. Another benefit of non-resistance to pain is that it brings an unexpected miracle: the pain doesn't hurt as much. Even if the pain is still there in all its intensity, we don't suffer as much from it. In this regard, pain is much like any other object of control. Control generates its own necessity, which intensifies over time at ever-greater cost.

Health means to be alive in pleasure *and* in pain. I have always thought of this book as a work of healing, and I hope that in reading it you feel more alive. Returning to the depression, anxiety, etc. discussed above, I cannot and will not ameliorate that suffering by assuring you that it's not so bad, that everything is okay. No, it is every bit as bad as you suspect, and even worse. Yet also, all is well.

When we awaken to the enormity of our crisis and the magnitude of our loss, often the first response is a crushing despair. I have been through that; I know what it is like. Yet on the other side of despair is fullness and an urgency to live life beautifully. We can choose a different world—the "more beautiful world our hearts tell us is possible" to which I have dedicated this book—but to choose it we must be familiar with what we are choosing. We must be fully cognizant of the world we have chosen up until now. Knowing the pain of the world fills me with energy and confirms the rightness of my life's direction. Otherwise what would stop me from occupying my hours with the trivial and the vain, staying comfortable for as long as possible until I died? We need not avert our gaze from ugliness and pain in order to live a happy life. Ignorance is not bliss. Quite the contrary: the more we insulate ourselves, the weaker we become, the less able we are to take on reality. The more we numb and defer the pain, the more afraid of it we are, until we willingly submit to confinement in the (temporarily) secure, predictable, controlled semblance of life our society offers.

Life in a Playpen

Our medical system with its War on Germs illustrates a general feature of control. Just as the unchallenged immune system fails to develop, yet also becomes oversensitive, and just as pain becomes more frightening in proportion to our efforts to numb and defer it, so also does our insulation from risk, challenge, and discomfort leave us weak and afraid of the world.

When we are deprived of the opportunity to explore our limitations, we become more fearful of them, more tightly bound to them, and less able to cope when, despite our strivings for control, reality presents us with a new challenge. This deprivation starts early in childhood with the playpen, the hovering parent, and more generally the regime of "safety first" that has infected modern society. Remember the latter part of Illich's definition of good health: "... it means to cherish but also to risk survival." Traditional societies allowed children productive, sanctioned ways of exploring their limits—real limits, not the phony freedom of the playpen's contained safety. Moreover, the consequences of their mistakes were real. The parent might say, "Don't poke that beehive or the bees will sting you," but would not through physical or verbal coercion try to prevent the child from doing it anyway. The result was that the parental voice gained an authority far beyond the fear-based coercive power today's parents exercise; at the same time, the child learned that consequences are real.

John Taylor Gatto has observed that one of the unspoken lessons of school is that actions have no real consequences. Children are essentially not permitted to fail, not at anything real. Teachers and parents too tend to praise shoddy work in order to "boost self-esteem", not understanding that the child herself knows the difference—at least in the beginning. Eventually, though, the child confuses praise from authority with the genuine article of satisfaction in the creative process, preparing for a life of doing it for the grade, for the client, for the paycheck. In this way, we early on become strangers to what we really love; our passions are lost to us and so we lose our passion for living. On the flip side of the coin, the absence of real punishments teaches children that socially destructive behavior has no consequences. But even punishment is only a substitute for real consequences.

When my son Matthew was four or five, he wanted a pocket knife just like his big brother. I decided to give him one, explaining carefully, "This knife is sharp, Matthew, and if you are not careful you will cut yourself." What happened? He was not careful, of course, and he cut himself. Not too seriously, but it hurt and there was blood. What did he learn from this? For one thing, he learned that knives are indeed dangerous—on their own merits, and not because one might get caught using one without permission. The second thing Matthew learned is that Dad is one smart dude. Dad was right about the knife. When Dad says something might happen, it's a good idea to listen.

No matter how deeply and thoroughly we frighten children with our power to invoke their survival anxiety, their natural curiosity and compulsion to test limits will eventually provoke them to "try it anyway," often in secret. When they find, as is often the case, that the consequences aren't as bad as their parents said they were, then parental authority loses all credibility. They find that no one loses an eye when they throw a paper airplane indoors, that they can smoke marijuana and not wake up in a crack house, that reading Harry Potter does not lead to Satanic ritual sacrifice. Now the stage is set for tragedy. On the one hand, they have always been insulated from the real consequences of their actions. On the other hand, the imposed substitute consequences (punishments) are no longer effective, because the wily teenager easily evades them by deceiving authority, not by abstaining from the behavior. The result is that the teenager acts as if he were immortal or invulnerable, and lies to his parents about everything he does.

Because human beings have an inherent need to explore boundaries and challenge limitations, today's obsession with safety forces teenagers into illicit, highly dangerous "risk behaviors". A lifetime of pent-up desire to know their limits explodes outward at the first taste of freedom, for instance, when they go away to college. Tragically, these behaviors nonetheless fail to challenge and expand significant boundaries. The inescapable fact is that although the exploration of one's boundaries is inherently unsafe, there is no other way to grow. When my five-year-old son says, "Daddy, watch how high I can climb this tree," I restrain myself from stopping him. And as a matter of fact, Matthew turns out to be quite prudent. It is not my authority ("You can't go higher than that branch!") but rather his own caution that limits his ascent. Imagine the consequences when time and again, parental authority halts a child's exploration before the limits of caution are reached. He will come to depend on external authority to define safety and danger on his behalf, while his own judgment atrophies. Never having had a chance to develop his own judgment, such a person is wont to take foolish risks. Yet paradoxically, because he is chronically dependent on authority to define danger, such a person is also easy to rule through fear. (I leave the political consequences to the imagination of the reader. Not much imagination is required though. Just read the newspaper.)

Of course I know that climbing trees is dangerous, and that scotch tape can't hold together wood, and that it's much too hot today for long sleeves, and scientific experts can tell you the most up-to-date optimum ways to eat, to exercise, to learn, to stay healthy, to be secure. But it is one thing to state my truth ("Matthew, you are going to be hot in that sweater") and another to presume to force someone to abide by it. As Gatto says, "The plans true believers lay down for our lives may be 'better' than our own when tested against some official standard, but to deny anyone a personal struggle is to strip humanity from their lives; what are we left with after our struggles are taken away but some unspeakable Chautauqua, a liar's world driven by the dishonest promise that if only all rules are followed, a good life will ensue?"[53]

I would like to add that I don't let my two-year-old run out into a busy street. I protect children from dangers that are beyond the horizon of their understanding.

The regime of safety is a facet of the "world under control" directly traceable to underlying assumptions about life, self, world, and purpose. It is an outgrowth of the survival anxiety implicit in our understanding of who we are (discrete subjects) and why we are here (no reason, just the random outgrowth of the struggle to survive and reproduce). As is well known, the safe is very rarely fun. What is safe is almost by definition predictable—free from the random variables that engender "risk". By the same token safety is inimical to creativity, which is about novelty, and therefore inimical to play. Hence the demise of unstructured, unsupervised play in favor of the contained, the controlled, and the programmed that has infiltrated younger and younger age groups.

The regime of safety, like the rest of the world under control, requires constant maintenance in order to quell the inborn human compulsion to transcend old boundaries, that is, to grow. The control starts out external and overtly coercive, then gradually becomes subtler and more deeply internalized. Usually the last visible hurrah of this drive to transcend is expressed in adolescence, and goes by the name of immaturity, teen rebellion, or youthful idealism. By the early twenties most of us have learned enough "self-control" to be trusted outside the confines of overtly coercive institutions such as schools and prisons. We are then dead in spirit, a condition which goes by the name of "maturity". Or if not dead, at least beaten, broken, subdued. Yet the fundamental energy to grow, frustrated though it is, is still latent and still a potential threat to a society built upon the diminishment of human creative energy. Society therefore channels this energy into various illicit or out-of-the-way outlets that do not threaten the status quo. This is life in the playpen, a contained environment where we can't make too much of a mess.

Three examples are especially illuminating. First is the self-destructive behavior of "getting wasted"—i.e. the abuse of drugs and alcohol—

along with high-risk activities such as sky-diving and speeding, and the imitation of such activities in amusement parks. When all other avenues for the transcending of limitations are denied, whether in fact or in perception, then self-destructive behavior is a logical result. If not any other way, I shall transcend my boundaries by dying.

Related in origin to violence against the self is violence toward the world. It is the desire to smash, to smash the world that seems to conspire to hold us stagnant. An enormous anger lies latent just underneath the veneer of our civility, an urge to break, to smash, to burn that manifests at the first sign of breakdown in the controlling authority. Ordinarily, society channels this violence toward victims who are insignificant to the preservation of the status quo: anything, essentially, that falls into the social classification of "other", which could be minorities, foreigners, other species, or the land itself. Previously I defined violence as "longing denied"; thinking along the same lines, Joseph Chilton Pearce traces it to frustrated transcendence.[54] Violence is what happens when we can see no possibility of ever realizing that more beautiful world and more beautiful life our hearts tell us is possible.

The third example of channeled desire for the expansion of boundaries is identification with sports teams, movie stars, and TV characters, which provide us with a second-hand counterfeit of the experience of striving after great things and pushing our limits. Of course, actual *participation* in sports (and in drama) provides genuine opportunities to test the limits of who we are and what we can do, and as such can be part of the unfolding of human potential, but most of the time we settle for watching other people do it. Another channel is the ersatz rebelliousness of impudent hairstyles, rebellious clothing, rude music, outrageous sneakers, and other statements of individuality via shopping.[55]

While these outlets or diversions might pacify us temporarily, the human spirit eventually recognizes the fraud and begins to seek the authentic article of transcendence. The untapped rage that results from the frustration of the natural desire to explore our limits and grow can only be contained by elaborate systems of control, both external and internal. And of course, the control only worsens the frustration, which aggravates the rage, which necessitates the intensification of the control in a never-ending vicious circle. As I have described, in childhood the control is established by the threat to survival implicit in parental rejection. Internalized early on, it requires constant reinforcement, for the human spirit is strong. Because we cannot discern the object of the rage (because it immerses us), we channel it toward sanctioned targets and through sanctioned means. When we do, by chance, hit upon the real target and threaten the status quo, the punishment is swift and sure. The lesson we learn when we lash out against the forms, institutions, and functionaries of authority is that resistance is futile, as when the high school student sets off a smoke bomb in the bathroom. Any challenge to their authority sends school administrations into paroxysms of panic—an independent student newspaper, a senior prank, a spontaneous symbolic rebellion where every student wears black one day. One of my students related a typical image: the principal pacing in front of students waving an underground student newspaper, livid, screaming, "Who is responsible for this? I want names!"

And so, exemplifying the vicious circle outlined above, the regime of control tightens inexorably in our schools, many of which now have video cameras, police patrols, chain-link fences, random unannounced locker searches, metal detectors, drug-sniffing dogs, networks of informants, undercover police posing as students, and a comprehensive system of passes so that there is a record of each student's authorized whereabouts at all times. What a perfect preparation for life in a prison or a totalitarian society! The result is much what we should expect from any series of technological fixes: more control has made the situation far more explosive and not any safer, justifying yet more control. It parallels the results of the Technological Program: life is not actually any more secure, leisurely, or comfortable, and the entire edifice teeters on the brink of catastrophe.

Totalitarianism is the inevitable destination of a society based on the Technological Program of achieving complete control over reality. As a practical matter, the engineering, managerial mindset naturally applies its methods—the methods of the factory—to governance as well as to manufacturing, promoting the complete inventorying, tracking, numbering, and classification of the population. Its technologies as well lend themselves to control: witness the Orwellian possibilities of biometrics and continuous automated surveillance in the computer age. On a more theoretical level, the falsity of the self-other distinction means that control over the world—the other—will result in the subjugation of ourselves as well. As Martin Prechtel puts it, "When the entire world is fenced and farmed, we will all be in prison."[56] Or as Derrick Jensen says, "When we imprison another we must also place one of our own in prison as a guard. Likewise, when we imprison a part of ourselves, other parts must move into that same dungeon."[57] Complete control over the world inevitably leads to complete control over ourselves, both on the collective, political level and on the personal, moral level. With each intensification of control, the individual and collective human spirit seeks new outlets, new doors to freedom which, when they are slammed shut, intensify the longing even more. The world under control is like a leaky pressure cooker: as each leak is filled, the pressure builds up to cause other, previously invisible seams to burst. The program of complete control aspires to seal off all possible leaks once and for all. I leave it to the reader to imagine what happens then.

Control of the world inherently involves separation from nature, just as the very idea of technology requires an objectification of the reality it seeks to manipulate. Control implies the circumvention or alteration of what would otherwise naturally happen. Of course, as I observed in Chapter Two, each part of nature exercises purposeful effects on its environment all the time—control is not the exclusive domain of human beings—so we might consider control and separation to be, in a sense, themselves "natural". As many an armchair philosopher has observed, human beings are a part of nature, so everything we do could be considered natural. The damage comes from the *mistaken belief* that we are separate from nature, and not from actual separateness. It is this misperception of separateness that allows us to suppose that we might be exempt from nature's laws. The very word "nature" as ordinarily used is a symptom of the problem, as if there were some other realm, non-natural, that were exempt. So when I speak of our separation from nature, what I really mean is a forgetting, a detachment, a delusion. It is everything that make us think that nature's laws and processes do not apply to us.

To say that all organisms exercise control over their environment smuggles in insidious biases about the nature of self and world. Control implies a reduction of uncertainty, a reduction of deleterious possibilities in favor of beneficial ones. It implies as well an imposition of power over the environment. But when self and environment are not so rigidly demarcated, and when an organism is seen not just as a discrete unit competing for resources but also as an integral organ in the functioning of the whole, then the whole concept of control loses its coherency. We could equally view the environment as *inviting* the effects that a particular organism has by offering a corresponding niche. The behaviors of all living creatures make a contribution to the functioning of the planetary ecosystem. There is no waste in nature. Nothing (except thermal radiation) is ever thrown "away"—there is no away. The error of separation is that we have convinced ourselves otherwise, not that we have actually succeeded in separating ourselves from nature.

If human behavior and technology were informed by such a non-dualistic view of self and environment, then our goal as conscious, self-aware beings would not be the supercession of nature implicit in the Olympian ideals of technotopia, but the discovery and fulfillment of our proper role. We would seek to conform technology to the rules and patterns that govern the rest of nature. There would be no waste or externalities. In the later chapters of this book I will advocate going "back to nature," not in the sense of abandoning technology, but rather to reconceive all activities of technological society in terms of natural laws and processes. To do anything else is folly—if our dualistic separation of self and environment is indeed a delusion. The next chapter, then, will return to the scientific issues raised in "The Way of the World," to describe the crumbling of the scientific underpinnings of the discrete and separate self in an objective universe.

CHAPTER VI

The Crumbling of Certainty

The End of Objectivity

The Age of Separation in all its dimensions has unfolded before us. The objectification and reduction of the world, the conversion of life into money, the program of understanding and control, the alienation of a discrete and separate self are all reaching their historical zenith in our time. All of these facets of separation are bound together in the official dogma of our civilization: the religion of science. However, starting perhaps a century ago, new countercurrents have welled up from within science itself that are contributing to a gathering sea-change. This momentous shift both drives and reflects a wholesale transformation of all the dimensions of separation that comprise the ascent of humanity.

Like most religions, science encompasses an ideology, a program, and a method. The ideology of science comprises our fundamental story of the world and how it works, our parsing of the possible into the real and the imaginary, and our definition of what types of knowledge are valid. The program of science is what I have named the Scientific Program. It is the ambition to become Descartes "lords and possessors" of the physical universe; that is, to bring all phenomena into the realm of measurement, predictability, and control, so that all knowledge rests on a firm foundation of experimentally verifiable, objective truth.

The third element of the religion of science, the Scientific Method, draws its validity from the ideology of science and its motivation from the program of science. The Scientific Method depends on the replicability of experiments, the testability of hypotheses, and ultimately on the assumptions of determinism and objectivity. Scientific inquiry in general—as well as a scientific, rational approach to life—assumes that there is a *reality out there* that we can query, test, understand, and in due course predict and control. Herein lies the quest for certainty, in which understanding arises from a foundation of facts. By sticking with the facts and reasoning from there we remain objective, and obtain knowledge of superior reliability.

The assumption of determinism encodes the conventional notion of causation and validates the Technological Program of control. It says that nothing happens that is not caused to happen. Events do not arise spontaneously, mysteriously, magically without cause. Events follow predictably from causes. If we can learn the causes of phenomena, and then master those causes, we become the masters of reality. With sufficient understanding, there is nothing that might not some day be brought into the human realm.

Determinism is deeply woven into our logic and beliefs. Do the same thing in the same way, and you *have* to get the same result. You see someone strike a flint and get a spark. You try it and it doesn't work. Do you then simply conclude, "Sometimes it works, sometimes it doesn't"? No, you assume that you must have done something different. You examine your stone—is it the same kind? You examine your striking motion. You do your best to recreate the conditions under which it worked the first time. Determinism is absolutely fundamental to a rational approach to the world. I emphasize that so that the import of the failure of determinism will be clear. We are still very, very far from having fully integrated the psychological shock of it.

Objectivity is equally crucial to our understanding of and relationship to the world. It says there is a reality out there that can be observed, measured, quantified, and controlled. It is the same for you and for me. Any apparent differences arise merely from different perspectives or interpretations of an independently existing universe. Its laws are invariant: God does not operate the world according to some changeable whim, nor do its laws operate differently for me than for you. We take statements like, "The unicorn was there—really there—for me, but not for you" to be the very epitome of irrationality. Come on—was it there or wasn't it? The same holds for "The computer works for you but not for me, even when I do nothing different." Reason as we know it insists that something must have been different, either between us or in the environment, to make the computer behave differently this time.

The aspiration toward objectivity affects nearly every realm of human endeavor, anything that tries to be scientific. It is in fact very hard to define such terms as "scientific" or "rational" without resorting to some variation of objectivity. In science, the experimenter is supposed to maintain an objective distance from his experiments, assuming that there is no necessary, ineliminable connection between himself and the system under study. In medicine objectivity is embodied in the controlled double-blind study, which seeks to isolate the objective effects of a therapy so that we know how well it "really works", independent of the attitudes and foibles of patient or doctor. In agriculture we might plant two (supposedly) identical fields with crops differing in only one significant variable, and measure the difference in yield. In jurisprudence the judge is supposed to maintain impartiality and consider only "the facts". In journalism the belief in objectivity implies that a reporter is just that: someone who "reports" whatever facts are already out there. She is not supposed to actually be taking part in those events, for then she would no longer be objective.

Together, determinism and objectivity promise technologies that apply generally and generically. Their standard application produces a predictable result. The person applying them is interchangeable, just as a scientific experiment is supposed to be replicable by any competent experimenter. Machine civilization depends on this interchangeability. Power over the physical universe comes via method and structure. Follow the prescribed procedures and you will get the predicted result, reliably. No matter who prescribes them, the right dose of antibiotics, taken according to objectively determined instructions, will cure strep throat. It does not matter the intentions of the canoneer: the cannonball will follow the same trajectory no matter what, as long as the initial angle and propulsive force are controlled. No less than for a caveman striking a flint, technological civilization's mastery of the physical world depends on having reliable, generalizable ways to control it. Or so it would seem.

This is the founding philosophy that galvanized the Scientific Revolutionaries and motivates still the program of understanding and control. For several centuries after Galileo and Newton, the Scientific Program extended the foundation of control by gaining an ever-finer understanding of the "reasons" and the "reason" of the world, making ever-finer observations of the reality out there, until it got down to the base level of the subatomic realm. Here were to be the building blocks of the determinism and objectivity that embody scientific reason. And then calamity struck.

The calamity for science and reason is simply that at the subatomic level, the very bedrock of the whole edifice of science, determinism and objectivity do not hold. At the most fundamental level of reality, our scientific intuitions (embodied in the above statements about unicorns and computers) are simply wrong.

As a result, the last eight decades have seen a proliferation of interpretations of quantum mechanics that attempt to reconcile the indeterminacy and observer-dependence of the quantum realm with the determinism and objectivity that we "know" characterizes the world of everyday experience. None of these attempts have been successful—a marked contrast to Newtonian mechanics, which provoked little serious dispute about what it all meant because it fit in with the tide of the times. The present lack of agreement about the interpretation of quantum mechanics—which after all lies at the basis of physics—testifies to its incompatibility with our fundamental ontology.

It is beyond the scope of this book to give a thorough summary of precisely how quantum mechanics violates determinism and objectivity. I refer the reader to the vast non-technical literature on the topic, in particular the works of Paul Davies, Nick Herbert, David Wick, Roger Penrose, Fritjof Capra, David Deutsch, and Johnjoe McFadden. I particularly recommend the last two: Deutch's *The Fabric of Reality* for its lucid exposition of the many-worlds interpretation that is currently in vogue, and McFadden's *Quantum Evolution* for its elegantly clear introduction to the basic paradox of measurement.

Quantum mechanics' violation of determinism is somewhat less challenging to conventional beliefs about self and world than is its violation of objectivity. Determinism holds that initial conditions completely determine final conditions: if you do the exactly the same experiment twice, you'll get the same result. This is a key assumption for the requirement of repeatability used to determine scientific fact. But in quantum mechanics the assumption is false. Fire a stream of electrons, photons, or indeed any particle through a slit onto a detector screen, and the final detected position of each one will be different. The overall *distribution* of particles is fully described by mathematical equations, but the fate of each individual particle is random. One might veer left, the next right, the next straight through, and there is no *explanation* for that behavior.[1] It is acausal, which violates a central assumption of the Scientific Program that with sufficiently diligent querying of nature, the reason for everything can be found. Here, at the very basis of the reductionist pyramid, matter behaves acausally, unreasonably, a state of affairs so troubling to scientific orthodoxy as to incite Einstein's famous protest, "God does not play dice with the universe."

If you are less troubled than Einstein was, perhaps that is because you have not pondered it as deeply as he did. So think about it. There is no reason for the photon's behavior. Why did it take a particular path? The only answer is because it did. Nothing *made* it veer left, or right, or go straight.

Quantum uncertainty provides us with a new source of metaphor and intuition for human life. Newtonian determinism contributed to the feeling that we too are mere masses, the trajectory of our lives wholly determined by the forces bearing upon us. But perhaps we are more like a quantum particle, whose path is constrained or influenced by outside forces, but behaves as if it made its own choices. The metaphor of quantum mechanics is one of choice, autonomy, self-determination. Forgive me if the following metaphor is a little corny, but perhaps we are hurled through the aperture of our circumstances toward a highly probable destination; we then have the power to choose that one, or one far divergent from it. And no one can predict where our path will take us, and no outside power can dictate our choice.

In the quantum metaphor, choice is the human counterpart of quantum randomness. Both are irreducible, inalienable properties of the subject in question. We can give reasons for our choices, justifications and excuses. We can explain why we "had to" do what we chose to do, but the fact is that there is always a choice. By falling back on justifications we give away our power. And I wonder if the correspondence between quantum randomness and human choice is mere metaphor. I suspect that if we described photon diffraction to an aborigine, he would say that the particle too chooses its path. Random? Feh! Randomness is just a feeble attempt to rescue the world of generic masses and uniform building blocks. What if they are all different? What if each bit of matter is unique? What if the sameness that we impose upon it is a mere projection of our own lot, as the standardized consumers and functionaries of the Mumfordian megamachine?

The same indeterminacy that characterizes the path of a particle through an aperture also characterizes the decay of a radioactive atom, the polarity of a photon or electron, and many other properties. But the challenge to our conventional worldview runs much deeper than that, because not only are these measured quantities random; apparently, until they are measured they don't have any definite status at all. Interference experiments such as the double-slit experiment, the Stern-Gerlach experiment, and countless others demonstrate that in the absence of measurement or observation, particles behave as if they occupied all possible states at once. What's more, the very presence of observation can affect the evolution of the system being observed,[2] even in the absence of any physical force operating between system and observer.[3] In other words, there is no independently existing universe "out there" separable from we who observe it. Observer and observed are intimately linked; the distinction ultimately does not even make sense. The discrete and separate self is an illusion. And, the Galilean "primary qualities" that we measure with our instruments are not primary at all, but created through the very act of measurement. Such properties as distance, time, and form are properties of a *relationship* between self and universe, not of an independently existing objective universe. The naïve concept of existence represented by the disembodied fork floating in nothingness that we visualized in Chapter Three is incompatible with modern physics.

Like the failure of determinism, the crumbling of objectivity opens the door to a profoundly different set of intuitions and metaphors. Seeing ourselves as isolated subjects in a vast, indifferent universe, we easily succumb to feelings of powerlessness, alienation, and despair. No more. Just as quantum randomness is the metaphorical counterpart of human choice, quantum measurement is analogous to our stories, our interpretation of experiences. Like a quantum measurement, these interpretations take on a creative significance. In interacting with the world and taking its measure, we collapse a plenum of possibility into a single actuality. We are not merely interpreting a reality separate from ourselves; we are, through the act of interpretation, actually causing that reality to come to be.

This is as we would expect if the foundational myth of our civilization, the discrete and separate self, is indeed only a myth. If that separation is an illusion, then of course the inner world of our interpretations, thoughts, beliefs, and attitudes will have an effect on the outside world that is not really outside. This is very close to the magical thinking of primitive animists, whose beliefs in the creative power of word and ritual take on a new significance. Could the metaphorical implications of quantum mechanics propel us toward a reunion of those long-separated inner and outer worlds? Let us begin imagining what a reunited world would be like for humanity, not as it was 50,000 years ago, but in the context of our long-accumulating technology and culture.

The consequences of failure of determinism and objectivity have so far been too huge for our culture to digest, so antithetical are they to the reigning orthodoxy. The quantum measurement "paradox" is the inevitable product of attempting to weld the observer-dependence of the quantum world with the supposed objectivity of the world of everyday experience inhabited by our discrete, separate selves. Quantum mechanics invalidates the discrete and separate self. Because quantum mechanics represents such a profound challenge to our very self-conception, for eighty years scientists and philosophers have gone through paroxysms of interpretation to somehow mediate the two realms of non-objective, acausal quantum events and the classical reality we think we experience.

Again, none of these attempts have been successful. On the practical level, most deny the extension of indeterminacy and observer-dependence into the macroscopic world by essentially claiming that quantum uncertainty tends to routinely cancel out, approximating classical mechanics on the scale of everyday experience. Thus while there is a non-zero chance that a marble flung through a hole will diffract onto a non-classical path, this chance is so close to zero that it can be ignored. While this solves the practical problem of why classical mechanics works so well for designing machines and bridges, it doesn't deal with the ontological problem: what is the fundamental nature of reality? Moreover, as the founders of quantum mechanics, particularly Shröedinger, realized, the ontological problem does not go away, but becomes especially pressing when quantum events are magnified into classical observations (which is essentially what a quantum measurement does).

Some unconventional thinkers such as Roger Penrose and Johnjoe McFadden argue that quantum effects are projected into macroscopic reality routinely in living systems, and not just in the contrived conditions of a physics lab. Some, more radical, even cite indeterminacy as an escape clause from mechanism that allows free will; others cite quantum phenomena as evidence to support various approaches to healing and spirituality. Such speculations range from the ignorant to the highly sophisticated, but I believe that someday science will establish a quantum explanation for many presently unexplained (and for the most part, unacknowledged) phenomena. However, a detailed discussion of the measurement paradox, and the dominant misunderstanding of decoherence, will have to await a future book. If you read my words carefully, you will see I claim no direct link between quantum phenomena and the world of human experience. For example, I am not claiming that quantum indeterminacy proves we have free will. What quantum mechanics has given us is, at the very least, a new way of thinking, a new kind of logic, and a new source of metaphor. These already may be powerful enough to transform our civilization.

The counterintuitive aspects of quantum mechanics I have described are only counter to those intuitions that are contingent on the modern conception of self and world. To people before the Age of Separation was well underway, descriptions of quantum phenomena such as "It occupied two positions simultaneously," or "It wasn't there until you looked for it," or "It was there for you but not for me" may not have seemed paradoxical at all. To them, there was no absolute distinction between observer and observed, imagination and reality, human and nature, self and other. To the extent that such distinctions existed, their provisional nature was recognized, perhaps as a play, a creative artifice. Hence the original identity discussed in Chapter Two between ritual and reality, and in the Original Language between the name and the thing named.

The mind of the primitive is often irksomely irrational to the Western visitor. I must admit having suffered the same annoyance in my early encounters with New Agey people who would (it seemed) taunt me with such statements as "It's true for you but not for me." I would say, "I believe that if 'qi' really exists we would have detected it with scientific instruments" and my friend would respond, "That belief is why you cannot detect it with your instruments." I would say, "I don't believe out-of-body experiences are possible," and he'd say, "Then for you they are not possible."

It was maddening. "I don't mean 'for me', I mean not possible for real."

"Then for you, it is not possible for real."

Aargh!

What I meant by "for real" was "objectively". One friend, the healer and musician Chad Parks, tried to explain a psychic invisibility technique taught him by some (to me) dubious New-age guru. People choose not to look at you or they simply don't notice you. "But surely *if they looked*, the light rays bouncing off your body would still reach their eyes," I said, "so you're not *really* invisible."

"To them I am."

A similar situation arises in one of Carlos Castenada's books, in which the narrator, trying to get a grip on Don Juan's shamanic powers, challenges him, "But what if someone was waiting in ambush on your path—surely you couldn't stop a bullet, could you?"

"No, I could not stop a bullet. But I would not take that path."

Castenada could have continued, "But what if the situation required you to take that path?" and Don Juan could have replied in kind, "Then I would not enter that situation."

In the prior example, I could have proposed to Chad an experiment: "Okay, make yourself invisible—I bet I can still see you." He would have said, "It won't work, I am already here for you." His invisibility is essentially untestable because the very grounds for objective testing embody a conflict of assumptions. It is testable only in an objective universe, and it only works in a non-objective universe. The whole idea of certainty of knowledge, built through objective reasoning, is only as sound as the objectivity at its basis. Question that, and we question the soundness of the entire edifice of experimentally-derived knowledge.

The reason that primitive and New Age logic seems irrational is that it *is* irrational. Reason, according to David Bohm's definition that I quote in Chapter Three, is the application of an abstracted relationship onto something new. A non-objective world defies such abstraction. If the world "out there" reflects in some way the inner world, then reason is but one of several cognitive tools for creating and defining our experiences. Reason is still a valid and useful tool; it is only when it becomes a reflexive habit rather than a conscious instrument that it is limiting.

Professional skeptics are fond of railing at the abysmal stupidity of their opponents, who seem dispossessed of that key function of higher cognition, reason. Like a fish unaware that it is wet, these critics rarely perceive their own immersion in assumptions of self and world that constitutionally limit them to certain narrow modes of cognition, those that we call rational. These are powerful in a certain domain, having enabled us to build the towering edifice of our civilization; they are behind the vast program to engineer the world and remake nature. As this program falters, we open to the possibility of other modes of cognition and relationship.

As quantum mechanics slowly replaces our Newtonian-Cartesian intuitions with those that are non-dualistic, all of the fruits of separation will lose their deepest rationale. For even if conventional philosophy is right that quantum indeterminacy and observer-dependence have no practical consequences for consciousness, mind, and self; even if no one ever proves that our level of matter departs appreciably from the classical description, there still lurks at bottom an implacable exception to the claim that "the universe is just like that." If only by way of metaphor, quantum mechanics confers upon us a new logic, a new framework of possibility. No longer will the discrete and separate self be the only conceivable, the only cogent way of understanding the world.

Quantum mechanics heralds a momentous shift in our intuitions that will rapidly accelerate as the failure of the old ways of life and thought becomes increasingly obvious. Just as the regime of separation both set the stage for, and was reinforced by, its apotheosis in the science of Newton and Descartes, so also will quantum mechanics quicken the emerging realization of our interconnectedness with each other and all of nature, which will in turn allow us to more fully digest quantum theory's profound ontological consequences. Quantum theory is both a cause and an effect, a harbinger and a symptom, of a larger shift in consciousness.

Armed with the intuitions, or at least the metaphorical possibilities, that quantum mechanics foretells, the beliefs of primitive humans will take on a new vitality, relevance, and import. Already we feel their pull, as the popularity of "Native American spirituality" testifies. (That this form of cultural capital is rapidly coopted and converted into money does not alter the kernel of its appeal.) Already, we are becoming more willing to believe that our thoughts, words, and actions have a power beyond their classical physical description as a mere shifting of masses and flux of chemicals. Already we grow more at ease with the idea of a fluid reality, not separate and absolute, but defined by our relationship to it and molded by our beliefs. Little do we realize that the stage is being set for a wholly different science, and a wholly different technology, no longer based on the premise of separation and no longer reinforcing that premise. And no aspect of human life will remain unchanged.

Truth without Certainty

If particle physics is the foundation of the reductionist program described in Chapter Three, then mathematics and formal logic is the bedrock upon which that foundation rests. To truly fulfill the Technological Program of complete control, we must first achieve a certainty of knowledge, so that results follow undeviatingly from expectations, conclusions from premises, and function from design. In this view, if a machine doesn't work or a design fails, it can only be because some variable was left uncontrolled; i.e., the knowledge of initial conditions was incomplete. When every factor is accounted for, every variable measured, every force captured in a mathematical equation, then the predictability of a physical system is no less dependable than the mathematics that underlies it.

But how dependable is this mathematics? In the imaginary future of Leibnitz and Laplace, where all linguistic meaning is fully precise and all of science fully mathematized, any dispute can be resolved by straightforward calculation, without doubt or controversy, and all the truths of nature will be laid bare. Mathematics, after all, is the epitome of certainty, in which conclusions are reached not by persuasion but by formal, deductive proof, indisputable unless logic itself is violated. But how do we know that mathematics is sound? How do we know that there are not hidden contradictions buried in the axioms of arithmetic? And equally important, how do we know that all truth can be reached starting from those axioms? As physics was placed on an axiomatic footing and more and more fields of knowledge appealed to mathematics for their legitimacy, these questions took on increasing urgency around the end of the 19th century.

The axiomatic method, which originated with Euclid, is implicit in today's notion of scientific rigor. It starts with explicit definitions of terms to be used, and the assumptions one is operating from. After all, how can reasoning be sound if its very terms are ambiguous? You start from basic definitions and premises, and build from those. In mathematics the necessity of the axiomatic approach was highlighted by various paradoxes in set theory that demonstrated the ultimate incoherency of naïve (non-axiomatic) definitions of a set. For example, consider Russell's Paradox: the set of all sets that are not members of themselves. Is that set a member of itself? By definition, if it is, then it isn't, and if it isn't then it is. Hence the necessity of axioms implicitly defining what is and is not a set. Axioms were also formulated for arithmetic: naïvely, we think we know what addition and multiplication are, but do we really? The axioms of arithmetic define them formally.

The program to put all of mathematics (and by implication, eventually all of science) on a firm axiomatic footing was articulated by the French mathematician David Hilbert, and its culmination was to be a proof that such axioms systems (particularly arithmetic) were both sound and complete. Sound, in that no contradictory results could arise from them; complete in that all true statements could be proven from them. At the time it seemed intuitively obvious: surely anything true is also provable. Else how would you *know* it were true? How could you differentiate it from any other assertion? In a sense, the entire Cartesian ambition to become the "lords and possessors of nature" hinged upon the completeness proof, for it would establish that no mystery, no truth, is beyond the purview of human logic. Starting with a formal axiom system, one can proceed according to the rules of logic to derive, mechanically and unthinkingly, all possible proofs from those axioms. A computer could do it. And even though computers didn't exist in 1900, the mechanical nature of axiomatic proof seemed to promise that Leibnitz' vision would come true. To settle any dispute, and indeed to eventually arrive at *all truth*, we would but need to say, "Let us calculate."

Imagine then the sense of bewilderment that followed Gödel's famous incompleteness theorem of 1931, which destroyed any hope of ever completing Hilbert's program. Usually presented in popular literature as demonstrating that "there exist true statements of arithmetic that cannot be proven from the axioms," Gödel's Theorem is actually a bit more subtle than that. I will present the theorem here in a little more depth, because its subtleties have both direct and metaphoric consequences for the Scientific Program and Technological Program.

The divergence of truth and provability in Gödel's Theorem can only be understood in the context of the distinction between a formal theory and an *interpretation* of that theory. A formal theory is the set of all theorems deducible from a given set of axioms according to the usual rules of logic. Its interpretation is whatever real or abstract system the theory describes.

To take a familiar example, the geometrical axioms of Euclid generate a theory, one interpretation of which is the idealized lines, points, angles, and so forth drawn on a flat surface. One interpretation of the theory of arithmetic is the set of natural numbers we use to count, add, and multiply. Provability is a property of sentences written in the formal language of the theory; truth is a property of their counterparts in the real world. For example, in the formal language the sentence "$\forall x \forall y\ x{*}y=0 \Rightarrow x=0 \vee y=0$" is provable from the seven basic axioms of formal arithmetic (named Q), and its interpretation "If the product of two numbers is zero, then at least one of those numbers must be zero" is true in real-life arithmetic. That assertion seems quite obvious, but how can we be sure? How, in other words, can we prove it? Only by abstracting from the real-world example a theory, a set of definitions ("here's what addition really means") embedded in axioms.

But then the question arises, How can we know whether the theory really corresponds completely to the interpretation? Indeed, the seven axioms of Q are so minimal that it is impossible to prove basic arithmetical facts such as "$\forall a \forall b \ a+b=b+a$". It is impossible to prove from Q that you can get to every number, eventually, by counting from zero. Such inconveniences are easily dealt with, however, by adding them in as new axioms. The ultimate goal of Hilbert's Program, and indeed of the Scientific Program of complete understanding, would be to add in as axioms all unprovable statements whose interpretations are true. You would then have a complete axiomatization of reality, the ultimate conversion of nature to number. The separate human realm would finally encompass all of reality.

What Gödel proved was that this is impossible, that there is no way of adding enough axioms to prove every true sentence (i.e. every sentence whose interpretation is true). An infinity of axioms would be required—but even this is not the deepest problem. The problem is that there is no "effective procedure" to generate that infinity, no finite means to make the theory complete. You cannot say, "Let every sentence that is true in such-and-such an interpretation be an axiom," because there is no way to tell what those sentences are.

In other words, there is necessarily something missing in any mathematical description of reality. There is no finite means to encompass all truth in a system of labels and quantities (which is essentially what a "formal system" is). Even without quantum indeterminacy, the Scientific Program is doomed to failure from the start. Doomed as well is the whole notion of reductionistic rationality as a sure guide to truth, the approach mentioned above of starting any problem by laying out rigorous definitions. By limiting knowledge to what can be proven, we exclude large swaths of the truth.

And it gets worse. The above state of affairs would be acceptable if the truths left inaccessible were unimportant ones, contrived sentences of arithmetic like the one Gödel constructed for his proof. But as soon became apparent through the work of Turing, Post, and more recently

Gregory Chaitin, it is not just a few recondite corners of mathematics that are impervious to proof, but the vast majority of all mathematical facts.

The very idea of rational understanding is to reduce the complex to the simple, to find the "reasons" underlying things. The quintessential example of this is the reduction of the complex paths of planets in the sky to Newton's universal law of motion. What Turing proved is that there are important mathematical questions that cannot be answered that way, but only along the lines of "because that's the way it is." His famous Halting Problem showed that there is no general means of determining whether a random Turing Machine (an idealized computer) will eventually halt given a certain input—no means, that is, except to actually try it out.[4] There are specific methods for some Turing Machines, but no universal method, no finite formula or set of instructions, nothing that could be programmed into a computer. There can be no general theory of why Turing Machines halt. Chaitin has extended this result even further to observe that almost all mathematical fact is unprovable, and even worse, that mathematical truth is random in the sense of algorithmic information theory.[5] There is no rhyme or reason to the truth; nothing that could be understood in the finite, standard terms required to bring reality into the human realm of control.

Mathematics has sealed the fate of the age-old attempt to substitute chaotic, unpredictable reality with a controllable artificial version of it. However fine our mapping of reality, however sophisticated our modeling, something will always be missing, and this limitation is inherent in the map itself. Of course a map, a set of definitions, an axiom system can be useful, but when we mistake it for the real thing then we are marooned in a finite world of our own making, a projection of our own assumptions, a tiny subset of Truth delimited from the very beginning by what we hold to be self-evident. As Chaitin puts it, "In other words, the current map of mathematics reflects what our tools are currently able to handle, not what is really out there."[6] The danger is that blinded by our assumptions, we reject actual experiential data with the thought, "It isn't true because it couldn't be true." This is precisely what has happened in many branches of science, which have become so mired in their principles that they cannot countenance "anomalous" phenomena no matter how well-documented.

We make an analogous mistake in everyday life whenever beliefs blind us to experience. Consider for example the shy teenager who is so convinced she is unattractive to boys that she is oblivious to their attentions, interpreting secret-admirer notes as mockery and compliments as sympathetic attempts to cheer her up.

The results of Gödel, Turing, and Chaitin imply a different way of pursuing knowledge that forgoes certainty in favor of utility. In the absence of irrefragable truths that exist "out there", the inquirer's relation to the world becomes paramount. In mathematics, the researcher can add new axioms onto a formal system, justifying them either with computational evidence or merely by the appeal of the results they can prove. In this way, subjectivity creeps back into mathematics. One body of theory derives from the axioms of set theory with the Axiom of Choice; another body of theory from the axioms without it; one body of theory derives from the addition of the Continuum Hypothesis as an axiom, another from the addition of its negation. In computation, many results follow from assuming that there is no polynomial-time algorithm for solving NP-hard problems like the "traveling salesman" problem, an assumption widely accepted based on computational evidence and the repeated frustration of mathematicians' best efforts to find a polynomial-time algorithm.[7] In geometry, the inclusion of non-Euclidean axioms provides a tool for understanding curved space-time, just as the Euclidean axioms describe geometry on a flat surface. We can, playfully, try out different axiom systems to come up with different descriptions of reality, parts of reality, or aspects of the universe.

If, my dear reader, you have become lost in these complexities, let me emphasize the key point. It is simply that in mathematics as well as physics, there is not always a *reason why*. Sometimes things are true just because they are. Have you ever heard someone say, "Oh yeah? If it is true, then prove it!" Unconsciously we have learned to equate truth and provability, just as we have learned to value reason over intuition. Hilbert's Program was just one manifestation of our craving for certainty, for an indisputable source of truth outside of ourselves. The same impulse underlies the ubiquitous elevation of "experts" in our society, and the giving over of more and more of our autonomy to external authorities. It also underlies many religious cults, in which certainty comes from the guru, as well as Christian fundamentalism, which looks to yet another external authority, the Bible, for an indisputable source of truth. The doctrine of Biblical inerrancy is the religious counterpart of the scientific ambition to axiomatize reality. Here is certainty! No longer is it necessary to look within oneself to know truth—it is all laid out in black and white.

We are no longer divine creators of our world, only receivers, only consumers.

Today all hope of ever achieving such certainty is ended. Of course we can, like the Christian fundamentalist, the cult follower, or the dogmatic scientist, choose to remain ensconced in our axiom system and refuse to explore any truth that lays outside it. But such certitude comes at a high price: insularity, stagnation, and cut-off from new worlds of knowledge and experience. In fact, the crumbling of certainty is incredibly liberating. Its effects are similar to the effects of the failure of determinism and objectivity in the realm of physics. Truth, like being, ceases to be an independent quality separate from ourselves. Both begin to make sense only as a relationship. Divorced from logical certainty, divorced from proof, what can truth mean? The only satisfactory answer that I've found is that truth is a state of integrity. When faced with two different interpretations of an experience, instead of gathering more and more evidence to decide which is true, the new metaphor calls us to simply choose one or the other, depending on which fits with greater integrity into all that we are and, more importantly, all we strive to be. We create who we are through the truths we choose.

Does that statement sound dangerous to you? Does it seem to give license to play fast and loose with the truth, to ignore the evidence and blindly maintain a self-serving interpretation of reality? Does it allow us to justify anything we want by saying, "It's my truth"?

Actually it has the opposite effect. When we choose truth consciously, in knowing self-definition, that choice takes on a gravity absent from ordinary reasons and justifications. If we understand truth as a creative choice, we will be all the more conscientious in choosing. Returning to our mathematical metaphor, truth is a property of an interpretation, not of proof, not of reasons. So in choosing a truth we are choosing an interpretation of our world; that interpretation, in turn, generates new experiences consistent with it. Our choice of the truths we live by has world-creating power.

Whether inside or outside of science, we might see the quest for truth not as an encompassing of more and more facts, not as a growing certainty about the world, but rather as a path of self-understanding and conscious creativity. When truth is, as in mathematics, often beyond certainty, beyond even reason, then how do we recognize it? How do we choose between a belief and its opposite? We are left with integrity, which we can clarify by asking, "Is that me? Is that the universe I choose to live in? Is that the reality I wish to create?" We are not discrete observers separate from the universe that we observe, able only to discover what is. We are creators.

Let me give you an example. Through a few personal experiences and extensive reading I have come into contact with many phenomena that conventional science does not accept. At some point I was faced with a crisis of belief, a choice between two different interpretations, each logically coherent. They went something like this: (1) My physical experience of *qi* was a fantasy induced by the unusual circumstances of the dojo and the culture shock of being in Taiwan. The hundreds of apparently normal, sincere, and humble practitioners were similarly deluded, except for those consciously conspiring in a hoax. Formerly respectable or even eminent figures like John Mack, Roger Woolger, and Barbara Brennan, whose books I'd read, had succumbed to some form of dementia. The many people of apparent integrity who'd shared stories with me of miraculous coincidences, inexplicable experiences, ghosts, and so on were actually putting me on, trying to seem special, hungry for attention, and I must be a poor judge of character. My life is full of dupes, frauds, hoaxers, liars, and the mentally unstable. Even my own wife lies to me for no apparent reason about extraordinary experiences that happened to her years ago. If I try hard enough, I can cobble together a belief system in which none of these "unscientific" occurrences ever really happens. (2) These experiences that I've had, that I've been told, and that I've read about are as real as any other. The people in my life are generally as they seem, and not pathologically plagued by confabulation, selective memory, and compulsive lying. John Mack did not write his books about alien abduction because being a Harvard professor of psychiatry wasn't good enough and he wanted fame and notoriety. Eminent scientists do not usually throw their careers down the sewer in some vain pursuit of purely imaginary psi phenomena. And in my own experiences, I saw what I saw and felt what I felt. And, finally, believing all this I must also believe that the entire corpus of science is fundamentally incomplete.

Neither of these two interpretations suffers any internal logical inconsistency. Just like the shy girl, I can fit all the data into one of many interpretations, many universes. Even Occam's Razor cannot always rescue me—it is usually "simpler" to discard inconvenient facts on some pretext. By choosing a truth, I am choosing what universe I will live in and making a statement to myself and the world about who I am. Sometimes I might even do this playfully, in a spirit of exploration and discovery, like when I spend time immersing myself in "skeptics'" writings, and notice how it changes my state of mind, emotions, and relationships. Usually, though, the progression from one belief-state to another is unconscious, subject to a logic and a process beyond my understanding. A truth that served me well in one stage of existence becomes obsolete as I move on to another. And so it is with all of us.

The work of Gödel and Turing has ended forever the Babelian program of taking nature by finite means. It has shown us the limits of reducing reality to label and number, and the impossibility of ever subjugating truth to certainty. Understood metaphorically, mathematical incompleteness hints at a new way of understanding truth, knowledge, and belief as a way of relating to the universe and defining who we are, a process of cocreation of reality, and not a mere unveiling and cataloging of what is already objectively out there. Certainty is gone, but in its place we have freedom.

Order without Design

Certain theologians and scientists have made much of the amazing fact that the physical constants of the universe seem to be precisely calibrated to allow the existence of life. The value of the strong nuclear force, the fine structure constant, the gravitational constant, the mass of the electron, and so on would, if different by even a few percent, no longer allow the possibility of life or even, in some cases, stars or solid matter.[8] Some cite this as evidence of a creator setting the stage for life; others invoke a multiplicity of universes, each with different constants. To me the most intriguing is the possibility that the constants actually covary with each other and are subject to feedback mechanisms that eventually brought them into the stable attractor configuration they occupy today. In any event, the universe's pregnancy with order, beauty, and life does not arise from this set of (possibly arbitrary) physical constants alone. Order and beauty are woven into the fabric of reality even more deeply than that. They emerge on every level, in every non-linear system of sufficient complexity. We shall look at a few of these systems at different levels to get some sense of the ubiquity of order—order without design—in the continuing miracle in which we live.

The metaphorical and practical implications of self-organization are staggering, perhaps even more so than those of quantum mechanics. I first became aware of self-organization sixteen years ago when, just out of college, a friend handed me *Order out of Chaos* by Ilya Prigogine and Isabelle Stengers. This book blew my mind. Amid numerous examples of self-organization in chemical systems, the book gave me my first glimpse of the Mandelbrot Set, an extraordinarily complicated fractal generated by an extremely simple recursive formula. To generate it, take a point "C" in the complex plane and apply the formulas:

$Z_0=0$

$Z_{n+1}=Z_n^2+C$

After a given number of iterations, the sequence of $Z_0, Z_1, ..., Z_n$ will either exceed absolute value 2, after which it quickly diverges toward infinity, or it will stay in the neighborhood of the origin. If C generates a sequence that stays tame and never diverges toward infinity, it is in the Mandelbrot Set. Unfortunately there is no general, finite way to decide whether that will happen, because even after a billion iterations, it could still start diverging on the billion-and-tenth. In mathematical language, the set is not recursively enumerable, and therefore not decidable either.[9] And this means that there is no shortcut or reason for the set to have the structure it does, no reason except: "That's the way it is." The only reason you can (generally) give why a point is in the set is simply to quote the definition: "After N iterations, it still has not diverged." *Our complete reductionistic description of the M. set tells us nothing about it.* In effect, here is a reality that cannot be reduced.

We could say the same thing about any random set of points[10], but what makes the Mandelbrot set special, and a fecund source of metaphor, is that it is evidently not some disorderly scattering of points on the plane, but seems to possess order, structure, and even beauty. Yet a full finite description is impossible; any such description leaves out an infinity of structure. (Notice the parallel with the inescapable reduction of reality entailed in representational language and measure described in Chapter Two.)

The M. set is not unique in the spontaneous emergence of structure—the same thing happens in other mathematical systems such as cellular automata, neural networks, and population dynamics. Stuart Kauffman calls it "order for free", because it is not evident in the defining equations. Studying boolean networks, Kauffman has identified general parameter values that virtually ensure order will emerge (even though the only way to know what that order looks like is to calculate it out).[11]

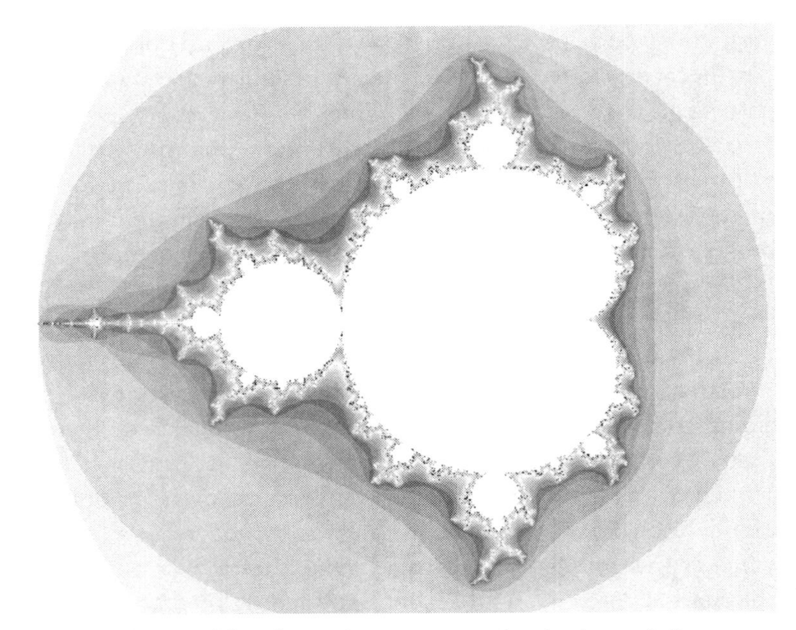

Figure 1: The Mandelbrot set. The white regions represent points in the set, fading to grey and black depending on how quickly the points diverge.

Figure 2: Different regions of the Mandelbrot Set under magnification. By zooming in one can often find distorted replicas of the whole set.

Let me offer one more example. The cellular automaton called "Langton's Ant" couldn't be any simpler. Imagine an ant crawling on a grid where each square must be either black or white. If the ant lands on a black square, he colors it white and turns right. If he lands on a white square, he colors it black and turns left. The ant wanders aimlessly about patterning and repatterning the plane, until at some point (depending on the initial setup, but usually after tens of thousands of steps) something very strange happens: the ant begins building a "highway" out to infinity.[12] Numerous computer trials confirm that this happens no matter what the initial setup, yet this is an empirical fact only. It is probably impossible to prove it analytically.[13] In other words, the only reason or explanation for "Why does the ant build a highway every time" is "Because it is true for setup A, setup B, setup C,..." which is really no explanation at all. Such an explanation amounts to, "Because it does." As with the M. set, we have a complete reductionistic explanation of the ant's every move, yet it tells us nothing about the large-scale structure.

In other words, nothing in the simple defining equations of Langton's Ant would indicate that highway-building behavior will result. There is no finite explanation, yet it happens just the same! There is no simpler "why". It just is. Neither, therefore, can we be certain that highway-building always happens. We could demonstrate it for a billion starting setups, but it could fail for the billion-and-first. Analogous situations are ubiquitous in the world of cellular automata and related fields, leading researchers such as Stephen Wolfram to advocate a "new kind of science" based on empirical discovery and not analytic proof. Since science is about *understanding* the world, what they are advocating is really a new conception of understanding, one which is no longer subject to certainty. It becomes necessary to accept that complex systems have emergent properties for which we will never find a reductionistic "why".

Not only certainty, but also predictability and control fall by the wayside. We must accept that no matter how precise our control over initial conditions, the results might surprise us. With Langton's Ant we know highway-building will result, but not when or where. Changing merely one square in the initial grid can utterly alter the "when and where". In chaos theory this is known as "sensitive dependence on initial conditions", which afflicts not just mathematical systems but physical, chemical, and biological systems as well—anything with non-linearity and feedback . We might empirically infer general information, analogous to the ant's highway-building, but the details will elude us. For example we can look at earth's meteorological data and predict that hurricanes will blow through the Caribbean, and explain why, but we are helpless to predict their timing or precise paths. Despite vast improvements in data density, weather forecasting is still inaccurate beyond a few days. Exponential increases in data density bring at best linear improvements in forecast period.

These examples show that determinism and practical predictability are actually two different things. Idealized Newtonian two-body mechanics possesses both: past, present, and future are contained in a single equation, and the evolution of a system at any point in time is easy to calculate. But with few exceptions, this only holds for linear systems. A free pendulum is infinitely predictable; a driven pendulum (where a motor drives the fulcrum back and forth) is chaotic. In a linear system, uncertainties in measurement result in only linear divergence in predicted results—"noise" stays manageably small in comparison to "signal". But in non-linear systems, the noise quickly overwhelms the signal. A tiny difference in initial conditions results in totally different behavior.

The word "design" implies an intentional awareness of end results that is built in to the initial setup. The engineer desires a certain outcome, and so builds that into his design. Design implies some kind of predictability, properties of the whole that can be predicted from the properties of the parts. When people speak of "intelligent design" as a cosmological theory, this is exactly what they have in mind: an external Creator God who calibrated the initial conditions and laws of the universe in precisely the way that would result in intelligent life. I am suggesting a different conception of God, not as creator but as Creativity itself, not outside the universe but an inseparable property *of* the universe. Why should the M. set be so ornately beautiful? We hunger for a "why", but there is none. No reason except, "That's the way it is." Reality is just like that. In the Mandelbrot Set I see God, a miraculous mystery that I can never fully grasp, not because of the limits of my mind but because it is constitutionally ungraspable, impervious to reduction. There is nothing to "understand", nothing to grasp. It is just like that. It is because it is. It is the eternal "I am". Just as 2+2=4, the M. set could be no other way. Order for free. Beauty for free. Not designed that way, but an irreducible property of reality. I wanted to say "built-in" but even that smuggles in a dualistic reduction of its self-sufficient miracle.

It is ironic indeed that mathematics, which through its "digits" sought to manipulate, tame, and reorder nature, has brought us back to the indubitable realization that reality at its most fundamental level is forever beyond our grasp. And because this is not an oddity of abstract computational mathematics, but broadly true of any non-linear system, the Technological Program of complete control can never come to pass. Time and time again, experience has shown us that however precise our measurement and control, the end results are unpredictable in a non-linear system. The putative successes of the ascent of technology owe themselves to the linearity of machine technology. In linear systems it works: the more precise the control of initial conditions, the more precisely predictable the end result. Our civilization has developed linear technology to its maximum. The more complex a machine becomes, the (exponentially) more relationships among its parts must be considered to keep it "under control". Control thus comes at an escalating price, a price which, in human terms, is nothing less than the subjugation of life itself to its demands. The exponential conversion of all things to money described in Chapter Four is just one aspect of this. Our present mode of technology only works when its objects are kept simple, linear; when variables are controlled or eliminated. We can control reality only through reducing it: reducing complex ecosystems to the managed forest, the monocultured farm, the suburban lawn; reducing complex chemistries in herbs and food to just so many "active ingredients" and vitamins; reducing the complexity of human social relationships to the orderliness of a planned society.

Modern technological society is an example of a system possessing order (and even hints of beauty) without design. It is a brutal fact that all human attempts to design society from the top down, as opposed to letting it grow organically, have been abject failures. (Witness the present contrast between North Korea and Cuba, both experiencing considerable isolation from the rest of the world. The former—a centralized command economy—is in dire straights while the latter, though poor, gets by with an adequate quality of life thanks to the resilience springing from the high degree of local autonomy there.)[14] The complexity and scale of technological society defies human attempts at design. Any new technological invention, any new law or social innovation, has effects that ripple through the system in unpredictable ways.

In a non-linear system such as a society or an ecology, the effects of any technological fix are wholly unpredictable. Introduce a new species to control a pest, and the end result might be the emergence of an even worse pest. Elevate serotonin levels with a reuptake inhibitor, and the end result might be a reduction in the number of receptors for serotonin *and* other neurotransmitters, resulting in depression even worse than before. Build a new road to ease congestion, and new developments might end up making congestion all the worse. Cut down all the slow-growing trees and replace them with a fast-growing lumber crop, and diseases might sweep through this weakened ecosystem that result eventually in less lumber, not more. Introduce computers to reduce the drudgery of office work, and people end up spending more time at their desks than ever. Add nitrogen fertilizer to increase soil fertility, and the resulting changes in soil ecology end up reducing its fertility. All of these unpredictable and even contradictory results arise from non-linearity. We can pretend the world is linear and enforce our pretense to a certain extent, but only at increasing cost to freedom and life, because outside a very narrow realm, the world is not in fact linear. Our civilization is built on that tiny segment that *is* linear, or that can reasonably be viewed that way. Today, that pretense is crumbling. The price has become unbearable. Yet unbearable though it may be, we will continue to bear it as long as we are ignorant of any alternative. We will continue to remedy the failures of control with even more control, intensifying the plundering of earth, life, beauty, and health until all are utterly consumed. That is why it is so important to be aware of another way, suggested in part by the scientific developments I describe in this chapter.

We are inclined to see the unintended consequences of technology as the result of lack of foresight, and to hope that better social engineering might solve the problem. In fact, the problem is insoluble. It is insoluble, that is, by the types of solutions we call "design". Unable to conceive of other types of solutions, we desperately try harder until we succumb to despair. What other types of solutions might there be that are not designed? How about those that are allowed to grow? The technology flowing from that paradigm might be so different from what we have today as to be unrecognizable.

Someday soon, when we fully digest the futility of control arising from the non-linearity of our universe, we will begin to adopt a wholly different approach to technology, one that does not attempt nature's reduction but seeks rather its fulfillment. This is not simply a decision about the "appropriateness" of such-and-such a technology; it is the recognition that we are incompetent to determine its ultimate effects. Technological development will come from a place of humility.

The Technological Program of complete control follows from the Scientific Program of complete understanding, in the Newtonian sense of reductionism and predictability. A new technology that is not of control depends therefore on a new mode of understanding. To understand something need not mean to have taken its measure, to know its inner workings, to reduce it analytically. We have now seen the limits of this mode of understanding in physics and mathematics. The next two sections will reveal identical limits in biology and ecology. Some things, perhaps most things, can be understood only as wholes and through their interaction with the rest of the world. Such understanding comes through relationship and experience, even, we could say, through intimacy. No longer can we expect science to make us "lords and possessors of nature"; instead we will see its purpose as being to bring us into more intimate relationship to nature. The spirit of discovery, curiosity, and wonder that is the deepest motivation for science will remain intact.

In the absence of certainty, reason will lose its primacy as the royal road to truth and resume its rightful place as one of several modes of knowledge, each suited to its proper domain. Returning to the alchemical model of Chapter Three, it is the function of the head to reflect, of the heart to know, and of the viscera to transform. The first is cool, the second warm, the third hot. The head—cool reason, cold hard science—has its place; the problem today is that it has usurped the functions of the others. We have built a civilization in which we look to reason, to science, to specialists of all stripes to guide our society and create our future. We have given them a task for which they are constitutionally incapable. Analysis and reason cannot *know*. They can explore, break apart, reflect, but it is not theirs to know or to choose. They cannot apprehend all the qualities listed in Chapter Three—spirit, beauty, life, consciousness, meaning, purpose, love—that "when you take it apart, are not there." As a result, all of them are notoriously absent from "cold, hard" science, which, following Galileo's lead, concerns itself only with the measurable—that which can be reduced to number. Another way of putting it is that science and reason are blind. Without the heart's guidance their methods are equally capable of good and of evil. Remember that the Holocaust was conducted according to a terrible logic with the utmost rational efficiency. According to a particular choice of first principles, the whole thing was perfectly reasonable. Similarly, enormous scientific and engineering efforts today go toward the manufacture of armaments, harmful chemicals, trashy consumer goods, and all the other instruments of planetary despoliation and cultural impoverishment—again, all in accordance with an implacable economic logic.

Reason cannot evaluate truth. Reason cannot apprehend beauty. Reason knows nothing of love. Living from the head brings us to the same place, whether as individuals or as a society. It brings us to a multiplicity of crises. The head tries to manage them through more of the same methods of control, and the crises eventually intensify. Eventually they become unmanageable and the illusion of control becomes transparent; the head surrenders and the heart can take over once again. We are now very close to this point.

Reason can know nothing of beauty, truth, and love, but it can be put into their service. Reductionistic techniques cannot apprehend the emergent phenomena of wholes, but they can be tools for exploring them. Can you imagine a science dedicated to discovering humanity's role and function in that vast cosmic Whole of which we are a part? Can you imagine a technology dedicated to the fulfillment of that role? Perhaps that is humanity's destiny: not lordship over the cosmos, nor meaningless life and death in an indifferent purposeless universe, but conscious participation in a cosmic evolutionary process whose grandeur we can only now begin to imagine.

The Nature of Purpose

The absolute, inherent limits of reductionist logic, both as a means of understanding and as a road to control, invite us into a new mode of relationship with nature. Rather than attempting to explain the workings of the whole according to the workings of the parts, emergent phenomena demand that we sometimes must explain the workings of the parts according to the purpose of the whole. Instead of looking for the underlying *causes* or *reasons* for things, this mode of relationship looks for the *purpose* of things. The word for this type of understanding is teleology, which is a kind of causality that pulls events toward a future purpose.

These two levels of explanation are complementary, not contradictory. Here is an example, courtesy of Stuart Kauffman: imagine a bacteria swimming up a glucose gradient toward a food source. A reductionistic understanding would ask, "What is really making it move in that direction?" and would answer that question by observing how the glucose receptors work, what protein messengers they trigger, how those in turn trigger deformations of the cell membrane to generate movement, and so on. A teleological approach would understand the same molecular-level phenomena from the opposite direction: they are happening *in order that* the bacteria can move toward food. Reductionism essentially asks, "What *makes* it do that?" while teleology asks, "Why does it do that?" As is obvious from this phrasing, reductionism lends itself naturally to the mentality of engineering and control (*making* matter behave in a certain way), while teleological questioning assumes a purpose awaiting discovery, and would therefore foster caution and respect.

Ideology aside, scientists use teleological thinking quite routinely. When we think in terms of empty niches attracting species to fill them, that is teleological reasoning, even if we adduce reductionistic mechanisms to explain how this occurs. Similar teleological reasoning informs our explanations of why birds fly south in the winter, why salmon swim upstream during spawning season, and why bears hibernate in the winter. In the history of technology, too, we can see teleology in action. Engineers develop components in order to meet a need in a larger whole. In fact, many components came into existence only after the larger whole was already there—the bicycle derailleur, for example. Why was it invented? I suppose you could answer that reductionistically, but the clearest answer is "So that bicycles could shift gears." To say that this isn't the "real" reason is mere dogma. Reductionism and teleology are two lenses through which to view the same reality. For many phenomena, understanding comes only through the teleological explanation—What is its purpose?—even when a reductionist explanation is possible (and in nonlinear systems it often is not). Why, then, do we so emphasize the reductionistic explanation? Why do we arbitrarily declare those properties that we can reduce, abstract, and measure to be *more real* than those that are emergent properties of wholes?

One reason is that teleological understanding brings neither absolute certainty nor control. We often understand what something does without knowing how it works, but only knowing how it works can we eventually replace it with an artificial substitute. We might observe that garlic has antifungal properties. When we figure out a reductionist *why*, an "active component," then we can isolate it and perhaps even synthesize it, no longer even needing garlic. We observe that apples taste good. When chemists isolate the flavor factors, we can reproduce the flavor of an apple without the need for apples. In this way we transcend nature and fulfill the Technological Program. We observe that soil fertility depends on bacteria. If we knew exactly why, then we could perhaps eliminate the bacteria by implementing their function through technical means. We observe that forest health depends on the presence of top predators. If we can reduce them to a precise description of their functions (deer kill distribution and so on) then we can replicate those functions and we won't need the predators anymore. The ultimate fulfillment of this ideology would be to replicate every natural function in a wholly artificial reality, the totalization of the separate human realm.

A teleological explanation does not give us this power. When we see the logic of the whole guiding the evolution of the parts, then we must recognize that each part—of a body, an ecosystem, the biosphere—has a purpose that we may not ever fully understand, not even if we measure its every component. With awareness of a purpose to all things, no longer will we so cavalierly try to improve on nature. Lack of purpose, on the other hand, makes *us* the masters of the universe. There is no higher plan in which we might interfere, no natural order that we might disrupt.

Today we are discovering such hubris to be sorely misguided. The pieces of nature or the body we once thought useless or redundant actually have unsuspected functions in maintaining the health and integrity of the whole. Our "improvements" come with hidden costs, stemming from these unseen functions, that we scramble to cover with yet more improvements. From a teleological perspective, the failure of the technological fix and the program of control is inevitable.

The teleological question, "What is it for?" is inherently respectful, both toward the subject of that question and toward the universe that embeds and purposes that subject. It makes of it more than just a thing, and of the world more than just a collection of things. "What is it for?" invests the objects and processes of the world with individuality, selfhood, uniqueness, life—something beyond their descriptions as standardizable, generic building blocks. "What is it for?" changes the world from an *it* into a *you.*

The consequences of applying "What is it for?" to other people and society are equally profound. It subverts the Machine agenda of rendering human beings into standardized, replaceable parts; it resists the conversion of the social capital of unique, local relationships and traditions into generic money; it confounds the social engineer's dream of molding human beings like so many building blocks to construct the perfect society. People are not "human resources", nor are they "assets". The world and everything in it is not a pile of instrumental stuff put here for my use. It has a purpose already, an organic necessity.

On a personal level, teleology fosters respect. In contrast to the world of the selfish gene, wherein each survival machine manipulates competing survival machines to maximize its own benefit, in contrast to the economic view that analyzes relationships according to transactional models of loss and gain—what's in it for me—we see other people as, again, "you's" and not "its", possessing their own purpose and destiny. We see them as other selves, not resources to manipulate. We no longer try to resist and control reality, but to align ourselves with the indwelling purpose that can only be discovered through relationship. We seek to know people and not to use them.

As for others, so for oneself. It is teleology that inspires the eternal human question, "Why am I here?" The science we have inherited provides no answer and it is incapable of providing one. Any explanation couched in terms of genetics and human origins is a non-answer; in fact it is an anti-answer. You are here because your ancestors survived—and no further answer is possible because there is no purpose, only cause. Thankfully, that emptying ideology is crumbling. I just wrote, "We seek to know people and not to use them." Applying that to oneself, we seek through self-knowledge to learn our purpose, instead of exploiting our gifts for the temporary aggrandizement of an illusory separate self. To even believe such a purpose exists or could exist, to even believe that "Why am I here?" has an answer, already empowers life. It generates an urgency to discover what that purpose is. It makes any other life intolerable, including the standardized, compressed life offered by the Machine. Self-respect demands that we live in accordance with our purpose and seek to fulfill it.

"We are not just here; we are here for a purpose." I don't think I've ever penned a more banal spiritual cliché, but it is true nonetheless. Can you feel that in your bones? Young people know it most certainly; we call that knowledge idealism. They know that there is a way the world is supposed to be, and a magnificent role for themselves in that more beautiful world. Broken to the lesser lives we offer them, they react with hostility, rage, cynicism, depression, escapism, or self-destruction—all the defining qualities of modern adolescence. Then we blame them for not bringing these qualities under control, and when they finally have given up their idealism we call them mature. Having given up their idealism, they can get on with the business of survival: practicality and security, comfort and safety, which is what we are left with in the absence of purpose. So we suggest they major in something practical, stay out of trouble, don't take risks, build a résumé. We think we are practical and wise in the ways of the world. Really we are just broken and afraid. We are afraid on their behalf, and, less nobly, we are afraid of what their idealism shows us: the plunder and betrayal of our own youthful possibilities. The recovery of purpose, the acceptance of teleology into the language of science, promises whether directly or metaphorically to undo all of that.

Unfortunately, teleology is still anathema to most scientists, to whom it is tantamount to magic and superstition. Yes, they might agree, we can provisionally say the bacteria wants to move toward food, but we actually know that what "wants" really means is a set of biochemical states. The primary reality is at the molecular level. The teleological explanation is not the real explanation, just a heuristic convenience. I have already explained in this chapter why this view is no longer tenable. Ironically though, it is perfectly consistent with the very religious views that science spurns, for both agree that purpose is not an intrinsic, real property of matter.

James Lovelock, co-originator of the Gaia theory, has gone to great lengths to shake off the "teleology" epithet that critics apply to his theory. One of Lovelock's observations in support of Gaia is that earth has maintained a roughly constant surface temperature for over three billion years, even as incoming solar energy has increased by some thirty or forty percent. Similarly, earth maintains fairly constant levels of oxygen in the atmosphere and salt in the oceans. All of life, and inorganic processes too, contribute to this homeostasis. According to the reductionist paradigm, this is something of a coincidence: after all, there is no annual committee meeting deciding what each species and ecosystem must do this year to make everything work out. There is no board of directors, no supernatural force coordinating it all. Lovelock's response was Daisy World, a simplified computer model of temperature homeostasis in which rising temperatures cause white flowers to cover the planet and reflect more radiation to cool it, while falling temperatures support dark flowers that absorb radiation and warm the planet. See, he said, no coordinator is necessary!

But has Lovelock really escaped teleology? Yes, he has escaped the necessity of an external coordinating force, but only in our dualistic mindset does this equate to teleology. For in fact, Lovelock's critics are right: the model of understanding that Gaia theory provides *is* teleological, because it invites us to understand life processes in terms of their purpose, not their cause. Why do algae emit dimethyl sulfide gas? As a circumstantial byproduct of their metabolism, or in order to seed rain clouds? Why do bacteria on legume nodules fix nitrogen? Because ammonia happens to be a waste product of their ATP conversion pathway, or in order to allow plants to grow? Why do other bacteria accelerate rock weathering? Because they need the minerals to survive, or in order to speed the removal of carbon from the atmosphere? The first answer to each of these questions is not invalid, but it provides a limited understanding.

Does random chance explain our great good fortune that all of Gaia's essential functions have been filled? Are we just lucky that aerobic bacteria arose just in time to rescue life on earth from poisoning itself on oxygen? Do we owe our very existence to an absurdly unlikely series of chance events, one after another? If so, we are surely alone in the universe, poised always on the edge of oblivion should our run of luck end. Hence, the necessity of mastering nature.

The mentality of control draws intimately from a fundamental scientific worldview that denies purpose. Until that changes, the regime of control will persist and continue to proceed toward totality. The random, indifferent universe of our ideology invites—nay demands—control. And always is our position a precarious one, and anxiety woven into the fabric of life. That is why the gathering sea-change in science is so significant. It is more than a mere shift of opinion; it portends a new relationship to the world and a new state of human beingness.

When you read the words, "Does random chance explain... are we just lucky that... an absurdly unlikely series of chance events..." did you expect I was about to unveil God as the alternative? Theologians have long cited the uncanny coordination of nature as evidence of God, noting as I have the hubris, despair, and license to plunder implicit in a mechanistic worldview. "See what happens when you don't believe in God?" they ask. Viewing the world of matter and the life of the flesh as devoid of sacredness, religious believers invoke an external God to invest life with meaning. Similarly, unaware that order and beauty can arise spontaneously, they invoke God the Creator or God the "Intelligent Designer" to infuse this universe of inert, randomly interacting matter with organization, beauty, and life.

For a long time we have seen their external, supernatural God as the only alternative to an indifferent, mechanical, purposeless universe. No longer. The spontaneous emergence of organization in mathematical, chemical, and biological systems offers a third alternative. What if the coordination, purpose, and beauty we observe are inseparable properties of matter? What if divinity is an organic property of reality?

So steeped are we in a dualistic mindset that we can hardly conceive of a purpose to nature except as externally defined by a supernatural intentional force, namely God. External to matter, such a God is by definition outside the subject area of science. Remember, though, that the whole conception of God as a force above and apart from nature is only as old as agriculture, reflecting the other dualistic separations that gained momentum at that time. Yet human intuitions of a meaning or significance to life surely predate agriculture with its supernal God, indicating that an external imposer of meaning, an external designer of purpose, is necessary only to the dualistic mind.

The Fundamentalist movement, along with most of the New Age as well as transplanted spirituality from the East, is in part a response to the loss of sacredness that science seems to imply, and in part a reaction to the moral and social fragmentation that shares a common root with science in the Age of Separation. Agreeing with science that the ordinary world of matter is devoid of sacredness, they imagine a spiritual realm apart from the world or, in the case of Eastern transplants, retreat inward to the depths of the mind, ritual, yoga, and other "spiritual practice". Religion thereby abets the mechanistic assumption of science: the world of matter is just matter. Seen in this light, the doctrine of Intelligent Design, intended to bring God back into science, actually ends up confirming the soullessness of the physical world—the world of life and human life.

Scientists typically cringe at any mention of an intention or purpose to evolution, tenaciously defending the Darwinian dogma that evolution arises solely from random variation and selection based on survival and replication.[15] Alternative theories of biogenesis and evolution subvert deep-seated cultural assumptions of self and world, thus offending our intuitions. On a more superficial level, scientists' antipathy toward dissenting theories stems from their common association with the anti-scientific attacks of various proponents of creationism. Therefore, before I draw out the cultural and spiritual implications of some alternatives to Darwinism, I want to make clear right now (doubtless eliciting a sigh of relief from the scientific reader) that I am not going to espouse any form of creationism or "Intelligent Design" (ID) in this book.

I sympathize with the motivation of the proponents of ID and creationism, however, and my sympathy is not of the patronizing variety which imputes to them a childish refusal to accept the obvious facts of the universe. No, their ultimate motivation is a sincere protest against a world that has been shorn of sacredness and purpose. They intuit that sacredness and purpose are real properties of the universe and not human projections, not the superstitious wishful thinking of people unable to cope with a universe that, to quote Richard Dawkins again, has "precisely the properties we should expect if there is, at bottom, no design, no purpose, no evil and no good, nothing but blind, pitiless indifference."[16] Unfortunately, the proponents of ID are themselves subject to most of the same deeper, hidden assumptions about the nature of reality that grip the scientific mainstream. This is no wonder, for these assumptions are written into the very culture we live in, into the conceptual vocabulary we use to describe the world, into the very structure of modern thought and language.

The proponents of Intelligent Design are unwittingly replacing a greater miracle with a lesser miracle, a greater God with a lesser God, and a greater mystery with a lesser mystery. They are also, again unwittingly, contributing to the very desacralization of the universe that is so unsettling about the scientific cosmology that dominates today. Their opposition to the atheistic cosmology, genesis, and evolution that is today's orthodoxy is actually a superficial opposition; on a deeper level, *proponents of Intelligent Design exacerbate the very dualism that has robbed the modern world of sacredness and meaning.*

The typical ID argument first cites some of the well-known, and very real, difficulties of the Neodarwinian orthodoxy. Then, citing the wild improbability of sophisticated structures such as the eye or the cellular metabolism having ever developed by chance, it goes on to conclude that such structures must therefore have been planned out and guided by a suprahuman intelligence. Something as complex, as tightly coordinated, as miraculous as a living being could never have emerged spontaneously; therefore it must have been consciously designed.

But the very concept of design already smuggles in Newtonian concepts of reductionism, dualism, and mechanism. How do we go about designing something? Through a reductionistic process of assembling components, each of which themselves may also require designing. Design is hierarchical and modular—a problem is broken up into manageable pieces arranged by an outside intelligence. Spontaneous processes of growth are not considered "design". Organisms do not design themselves, nor do parents design their children. Trapped in a dualistic mindset, we are hard pressed indeed to see how complex, highly ordered, tightly coupled systems could have come about by themselves, without external guidance. Our logic tells us that if it is purposeful, it must have been designed with a purpose. We cannot conceive of a purpose without a purposer, a design without a designer, beauty without an artist, because that has been our relationship to the world ever since agriculture supplanted the spontaneous goodness of nature with the imposed goodness of cultivation. In the mindset of agriculture, goodness comes forth only with work; else the field runs to weeds and the livestock goes feral. The logic that inspires Intelligent Design is also the logic of engineering and control. It is the logic that defines the Technological Program whose failure is increasingly evident.

To me it is a far greater miracle that reality have the property of self-organization. Why should it? Why should reality not be boringly linear instead? Intelligent Design carries a deep hidden assumption that the universe is not inherently beautiful, alive, inspirited, but that it had to be made that way by an outside deity. The universe is just dead matter until God makes something of it. Life could have only sprung from the primordial soup if God had made it happen. The cell, the organ, the organism, the ecosystem could not work so perfectly by themselves; matter has not that property. Intelligent Design posits a universe that is less spiritual and not more. The mother of all religions, animism, had no such dualism in it. Animism, often misunderstood as the belief that all things are possessed of spirit, actually holds that all things *are* spirit. It holds that all things are sacred in themselves, and not conditionally sacred because of something they may (and by implication, may not) possess.[17] One of the goals of this chapter is to translate animism into the language of modern science.

I reject Intelligent Design more out of a religious sentiment than a scientific sentiment, for the wonder, the magic, the ongoing miracle of a living universe pregnant with creativity, order, and beauty is far more stupendous than a conditional Creation contingent on a separate creator God. The quasi-religious awe a biologist feels upon truly appreciating the complexity of a living cell is diminished, not enhanced, by the explanation, "It's like that because God made it that way," just as it is also diminished by the conventional explanation that it arose through a purposeless concatenation of improbable events. The spontaneous arising of order, beauty, and life that is written into the laws of the universe, and even more deeply, into the structure of mathematics, that is repeated in every non-linear system with certain very general characteristics, and that is like that only because it is like that and could be no other way, is far more awesome. I offer the reader not a mundane universe in which nothing is sacred because there is no God, nor a split universe in which some things are holy, of God, and others just matter, but rather a universe that is fully sacred, pregnant with meaning, immanent with divinity, in which order, organization, and beauty arise spontaneously from the ground up, neither imposed from above by a designer nor projected from within by the observer, and of which God is an inseparable property. The marvelous complexity and beauty of nature is not some consolation prize for science's denial of the sacred, but evidence that the universe is itself sacred. The awe of the cell biologist is not *quasi*-religious, it *is* religious.

Note the obstacle of language apparent in the above discussion. By using words like sacred and spiritual, I draw on meanings that imply a divided universe containing also the non-spiritual and the profane. Dualism is built into our language and thought. What could spiritual or sacred mean in the absence of a separate realm? For me as for the animist it means that every living being, every process of nature, and every bit of matter is a unique individual that is nonetheless not separate from myself. Not even electrons are generic identical particles; each is unique and its behavior forever irreducible to causes. There are no "its" in the universe, only you's or thou's, each utterly unique, and yet I and thou are one. It is impossible to put this into words without generating a paradox. Impossible. Why? It is in the nature of words as labels, representations, symbols to categorize, abstract, and divide the world. So don't rely on words to understand how each "thou" is unique and yet I and thou are one. Look into a lover's eyes instead, or let the birds sing it to you.

Each electron, each drop of water, each pebble, each everything is therefore qualified to receive our love. We cannot love the standard and generic except in the abstract. Love is personal. It sees *you* as unique and fully you, and it knows a connection so profound as to blur *your* distinctness from *me*. The purpose possessed by all beings in the universe, and indeed by the universe itself, is an aspect of this uniqueness, this sacredness of all being. The experience of it suffuses our lives with love, awe, and respect.

Perhaps the import of the gathering paradigm shift is now becoming clear. It marks the end of the Age of Separation and the beginning of a new age that I call the Age of Reunion. It marks the end of civilization as we know it, and the birth of a new kind of civilization. The civilization we have known has always been built upon the escalating domestication of nature. The ascent of humanity has always been a project of imposing human purpose and human design onto the raw materials of an inert reality. That has been the grand project of civilization, to bring order to chaos. And now we are discovering that order arises naturally from chaos anyway, and chaos from order, and from that chaos new order of ever-higher degree—an ascending spiral of Yin and Yang. The age of the frontiersman conquering nature and bringing it to order is over, as we turn toward seeking the right role for human consciousness in the continued unfolding of order in the universe. The new human relationship to the world will be that of a lover to his or her beloved. For thousands of years the relationship has been one of control, based on fear. Our technology, the concrete manifestation of that relationship, is therefore mostly a technology of fear as well. Can we even imagine what a technology of love would look like? One thing is certain: the devotional relationship of which I speak does not mean the diminishment or fading of the human race, or a return to the Stone Age. In his devotion to the beloved, the lover grows himself. Love is not a sacrifice; it is a mutual fulfillment. Love itself denies the logic of the separate self.

A universe that is inherently creative or inherently purposeful has no need of intelligent design nor of unlikely chains of coincidences, and has broad implications for our individual lives as well as the collective function and destiny of the human species. This will become apparent in the second half of this chapter as we come "back to earth" to examine some elements of the neo-Lamarckian paradigms that are gradually creeping into biology. Growing evidence that mutation is not in fact random calls into question the very cogency of the "selfish gene" theory of biogenesis, evolution, and behavior. In place of this competition-based world-view, a new paradigm is emerging that emphasizes symbiosis, cooperation, and the sharing of DNA across species boundaries, calling the integrity of the discrete biological self further into doubt. In its place a new concept of self arises, defined through its relationships and constitutionally impervious to the program of reduction and control. Remember, the entire edifice of civilization arises from our sense of self as discrete and separate. Now we are building the foundation of a new sense of self, and therefore of a new kind of technology, a new kind of money, a new kind of civilization. What will a technology of love look like, in the Age of Reunion?

Nature provides a clue.

The Purpose of Nature

The Neodarwinian account of the evolution of life rests on two key pillars, each of which is deeply woven into our worldview: random mutation and natural selection. Randomness is the opposite of purposiveness, and a necessary adjunct to a world of standard, generic building blocks. Natural selection is the projection onto biology of the ambient anxiety and competitiveness of our culture. These associations ensure that Neodarwinism will be orthodoxy in the Age of Separation.

But what exactly do random mutation and natural selection act on? What is the biological self that suffers mutation and that competes for survival against other selves? A coherent answer to this question is crucial to the coherency of the entire Neodarwinian paradigm. If we find instead yet another projection of our culturally-constructed conception of self, then the entire edifice of Neodarwinism is a projection as well.

The standard answer is that the subject of natural selection is the gene, biology's version of the discrete and separate self. Through their manipulations of the environment (via the organisms they "code for", command and control), genes enact the struggle for survival that drives evolution. By necessity, their primary relationship to each other is competitive: they are here because over the eons they out-survived and out-replicated all their competitors. Genes have no desire to evolve, but sometimes they mutate by chance into new genes that are even better at survival and replication. In this way, without any intention, purpose, or program, the genes evolve. The only direction to evolution is toward genes that are better and better at surviving and replicating. To the extent that the word "purpose" has any meaning at all, this is the genes' sole purpose. As for the organisms—such as you and I—that the genes code for, these are merely the instruments by which genes survive and reproduce. Insulated like the Cartesian soul from the environment, the interests of the genes are only coincidentally aligned with the interests of the organism. The struggles, desires, and adaptations of the organism do not affect the genes, nor do the needs of other organisms. The relationship between genes and environment is one-way: the genes alter the environment through the organisms they program, but the only effect the environment has on the genes is either to prevent them from being passed on, or to randomly alter them through mutation. The gene is the master, the organism the servant, the environment a reservoir of resources and a source of threats.

In parallel, we see humankind as the master, technology as the servant, and the environment again as a reservoir of resources and a source of threats. On the individual level, economics and other transactional models of human behavior utilize a structurally equivalent scheme. Life is about me, what I can get, and how I can get it.

The integrity of the Neodarwinian worldview depends on all of the elements described above. Remove one, and the whole fabric unravels. Along with it goes a whole constellation of intuitions about the nature of the self and the way of the world.

Today, indeed, the entire fabric of Neodarwinism is unraveling, contributing to and impelled by the unraveling of the larger paradigms of world and self that embed it. We could start with any of Neodarwinism's frayed threads and it will lead us to all of the others. So let us start with the core of the biological definition of self, the genes.

Contrary to prevailing dogma, the genes are not in fact the command and control center for the cell, nor do they stand in isolation from the non-random (and hence in some sense purposeful) effects of the environment. In his 1990 paper, "Metaphors and the Role of Genes and Development," the eminent biologist N.H. Nijhout argues that two central metaphors of biology, that "genes 'control' development, and the genes embody 'programs' for development," are highly misleading. He writes, "The simplest and also the only strictly correct view of the function of genes, is that they supply cells, and ultimately organisms, [and ultimately the environment?] with chemical materials." And, "When a gene product is needed, a signal from its environment, not an emergent property of the gene itself, activates expression of that gene."[18] If the gene is the biological self, it has few of the attributes of Cartesian selfhood. These are projections. The gene is not the control center, not the brain of the cell. In fact, enucleated cells can survive for months without any impairment of their ability to move, digest food, excrete wastes, exchange gases, communicate with other cells, and otherwise respond to their environment—all without genes.[19] Only when their protein parts begin to wear out do these cells' functions begin to deteriorate.

Applying the gene metaphorically as an analog to the self, we exist to contribute our gifts to the larger wholes of which we are part. Something in our environment calls upon us (activates expression of the gene), and we provide our unique gift that in turn activates the contribution of another. It is not about maximizing self-interest at all!

The microbiologist Bruce Lipton offers a conceptual model of the cell that makes the role of the genes clear.[20] The brain of the cell, he says, is not the nucleus but actually the cell membrane, which senses the extracellular environment via its receptor proteins and then, via effector proteins, translates this data into instructions for the interior of the cell—including the DNA—to carry out. This, he observes, is on a schematic level exactly what a brain does. The genes cannot turn themselves on and off; it is the cell membrane with its chemical messengers that does that. The nucleus and ribosomes are merely the manufacturing facility and not headquarters, the hard disk and not the CPU. It is in fact the cell membrane, the mediator for the environment, from whence the instructions actually originate. So in contrast to a lordly self manipulating an external not-self from its headquarters, we have a two-way interaction in which environment defines self just as much as self molds environment, blurring the distinction between the two. No longer does biology lend itself as a metaphoric model for some Cartesian "seat of the soul", a center of awareness and free will looking out upon a mechanistic world of dead matter.

Further eroding the self/other distinction is the fact that not only does the environment turn DNA on and off, it can even change DNA in a non-random way. Cells and organisms—and through them, other life forms and the environment as a whole—can modify their own DNA to produce necessary traits. What's more, these modifications can affect germline cells, so that the associated acquired traits can be inherited.

Hold on. Haven't I just asserted Lamarckism, the thoroughly discredited theory of inheritance of acquired characteristics proposed by Jean Baptiste Lamarck in 1806? Lamarckism, of whose founder George Bernard Shaw wrote in 1921, "poor Lamarck was swept aside as a crude and exploded guesser hardly worthy to be named as [Darwin's] erroneous forerunner"?[21] Despite 200 years of unremitting ridicule, Lamarck's basic idea is receiving increasing vindication. Much more than the mere mechanism of evolution is at stake, because Lamarck's theory was not so simplistic as the usual derided example of "each generation of giraffes stretched their necks to reach higher leaves, and the longer necks were passed on to the next generation." The crux of his thinking is laid out in the following:

> It is not the shape either of the body or its parts which gives rise to the habits of animals and their mode of life; but that it is, on the contrary, the habits, mode of life and all the other influences of the environment which have in the course of time built up the shape of the body and of the parts of animals. With new shapes, new faculties have been acquired, and little by little nature has succeeded in fashioning animals as we actually see them.[22]

Of course, we must interpret Lamarck's theory through our knowledge of the genetic basis of inheritance and morphology. To make the theory work, the acquired characteristics must arise from changes in the genes that can be passed on. Lamarckism in its modern form simply states that these changes can be acquired in the course of a lifetime through purposeful adaptation; Neodarwinism, on the other hand, says they arise only from random mutation. Thus the experiment that purported to disprove Lamarckism once and for all is not really a test of Lamarckism at all. Weismann's experiment with mice, in which he cut off the tails of each generation and found that the next generation always was born with tails of normal length, demolishes at best a reduced and distorted caricature of Lamarckism.

If acquired characteristics may be inherited, the door is open for this inheritance to reflect purpose, both of the organism and the environment. Shaw interprets Lamarck along these lines: "How did he come by his long neck? Lamarck would have said, by wanting to get at the tender leaves high up on the tree, and trying until he succeeded in wishing the necessary length of neck into existence." Evolution happens, in other words, through wanting, through intention. Shaw goes on to write:

> Another answer was also possible: namely, that some prehistoric stockbreeder, wishing to produce a natural curiosity, selected the longest-necked animals he could find, and bred from them until at last an animal with an abnormally long neck was evolved by intentional selection, just as the race-horse or the fantail pigeon has been evolved. Both these explanations, you will observe, involve consciousness, will, design, purpose, either on the part of the animal itself or on the part of a superior intelligence controlling its destiny. Darwin pointed out—and this and no more was Darwin's famous discovery—that a third explanation, involving neither will nor purpose nor design either in the animal or anyone else, was on the cards. If your neck is too short to reach your food, you die. That may be the simple explanation of the fact that all the surviving animals that feed on foliage have necks or trunks long enough to reach it. So bang goes your belief that the necks must have been designed to reach the food. But Lamarck did not believe that

> the necks were so designed in the beginning: he believed that the long necks were evolved by wanting and trying. Not necessarily, said Darwin... suppose the average height of the foliage-eating animals is four feet, and that they increase in numbers until a time comes when all the trees are eaten away to within four feet of the ground. Then the animals who happen to be an inch or two short of the average will die of starvation. All the animals who happen to be an inch or so above the average will be better fed and stronger than the others... And this, mark you, without the intervention of any stockbreeder, human or divine, and without will, purpose, design, or even consciousness beyond the blind will to satisfy hunger. It is true that this blind will, being in effect a will to live, gives away the whole case; but still, as compared to the open-eyed intelligent wanting and trying of Lamarck, the Darwinian process may be described as a chapter of accidents. As such, it seems simple, because you do not at first realize all that it involves. But when its whole significance dawns on you, your heart sinks into a heap of sand within you. There is a hideous fatalism about it, a ghastly and damnable reduction of beauty and intelligence, of strength and purpose, of honor and aspiration... To call this Natural Selection is a blasphemy, possible to many for whom Nature is nothing but a casual aggregation of inert and dead matter.

Ironically enough, it is now neo-Lamarckism that offers a "third explanation" besides intelligent design and random chance: "consciousness, will, design, purpose" can guide evolution even without an external designer or purposer. And just to make the full extent of the heresy plain, let me emphasize that it is not just organisms or cells that modify their own DNA—the blurring of self/other goes further than that. The extra-somatic environment also participates in the modification of genes, so that they serve the purposes not only of the organism but of the community, the ecosystem, and the planet... maybe even of the cosmos.

In other words, it is not just that the giraffe wants to reach the higher leaves, or that protohumans wanted hands that could grasp tools. An empty niche in the environment exerts an evolutionary pull. An unfolding pattern draws creatures to occupy the necessary roles. That is to say, the world wanted it too! Speaking metaphorically, our gifts and potentials are called forth by life's opportunities and by the world's needs. They are not independently preexistent, to be imposed by force upon an unconscious universe. Evolution, whether of the soul or of the species, unfolds in coordination with the evolution of all. "Lamarckism" stands as a term of disparagement because its doctrine is inescapably teleological.

That heritable change is more than a haphazard triage of randomly generated possibilities is problematic to our culture's program of control, for it undermines the ideology that leaves us the "lords and possessors of nature." While in Darwin's scheme the ultimate purpose of life is to survive, Lamarck's admits to a higher sort of purpose associated with will or intention. And if nature's forms and systems express a purpose, then we must doubt our absolute suzerainty over nature; we must doubt the assumption that we can engineer nature endlessly with impunity, especially when we do so in ignorance of its purposes. In a blind, purposeless universe we are at perfect liberty to do our will, for there is no natural order on which we might infringe, no destiny to interfere in, no destiny at all, in fact, except that which we create. But if there is a purpose inherent in the way of the world, then the whole bent of science must change from understanding for control's sake, to understanding for the sake of according more closely to nature's purpose. Asking of ourselves, "What are we for?" we will seek out our proper role and function on the planet and in the universe. This transition, which I believe will inevitably flow from the new scientific paradigms described herein, represents our abdication from the pretense to lordship over nature, to become nature's humble student.

Observing nature through our cultural lenses of separation, competition, and survival anxiety, we have long been blind to the fact that nature doesn't always work that way. Nonetheless, evidence for the non-randomness of genetic change is mounting. It is now widely accepted that rates of genetic mutation increase when organisms are under stress, presumably to increase the likelihood that a mutation will arise that is better adapted to the stressor.[23] Reflecting consensus, a recent article in *Nature* opines, "It makes sense for stress responses to cause mutations; it may be a 'selected' feature that increases genetic variation, thus increasing 'evolvability' under stress when organisms are suboptimally adapted to their environments. Most of the mutations would be harmful or neutral, but rare adaptive mutations would also occur."[24] Recent research implies that this "stress" can be the quite common occurrence of running low on food, implying that accelerated mutation is an ordinary feature of bacterial life. Even though such mutations are thought to still be random, stress-induced mutagenesis is immensely problematic to standard selfish gene theory. The unit of selection is supposed to be the gene, not the organism! Genetic characteristics, such as mutability, that benefit the organism in a way that does not promote the gene's replication in future generations would be selected *against.*

More controversial by far is a concept sometimes called *adaptive mutation* that has cropped up intermittently in biology under various guises ever since the days of Jean-Baptiste Lamarck. Reintroduced in 1988 by John Cairns, it claims that mutations are not fully random but are somehow biased toward an adaptive purpose.[25] Cairns disabled a gene in *E. coli* that produces an enzyme to digest lactose, then put the *E. coli* in a lactose-only medium. To his surprise, the disabled gene mutated back to the original enzyme-producing version—not in every bacterium, but at a higher rate than in a control culture where the bacteria had food sources other than lactose. While the consensus is that an overall elevated (but still random) stress-induced mutation rate can explain this finding, other research supports the idea that it may be truly adaptive. B.G. Hall found that a similar *E. coli* mutation happened to permit utilization of the sugar arbutin in the presence of starvation—but only when there was arbutin in the medium![26] The mutation happened at a much higher rate specifically when it was useful. Other research has found that increased mutation rates under stress are not evenly distributed across the genome, but are much higher in precisely those areas which may generate beneficial mutations.[27] The controversy rages on today, with orthodox opinion maintaining that stress-induced hypermutability can fully explain adaptive mutation, and thus preserve Darwinism.

There are several reasons for the mainstream aversion to true adaptive mutation. Some commentators seem to feel it would entail "spooky foreknowledge"[28] of what mutations would be useful. Personally, I love the idea of spooky foreknowledge, but in the case of adaptive mutation it is unnecessary. All that is needed is a way for the DNA to sense and internalize signals from the environment; in other words, a way for the environment to speak to the DNA, to alter it purposefully just as the DNA, mediated by the organism, alters the environment. Adaptive mutation implies a tight coupling between gene and environment that calls into question the very definition of the biological self. To say that the gene responds to the environment is a conceit; just as easily we could say that the environment molds the gene.

The precise mechanisms by which this happens are unknown and mostly unexplored due to the widespread belief that nature doesn't work that way. Nonetheless, there are several confirmed examples already, along with some interesting speculation. Mainstream biology has long accepted that reverse transcriptase allows RNA to rewrite DNA, and that antibodies in immune cells help to modify the DNA that produces them.

More radical, but now also widely accepted, is the observation that epigenetic proteins, which envelope the DNA double helix, also carry heritable information that affects the way genes are expressed. What's more, these epigenetic proteins are also subject to heritable environmental modification. More radical still are proposals that invoke electromagnetic or other vibrational induction of genetic change. Finally, I must mention Johnjoe McFadden's very clever effort to bring "spooky foreknowledge" back into evolution through the Quantum Zeno Effect.[29] His full argument is beyond the scope of the present chapter, but essentially he believes that genes take advantage of quantum superpositions of states to scan the search space of possible mutations, whose viability is confirmed by the observer effect of their own progeny.

Whatever the mechanism, evidence is mounting that the environment influences genes through mechanisms beyond mere natural selection. The evidence has been mounting, in fact, for a long time, as Charles Darwin, himself remarkable free of the dogma of Darwinism, acknowledged in 1888:

> In my opinion, the greatest error which I have committed has been not allowing sufficient weight to the direct action of the environments, i.e. food, climate, etc., *independently of natural selection....* When I wrote the "Origin," and for some years afterwards, I could find little good evidence of the direct action of the environment; now there is a large body of evidence..[30] [emphasis mine]

If Darwin was aware of it in 1888, why do most biologists still ignore this "good evidence" today? Indeed, the evidence today is far greater. The institutional insistence on the primacy of the DNA as the "controller and program" for morphology and behavior, and on the randomness of its evolution that is driven solely by natural selection, derives not from the evidence, but from deeper cultural assumptions about who we are.

The challenge that adaptive mutation presents strikes at the heart of our culture's dualistic conception of world and self. Purpose, once the sole province of the conscious self of psychology (or at least the sentient self of biology), escapes the confines of individual actors—purposers—to become a property of the world as a whole. No longer do the genes necessarily act in their own self-interest, nor even, as we shall see, the selfish interest of the organism they are part of. Rather than being the source of purpose (which is only to survive), they are a means by which organism and environment enact purpose. And in that case, in what sense are the genes the kernel of biological selfhood? Yes, they carry heritable instructions for building bodies, but it is outside signals that activate them. Nor are these instructions inviolate; they are readily modified according to need. Wouldn't that-which-modifies-them be a more appropriate choice for selfhood? Our Cartesian intuitions tell us that ultimate authority must reside *somewhere*. But the invoker or modifier is not a discrete entity either: it is the cell, the organism, even the environment. The self is no longer so rigidly defined; therefore neither is self-interest. Who are we? The question has no definite answer. Different answers are expedient for different purposes, but none is complete. Natural selection depends for its primacy on a discrete self to act on, but we are now learning that the self of biology, the subject of natural selection, is not discrete but fluid, merging by degrees into the genetic plenum.

Direct evidence for adaptive mutation is still ambiguous. However, if natural selection on random mutations of replicating DNA is not the prime mover of evolution, then we must construct a coherent understanding of life, its origins, and its evolution that does not rely on competition among discrete genetic subjects. Fortunately, just such a story is emerging today and gradually infiltrating mainstream biology. It is a story in which cooperation just as much as competition defines relationships among living beings, in which symbiosis and merger across fluid genetic boundaries drive evolution, and in which purpose arises from both within the organism and without. No longer is biology a study of separate and competing selves, and no longer will that idea inform our intuitions about the way of the world.

Life without a Replicator

A creation myth can be a powerful window into a culture's most important belief systems. In Chapter Three I described the creation myth of biology, the conventionally accepted story of how life originated with an initial replicating molecule, which eventually mutated and evolved into the diversity of forms we see today. Lost amid all the speculation about possible RNA or peptide candidates for the "first replicating molecule" is the fact that no such molecule exists today. There is no gene, no sequence of DNA or RNA, that can replicate itself—not without an awful lot of help from other genes. In fact, what genetic reproduction entails is many, many genes cooperating in their mutual replication. Each gene plays a very limited—though sometimes indispensable—role in this collective replication. Perhaps it provides instructions for the production of a protein important in sex hormone production; perhaps it turns on another gene that initiates cell differentiation at a crucial stage of development. Each of these and other functions certainly are necessary for a gene's replication, but they are not sufficient. Even in the simplest organism, even in a virus, no gene can autonomously replicate itself.

In mitotic cell division, a diversity of enzymes and signaling proteins contribute indispensably to the whole process, each coded for by a different gene and each useless by itself. Helicase, polymerases, cyclin, and CDK are each produced from a distinct gene, and if even one is missing, mitosis fails utterly. Genes do not replicate themselves. At its most fundamental level, reproduction (and survival) is a cooperative effort.

Why, then, is biogenetic theory overwhelmingly focused on explaining the origin of the "first replicator"? Why do we look for this mythical molecule capable of self-replication, when life does not work that way? In Chapter Three I described the cultural biases that motivate and reinforce the selfish gene conception of life. A world of clearly demarcated, competing selves, in which cooperation is incidental and not a necessary or fundamental feature of life, is consistent with the philosophies of the Scientific Revolution and our self-imposed separation from nature and each other. We find it as well in our systems of money, education, medicine, law, and religion. Projecting our own culture onto life, we see it as red in tooth and claw, driven from the outset by a ruthlessly selfish competition to survive. And the subject of this "self"ishness is the gene, the replicator.

Because it is an elegant and parsimonious theory, and because it fits in so well with our culture's conception of self and world, the Neodarwinian synthesis has persisted as the dominant paradigm for a long time, despite several formidable problems that have never been resolved. Chief among these is the survivability of intermediates, and the closely related problem of irreducible complexity. At each level of organization, standard evolutionary theory runs up against enormous leaps in complexity that are difficult to reconcile with undirected, random mutation. And the solution points toward not only a different conception of the self that is relationally defined, but also a different understanding of the relative importance of cooperation and competition, a different conception of the nature of life that is not nasty, brutish, and short, a different attitude toward the program of control, and a vastly different understanding of progress, human society, and human relationship.

The first such leap of complexity that evolutionary theory must contend with is, of course, the origin of life. The original version of the problem was the chicken-and-the-egg situation in which DNA requires proteins to catalyze its replication, while proteins can only be produced by DNA. Which came first? A possible solution came with the discovery in 1982 of "ribozymes": RNA molecules that can perform both catalytic and information-storage functions, implying the possibility of an "RNA world" without the need for proteins. However, this solution only pushes the chicken-and-egg problem into another realm, as the next paragraph describes (you can skip it if you want). Despite these enormous difficulties, research on the "RNA world" remains, in the words of one commentator, a "medium-sized industry" unto itself.[31] Motivating this effort is the conviction that there must have been a "first replicator" somehow, because our entire conception of what life is, what the self is, and how the universe works is dependent on it. It is the original discrete and separate self. RNA world theories are full of all kinds of highly contrived, *ad hoc* pre-conditions, based on the assumption that "something like this *had* to have happened."

First there is the problem of getting an initial prebiotic soup full of β-D-ribonucleotides, about which two leading researchers, Joyce & Orgel, write, "We conclude that the direct synthesis of the nucleosides or nucleotides from prebiotic precursors in reasonable yields and unaccompanied by larger amounts of related molecules could not be achieved by presently known chemical reactions."[32] Worse yet, even if such synthesis were achieved, there is no plausible mechanism to resolve the left-handed L-isomers from the right-handed D-isomers, which is necessary because the presence of the L version (which does not exist in biology) inhibits the polymerization of the D version.[33] A further problem is the synthesis of polynucleotides with the correct 3',5' linkages, which would be in the minority in the absence of some specific catalyst.[34] Thirdly, even if some pre-replication polymerization mechanism existed, there is the difficulty of traversing the enormous search space of possible polynucleotides for one that engages in self-replication, yet without the evolutionary search mechanism of error-prone replication—another chicken-and-egg problem. [35] To give you some idea of this difficulty, an RNA polymerase, which carries out only one of the functions an RNA replicase would need, has been created in the lab; however, it can only perform polymerization of strands three nucleotides long and itself contains a hundred nucleotides, implying a search space on the order of 4^{100}.[36]

In other words, the emergence of even the simplest imaginable replicating molecule is only plausible within a highly complex chemical system—a system that includes molecules generally only produced by living systems. It would seem that life is a prerequisite for life. Since the disproof of "spontaneous generation" in the 19th century, such a maxim has indeed corresponded to our every observation. On the other hand, at all stages of evolution life has undergone sudden leaps in complexity, each of which admits to the same chicken-and-egg explanatory dilemma. The question of the origin of life is a special case of a more general question: what is the origin of complexity, order, and organization?

One theory that tries to explain the origin of life without a "first replicator" is known as complexity theory. Elucidated by Stuart Kauffman in his book *Origins of Order,*[37] this theory is an important step toward a biological paradigm no longer based on our culture's present conception of self. In describing the origin of life, Kauffman takes advantage of the "order for free" concept described above in the "Order without Design" section. Complex evolving structures appear in many mathematical, physical, and chemical systems in the presence of certain basic conditions such as feedback. Could life be one of them?

The key to Kauffman's account of biogenesis is the idea of an autocatalytic set. Many of the problems with the standard selfish gene theory come from the absurd unlikelihood of obtaining such a molecule from the building blocks likely to be found in the prebiotic soup. In Kauffman's theory it is not necessary for a molecule to appear that can catalyze its own formation. All that is necessary is to have a set of molecules each of which catalyze a step in the formation of one or more other molecules in the set. The final step in the formation of each molecule in the set must be catalyzed by another member of the set, a condition called catalytic closure. Based on combinatoric reasoning, Kauffman argues that the emergence of autocatalytic sets is highly probable if not inevitable when molecular diversity crosses a certain concentration threshold.

Let's look at a simplified example, an autocatalytic loop in which A catalyzes the formation of B, B catalyzes C, C catalyzes D, D catalyzes E, and E catalyzes A. One way to look at it is to say that A is the replicator and uses B, C, D, and E as tools to achieve replication. However, that is an arbitrary designation, because we could say the same thing about any other member of the set. Each is dependent on the others.

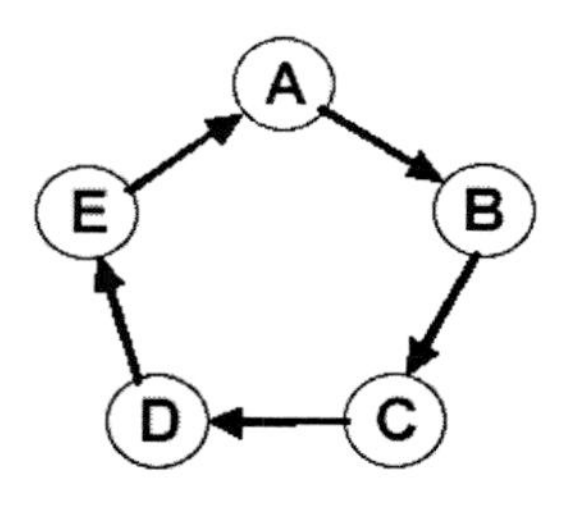

Figure 3: An autocatalytic loop

An autocatalytic loop like this lies at the heart of Kauffman's autocatalytic sets, but it gets far more complicated than that. Imagine a system of multiple loops and chains, loops within loops, mutual cross-feed relationships connecting them, inhibitory connections, preferential reactions given different substrate concentrations... very soon the picture starts looking very much like a metabolism or an ecosystem. There is still no unequivocally identifiable unit that might be said to be *the* replicator, but we may impose somewhat arbitrary boundaries on various subsystems within the system and call these parts alive, recognizing that while they might be dependent for certain of their reagents on other subsystems, they are able to maintain a constant internal environment.

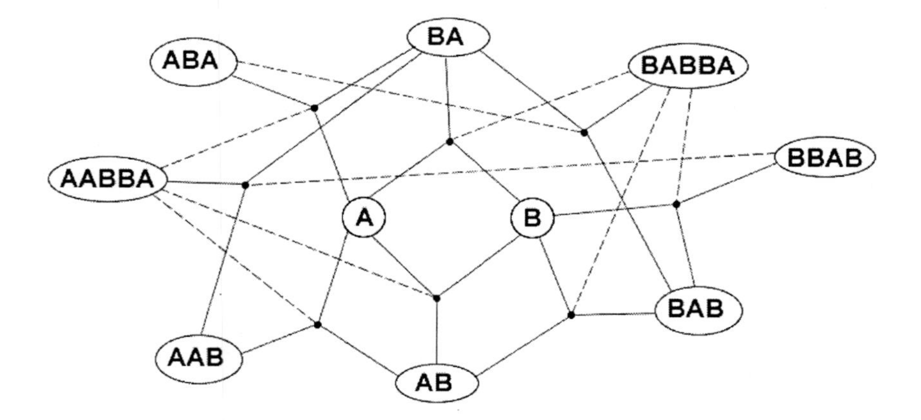

Figure 4: A very simple autocatalytic set. Each node represents a ligation/cleavage reaction comprising three elements. The dotted lines represent the catalytic action of a fourth element.

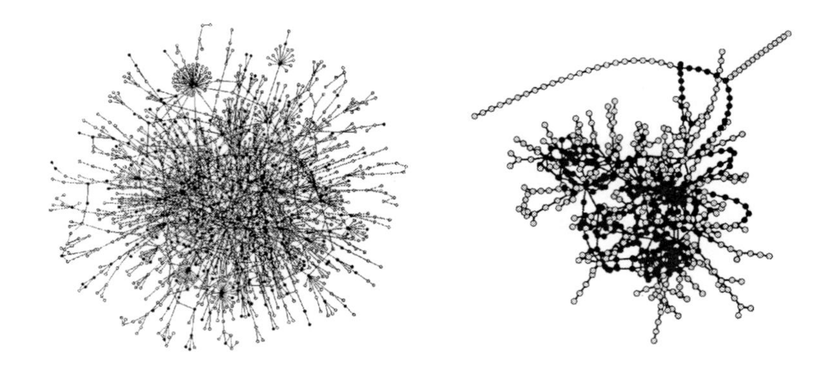

Figure 5: Map of protein-protein interactions. Left-hand diagram by Hawoong Jeong, right-hand diagram by Erzsebet Ravasz.[38]

The resemblance of autocatalytic sets to metabolisms and ecosystems is surely not coincidental. When ecologists draw a nodes-and-arrows graph of the interdependencies among species, it looks like a vast elaboration of the autocatalytic set pictured above. Moreover, as Kauffman points out, the modern cell is an autocatalytic set in which DNA, RNA, proteins, and various intermediate products all contribute to each others' synthesis. Just as in our autocatalytic loop above, it is arbitrary to say that the DNA is the replicator and the RNA, proteins, etc. just its tools. Instead of saying that our genes act collectively, by coding for proteins, to replicate themselves, why not say that the proteins act collectively, through catalyzing the chemical steps in DNA transcription and RNA translation, to replicate *them*selves?

Stuart Kauffman suggests a self that is cooperative at its very origin, yet his model still carries a subtle projection of separation. Is a cell really autocatalytic? What about a human being? No. At best we can say that each contains autocatalytic systems and systems-within-systems. Each requires a "food set" of molecules that it cannot produce itself. A human being cannot produce sugar from sunlight, nor the free molecular oxygen our metabolism requires, nor a number of essential amino acids, fatty acids, and vitamins. These substances and their sources are therefore excluded from self.

When we wall off part of an autocatalytic set to define an organism, we recognized a certain arbitrariness to that definition. We could just as easily have named the whole as the unit of life, or a smaller part, just as we could say in a human that our organs are alive, our cells are alive, our mitochondria are alive. True, none are viable on their own, but *neither are human beings*. Nor is any life form, at any level.

The biological definition of the organism, and therefore of the self, usually draws on the concept of the phenotype: the "expression" of the nuclear DNA. (Already this definition is problematic because DNA is expressed very differently under different environmental conditions.) However, the resulting organism is often no more viable—that is, no more capable of survival and replication—than an isolated human organ or cell. Most life forms are so utterly dependent on symbiotic relationships with other life forms as to call into question the validity of the phenotypic definition. Without the bacteria in their rumens, for instance, cows would be unable to digest cellulose and would quickly starve. Is the bacteria part of the cow, or a separate organism? Or is our identification of discrete organisms as the atomic elements of an ecosystem a source of confusion?

It may very well be that contrary to the selfish gene theory, in which the organism came first and eventually mutated, evolved, and generated the ecosystem, *it is in fact the ecosystem that is primary*, while organisms are merely semi-autonomous off-buddings that arose after the ecosystem developed to a certain level of complexity. If so, the quest to find the first "replicator" is a diversion, an artifact of a cultural prejudice about the nature of the self as a discrete, independently existing entity. If our boundaries of self were more fluid, as they were in other cultures, more open, less rigid in their demarcation of the universe into self and other, then perhaps we would not be so fixated on finding the replicator and better able to understand a different genesis story.

Kauffman's work suggests a story of life based much more on cooperation than on competition. In the simple example of the autocatalytic loop, each element is indispensable to the viability of the whole. Remove one and the whole system disintegrates. A more complex system such as an actual ecosystem is more robust—remove C, and there will be other pathways to get from B to D, or if not, from B to E—but the loss of one element, or one species, will typically still initiate a cascade of other losses causing the entire system to collapse into a simplified, but still viable, subset of the original. The system is cooperative in a another way, too:

the whole is greater than the sum of the parts. Split an autocatalytic set into two pieces, and both may be non-autocatalytic and hence incapable of maintaining themselves away from chemical equilibrium. Autocatalysis, and therefore life, is an emergent property arising out of complexity. Life on all levels is a collective.

This is an example—quite literally a living example—of the limits of reductionistic logic. As in any system with emergent order, there is something about the whole that eludes analysis, that cannot be understood by taking it apart. Poets have long known that something of a flower is lost when it is reduced to just so many botanical properties of stamen, sepal, anther, and petal, just as life is more than the collection of enzymes, fatty acids, DNA, proteins and so forth that make it up. The ideology of science has long been that higher-order properties, such as beauty, meaning, and love, are either human projections that are not really there, or simply collective terms for a number of lower-order "real" phenomena. What is happiness, *really*? Merely a set of biochemically determinate conditions: hormone levels, neurotransmitter levels, activation of brain regions, etc. Today, however, we can demonstrate mathematically that irreducible higher-order properties exist, and we can see them as well in living systems.

But my dear reader, when I say "life is more than a collection of enzymes, fatty acids…" please don't think I am proposing to add yet one more ingredient, an immaterial spirit to inhabit and animate the body. The truth is far more marvelous. I do not believe in an immaterial spirit, but I do believe in a material spirit! Spirit is not separate from matter, it is an emergent property of matter. On the one hand, yes, there is nothing more than the masses and forces of physics and chemistry,[39] but that does not mean emergent higher-order phenomena are unreal. Spirit is just as real as Langton's ant highways. It has real effects and real explanatory power, but when we take things apart to look for it, it isn't there. The soul is not just another ingredient in a living being, which is why experiments purporting to prove the soul's existence by weighing the body before and after death are misguided.[40] That doesn't mean it isn't real though! The same goes for other emergent phenomena like happiness, consciousness, beauty, and selfhood, which puts them forever outside traditional paradigms of engineering and control. As in any complex, non-linear, feedback-ridden system, monkeying with the parts can have unexpected or even contrary effects on the whole, which is why attempts to "manage" ecosystems are so problematic.

A technology of wholes rather than parts will look very different from what we have today, running contrary to the scientific intuitions of the last several centuries. If posed with an herb that has a healing effect, science would take it apart chemically in search of the "active ingredient". Happiness was to be explained not as a holistic state of the entire person, but isolated as a limited set of neurotransmitter levels, hormone levels, and related responses. Consciousness was similarly to be identified in a certain part of the brain, the "seat of consciousness". Soil fertility was to be explained reductionistically as well, in terms of percentages of a finite list of minerals and other ingredients. Science was based on distilling, purifying, extracting the active principal and separating out the dross.

Closely related to this goal was the program of control. For once we had grasped the active ingredient in pure form, we could wield it to manipulate reality to our liking. We could make the soil fertile, we could make the patient happy, we could make a medicine in pure form more precisely effective than the original herb containing it. Future technology will emerge from our understanding of all the things that vanish when you take them apart, and that are real nonetheless. It will be a technology of the emergent.

When René Descartes pondered the nature of the self, he took the very same approach of isolating the essential principle. In fact it is our concept of self, enunciated so clearly by Descartes, which has set the template for our investigations into the world as well. Separating out the dross, Descartes arrived at what he believed to be the essential, purified kernel of selfhood, a mote of am-ness separate from the body, the emotions, the sense-impressions, and the thoughts, but gazing down upon them, experiencing them but apart from them. Descartes' self is the audience for what Daniel Dennett calls the "Cartesian theater", viewing the play of thoughts, experiences, sense data, and emotions on the stage of the brain.

But like the original replicator, like the vital principle of life, when we take the self apart to find its essence, we discover that it too is not there. This has certainly been the case in neuroscience, which has found that properties such as consciousness and memory are not localized anywhere in the brain. Buddhism has arrived at an identical finding, that the self has no objective, discrete reality, but emerges from relationships spanning the entire cosmos; there are no separate individuals, all are interconnected, interdependent, interdefined.

The vision of life, the organism, and hence the self as an arbitrarily-bounded open subsystem itself composed of numerous interdependent sub-subsystems, and therefore without a discrete objective reality, conflicts with many of the founding assumptions of modern philosophy, medicine, economics, religion, law, and psychology. The new conception of self will give birth to momentous changes in all these areas. The Cartesian self is fundamental to the dualism that informs so much of modern thought: I think, therefore I am. Descartes believed in an irreducible kernel of selfhood or am-ness that is the true "I", discrete and separate. That conception of self is wholly consistent with the intuitions of the Age of Separation. Now we are discovering that at the very basis of life, no such self exists, and we shall observe that fact again and again, level after level, through the cell, the organism, the ecosystem, and the whole planet. The progressive alienation of ourselves from the community of life, our progressive distancing from nature, is based ultimately on an illusion. The Scientific Revolution gave voice to this illusion; today science is undermining its very foundations.

The Community of Life

While complexity theory has some serious flaws as an explanation of life's origins, many of its features strongly correspond to a newly emerging understanding of life and evolution. Recent discoveries in ecology and genetics confirm two of the basic implications of complexity theory: (1) That life is at its basis cooperative, and (2) that there is no absolute integrity of biological self.

These new discoveries imply a different mode of understanding that subverts Neodarwinism at its very foundation, calling into question the primacy of natural selection and random mutation as agents of evolution. While competition certainly may exist among members of an ecosystem or an autocatalytic set, each also depends on the others to survive. Each has its niche, its essential role and function, that the others would destroy at their own peril. There is no discrete replicating unit, and therefore nothing on which Darwinian laws can operate until much later, when semi-autonomous offbuddings of the set gain enough independence to compete as discrete individuals. But let me emphasize the modifier "semi", because no life form is ever fully independent of the rest of nature.

When we look at contemporary living systems, we find a situation similar to that of the autocatalytic set. Of course there is competition, but its primacy as the determining factor in behavior and evolution is exaggerated. We tend to find what we look for, and now that the cultural blinders of separation are beginning to fall away, scientists are discovering more and more the overriding importance of cooperation in biology. The extent of these cooperative relationships calls into question the very coherency of the concept of an individual organism, threatening, as Kauffman's sets do, to deprive Neodarwinism of the subject of natural selection.

Let's look at some examples of cooperation in nature, starting with our own cells. Each of our cells are inhabited by mitochondria, which have their own DNA. These bits of protoplasm are parts of our selves (and of all other animals) that are *not* "coded for" by our nuclear DNA. In a sense they are separate organisms—in fact most scientists now accept Lynn Margulis' theory that mitochondria were originally aerobic bacteria that merged with other cells. Yet without them we'd be dead, for they provide at least 90% of our energy. Even on the cellular level, we are cooperative beings.

On the coast of Brittany there is a flatworm, *Convoluta roscoffensis*, that has no functioning mouth or digestive tract. Instead, its transparent body hosts trillions of green algae who provide the worm energy through photosynthesis. In that protected environment, generations of algae live and die. They even process the worm's metabolic wastes! Another worm, a roundworm that lives near undersea vents, also has no digestive tract but harbors bacteria in a special organ called a trophosome. The bacteria produce energy from hydrogen sulfide gas collected by the worm. What kind of bacteria? No one has named them, because they are impossible to culture in a lab. They can only survive in the worm. Bacteria and worm are each wholly dependent on the other.

I have already mentioned the dependency of ruminant animals on bacteria to digest cellulose. Recently, evidence has accumulated that humans too require a complex intestinal ecology to thrive. Hundreds of species of yeasts, bacteria, and other organisms inhabit the healthy human gut in numbers far exceeding our own cells. They produce K and B vitamins, protect us from colonization by pathogenic bacteria and fungi, assist digestion, and interact with the immune system in ways that are not fully understood. There is even evidence that human beings are meant to host certain species of "parasitic" intestinal worms, which have been used to successfully treat hundreds of cases of ulcerative colitis and Crohn's Disease.[41] Not just our intestines, but all of our mucosa, skin, and even eyelashes are host to bacteria, yeasts, and microscopic insects that are not competitors drawing down our resources, but partners in health. Since we are partly or even wholly dependent upon them, no less than we are on our heart and liver, by what right do we exclude them from the definition of self?

Stephen Buhner gives many amazing examples of cooperation in nature in his beautiful book, *The Lost Language of Plants*.

> In Central America and Africa certain species of *Acacia*, a large shrub or small tree, is covered with thorns, some of which are hollow and house ants. Much like coevolutionary bacteria, *Pseudomyrmex* ants recognize new shrubs as coevolutionary partners and colonize them. The trees produce special nectar along the stems for the ants to eat. Like the compounds released from plant roots, this nectar contains a rich mix of fats (lipids), proteins, sugars, and other compounds necessary for the ants to remain healthy. The ants remove vegetation from around the base of the plant, remove leaves of other plants that shade the tree, kill any vines that try to grow up the tree, and attack any herbivore that tries to eat the plant.[42]

Which is the organism, the tree itself, or tree plus ants? Most plants could not survive without mycorrhizal fungi on their rootlets, which in turn depend on sugars provided by the plant. A similar relationship exists between legumes and nitrogen-fixing bacteria, except not only the organisms involved but the entire plant community depends on the fixing of nitrogen. This is typical. Cooperation in nature extends far beyond mutual dependency between plant and pollinator, flatworm and algae. More often, the dependencies are not one-on-one but highly distributed: thousands of species in mutually dependent partnership. Just as a mammal cannot survive without a heart or lungs, which justifies including these organs as part of "self", so also are these plants, insects, fungi and so forth unable to survive without their symbiotic community. Remove one element, and the whole system might collapse. Life does not thrive in sterile isolation.

Buhner gives a striking example of a community of life in his description of the ironwood tree of the Sonoran desert:

> Smaller plants begin to appear. Continually shaded from the desert sun, cooled by transpired water, and watered daily by hydraulic lift, some 65 species of plants will come to grow under ironwood... Thirty-one of these will grow nowhere else. This emerging plant community connects to the mycelial network and plant chemistries flow throughout the

> network. Wherever plant roots touch, they can share their chemistries directly. All the plants exude volatile aromatics. Some aromatics call in pollinators, others fall in a continual rain over the plant community and to the Earth below. The soil takes them up; the companion plants under the ironwood breathe them in. The smaller community plants cover the ground, keeping the soil moisture high. They all release their own unique mixtures of phytochemicals that blend together with ironwood's in maintaining the microclimate and soil community under the tree.
>
> As leaves, bark, and limbs age, they fall to the Earth, forming a layer of decaying matter. Over the centuries, the tree and its community build up a mound of detritus around its trunk and under its canopy, in effect becoming an island or archipelago of life and richness amid the desert—a facilitative nucleus of life. Scores of insects, birds, and animals come to the archipelago. They pollinate, spread seeds, build nests from archipelago plants, dig burrows, mate, aerate the soil, use plant chemistries in their growth, as their medicine, as their food, and contribute, over the years, tons of their own "night soil." Ironwood increases the abundance of life by 88 percent and species richness by 64 percent in any area in which it grows. Plants such as the endangered saguaro cactus can rarely germinate outside the kind of zone that trees such as ironwood create. Ironwood, and similar trees, literally create the ecosystems in which they and other beings live.

The ironwood communities are unusual only in their relative isolation. In other ecosystems, the same kind of relationships exist among keystone species, the nurse plants which help them get established, and subordinate plants which regulate the flow of life and energy through the community. Of course, the community as a whole, which also includes insects, bacteria, birds, mammals, and fungi, does exercise limitations on the proliferation of each member, and in the establishment and maintenance of these, competition surely comes into play. But it is a mistake to try to understand the functioning of the whole community in those terms. Instead of seeing competition as primary and symbiosis somehow arising out of it, it might be more illuminating to see competition as one of several ways by which resource flow is optimized within a living community.

Understanding of human communities is crippled in a similar way, by insisting on viewing all reciprocity in terms of competitive economics. Some anthropologists have tried to do just this by analyzing trade between individuals or tribes in terms of net caloric gain or net time gain. However, this kind of analysis is motivated by the implicit assumption that calories or time are scarce commodities (an uneven trade would be no cause for concern for people without a concept of time scarcity or food scarcity), and therefore harks back to the assumption that primitive life was a struggle for survival. Orthodox biology holds an identical assumption, and like anthropology, cannot fully apprehend systems in which competition is not primary. Something is always missing.

Of course, none of the communities of life described above are autonomous. All organisms and microsystems depend on the health of the ecosystem in which they live, and each ecosystem depends on other, distant ecosystems. And all higher life forms depend on the bacteria which maintain a life-supporting atmosphere. While life on earth can sustain the loss of some species, each species depends on the whole. None can exist in isolation on a bare and lifeless planet. We may thus also consider the only viable unit of life to be the entirety of all life, along with inorganic processes relating to the water cycle and carbon cycle. The conception of the organism as an autocatalytic set is misleading, because no organism is fully autocatalytic. Like the human depending on essential amino acids, etc., all life forms depend on the rest of life for their long-term survival. The only autocatalytic set is the entire planet. If that.

These examples of flatworms and mycorhizza are not anomalies, not some odd curiosity of nature. They are ubiquitous. Cooperation is everywhere. Life depends on it. Only the cultural blinders of Darwinian survival-of-the-fittest, the dog-eat-dog world, prevent us from seeing it. We live in a cooperative biological world, a living entity which we call Gaia.

Some might quibble with the characterization of earth as a living organism, but it does possess many features of one, most notably homeostatic regulation of temperature, gases, salinity, and other variables. Each species contributes some way to the metabolism and homeostasis of the planet. Coral creates lagoons that help remove salt from the ocean, which would otherwise double in salinity in just 60 million years. Photosynthesizing algae and rainforests produce the oxygen that sustains animal life, while other mechanisms prevent oxygen levels from going too high and sparking devastating planet-wide forest fires. Bacteria accelerate rock weathering to bring carbon out of the atmosphere, while marine animals eventually turn that carbon into shells ultimately sequestered on the ocean floor. And something, some combination of organic and inorganic processes, has kept earth's surface temperature stable as the sun's apparent brightness has increased by 30 or 40 percent over three billion years. Gaia maintains homeostasis, responds to external stimuli, and grows, if not in size at least in complexity.[43] The only attribute of a living being that Gaia does not possess is, we are told, the ability to reproduce.

Evolution is often depicted as an arms race, with plants developing ever more sophisticated chemical defenses against predation, while the insects that feed on them try to adapt to those defenses or face starvation. While examples of this do occur in nature, as between cheetah and gazelle, it is actually an unusual situation that we take as typical only because that's what we look for. Far more typical is the relationship between the Douglas fir and the spruce budworm. During periods of very light infestation the tree produces no response, but when budworm numbers begin to grow the trees alter their terpene releases in a way that interferes with budworm feeding and reproduction.[44] The trees don't try to eliminate the budworm and other pests and take over the earth, but merely help maintain budworm populations at the right level. Many other plants do the same thing, tolerating moderate foraging but responding aggressively to severe infestation. Others possess compounds that are toxic only in large quantities, such as phytoestrogens that interfere with grazers' reproduction if eaten to excess.

With few exceptions, modern human beings are the only living beings that think it is a good idea to completely eliminate the competition. Nature is not a merciless struggle to survive, but a vast network of checks and balances that ensures each species occupies its proper place. Indeed, the extinction of any species usually has negative consequences that spread throughout the ecosystem, often to the detriment of even its former prey. Are the deer better off when they are finally free from the tyranny of the wolves? Only if you think starvation, bark stripping, and the degradation of the entire forest ecology are an improvement.

The study of ecology leads toward a view of nature as a vast gift-giving network, rather than a competitive, accumulative network. In his classic work, *The Gift*, Lewis Hyde observes that in primitive cultures it was in the essence of a gift that it had to be passed on or consumed—many cultures actually used the word "eaten". Gifts were not accumulated. Similarly, each species, each organism, has something to give to its environment, through which resources flow freely. Even in the case of predation, locutions like, "The deer gave itself to the wolf" reveal an unconscious insight, that underneath the very real life-and-death struggle there is a fateful and intimate connection between predator and prey.

Just as hunter-gatherers did not accumulate possessions, animals and plants don't try to make the world theirs by taking over ecosystems and wiping out other species. (Opportunistic weeds may "take over" an area for a time, but soon give way to more complex ecosystems. Their rapid initial takeover might be a gift to the community as well, for example by stabilizing denuded soil and preventing erosion.) In a gift-based world, the needs of the rest of the community define a purpose to life. Instead of a struggle to survive, life is an aspiration toward excellence in the role presented to each organism, or each person.

It was only with the advent of agriculture that human beings began to think in terms of eliminating the competition: weeds, wolves, and insect pests. What of the deer problem? We will cull the herd and manage deer populations. What of the diseases that afflict monocultures? We will manage them with chemicals. The project of eliminating the competition coincides with the ambition to order and manage nature, culminating in the total mastery of nature that is the fulfillment of the Technological Program. Or as a *Scientific American* cover once put it, "Managing Planet Earth."

Today, as the convergence of crises renders life increasingly unmanageable, we are realizing the bankruptcy of this ambition. We need only to look to nature to see the false premises on which it is based. Projecting our own alienation, competitiveness, and survival anxiety onto biology, we have seen life as a Steinerian war of all against all, but nature, like primitive society, is not like that. Yes there is competition and there are times of anxiety, hunger, and life-and-death struggle, but these are merely a few of the strands of existence, not its warp and woof. The entirety of life, Gaia, makes room for each species until its time has passed. We live, as Lynn Margulis puts it, on a symbiotic planet.

All of this suggests a very different way of relating to nature, which is another way of saying that it suggests a different mode of technology. Technology, ever directed at the management, subjugation, and supercession of nature, will be turned toward fulfillment of our proper role and function *within* nature. I will share the outlines of such a mode of technology in the next chapter, fleshing out a vision of the future that holds for humanity a beautiful destiny, an *ascent* which nonetheless explicitly rejects today's longstanding consumptive, linear view of "progress". Perhaps a better word for this ascent is "fulfillment".

The Genetic Plenum

A major flaw of complexity theory as an account of life's origin is that it never works in a test tube. Critics of Stuart Kauffman point to the failure in the laboratory to create anything like the expanding, evolving autocatalytic sets he describes; invariably, instead, we end up with uninteresting tarry gunk sitting on the bottom of the flask at chemical equilibrium. Computer simulations of evolution, where various mutating organisms compete for memory or some other resource, similarly fail to generate genuinely new genes. Artificial life programs like Avida and Tierra appear to simplify and prune genomes rather than generating novel ones. Interesting variations on the original creatures emerge, but not to my knowledge truly new organisms at a higher level of complexity.[45] The only exceptions I could find are systems like Ray's Tierra when length is specifically rewarded in the program environment, and Dawkins' simulations where fitness is defined by proximity to a predetermined form. Ironically, such systems model Intelligent Design, not the undirected evolution Dawkins and Ray believe in.

Yet leaps in complexity do happen, not only in nature but in human civilization. In direct analogy to Lamarckian adaptive mutation, technological development is not a random search of the entire possibility space of new inventions, that are then selected according to fitness; rather, an awareness of purpose, a goal, an intention guides trial-and-error and narrows the search space. Scientists believe that such intentionality is only the province of human beings, not genes, not ecosystems, and not the planet as a whole. Creationists believe the same thing, except they ascribe intentionality as well to a supernatural intelligence. No one says that because human society is so complex, interdependent, and tightly coupled, it could never have evolved but must have been created this way!

The chicken-and-egg problem of irreducible complexity that plagues selfish gene biogenesis repeats itself on all levels of evolution. Generally speaking, a single mutation in an existing gene cannot produce a new gene with a different function. If two genes differ by only a single base pair, then a single point mutation could convert one into another. Usually, however, these two genes will be regarded merely as variants of the same gene, and will have an identical function unless the mutation is at some critical spot that renders the second gene non-functional. The same goes for other basic alterations, such as the deletion or repetition of an existing sequence. To get a gene with an entirely new function usually requires many, many alterations, a concatenation of several unlikely steps—a series of just the right mutations happening either all at once, or one after another. Unlikelihood multiplies into impossibility. If I guess the next card you draw from the deck, you'd be impressed but not amazed. If I guessed ten in a row, you'd suspect a trick because that would happen only about once every fifty quadrillion trials.

Well, maybe it isn't so unlikely given billions of years for evolution to happen, right? Wrong. To give you some idea of the numbers involved, in 1997 a gene for an anti-freezing protein was discovered in an Antarctic fish that is extremely similar to another gene in the same fish that codes for a pancreatic enzyme—an entirely different function.[46] The two genes are so similar that indeed, a mere handful of mutations (a deletion, a duplication, a frameshift, an insertion of a short intron, and the amplification of a formerly non-coding spacer sequence) can take us from one to the other. However, if we assume that these mutations were totally random, the number of possible genes that can be made from the original by those steps is astronomical—one in 10^{370}, according to one author.[47] And remember, these two genes are extremely similar: for most genes the number of steps to convert one to another is much greater.

The Neodarwinist solution to the problem is to postulate that each of the intermediate genes has some kind of function that benefits the organism's survival, allowing a gradual evolution from a warm-water fish to an Antarctic fish. In general, though, getting halfway to a new gene won't give you half of its new function. It's usually all or nothing—or less than nothing, as intermediates might be useless for both the old and the new functions. The authors of the antifreeze paper were aware of this problem, and in a remarkable (though certainly unwitting) appeal to teleology wrote, "The selection of an appropriate permutation of three codons... was likely shaped by the structural specificity required for antifreeze ice interaction to take place."[48] In other words, they realize it couldn't have happened purely by chance, nor is it plausible that each intermediate conferred a survival advantage.

The problem is actually even worse than that. In most cases, a new gene with a new and genuinely useful function is only useful in the presence of a number of other new genes that must be expressed at the same time. For example, whatever mutation gives a giraffe a long neck would be fatal in the absence of some other gene to give it a special nasal structure and blood vessel network to cool the brain, and another to create the vascular adaptations that regulate blood flow to the brain when the giraffe lowers its head to drink water.[49]

The Intelligent Design proponent Michael Behe gives similar examples in his book *Darwin's Black Box*. Most striking is the example of the synthesis of AMP, which is essential to all life. Behe describes thirteen steps of AMP synthesis requiring twelve separate enzymes, each of which is basically useless without the other eleven. Even if one of the requisite genes emerged through random mutation, in the absence of an adaptive advantage, why would it remain in the gene pool instead of any of trillions of other, equally likely mutations? A parallel situation pertains to blood coagulation, which requires numerous proteins and other substances that are highly specific to their tasks and useless outside of them. Both exemplify irreducible complexity, of which Darwin wrote, "If it could be demonstrated that any complex organ existed which could not possibly have been formed by numerous, successive, slight modifications, my theory would absolutely break down."[50]

The extreme difficulty in using gradual, random genetic mutation to account for the complex, tightly coordinated systems of biology, added to the fossil evidence for sudden evolutionary jumps, is slowly eroding support for the conventional Neodarwinian synthesis. A new paradigm is emerging to replace it, one which subverts many of our cultural assumptions about the nature of life. Part of this paradigm shift relies on the adaptive mutations described earlier. However, irreducible complexity poses difficulties even for adaptive mutation, because not one but many genes for complex interdependent processes must all appear at the same time. Alternatively, they can appear at different times, but then they must somehow be preserved in the genome without help from conventional selective mechanisms (because they have no beneficial expression). Either way, some coordinating influence is necessary to guide each adaptive mutation toward an outcome that meshes with all the others, whether this force is supernatural (Behe), an environmental purpose (Lamarck/Buhner/Lipton), or the observer effect of the system's own evolutionary future (McFadden).

Perhaps the difficulty again rests upon our assumptions about the nature of self. Adaptive mutation, even when triggered by environmental "purposes", still preserves the gene as a unit of selfhood, albeit an evolving unit. But there is more. While adaptive mutation is certainly part of the story, perhaps the sheer complexity of the cooperative systems described in the last two sections demands an account of evolution that builds cooperation in to its fundamental mechanisms—a cooperative account of evolution to complement the cooperative nature of life.

According to Lynn Margulis, cooperation (i.e. symbiosis) is not only essential to the survival of all life on earth, it is also the driving force of evolution. Her theory of serial endosymbiogenesis explains the evolution of the modern eukaryotic cell as the progressive incorporation of simpler organisms. In her view, genuine novelty in evolution comes through the merging of simpler organisms, and not through the mutation of DNA. Organisms—including higher organisms—evolve through the incorporation of external DNA. This has been well-documented for half a century in the case of bacteria, which have been closely studied to determine how they achieve resistance to antibiotics. While occasionally a point mutation will confer resistance, for example through the alteration of some surface protein, normally bacteria import resistance-encoding genes from other bacteria via viruses, conjugation, and other means. And it's not just resistance. Recent studies have demonstrated that the genes for photosynthesis are also transferred horizontally among bacteria.[51,52] This phenomenon, named horizontal gene transfer (HGT), was observed in plants as long ago as the 1940s in Barbara McClintock's pioneering studies of corn. More recently, evidence has mounted that HGT is also common in multicellular animals, even in humans. The following press release describing bacterial photosynthesis might therefore be generalized to all evolution:

> The analysis revealed clear evidence that photosynthesis did not evolve through a linear path of steady change and growing complexity but through a merging of evolutionary lines that brought together independently evolving chemical systems — the swapping of blocks of genetic material among bacterial species known as horizontal gene transfer.[53]

A major piece of evidence for HGT in nature is the presence in unrelated organisms of similar DNA sequences that, based on molecular clock evidence, cannot be explained through common ancestry. While viruses are probably the main vector of gene transfer, there is also laboratory evidence for direct bacterial transmission of DNA into mammalian cells.[54] Other possible vectors include, in the words of one researcher, "External parasites, infectious agents, intracellular parasites and symbionts (especially those in the germline), DNA viruses, RNA viruses, retroviruses, even hitchhiking in other transposable elements."[55]

The primary vectors of HGT, whether viruses or other parasites, are not merely germs or invaders but carriers of genetic information from organism to organism and species to species. In one study, researchers found that DNA from *T. cruzi*, a protozoan which causes Chagas disease in humans, had been incorporated into the nuclear DNA of laboratory rabbits and passed onto their offspring, thus having entered the germline.[56] Some scientists have even put forth evidence that the incorporation of viral DNA initiated the speciation event in which human beings and chimpanzees diverged six million years ago.[57]

These are not controversial findings by fringe scientists, but are reported in prestigious mainstream journals, as a cursory examination of this section's footnotes will confirm. It is becoming increasingly apparent that eukaryotic genomes, including humans', are riddled, perhaps even dominated, by the remains of viral DNA incorporated into the germline millions of years ago.[58,59] As Lynn Margulis puts it, "We are our viruses."

But let's not limit ourselves to the uncontroversial. Perhaps the integration of exogenous DNA into the human genome is on-going and purposive. The great epidemics of agricultural civilization may have contributed important genes bearing adaptive utility for humans living in dense populations and carrying out civilized lifestyles. In particular, nonlethal viral diseases like measles, mumps, and chicken pox, which were pandemic until the era of vaccines, might comprise an important alternative transmission system for genetic material. Perhaps they are our symbionts, having adapted with us into a mutually beneficial relationship. If so, eliminating them could be dangerous. The precise consequences are hard to predict, but perhaps it has something to do with the alarming rise in autoimmune diseases and allergies over the past thirty years.

Orthodox genetics maintains, because of the slowness of evolution, that we are essentially the same species as our Stone Age ancestors 20,000 years ago. But if viral DNA is regularly incorporated into the germline, it is conceivable that the diseases of civilization are an agent of speciation, triggered by the population concentrations that come only with agriculture. Could it be that humanity is undergoing speciation right now? And if, as Bruce Lipton argues, our emotions, thoughts, and beliefs can modify our DNA, could speciation even be a matter of choice? Do we have the opportunity to literally become what we choose to be? I realize I have entered the realm of speculation, but even so, the metaphoric significance of this possibility is profound. We are not trapped into what we are by our biology. We can acquire new biology in the course of a lifetime. Neither are we trapped by our other inheritances. We are free to create ourselves.

On the level of the organism, HGT offers a model of evolution that does not primarily depend on random mutation, but rather on the integration of already existing external DNA. It offers a solution to many of the problems related to irreducible complexity, because it does not rely on the simultaneous occurrence of many highly unlikely mutations at just the right time. The necessary genes need only be imported, even sequestered in non-expressed form until needed. The ubiquity of HGT also suggests a new way of relating to other life forms such as bacteria and viruses (a matter I will explore in Chapter Seven) as well as a new model of progress outside the biological realm.

Horizontal gene transfer removes the biological underpinnings of the ideology of the discrete and separate self. It suggests a new self, a new identity that might be described as "interbeingness". This is a much more intimate relationship than mere interdependency among life forms. Thanks to HGT, we are all incorporated into each others' being. As we shall see in the next chapter, the spiritual and metaphorical implications for human technology, medicine, and economics are profound.

Horizontal gene transfer may explain macroevolution, but the origin of novelty on the genetic level remains a mystery. The transferred genes had to come from somewhere. If HGT is primary, it implies that many genes existed millions of years before they were ever expressed. Indeed, single-celled choanoflagellates have been found with genes for proteins previously thought to exist only in multicellular animals, and with no known function in the choanoflagellates.[60] Another, more general mystery is the appearance of homeotic genes (which coordinate embryological development) long before the genes for the structures of the features they coordinate. For example, genes that coordinate the development in embryos of the eye are older than the genes for the proteins that make up the eye. Brig Klyce sums it up quite nicely: "It is difficult for Neodarwinism to explain the appearance of embryological coordinating genes before the appearance of the embryological steps they coordinate. It's like saying that the blueprints for automobile manufacturing plants were on hand before the invention of automobiles."[61] In both the above cases, there is nothing on which Darwinian natural selection could operate, no reason for the genes to persist. What survival advantage could they confer when they have no function?

That genes persist in the genome despite having no beneficial expressed function flatly contradicts Darwinian evolution: what would select such genes over trillions of useless variants? Oddly, ultraconserved elements—strings of 200 or more DNA base pairs—have been discovered that are identical across the genomes of rats, mice, humans, dogs, and fish, but which do not code for any protein and do not exhibit any regulatory function. Indeed, when these sequences are removed from mouse genes, the mice appear to be normal in every respect.[62] What selective mechanism, then, could preserve these sequences, virtually mutation-free (far below the "background" mutation rate), for hundreds of millions of years? Why are they so important if they don't affect survival and reproduction? Clearly, other selective mechanisms are at work. The purpose of life is not, as Darwinism implies, to survive.[63]

Brig Klyce speculates that such genes are actually "high-level software capable of recognizing, installing, assembling and activating horizontally acquired genetic programs."[64] As such they are agents of evolutionary purpose serving needs transcending individual organisms and their genes.

Klyce is an advocate of Panspermia, the theory pioneered by Fred Hoyle and Chandra Wickerhamsinghe that says that life on earth was seeded with biological material from outer space. Aside from the mysterious origin of genetic novelty, there are indeed indications of an extraterrestrial origin for life, such as the discovery of complex organic molecules on meteorites, and spectroscopic analysis of interstellar dust consistent with the presence of bacterial spores. Of course, on the cosmological scale Panspermia is not a satisfactory explanation for the origin of life either, since it just pushes the question back in time to before the earth was formed. If life didn't start on earth, it had to have started somewhere... right? Very quickly (only three or four earth lifespans) we run up against the Big Bang, posing the same combinatorial plausibility problems as before. That is why Panspermia advocates usually subscribe to alternative cosmological models such as the steady-state universe. Life in that case never began; it has always been here, a property of the universe.

To the linear mind, that life has always existed, beginningless, is a mind-blowing concept. (At least, it blows my own linear mind!) Unlike primitive people, who thought in terms of cycles and lived in a timeless world, the modern mind understands all things, including the universe, as having a beginning and an end, just as our way of life is based on the linear consumption of raw materials that end up, finally, as useless waste. It is surely no accident that standard cosmology posits precisely this model: the universe began in a state of low thermodynamic entropy and ends in "heat death" when all usable energy has been consumed. The Second Law of Thermodynamics, that entropy always increases, is the ultimate statement of the ideology of domestication: nature is bad, a force to be resisted. We establish a separate human realm, a realm of order, maintained only through constant and eternal effort against the forces of chaos. Can you feel the despair implicit in the Second Law? No more possible is it to overcome entropy, than to build a tower all the way to Heaven.

How different a universe it is that is continually created, in which new matter and new order is continually being born in perfect counterbalance to the entropic tendency toward death. It would be a living universe, a fecund universe, a universe consistent with the hunter-gatherer's view of nature as endlessly provident. We don't need to establish a separate human realm. We can instead turn to the extension of the natural realm. No longer need we strive to build a Tower to the sky, when we realize that the sky is all around us already. Order, creativity, birth are in nature, woven into mathematics and physics and biology. Instead of building toward an unattainable and ever-receding Heaven, we turn instead toward another kind of edifice designed for beauty instead of height.

If humankind is now shifting toward a way of life that no longer denies nature's cycles, then perhaps the Big Bang will lose its intuitive appeal. (There are already cracks in the Big Bang facade, most notably in the work of astronomer Halton Arp and plasma physicist Hannes Alfven.)[65] Deep cosmological and philosophical issues are embedded in the debate over life's origins and evolution: Is the universe finite or infinite? Created or eternal? Random or purposeful? Or could it be, perhaps, that even these dualities will ultimately collapse?

My problem with Panspermia is that just as Intelligent Design postulates a divine plan (i.e., from *outside* nature) for evolution, Panspermia explains away the miracle of genetic novelty by saying it arrived from outside the earth. Both agree that such an irreducibly complex phenomenon as life could not arise spontaneously. Both agree that the genes for each jump in evolution already existed, whether in outer space or from the mind of God. But following their logic, are we to believe also that the emergent complexity of human society, economy, culture, and the noosphere couldn't have *just happened* either? As Ilya Prigogine observed in the 1960s, order and even organization—new structures—emerge spontaneously in any non-linear system. Why not life? Like the created universe of ID, Panspermia's "steady state" universe is one which is not inherently creative, but in which creativity is an already-accomplished fact. That conflicts with my deepest spiritual intuitions. Is not life the very essence of creativity?

On the other hand, Panspermia has much in common with the complexity theory arising from Prigogine's work. The universe, they both believe, is abundant with life. Like Stuart Kauffman's theory, Panspermia implies that life is extremely likely, if not inevitable, on any planet within a fairly broad range of parameters. If the universe is replete with life, then the chances of its seeding earth are also very high. If on the other hand Stuart Kauffman is wrong, and life is an incredibly unlikely fluke, then the chances of its having originated somewhere else and found its way to earth are equally unlikely. Both theories agree that, one way or another, ours is a living universe.

In this statement lies the solution to the serious difficulties that complexity theory encounters in the lab. Just as order arises out of chaos, it is equally true that, as I wrote in the context of symbiosis, "Life does not thrive in sterile isolation." It is in the nature of a controlled experiment that it must be a closed system—you cannot leave the beaker open and claimed to have created new life inside. Perhaps what is missing from laboratory and computer experiments aimed at creating life is a seed, a seminal power that originated at the beginning extraterrestrially. Perhaps, indeed, life cannot originate or persist in a closed system; that in addition to an input of energy it needs an input of information; that no one part of the universe can be alive unless the entirety is also alive. Just as we humans are neither discrete nor separate from the rest of life on earth, perhaps Gaia herself is a dependent, semi-autonomous off-budding of a living universe.

It is possible that both theories are true, that life on earth is of dual parentage: a female parent, the earth, and a male parent, the sky. Ours is a living universe. No less than a gene, an organism, or an ecosystem, perhaps Gaia herself is also dependent on an open, fluid interaction with the outside. By sealing the beaker in origin-of-life experiments, we create conditions that are fundamentally anti-life.

Metaphorically, we have also replicated the controlled conditions of the laboratory in the modern "world under control." Believing ourselves fundamentally separate from a mechanistic universe of inert matter, we have sought to insulate ourselves from its vagaries, to bring it under control through the Technological Program. That this program, too, is anti-life is becoming more and more apparent. The reduction of life—

literal in our devastation of the ecology, figurative in the conversion of human life into money—is reaching its apogee. Fortunately, the new paradigms that this chapter has gathered from every corner of science invite us into a new way of relating to the universe that is not anti-life. What kind of society will emerge with the fall of the Newtonian World-machine? How will the human spirit blossom, when we no longer enforce the artificial boundaries of the discrete and separate self? What will human life look like, when we cease seeing it as a competition for survival in a objective universe devoid of purpose, sacredness, and meaning? The second Scientific Revolution that I have outlined is but part of a vaster sea-change in our own self-conception, a symptom as well as a cause for a new way of relating to the world. It marks a turning point, the end of the Age of Separation and the beginning of a new Age of Reunion.

CHAPTER VII

The Age of Reunion

The Convergence of Crises

For people immersed in the study of any of the crises that afflict our planet, it becomes abundantly obvious that we are doomed. Politics, finance, energy, education, health care, and most importantly the ecosystem are headed toward near-certain collapse. During the ten years I've spent writing this book, I have become familiar with each of these crises of civilization, enough to get some sense of their enormity and inevitability. Every year I would wonder whether this might be the last "normal" year of our era. I felt the dread of what a collapse might bring, and visited the despair of knowing that our best efforts to avert it are dwarfed by the forces driving us toward catastrophe.

One of the main purposes of this book is to speak to that despair. In answer, I offer a plausible and unexpected optimism. It is not a blind optimism that ignores the magnitude of our crisis, but a practical one that sees and integrates all the ugly facts of our world. It is optimism fully aware of the horror and suffering that are as old as civilization and that are approaching their feverish crescendo in the convergence of crises that is almost upon us.

First and foremost, I am aware of the environmental crisis: climate change, desertification, coral bleaching, tree death, topsoil erosion, habitat destruction, irreversible loss of biodiversity, toxic and radioactive waste, the PCBs in every living cell, the vast swaths of disappearing rainforests, the dead rivers, lakes and seas, the slag heaps and quarry pits, the living world reduced to profit and pavement.

I am aware of Peak Oil and the dependency of all aspects of our economic infrastructure and food supply on fossil fuels. And I realize that no conventionally-recognized alternative energy source can possibly hope to replace oil and gas any time soon.

I am aware of the impending health crisis: the epidemic rise of autoimmune diseases; the causes and effects of heavy metal poisoning, electromagnetic, chemical, and genetic pollution. I am aware of the degeneration of the modern diet, the toxicity and impotence of most pharmaceutical drugs, the suppression of alternative therapies.

I am aware of the fragility of the global financial system, a Ponzi scheme five hundred years in the making, and the hyperinflation, currency collapse, and depression waiting for the day when the American debt pyramid can no longer be sustained.

I know about the political trend toward fascism and the surveillance state, the barely-concealed contingency plans for martial law, the concentration of media, and the ubiquitous propaganda machine—so successful that even its operators are unconscious of it—that legitimizes the hijacking of government for profit in the world's most heavily armed nation. I know of the machinations of the global power elite, its addiction to the arms, narcotics, and prison industries, and its increasing desperation to hold onto power as Things Fall Apart. Nor do I imagine that we can cleanse the system by removing a few bad apples, because I understand that nearly every important institution of our civilization is complicit in this plunder.

I am aware that the bloody terror that has long stalked civilization remains essentially undiminished today. My optimism does not ignore the Tutsi babies smashed to death against walls in front of their mothers. It does not ignore the genocide repeated throughout history, with no end in sight. Nor I am ignorant of the widespread torture happening in prisons and police stations around the world, including some my own country operates.

My optimism is cognizant of the ongoing destruction of the world's languages, cultures, and communities at the hands of the all-consuming global corporate-consumer monoculture. I know about factory working conditions in the Third World, and the utter destitution and despair that prevail there. I have met people working eighty-hour workweeks, barely subsisting, so that the developed world can suffocate under useless piles of plastic junk that contribute to its own brand of misery.

I am aware that even the winners, the most privileged under the present global regime, are actually among its most pitiable victims. I have witnessed the alienation, despair, and loneliness of the very rich, whose acquisition of mansions, money, sports cars, wealth, prestige, and power do nothing to fill the inner void.

I am as aware as the most outspoken radical that our schools have become much like prisons, complete in some cases with barbed wire barricades, metal detectors, random locker searches, uniforms, prohibitions on personal articles, drug testing, armed guards, undercover police, video surveillance, and chemical control over those who will not submit. I know of parents threatened with legal action for failing to medicate their children.

Nor does my optimism depend on the god Technology come to save us, for I am aware that technology has abetted the despoliation of the planet, the acceleration of the pace of life, and the dehumanization of the world as it is converted into money, property, and data.

So please do not think that I am an optimist only because I do not recognize the true dimensions of the crisis. I have integrated it all, and remain an optimist. The kind of optimism associated with a blithe disregard for the fact that many people today live in Hell and are creating more Hell is no optimism at all, but mere fantasy. It is the kind of optimism that paves over a toxic waste dump and forgets it was ever anything but a parking lot. Out of sight, out of mind.

As the parking lot metaphor suggests, false optimism of this sort is actually a recipe for disaster. My optimism is of a different sort, independent of the logic of the technological fix, which says that these problems swept under the rug don't matter—that science, after all, will find an answer, that technology will find a way. By now, I think most of us see through that, as the long-awaited technological Utopia recedes ever farther into the future.

It is not my purpose to persuade you that we indeed face an environmental, financial, political, energy, soil, medical, or water crisis. Others have done so far more compellingly than I could. Nor is it my aim to inspire you with hope that they may be averted. They cannot be, *because the things that must happen to avert them will only happen as their consequence.* All present proposals for changing course in time to avert a crash are wildly impractical. My optimism is based on knowing that the definition of "practical" and "possible" will soon change as we collectively hit bottom.

Another way to put it is that my optimism depends on a miracle. No, not a supernatural agency come to save us. What is a miracle? A miracle comes from a new sense of what is possible, born from a surrender of the attempt to manage and control life. In individual experience miracles often happen when life overwhelms us. For an alcoholic, to suggest "just stop drinking" is ludicrous, impossible, unimaginable. It takes a miracle. The changes that need to happen to save the planet are the same. No mainstream politician is proposing them; few are even aware of just how deep the changes must go.

When the above-mentioned crises converge, when we experience acutely and undeniably that the situation is out of control, when the failure of the old regime is utterly transparent, then solutions that appear hopelessly radical today will become matters of common sense.

And this *will* happen. The timing of each crisis is uncertain, but the forces driving them are inexorable and cannot fail to be expressed sooner or later. Processes set in motion long ago have accelerated past critical mass; we are just beginning to taste their effects. Even if we somehow stopped making new pollution right now, the cumulative effects of existing ecological damage are enough to generate catastrophe. The same inevitability applies to other realms as well: public health, education, finance, and politics. It is already too late. It is only a matter of how soon, how bad, how long. However bad you think it is, it is probably worse. Read books like *The Dying of the Trees* or *Boiling Point* if you don't believe me.

Like the Titanic, the momentum of technological society is so huge that even if we reversed the engines and steered hard right now, the short-term and mid-term course of events would not change much. We are on a collision course with nature that can no longer be averted. Yet not only have we done little to brake or steer away from the looming iceberg, we have maintained an oblivious policy of "full speed ahead!" In the United States, Republican policy has been essentially, "What iceberg?" while the Democrats try to change course by a few degrees—but not so quickly as to spill the drinks on the first class deck. The "practical" proposals and workable compromises on the table are woefully inadequate. One party repudiates the Kyoto Treaty and the other endorses it, but few acknowledge that even that is far too little, far too late. Outside the United States, "developing" countries such as India and China, abetted by Western institutions, stoke the Titanic's furnaces with their headlong industrialization using the old linear model of extraction, processing, consumption, and waste.

And meanwhile, on deck the party continues, as it will continue to continue even after the first crunch reverberates through the ship, even as the icy torrent consumes compartment after compartment. On the top deck the band will play on even as the ship lists and rolls, maintaining a desperate and deadly illusion of normalcy.

At this point the utter bankruptcy of the program of competition, security, and financial independence will begin to become so flagrantly obvious that no one will be able to ignore it. I once read a pessimistic book of the business genre forecasting a polarized society of crime-ridden slums and wealthy walled, gated, fenced, alarmed, guarded communities. The author's advice was to contrive to live in the latter! This is tantamount to climbing to the highest deck of a sinking ship. Everyone speaks of the intensifying competitiveness of the present era, evoking in my mind masses of rats struggling and clawing for the top—where they will perish but a few minutes later than the rest.

Yes, you can locate yourself as far as possible from the war zones, trash incinerators, toxic waste dumps, smog zones, bad neighborhoods, and other perils of an increasingly toxic world, but sooner or later the converging crises of our era will obliterate all defenses. No matter how diversified your investments, no matter how many guns in your walled compound or cans of food in your basement, the tide of calamity will eventually engulf you. Gates, locks, razor wire and guns can ensure security only temporarily, and a fraudulent, anxious security it is. Eventually we will abandon our bunker mentality and understand that the only security comes through giving, opening, and being at the center of a flux of relationships, not taking more and more for self; security comes not from independence but from *inter*dependence. The survivors will not be those who try to insulate themselves in a fortress, but who are able to give, to help, and to contribute to a community. They will form the basis of a new kind of civilization.

Our crises are converging within a ten or twenty-year timespan because they are related. Each helps precipitate the others, even as each arises from the common source described in this book. We can see harbingers of what is to come when politicians use terrorism as a pretext to abrogate civil liberties and intensify the recording and controlling of people's activities. In Asia the same thing happened during the SARS crisis, as it might here under contingency plans for influenza or smallpox epidemics. And historically, the connection between economic collapse and political fascism is well-documented. Already drought and ecological degradation are beginning to generate refugees; imagine the political instability that could result when today's localized disasters coalesce and swell into regional or global environmental degeneration. Nations are already at war over oil; imagine the consequences of shortages not just in energy but in food and water too.

The foregoing doom-and-gloom scenario might seem familiar in tone if not in details, but consider that it may be not just The End but a Beginning as well, a birthing crisis that will propel us into a new age based on a different sense of self. This is not to say we can sit back and wait for the birthing to happen. Despite the inevitability of our gathering crisis, the seemingly futile efforts of generations of activists to avert it are extremely important. If you are such a person, facing down despair to tackle impossibly huge problems, take heart that your work is not in vain. While it is true that no effort at renewable energy, wastewater recycling, local currency, wetlands preservation, or reform of any aspect of society is going to avert catastrophe, these efforts are sowing seeds for the planetary renewal that can happen after the present regime collapses, after the addict has hit bottom upon the exhaustion of his very last technical fix.

All the technical solutions for living sustainably and harmoniously exist already, and indeed always have existed. What is required is a shift in consciousness, a reconception of ourselves as individuals and as a species that will reverse the widening separation and deepening misery of the past millennia, but that, paradoxically, will only come as their result.

The shift in consciousness I speak of is not predicated upon any sort of technological invention, nor does it insist on a regression in our technological level. Once it happens, though—and it is already beginning to happen—vast technological consequences will proceed as a matter of course. Visionary people are pioneering these material and social technologies today, in response to the increasing futility of the old modes of management and control. That their inventions are not adopted on a wide scale simply means that the requisite shift in consciousness has yet to manifest. They are simply inconsistent with who we are today. We already know everything we need to know—it is just a matter of growing into it.

The mechanisms by which society suppresses the technologies of sustainability all rest on the same delusion that is at the core of this book: the discrete and separate self. In economics, this delusion manifests as interest and the externalization of costs. These in turn present an insuperable barrier to processes based on cyclical flow. In science, the same delusion blinds us to other conceptions of what is possible and practical, generating barriers to the acceptance of new understanding and new technologies. In medicine it focuses research within the old us-versus-them mentality that is constitutionally unable to grapple with the new autoimmune diseases, and classifies other modalities as unscientific. In the areas of politics, law, and education we have also seen how the paradigm of control eliminates solutions that do not extend the ordering, numbering, standardizing, and controlling of the world.

The fact that the regime of separation appears to be reaching new heights, the fact that the whole globe is falling into the grip of the monetization of life and the commodification of relationship, the fact that the numbering, labeling, and controlling of the world and everything in it is approaching unprecedented extremes, does not mean that prospects for a more beautiful world are receding into the distance. Rather, like a wave rolling toward shore, the Age of Separation rears up to its maximum height even as it hollows out in the moment before it crashes. This crash, inevitable eons ago, is upon us today. As for the world that we can build thereafter, we can see glimpses of it in all the "alternatives" presented today with so little effect.

The present convergence of crises was written into the future thousands of years ago. It is the inevitable culmination of the separation that began in the deep past, and that once initiated, could do no other than to build upon itself. From the very moment we began to see ourselves as apart from nature, our doom was sealed. While the impending crash is all too plain today, it was much harder to foresee centuries or millennia ago when the world was still large and we were still few and the effects of treating the world as an other were easy to escape. Nonetheless, throughout history perceptive individuals have seen the writing on the wall, the inevitable destination toward which our conception of self and world propels us. Long ago they saw the first stirrings of a gathering calamity written into who we are, and they couched their insights in the language of myth and metaphor.

Some of their metaphors are quite striking. Plagues of locusts symbolizing the ecology gyrating out of balance; plagues of boils symbolizing the diseases we have brought upon ourselves. Wars and rumors of wars. The Whore of Babylon, representing the commoditization of sexuality but also the prostituting of the sacred in general. The end-of-the-world prophecies so popular in American fundamentalist Christianity today tap into an authentic realization, except for the idea that our salvation will come from without. They imagine Jesus coming down from the sky to take away the True Believers in a rapturous escape from a world we have ruined. It is a thought form exactly parallel to that of the techno-utopians. Only the identity of their god is different.

The same awareness of gathering crisis manifests subconsciously in mass society as an all-pervading dread. Subtle when times are good, even the best of times cannot allay the ambient anxiety that pervades the ebb and flow of life and business—the very same anxiety that is embedded in our science and that drives the Technological Program. Infusing our entire culture, anxiety fuels our defining compulsion to control.

Accordingly, for thousands of years now people have been predicting the end of the world—and soon! Though their frequent postponement of the date of Armageddon detracts from their credibility, the basic psychic energy behind the loonies holding up placards on Times Square comes from a real source. They are tapping into a true insight: the edifice of civilization has an irremediable structural flaw that dictates its eventual collapse. We are on a collision course with nature and with human nature.

This long chiliastic tradition, going back to the originators of myths like Armageddon and Ragnarok—the battle at the end of the world—has consistently underestimated just how far separation could proceed, the depths of alienation we could reach, and the capacity of the technologies of control to patch up and shore up the teetering edifice of civilization. And perhaps I too am premature: perhaps we will continue to manage the proliferating consequences of past technology; perhaps the mad scramble to compensate for the lost functioning of ecosystem, polity, and body will be successful for a long time to come; and perhaps we will find as yet unimagined new realms of social, natural, cultural, and spiritual capital to feed the engine of perpetual growth.

Perhaps. For a time. But even if we find a way to hold off for another century the looming convergence of crises, everything I am telling you is still valid.

Notably, these world-ending myths had also in common that the ending is not of the world *per se*, but of the world as we know it. Some even described a vision of what was to come thereafter. Like the world-ending battle, the world thereafter projects into the collective unconscious as intuition and myth. Deep deep down, we all know that a much better world is possible, and more than possible, certain, someday. Ultimately it is this knowing, and not the ideology of the Technological Program, that generates the "Gee Whiz—The Future!" myth with which I opened this book. The Technological Program ideology merely co-opts this intuitively sensed future by claiming it will be brought about by more of the same rather than by its collapse.

The same knowing comes out also in the age-old myth of Heaven which, though idealized and abstracted under the regime of dualism into a realm separate from earth, nonetheless portends on a metaphoric level the necessity of the end of life-as-we-know-it. Even the procedure for entering Heaven usually involves some kind of transcendence of the customary self, a letting go of the old ways of being; as the Christian formulation has it, to be reborn in Christ. In the same way, the glorious estate to which humanity may ascend after the convergence can only come after the breakdown in our collective self-definition as distinct from nature and exempt from nature's rules.

The visions of Utopia that have recurred throughout the modern age are more than technologicalist propaganda, but hark as well to that universal certainty that a world is possible far more beautiful than what we have wrought today. Yet as the word Utopia, which literally means "no such place," implies,[1] we can never attain such a world through the types of efforts that have brought us to where we are today. Utopia will not be achieved by better science, more precise technology, finer control over inner or outer reality. It will not happen by trying harder to be good, and not by better controlling nature or human nature. Quite the opposite. The Hell we have created *originated* in the program to objectify and control nature. It is only by transcending that program and its accompanying conception of self that we can expect to create anything other than a further intensification of what we have today.

In parallel with Millenarian predictions of doomsday, New Age proclamations of the dawning Age of Aquarius have also proven premature. But that doesn't mean these visions are false. The Sixties hippies who knew beyond doubt that in ten years, war, money, laws, school, and so on would be obsolete were seeing a true future, a true inevitability that is not invalidated by the fact that most of them went on to become dentists. Once when my brother was standing in line at the Department of Motor Vehicles, the ex-hippie standing next to him said, "Man, we were through with all this. In the Sixties we were finished with lines and forms and ID's." To the hippies it seemed obvious that all these institutions were obsolete, that they would wither away in the light of expanded consciousness quickly overtaking the planet. A few short years, that was all it would take.

The hippies were not mistaken. Indeed, for some the end of the old civilization manifested, subjectively, in their own lifetimes as they dropped out of the matrix. Some are still living today in the interstitial spaces of our society, nearly invisible, and neither money nor laws nor war is part of their experience. They are akin to the Taoist Immortals of Chinese legend who fade away from normal society into remote mountains, invisible to anyone subject to the usual cultural blinders, and interceding but rarely in human affairs. For the rest, those that stepped back from their vision to become dentists and lawyers, the future they saw with such compelling clarity remains just that, a future. What they saw was true; only their temporal interpretation of that vision was off.

It is an inevitable future, yes, but also, paradoxically, one that we have the power to postpone indefinitely, to the day when every last vestige of beauty is consumed.

For many people, the convergence of crises has already happened, propelling them, like the hippies or Taoist Immortals, into a release of controlled, bounded, separate conceptions of self, away from the technologies of separation, and toward new systems of money, education, technology, medicine, and language. In various ways, they withdraw from the apparatus of the Machine. When crises converge, life as usual no longer makes sense, opening the way for a rebirth, a spiritual transformation. Mystics throughout the ages have recognized that heaven is not some distant, separate realm located at the end of life and time, but rather is available always, interpenetrating ordinary existence. As Jesus said, "The Kingdom of the Father is in us and among us." This is the esoteric meaning Matthew Fox ascribes to the Second Coming: not a single definable event in objective time, but the sum total of all our temporally separated, fitful, but inexorable awakenings to Christ consciousness.[2] What is special about our age is that the fulfillment of processes of separation on the collective level are causing this personal convergence of crises, and the subsequent awakening to a new sense of self, to happen to many people all at once.

The promised recovery of a long-lost Golden Age reverberates through countless myths. The heart-chord it strikes has inspired visionaries and idealists from time immemorial, and explains the unquenchable hope that springs, as the saying goes, eternal from the human breast. As well, it fuels a healthy discontent—the flip side of modern anxiety—that refuses to believe that this is the best we can do. It is an indignation, a muted outrage that can be allayed temporarily by comforts and luxuries, that can be subdued, temporarily, by survival anxiety, that is always strongest in the young, and that lies latent in all of us, ever-ready to be roused into a crusading idealism, though often co-opted toward the perpetuation of the very conditions that give it rise. It is my purpose, dear reader, to give voice to your indignation and to reaffirm your intuitive knowledge that life is meant to be more. I'll conclude with a lyrical description of the lost and future Golden Age by the visionary cartoonist Patrick Farley:

> Could you or I believe that—despite all our hard work to ensure the contrary—our descendants finally figured out a way to live without hurting each other?
>
> Could you or I believe how fantastically wealthy they all became?
>
> Could you or I comprehend, even for a moment, how fiercely my great-granddaughter and her friends loved being alive, and that this love was not an evanescent mood, but a never-ending power which pervaded every sleeping and waking moment of their lives?
>
> Could you or I believe it? Could we stand to believe it?[3]

This chapter will describe the technologies—social forms as well as paradigms of material production—of a future in love with life, which encompasses the love of being alive as well as the love of living beings. All of them arise from and embody a different understanding of self and world. Just as present-day social forms and technologies both spring from and reinforce separation, we will see how 21st century technology will be both a cause and an effect of separation's reversal—a very different understanding of the universe articulated on every level from psychology to cosmology. Now I will share my vision of what today's most promising social and material technologies might look like in full flower, after the seeds germinating today have the chance to grow and blossom in (and in*to*) the Age of Reunion. As our crises intensify we will be faced with new choices and new possibilities. Let us recognize the full ramifications and full power of the choices that will soon open up to us.

The Currency of Cooperation

Prosperity is relating, not acquiring.
-- Tom Brown, Jr.

The "irremediable structural flaw" in our civilization that has generated subtle omnipresent dread and doomsday myths for thousands of years manifests in every human institution, from science to religion to business. None is independent of the others; none can change in isolation; yet when one changes, all will change. Let us begin where the structural flaw is the clearest and its effects most explicit: the institution of money.

Chapter Four described how our present system of money-with-interest generates the necessity for endless growth, how it embodies linear thinking, how it defies the cyclical patterns of nature, and how it drives the relentless conversion of all forms of wealth into money. As well, interest is the wellspring of our economy's ever-intensifying competition, systemic scarcity, and concentration of wealth. Yet more than an accidental artifact of history, interest is tied in to our self-conception as separate, competing subjects seeking to gather more and more of the world within the boundaries of "mine". The change in our fundamental ontology expressed in part by the new sciences will also, therefore, ultimately generate a new system of money consistent with a different conception of self and world.

A society's system of money is inseparable from other aspects of its relationship to the world and the relationships among its members. Money as we know it today both reflects and propels the objectification of the world, the paradigm of competition, and the depersonalization and atomization of society. We should therefore expect that any authentic change in these conditions would necessarily also involve a change in our system of money.

As a matter of fact, there are money systems that encourage sharing not competition, conservation not consumption, and community, not anonymity. Pilot versions of such systems have been around for at least a hundred years, but because they are inimical to the larger patterns of our culture, they have been marginalized or even actively suppressed. Meanwhile, creative proposals for new modes of industry such as Paul Hawken's *Ecology of Commerce*, and many green design technologies, are uneconomic under the current money system. The alternative money systems I describe below will naturally induce the economies described by visionaries such as Hawken, E.F. Schumacher, Herman Daly, and many others. They will reverse the progressive nationalization and globalization of every economic sector, revitalize communities, and contribute to the elimination of the "externalities" that put economic growth at odds with human happiness and planetary health.

Given the determining role of interest, the first alternative currency system to consider is one that structurally eliminates it. As the history of the Catholic Church demonstrates, laws and admonitions against interest are ineffective if its structural necessity is still present in the nature of the currency. A structural solution is needed, such as the system proposed in 1906 by Silvio Gesell in *The Natural Economic Order.*[4] Gesell's "Free-Money" (as he called it) bears a form of negative interest called demurrage. Periodically, a stamp costing a tiny fraction of the currency's denomination must be affixed to it, in effect a "user fee" or a "maintenance cost". The currency "goes bad"—depreciates in value—as it ages.[5]

If this sounds like an outlandish proposal that could never work, it may surprise you to learn that no less an authority than John Maynard Keynes praised the theoretical soundness of Gesell's ideas. What's more, the system has actually been tried out with great success.

Although demurrage was applied as long ago as Ancient Egypt in the form of a storage cost for commodity-backed currency,[6] the best-known example was instituted in 1932 in the town of Worgl, Austria by its famous mayor, Uttenguggenberger. To remain valid, each piece of this locally-issued currency required a monthly stamp costing 1% of its face value. Instead of generating interest and growing, accumulation of wealth became a burden, much like possessions are a burden to the nomadic hunter-gatherer. People therefore spent their income quickly, generating intense economic activity in the town. The unemployment rate plummeted even as the rest of the country slipped into a deepening depression; public works were completed, and prosperity continued until the Worgl currency was outlawed in 1933 at the behest of a threatened central bank.

Demurrage produces a number of profound economic, social, and psychological effects. Conceptually, demurrage works by freeing material goods, which are subject to natural cyclic processes of renewal and decay, from their linkage with a money that only grows, exponentially, over time. This dynamic is what is driving us toward ruin in the utter exhaustion of all social, cultural, natural, and spiritual wealth. Demurrage currency subjects money to the same laws as natural commodities, whose continuing value requires maintenance. Gesell writes:

> Gold does not harmonise with the character of our goods. Gold and straw, gold and petrol, gold and guano, gold and bricks, gold and iron, gold and hides! Only a wild fancy, a monstrous hallucination, only the doctrine of "value" can bridge the gulf. Commodities in general, straw, petrol, guano and the rest can be safely exchanged only when everyone is indifferent as to whether he possesses money or goods, and that is possible only if money is afflicted with all the defects inherent in our products. That is obvious. Our goods rot, decay, break, rust, so only if money has equally disagreeable, loss-involving properties can it effect exchange rapidly, securely and cheaply. For such money can never, on any account, be preferred by anyone to goods.
>
> Only money that goes out of date like a newspaper, rots like potatoes, rusts like iron, evaporates like ether, is capable of standing the test as an instrument for the exchange of potatoes, newspapers, iron and ether. For such money is not preferred to goods either by the purchaser or the seller. We then part with our goods for money only because we need the money as a means of exchange, not because we expect an advantage from possession of the money. [7]

In other words, money as a medium of exchange is decoupled from money as a store of value. No longer is money an exception to the universal tendency in nature toward rust, mold, rot and decay—that is, toward the recycling of resources. No longer does money perpetuate a human realm separate from nature.

Gesell's phrase, "... a monstrous hallucination, the doctrine of 'value'..." hints at an even subtler and more potent effect of demurrage. What is he talking about? Value is the doctrine that assigns to each object in the world a number. It associates an abstraction, changeless and independent, with that which always changes and that exists in relationship to all else. Demurrage reverses this thinking and thus removes an important boundary between the human realm and the natural realm. When money is no longer preferred to goods, we will lose the habit of thinking in terms of how much something is "worth".

Whereas interest promotes the discounting of future cash flows, demurrage encourages long-term thinking. In present-day accounting, a rain forest sustainably generating one million dollars a year is more valuable if clearcut for an immediate profit of $50 million. (In fact, the net present value of the sustainable forest calculated at a discount rate of 5% is only

$20 million.) This discounting of the future results in the infamously short-sighted behavior of corporations that sacrifice (even their own) long-term well-being for the short-term results of the fiscal quarter. Such behavior is perfectly rational in an interest-based economy, but in a demurrage system, pure self-interest would dictate that the forest be preserved. No longer would greed motivate the robbing of the future for the benefit of the present. As the exponential discounting of future cash flows compels the "cashing in" of the entire earth, as illustrated in Chapter Four, this feature of demurrage is highly attractive.

Whereas interest tends to concentrate wealth, demurrage promotes its distribution. In any economy with a specialization of labor beyond the family level, human beings need to perform exchanges in order to survive. Both interest and demurrage represent a fee for the use of money, but the key difference is that in the former system, the fee accrues to those who already have money, while in the latter system it is *levied upon* those who have money. Wealth comes with a high maintenance cost, thereby recreating the dynamics behind hunter-gatherer attitudes toward accumulations of possessions.

In an interest-based system, security comes from accumulating money. In a demurrage system it comes from having productive channels through which to direct it. It comes from being a nexus of the flow of wealth and not a point for its accumulation. In other words, it puts the focus on relationships, not on "having". The demurrage system therefore accords with a different sense of self, affirmed not by defining more and more of the world within the confines of me and mine, but by developing and deepening relationships with others. In other words, it encourages reciprocation, sharing, and the rapid circulation of wealth. A demurrage economic system could evolve into something akin to that of the old Pacific Northwest or Melanesia, in which a leader "acts as a shunting station for goods flowing reciprocally between his own and other like groups of society."[8] These "big man" societies were not fully egalitarian and bore some degree of centricity, as perhaps is necessary in any economy with more than a very basic division of labor. Yet, leadership was not associated with the accumulation of money or possessions, but rather with a huge responsibility for generosity. Can you imagine a society in which the greatest prestige, power, and leadership accorded to those with the greatest inclination and capacity for generosity?

Consider the !Kung concept of wealth, explored in this exchange between anthropologist Richard Lee and a !Kung man, !Xoma:

> I asked !Xoma, "What makes a man a //*kaiha* [rich man]—if he has many bags of //*kai* [beads and other valuables] in his hut?"
>
> "Holding //*kai* does not make you a //*kaiha*," replied !Xoma. "It is when someone makes many goods travel around that we might call him //*kaiha*."
>
> What !Xoma seemed to be saying was that it wasn't the number of your goods that constituted your wealth, it was the number of your friends. The wealthy person was measured by the frequency of his or her transactions and not by the inventory of goods on hand.[9]

In the present interest-based money system, security comes from *having*—from accumulation—and its consummation is "financial independence". Yet the original affluence of the hunter-gatherer sprung from a security associated not with independence but with *inter*dependence. Remember the Piraha: "I store meat in the belly of my brother." A lone woodsman or woman can survive in the wild, but his or her existence is far more precarious than that of a cooperating group. Similarly, in a demurrage-based money system it is sharing and not personal accumulation that forms the basis of security. Demurrage recreates the hunter-gatherer's disinclination toward food storage or other material accumulation, perhaps inducing the same mentality of trust in a providential universe that existed in those days. The Age of Reunion is a return to an original psychology of abundance, yet at a higher order of complexity. It is not a technological return to the Stone Age, as some primitivists envision after the collapse, but a spiritual return.

Silvio Gesell, the originator of the demurrage idea, foresaw that it would spark a profound change in attitudes towards money:

> With the introduction of Free-Money, money has been reduced to the rank of umbrellas; friends and acquaintances assist each other mutually as a matter of course with loans of money. No one keeps, or can keep, reserves of money, since money is under compulsion to circulate. But just because no one can form reserves of money, no reserves are needed. For the circulation of money is regular and uninterrupted.[10]

No longer would money be a scarce commodity, hoarded and kept away from others; rather it would tend to circulate at the maximum possible "velocity". The government would ensure stable prices according to the "equation of exchange" (MV=PQ) by regulating the amount of currency in circulation (M) to correspond to total real economic output (Q). (The same result could be achieved by linking the currency to a basket of commodities whose level corresponds to overall economic activity, as proposed by Bernard Lietaer.) As Gesell concludes:

> It follows that demand no longer depends on the whim of the possessors of money; that price-formation through demand and supply is no longer affected by the desire to realise a profit; that demand is now independent of business prospects and expectations of a rise or fall of prices; independent too, of political events, of harvest estimates; of the ability of rulers or the fear of economic disturbance.[11]

Free-Money would eliminate a major source of our society's endemic economic anxiety. Can you imagine a world in which money were not scarce? How would your own life be different, if you felt no compulsion to accumulate money for security's sake? In a world where survival depends on money and where money is scarce, then survival too is hard, and security only won by outcompeting everyone else.

In a demurrage-based currency system, even though the total amount of currency would be determined by the issuer, its dynamics would ensure a sufficiency for all. The contradiction in today's economy that couples a surfeit of material goods with their grossly unequal distribution, so that some remain always in want, would disappear, as would the feedback cycle that leads to economic recession and depression.[12] Perhaps it would also address a fundamental paradox of the modern world: On the one hand, there are hundreds of millions of people who are unemployed or engaged in trivial, meaningless jobs. On the other hand there is much necessary, meaningful work left undone. This is a profound disconnect between human creativity and human needs. "With Free-Money demand is inseparable from money, it is no longer a manifestation of the will of the possessors of money. Free-Money is not the instrument of demand, but demand itself, demand materialised and meeting, on an equal footing, supply, which always was, and remains, something material."[13]

The greed that leads us to ignore good and necessary work in favor of narrow self-interest is not a fundamental pillar of human nature, but an artifact of our money system and of our misconception of self and world that underlies it. Our system's built-in scarcity has conditioned us to believe that we "cannot afford" to act from love, to do fulfilling work, to create beauty. Gesell's Free-Money represents a liberation from these constraints and from the delusions of self underlying them. It lays down a structural incentive for generosity and frees creativity to seek out need. In this regard Free-Money represents a return to the gift-based societies of yore. Notice its amazing congruity to Lewis Hyde's description of the dynamics of gift flow:

> The gift moves toward the empty place. As it turns in its circle it turns toward him who has been empty-handed the longest, and if someone appears elsewhere whose need is greater it leaves its old channel and moves toward him. Our generosity may leave us empty, but our emptiness then pulls gently at the whole until the thing in motion returns to replenish us. Social nature abhors a vacuum.[14]

Free-Money reverses the compulsion to constantly expand and fortify the accumulation of the private; that is, to expand and fortify the separate realm of self, me, and mine. Just as interest shrinks the circle of self until we are left with the alienated ego of modern civilization, demurrage, the negative of interest, widens it to reunite us with community and all humanity, ending the artificial scarcity and competition of the Age of Usury.

We live, after all, in a world of plenty, and we always have. The present money system and underneath it, the enclosure of the wild into the exclusively owned, has created artificial scarcity where none need exist. Half the world goes hungry, while the other half wastes enough to feed the first half.[15] It is not food nor any other necessity that is scarce; it is money, whose built-in scarcity induces the same in everything else.

A negative-interest currency is a step back toward the gift economies of yore described in Chapter Four, that literally create ties (obligations). Describing Lewis Hyde's theory of the gift, author Jessica Prentice writes, "Part of the sacred/erotic energy of gifts is that the receiver cannot accumulate them—either a gift needs to be passed on, or another gift needs to be given so that the gift-giving energy keeps moving. Gifts are about flow, and they are meant to circulate."[16] This is a perfect description of Free-Money, which like a gift collecting dust in the closet loses its value when kept unused. In a Free-Money system, monetary transactions become like the exchange of gifts, because money is no different from any other object.

Another type of money that addresses the scarcity problem (even more directly) is the mutual credit system, often going by the acronym LETS (local exchange trading system). In a mutual credit system, money is not created by banks or by a central issuer, but is generated by the transaction itself. Here's how it works: Suppose Jane needs someone to mow her lawn. Joe agrees to do it for ten LETS-dollars, and the transaction is recorded on a computer or other ledger: Jane's account is debited by ten dollars, and Joe's credited by ten dollars. If both started with a balance of zero, now Jane has minus-ten and Joe has ten dollars. Money is created out of nothing. Now Joe can use these ten LETS-dollars to buy some other good or service from Fred; Jane, meanwhile, owes the community ten dollars worth of some other service.

If the above scenario seems unsound, understand that its money-creation mechanism is in essence no different from the present bank-debt system. When a bank lends you $100,000 for a mortgage, it essentially creates the money out of nowhere. As in the LETS system, debt and money are, as Lietaer puts it, "two sides of the same coin." The difference lies in the way the creation of money is regulated. In our present system, money-creation is constrained by (1) margin reserve requirements and (2) the availability of credit-worthy borrowers. As described in Chapter Four, these constraints lead to a scarcity of money, because they are not directly connected to demand for a medium of exchange. They also can lead to a polarization of wealth when the means of wealth production (capital) is available at higher cost or not at all to precisely those people who already lack the means of wealth production (to pay back interest).

In contrast, LETS money-creation is typically governed by a community. In theory there is nothing preventing Jane from going deeper and deeper into debt with no intention of ever paying it back, and nothing to prevent Joe and Fred from recording fictitious transactions to build up Fred's credits as Joe skips town. But in practice, anyone who does this will eventually be refused service by the community. LETS currency represents a formalization of "I owe you one," where "you" is not the individual who performed the service but rather the community. It relies on social pressure, both overt and internalized in the form of duty and shame, to prevent abuse of the system. Someone is only permitted to run up huge debits if she and the community feel circumstances warrant it, as in the event of illness.

Because money is created by the transaction, there is never a shortage of currency in a LETS system. The simple willingness to perform a needed service or provide a needed good is enough to make a transaction take place, in contrast to the present contradiction-ridden system that often fails to connect the underemployed and the under-served. With a non-scarce currency, no one ever lacks the money to pay for what they need. Of course, "needs" are still determined by culture and community, which might be unwilling to provide what a person wants. Nonetheless, it is the community that makes this judgment, and not the impersonal, anonymous forces of economics. In this way, LETS brings back the ancient customs of reciprocity that once held communities together.

In addition to quantifying individuals' contributions to the community, LETS and other local-currency schemes draw communities closer together by keeping the economy local. Local currencies ameliorate and can even reverse the ruthless pressures of the global economy, which effectively pits everyone against everyone else. Few of the supermarkets in Pennsylvania sell the apples which grow abundantly here, importing them instead from the West Coast or even New Zealand. Local dairy farmers go bankrupt while consumers drink milk from Michigan or Wisconsin. Traditional economics says that is because these out-of-state producers are more "efficient", enjoying a "comparative advantage" arising from land, climate, or culture. The truth is often that this "efficiency" comes from a more effective externalization of costs. Subsidized inputs of water, transportation infrastructure, and environmental capacity to absorb waste artificially reduce the cost of the imports. Their price reflects none of these externalized costs.

While local economies can also suffer from externalities, it is less likely because those who pay the externalized costs are usually within the same community that supports the business. A business that relies solely on an impersonal, remote commodities market for all its raw materials and sales has little investment in its locality, and therefore little incentive to act responsibly.

Another reason besides externalization of costs for the higher efficiency of large external producers is that they exploit greater economies of scale than are possible on a local level.[17] Unfortunately, the flip side of economy of scale is uniformity and standardization. Today's economy is increasingly dominated by national or global corporations providing uniform products and services over vast areas. One sector of the economy after another has succumbed to the business model exemplified by Wal-Mart. Long gone are the independent grocers, bakers, hardware stores, clothing stores, and burger joints of yesteryear, replaced by corporate franchises. The result has been a homogenization of culture, cuisine, and urban landscape throughout North America and to some extent globally—the "landscape of the exit ramp." You know what it looks like: mile after mile of fast food restaurants, gas stations, automobile dealerships, big box stores, and shopping centers. No fast food franchise opens a store to contribute to the community, only to extract wealth from the community to be sent, eventually, to far-away headquarters or shareholders. Local currencies counter this dynamic because they can only be used locally. If a grocery store, say, decides to accept partial payment in local currency, it must find a use for that currency locally, by sourcing products from local farmers and perhaps by paying a portion of its salaries in local currency. A national chain is less likely to attempt this, opening the door to local competition. The result is a self-reinforcing "virtuous circle", because the local suppliers and employees paid in local currency must themselves use it locally. The more business and consumers who use it, the more useful it becomes and the greater the incentive for businesses to accept it.

The most profound effect of these currency systems is probably psychological. Currencies that are not defined by interest and scarcity resurrect the ancient hunter-gatherer mentality of abundance, in which sharing is easy and natural, in which there is no mad scramble to enclose the world in me and mine. As such they are consistent with a more open conception of self, defined by relationships and not by absolute boundaries of self and other. Consequently, while none of these currencies will ever truly supplant our present system unless our self-definition changes, they are nonetheless agents of change that can help to induce the reconception of self and world. A convergence of crises will precipitate this turning point, and its financial aspect (which we can see building today) will clear the way for all the ideas laid out above to crystallize into a radically different system. When this happens, money, ever a force for evil, will come to embody a new set of incentives aligned with the priorities of the connected, interdependent self: sustainability, beauty, and wholeness.

Is it too much to imagine an economy in which the best business decision is identical to the best ecological decision? In which there is a built-in economic incentive—as opposed to a regulatory compulsion—to protect the environment? In which the creative entrepreneurial energy of business is constitutionally aligned with the wholeness of nature and the health of society? As Gesell puts it, "Money, anathema throughout the ages, will not be abolished by Free-Money, but it will be brought into harmony with the real needs of economic life. Free-Money leaves untouched the fundamental economic law which we showed to be usury, but it will cause usury to act like the force that seeks evil but achieves good."[18]

As it stands, money is almost universally recognized as a force for evil or even "the root of all evil." The locution "cannot afford to" reveals just how often money is an impediment to our innate tendencies toward kindness, generosity, leisure, and creativity. Interest-money generates the greed that we mistake as human nature. It perpetuates the illusion that security and wealth come from gathering more and more of the world unto the self, from carving out a larger and larger exclusive province of "me" at the expense of every other living person, animal, plant, and ecosystem. As well it seems to directly contradict the teaching of karma, which says that what we do to the world, we do to ourselves. In our current money system, giving out to the world means less for me, not more! Money thereby contributes to the illusion of separateness that is considered in Buddhism to be the root of all suffering. Free-Money and LETS systems reverse this role and bring money into line with karma, reinforcing rather than denying its fundamental principle that by enriching the world we enrich ourselves. In so doing, they also subvert the dogma of the discrete and separate self that underlies the development of the present money system.

A harbinger of the convergence of crises occurred in the 1930's and began to generate the types of money systems I've been describing. Thousands of local currencies appeared throughout North America and Europe. Named "emergency currencies" in the United States, they rejuvenated local economies wherever they were instituted. They worked so well that one of the leading economists of the time, Irving Fisher, advocated the stamp-scrip (demurrage) system as the way out of the Great Depression: "The correct application of stamp scrip would solve the depression crisis in the US in three weeks!"[19] Other economists agreed, but pointed out possible decentralizing political effects to Treasury Undersecretary Dean Acheson. President Roosevelt responded by banning all emergency currencies, choosing instead the centralized solution of the New Deal. A similar story transpired in Austria with the Worgl currency and its numerous imitators, as well as in Germany a few years before. Today, though most countries tolerate complementary currencies, they still discourage them through a variety of means; for example, local-currency income for professional services is still subject to income taxes that must be paid in the national currency, forcing people to operate substantially in the non-local economy.[20]

The world was not yet ready for a shift of power to local communities. The centralized solutions represented by the New Deal, and by fascism in Europe, were more than just maneuvers for political power. They were part and parcel of the mentality of control, top-down engineering, and reductionistic thinking. The *zeitgeist* of the time dictated such solutions. In Newtonian science one does not simply allow a solution to grow, but engineers one from above. From Keynesian pump-priming to fascism's "We will *make* the trains run on time," the economic solutions chosen during the Great Depression all involved central authority taking control and imposing order upon a broken machine.

No matter that the centralized solutions were not, in fact, effective. While the New Deal did manage to rescue people from utter destitution, the depression only ended with the military buildup and conflagration of World War II. That wartime economy has been with us ever since.

A similar choice faces us today, and will become clearer as the financial crisis grows more severe. Again we will have to decide between, on the one hand, more centralized top-down control leading as it did in the 1930's to fascism and war, and on the other hand a release of control to a more organic solution. From the central government perspective, the latter is not a "solution" at all, not in the sense of a design or plan. Local currencies spring up spontaneously. Those of the Great Depression were preceded by currencies of similar design during the panics of 1873, 1893, and 1907.[21] We are habituated to thinking that a solution means more control, more detailed measurement of all the variables, a more comprehensive design—this is the Technological Program. But just as in other areas of technology, the illusion of monetary control, so compelling in the last sixty years due to the absence of major financial panic, is wearing thin. Soon its failure will be undeniable, and we will have to decide whether to deny it anyway at a higher and higher price, or to let go and trust the Wild—the spontaneous creativity of human communities.

The longer we hang on, the harder we scramble to apply one technical fix after another to our tottering money system, the more severe the crisis and its subsequent dislocation will be. The eventual result is assured, though, that a new system of money will emerge that is aligned with the priorities of the connected, interdependent self: sustainability, beauty, and wholeness.

Money in the Age of Reunion will be an agent for the development of social, cultural, natural, and spiritual capital, and not their consumption. It will be a mechanism for the sharing of wealth and not its accumulation. It will be a means for the creation of beauty, not its diminishment. It will be a barrier to greed and not an incentive. It will encourage joyful creative work, and not necessitate "jobs". It will reinforce the cyclical processes of nature, and not violate them.

Quoting again Patrick Farley's visionary future, "Could you or I believe how fantastically wealthy they all became?" we can now see what this might mean. When wealth is separate from accumulation but refers to a richness of relationships, each person's wealth makes everyone wealthier. Art, the creation of beauty, will no longer be limited by what we can afford, for money will be art's ally not its enemy. Business will be the seeking of ways to bestow wealth upon others rather than the stripping of wealth from others. No longer will our lives be full of cheap stuff and every transaction a rip-off. Work will no longer be bound to the search for money, but will seek out ways to best serve each other and the world, each according to our unique gifts and temperament. That will be the way toward riches, both spiritual and financial, for no longer will the two be in conflict.

The Restorative Economy

A money system does not change in isolation. Everything else must change with it, both as a consequence and as a precondition. My logic in the previous section is *not* "If only we could change our money system, then everything else would change too and our problems would be solved." No. It is rather part of a vast phase transition, a gestalt, a holographic repatterning in which we could take any aspect as a starting point—and indeed must, in a linear narrative such as a book. Together, the shifts in money, property, education, medicine, art, and others that I describe in this chapter all arise from and are part of a general reconception of self and world. I've started with money, but that doesn't make it primary.

As explained in Chapter Four, much of the destructiveness of our present economic system is due to the externalization of costs, which like interest attempts to impose a linearity onto nature. Externalization assumes an "out there" that is the start and end point of linear consumption. Whereas interest demands endless growth, economic externalization assumes an infinite reservoir of resources and absorptive capacity to accommodate that growth. Interest and externalization are, in fact, two sides of the same coin, a connection that has gone largely unnoticed by those who hope to halt our species' self-destruction through the internalization of costs. Nonetheless, proposals like green taxes, consumer leasing, and pollution permits offer considerable insight into what an industrial economy might look like that no longer sees itself as apart from nature. I borrow Paul Hawken's term for this, the "restorative economy," because it not only halts but also reverses the progressive distancing of ourselves from community, ourselves, and nature. One way it does so is by motivating technology that mimics, strengthens, and extends the processes of nature rather than seeking to deny them.

Writers such as Hawken, Herman Daly, Bill Rees, Lester Brown, Thom Hartmann, and Kenneth Boulding have made the seemingly obvious point that no economy is sustainable that consumes energy and other resources at a faster rate than they can be renewed. They view the consumption of fossil fuels, along with other finite resources such as the ecosystem's capacity to absorb waste, as a depletion of capital. Sustainability (an overused word which really means survival) is only possible if we live off our current income, not our capital reserves. This income includes "solar income" in addition to the earth's ability to convert pollutants into the building blocks of new life. We might also include the "cultural income" of new art forms, new ideas, and so on, which today are instantly co-opted and commoditized along with the "capital reserve" of our heritage.

Our economic paradigm separates us from nature through its linearity, its demand for endless growth, and its reduction of nature into "resources". It also separates us from each other through its consumption of social and cultural capital. An economy that reverses this separation will have a key defining feature: *there will be no such thing as waste.* The whole idea of waste, garbage, etc. is that there is an "out" to which to throw things. That idea can only exist within the mindset of separation.

If there is to be no waste, then everything we take from the biosphere must be returned to it in a non-toxic, non-accumulative form.[22] One proposal to implement this is Braungart and Englefield's "intelligent product system", which attempts to implement the ecological principle that "waste is food" through (1) A leasing economy for durables; (2) complete non-toxic biodegradability of consumables, and (3) permanent storage charges for any other waste—you can still produce dioxin if you want, but it is yours forever.[23] The first means that consumers lease large appliances such as refrigerators and washing machines instead of buying them. In effect they buy services (refrigeration, clothes-washing) instead of the machines themselves. This removes the structural incentive toward planned obsolescence while creating new incentives to make products easy to repair and recycle. The same logic applies to industrial durables, and is especially potent in the context of a demurrage system because it deemphasizes capital accumulation. The second principle is fairly self-explanatory, with the caveat that many supposedly biodegradable substances do not degrade because they are suffocated in landfills. The third principle is essential for making the first work, because it creates a huge financial incentive to make products non-toxic and recyclable. (The effect of an eternal storage fee is even more powerful in the context of a demurrage-based currency, in which future costs are not discounted.) As well, storage fees for toxic waste create an incentive to develop processes to render it non-toxic. The mycologist Paul Stamets has developed some amazing bioremediation techniques involving fungi that can detoxify many classes of toxic waste;[24] however, since the social and ecological costs of toxic waste are nearly all externalized, at present his methods have limited economic motivation. In a restorative economy, Stamets might well be the richest man on the planet!

To foster cyclicity in the industrial economy, author Paul Hawken advocates "green taxes", modeled after the proposals of economist A.C. Pigou. Exemplified by the fossil fuel tax, these would impose a large surcharge on the cost of natural resources reflecting the damage generated by their extraction and use. The tax on oil, for instance, would incorporate the cost of habitat damage in drilling the well, the apportioned costs of oil spills, the costs of air pollution from its burning, and so forth.

In practice and in principle this approach is fraught with difficulties. The practical difficulty is in calculating long-term, highly distributed costs such as economic damage caused by global climate change, to which oil burning contributes but not as a sole factor. The difficulty in principle lies in assigning monetary value to such things as species extinction, the destruction of beautiful views, and human mortality. Gesell's "monstrous doctrine of value" rears its ugly head again! The assumptions underlying green taxes once again involve the denomination of everything in monetary units. Nonetheless, green taxes are intended to eliminate the profit strategy of dumping costs onto an other. They are the financial embodiment of the realization that what we do unto an other, we actually do unto ourselves. We are not separate from the world. Today's economic system fosters the opposite belief, because damage dumped onto society or the environment does not usually end up as a cost on our own balance sheet. Pigovian taxes encourage a more expansive sense of self.

The immediate practical effect would be to encourage alternatives which are presently uneconomic only because the price of "natural resources" doesn't reflect their true cost. All of a sudden, the marvelous green technologies—bicycles, composting toilets, organic agriculture, photovoltaics, zero-emissions plants, grass roofs, hypercars, rail transport, fungal bioremediation, sailing ships, non-toxic industrial processes, and many more—would benefit from huge economic incentives. Today we attempt to force manufacturers to reduce pollution through regulation and control. Green taxes would eliminate the need for regulation by aligning ecological sense with business sense.

When humanity internalizes the damage of present industrial processes, those processes must change dramatically. Mining and quarrying will become prohibitively expensive, as will the refining of ore, when it is unacceptable to leave behind mine pits and slag heaps. Our main source of raw materials might be the junk of the 20th century: the pound or more of lead in every TV set and computer monitor, the vast amounts of iron and steel rusting away in junkyards, the cinder blocks of demolished buildings. What durables we produce will be designed to be easy to repair and maintain, so they might last hundreds of years. Enormous entrepreneurial energy will pour into such endeavors, as the objectives of business come into alignment with the restoration of nature. (In contrast, the regulatory approach to environmental protection puts it into opposition with business success.)

Material goods of plastic and metal will become so precious that to simply throw them away will be unthinkable. Not only containers, but the entire distribution system for their contents, will be designed in the interests of reusability—not because the government requires it, but because that will be sound business practice. Because such services as cleaning and refilling are by nature local, less capital will be exported to distant large-scale manufacturers, strengthening local communities and expanding the viability of local currencies.

The entire suburban landscape will change when gasoline, cement, and asphalt are no longer artificially cheap. The community-enhancing high-density principles of contemporary urban design will no longer fight an uphill battle against developers' economic interests. Without subsidized fuel, pesticides, and irrigation groundwater, local produce will become economically more competitive, and home gardening will gain an economic incentive in addition to its present aesthetic and spiritual benefits. Food production will become more local, and more people will become food producers.

For local travel, the bicycle (which is still undergoing technical improvements) will come into its own as the primary transportation technology of the future. Ultimately, I think that the construction and maintenance of bicycle trails is the maximum incursion into nature that is socially and environmentally sustainable. Just look at the carnage that goes into building a road, the bulldozing of trees and tearing into the earth. In the future, we will not be able to bear anything more than modest bicycle paths. Certainly, current levels of road kill—perhaps the most direct reminder that externalities equal death—would be intolerable to the conscience of any sane society. Is it mere coincidence that normal bicycle speeds—say about 20 miles per hour maximum (also the natural speed of a *woonerf* intersection)—also delimit the pace of mindful travel? At bicycle speed I can still notice what plant species grow on the wayside, I can react quickly enough to avoid running over rabbits, and I am moving slowly enough to enter briefly into the social universe of those I pass. There is time for a "hi", a token of acknowledgment, a flash of recognition. Anything faster entails some degree of oblivion to the people and places traversed—all the more when traveling in an enclosed vehicle. Automobiles, no matter how low their emissions, distance us from nature simply by their speed, of which roadkill and habitat fragmentation are inevitable byproducts. Cars (and even more dramatically, airplanes) allow us to go from point A to point B without truly experiencing the places in between; they allow us to travel without journeying. No wonder we are so oblivious to the "spaces in between" the matrix of human culture, and so numb to their destruction. Unexperienced, casually traversed, sped by and flown over, it is as if they did not exist at all. Externalities.

Unfortunately, the ecological economy envisioned by Hawken and others is fundamentally at odds with our money system. The former seeks to eliminate externalities; the latter requires and creates them. The former requires the cycling of all material; the latter has a built-in imperative of endless linear growth. A green economy can never exist, except marginally, in the context of interest, which demands endless "growth"—the one-way and therefore unsustainable conversion of all the world into money. On the bright side, the unsustainability of our industrial system and the unsustainability of our financial system are converging, presenting the opportunity for the transformation of both.

In place of today's linear (or exponential) economy will arise a cyclical industrial *ecology* that will be, in concept as well as in fact, an extension or a new dimension of nature. In no sense will it be unnatural. Like a plant or a whole species, each business enterprise and each industry will seek not the maximally efficient extraction of resources, but to serve an essential function in an interdependent web of relationships. Any step in the cycling of resources will be a viable business opportunity. Paul Hawken has compared the economy of the last few centuries to a field of weeds which, entering barren soil, do indeed seek maximum growth in a headlong competition for sunlight and other resources:

> At the dawn of the Industrial Revolution, a vast new world of apparently unlimited natural resources became available for the taking. By constructing an economy that demanded ever-increasing supplies of all resources... humans successfully mimicked the processes of a newly formed ecosystem. Like pioneer plants, we were aggressive and competitive. We emphasized untrammeled growth and didn't worry about efficiency, conservation, or diversity.[25]

Obviously, this can only be a temporary phase. Eventually, increasingly complex, interdependent relationships among species develop as the original opportunistic colonizers give way to a more diverse, mature ecosystem. "In immature systems, most energy is used to create new growth, so that bare soil is quickly covered. In a climax system, the greater part of energy is devoted to the continuation of the existing plant and animal communities."[26] We are at the cusp of just such a transition. Already, even with a money system that demands constant growth and even with our conceptual distancing from nature, nascent industrial ecologies are forming within a linear framework. When we integrate the parameter of cyclicity into the system, these will gel into an ecological economy, the equivalent of a diverse jungle ecosystem of hundreds of thousands of mutually reliant species, as opposed to a field of thistle and burdock competing to cover bare ground.

In such a system, each niche is self-limiting. Success lies not in taking as much as possible for "me", but in fulfilling one's role in the maintenance of the whole, a mission that calls forth the unique gifts of each participant. It defines each person not by what they can keep and possess, but by what they can give, ending the age-old contradiction between service and selfishness, greed and giving. Yes, there will always be competition, just as there is in the jungle; however, it will not be the destructive competition that seeks to enclose the commonwealth and fleece the chumps, but rather a constructive competition to excel in a given niche, to be a greater contributor to the common good. Already we have these ideals but must fight for them against the grain of economic self-interest. When the conception of self that underlies the economy changes, then so will economic self-interest realign itself toward the common good.

The restorative economy is not a matter of passing some new legislation, implementing some reforms, or seeing through the errors of conventional economics. It is nothing less than a facet of an all-encompassing spiritual transformation in our fundamental relationship to the world. Martin Prechtel explains that in Mayan culture, it was necessary to pay back to nature the debt incurred in the creation of any material object, and indeed in the living of a human life.[27] The greater the disruption of nature, the more ritual effort required to pay back the debt to the other world. Thus an iron knife would demand lengthy and elaborate rituals in consideration of the cost to nature of its forging: the digging of the ore, the burning of fuel, and so forth, so much so that no one would make a knife without very good reason. Such things as air conditioners would, in this world-view, demand so much ritual recompense as to be prohibitively "expensive". A limit to consumptive technology was thus built into the Mayan culture. And what would happen if you mass-produced knives, as we do, without paying the price in ritual? The knives would exact payment in their own way, either through direct physical violence or through the reduction of human life, hope, vitality, creativity, joy, and beauty. Can we say that this has not already come to pass?

In other words, when we produce and consume linearly, without regard for the consequences to the "other"—the planet and the rest of life—we unavoidably amass a debt that must, inevitably, be paid. One technological fix after another only postpones the day of reckoning, just as a few more drinks to prolong the buzz only brings a worse hangover in the end. The all-important myth of the Ascent of Humanity, the Technological Program, is that we can put it off forever. Yet the bankruptcy of this myth is increasingly obvious.

I am not advocating that we imitate Mayan rituals, but that we translate the idea into a context that makes sense today. The ritual price we must pay translates into devoting care, consideration, and thought into anything we make. For there to be no left-over debt requires that this consideration cover the origin of the raw materials, the effects on the ecosystem of producing and shipping them, the effects on people's lives, the effect the use of the object will have, and how it will eventually return to the earth. Translated into economics, it requires the full internalization of all costs. If we neglect any of these considerations and allow costs to remain external, then consequences will surely come to haunt us—a debt will be written into our future—because there *is* no "out there" to which to externalize.

Can you imagine, then, the enormous debt that our society has amassed over the last few millennia? While it may seem that we have escaped more or less scot-free and passed the debt on to future generations (which is bad enough), the truth is much worse. We pay the price ourselves too. The price comes out of our own flesh and, worse yet, out of our souls. The just reward for our ignorant, arrogant, deluded despoliation of planet and culture is the ocean of suffering that engulfs our species, the wars and atrocities, rape and genocide, the brutalized children, the slums of the Third World, the lives without hope, the diseases and famines, and in the rich countries, the depression and despair, anxiety and ennui, anomie and loneliness, the tragic reduction of human potential that leaves even the winners of the rat race among the sorriest rats of all.

An interesting parallel to Prechtel's view that the spirit world (underlying nature) takes its payment in the form of human violence and suffering can be found in the cosmology of G.I. Gurdjieff, who called it "food for the moon". Gurdjieff's prescription for human development—to act in the full consciousness of "self-remembering"—is also consistent with Prechtel's prescription for a ritualistic approach to life. True ritual is not a sequence of steps performed by rote, but a means to induce mindfulness and expand consciousness beyond the limited self that reached its extreme in Western culture as Descartes' "I am", Adam Smith's economic man, and biology's phenotype. Indigenous culture never reached anything near that extreme of separation, but they did recognize the tendency and took active steps to counteract it. Remember the Yurok story of the Wo'gey in Chapter Two, who "knew how to live in harmony with the earth." Before they departed, "because they knew that humans did not always follow the laws of the world, they taught them how to perform ceremonies that could restore the earth's balance." Ceremonies or rituals allow us to internalize, psychologically, the cost of our actions, in effect paying the debt right now instead of postponing it that it might be taken instead from our souls and the lives of our children. Here again is a parallel with Gurdjieff, whose primary prescription for recovering mindfulness was "intentional suffering", by which he meant an unwavering intention not to avoid or escape the consequences due. On a psychological level, this closely parallels the internalization of costs in economics. Both recognize that the Technological Program must ultimately fail, that the technological fix cannot work forever, that life will elude infinite control, and that the debt incurred through the destructiveness that arises from our separation must inevitably be paid. Every technological fix by which we generate yet more environmental damage in order to insulate ourselves from the consequences of previous environmental damage contributes to a vast Ponzi scheme, a debt pyramid, a bubble no more sustainable than our perpetual-growth economy. The longer we accumulate that debt, the greater the eventual crash, and the more complete our eventual bankruptcy.

The restorative economy I have described grows out of a recognition of this debt. It is motivated by a love of the world around us—the love of life and the love of our own lives—and a sincere desire to make amends. I have said little about the mechanisms by which the transition will take place, but I believe that the requisite changes are so deep that nothing less than a complete collapse of the current regime will do it, a collective hitting bottom. When the illusion of our separateness becomes utterly untenable, then new structures of money and economy will crystallize that enforce and embody our reemerging wholeness.

How far must we fall to hit bottom? That is up to us.

The Age of Water

In a future where all costs are fully internalized and in which no object is created mindless of its origins and consequences, it would seem that technology itself would be nearly impossible. Many thinkers, recognizing the long-term unsustainability of industry even at 12th century levels, and perhaps even of agricultural civilization,[28] can offer no alternative but a long descent back to the Stone Age, an erasure of the series of blunders that have brought us to our present condition.

Admittedly, Stone Age life seems pretty fine to me, but as I have hinted throughout this book, I believe that the errors of our mad, senseless age are nonetheless setting the stage for a new phase of human development that brings the past into the future rather than trying to bring the future back to the past. Machine Age technology is fundamentally unsustainable, yes, but let us not assume that all possible technology shares this trait. In this section I will outline a new mode of technology, already emerging, that is an extension and not a violation of nature.

In Chapter Two I wrote:

> The unsustainability of our present system derives at bottom from its linearity, its assumption of an infinite reservoir of inputs and limitless capacity for waste. A fitting metaphor for such a system is fire, which involves a one-way conversion of matter from one form to another, liberating energy—heat and light—in the process. Just as our economy is burning through all forms of stored cultural and natural wealth to liberate energy in the form of money, so also does our industry burn up stored fossil fuels to liberate the energy that powers our technology. Both generate heat for a while, but also increasing amounts of cold, dead, toxic ash and pollution, whether the ash-heap of wasted human lives or the strip-mine pits and toxic waste dumps of industry.

Fire is natural; in fact it takes biological form in any aerobic being as the liberation of stored energy through oxidation. Nature only lasts because there are other steps, powered by sunlight, by which the ash is reincorporated into forms of stored energy. An economy that is sustainable must do the same; it cannot be based on fire alone, in either its literal or metaphoric form. The restorative economy, then, will have a technological infrastructure that is not based on the technologies of fire.

From the Mayan perspective, we could view fire as a borrowing or a stealing from Gaia. When we burn wood, we replace the slow oxidative processes that would ordinarily sustain generations of insects and fungi with the rapid oxidation of fire, therefore reducing life's possibility. According to Martin Prechtel's logic, any fire-making, even the smallest campfire, should therefore be accompanied by some kind of ritual payment proportional to the quantity of fuel burned. A society that is truly sustainable would only use the technologies of fire with great circumspection.

The technology described in the previous section is not so different from what we have today, for it still seeks to impose human control onto nature. A bike path, for instance, requires keeping that strip of earth away from its natural state. As the Maya realized, any technology by which nature is put under control and bent to human purposes entails a price, that primitive cultures recognized and paid, and that we have attempted to evade. Even if no actual fire—no preemption of oxidative energy stored in biomass—is used, any technology that embodies fire's consumptive linearity and its cooptation of natural processes still runs on the fire model. Since almost all of our present technology derives somehow from fire, or seeks, like a bike path, to subdue nature, many ecologically-oriented futurists quite naturally assume that a sustainable future means a low-tech future.

This conclusion falls apart if we can envision a mode of technology not based on fire. I envision a high-tech future, but one whose technology is so dramatically different from our own that it is almost unrecognizable as "technology" to present ways of thinking. The end of the Age of Fire promises a reversal of the course of separation and domination that fire has fueled. Immersed as we are in the ideology of separation, it is hard to conceive of a mode of technology that does not involve the objectification, domination, and control of nature. Yet such technologies exist, even if we hardly recognize them as such. They are based not on fire but on earth, water, light, sound, and the human body. Rooted in an ancient past, they nonetheless carry the promise of a "new age". Who knows what unconscious wisdom has named it the "Age of Aquarius"? But I shall call it the Age of Water.

Water (to risk stating the obvious) carries metaphorical connotations very different from those of fire. Water denies linearity: cycling endlessly, it is also the agent of nature's cycles, nourishing both growth and decay. Similarly it resists separation: named the "universal solvent", it tends away from purity to partake of its environment. Water is also the nemesis of control. Seeking out the tiniest crack, nothing can hold it in. As waves in the ocean, it destroys any bulwark. Whereas fire burns clean and purifies what it touches, water makes a mess. Hence the key to preserving anything—houses, books, food, clothes, metal—is to keep it dry.

Water, with its cycles and flows, its unruliness and its ubiquity on earth, could be called the essence of nature. To keep everything dry, and therefore outside the cycle of decay, transformation, and renewal, is a concrete expression of our separation from nature. It is this cleavage between human and nature to which the Age of Reunion will put an end.

Earlier I asked, "Since human beings are themselves natural, then isn't everything we make and do natural too?" To be natural isn't a matter of who designed something or what materials it is made from. The products of the human hand are only unnatural to the extent they pretend to a linearity that defies the cyclical laws of nature. The industrial ecology described above *is* natural, because there is no waste. So I am saying more than "technology will be in harmony with nature". Technology will *be* natural. We need not abandon our uniquely human gifts and return to the Stone Age or before. Instead, we will recover a Stone Age mentality in the context of modern technology. Underlying the future technological economy will be principles of interdependence, cyclicity, abundance, and the gift mentality. Can you think of a better simile for all four of these principles, than that they are like water? Water, upon which all depends. Water, which moves in cycles. Water, abundant to ubiquity. Water, bringing the gift of life.

Our dependence on water—the fact that we are made mostly of water—denies the primary conceit of civilization, that we are separate from nature or even nature's master. No more nature's master are we, than we are the master of water!

Yet for centuries we have tried to persuade ourselves otherwise. In science our pretense of mastery manifests most fundamentally in the supposition that water is a structureless jumble of identical molecules, a generic medium, any two drops the same. To a standard substance we can apply universal equations. That each part of the universe is unique is profoundly troubling to any science based on the general application of standard techniques. The same is true of technology. Only a universe constructed of generic building blocks is amenable to control. Just as the architectural engineer assumes that two steel beams of identical composition will have identical properties, so does the chemist believe the same of two samples of pure H_2O.

That any two samples of H_2O, or graphite, or ethanol, or any other pure chemical are identical is a dogma with enormous ramifications. It implies that the complexity and uniqueness of objects of our senses is an illusion, that they are mere permutations of the same standard building blocks. Such a view naturally corresponds to the objectification of the world, which makes of it a collection of things, masses.

The opposite view sees every piece of the universe as unique. No two drops of water, no two rocks, no two electrons are identical, but each has a unique individuality. This is essentially the view of animism, which assigned to each animate and inanimate object a spirit. To a Stone Age person, the idea that water from any source had a unique character or spirit would have seemed obvious. Modern chemistry denies it and says any apparent differences are merely due to impurities—the underlying water is the same. Animism says no—to have a spirit is to be unique, irreducibly and intrinsically unique. To have a spirit is to be special.

With the dawning of the Age of Water, we return to our animistic roots and recognize the unique, enspirited nature of each drop of water and indeed every substance in the universe. Not even the field of chemistry is immune to this paradigm shift, as it becomes increasingly apparent that water does indeed exhibit structure on several levels. Chemists and materials scientists are now recognizing that structure maintained by hydrogen bonds and van der Waals forces is responsible for many of water's anomalous properties. Few, however, believe that this structure can convey information to biological systems. Yes, water has structure, they might admit, but there is no signal in the noise.

Now let's leave the mainstream behind and take a journey away from scientific respectability. Our first stopping place is the empirical science of homeopathy, which has been developed over two centuries to a remarkable degree of sophistication despite the absence of any cogent reductionistic theoretical underpinnings. In other words, no one really knows how it works. What is clear, however, is that it somehow uses water to convey the information embodied in natural substances to the body. Two different homeopathic samples of high potency may both be chemically pure H_2O, but they are far from identical in their effects—a contention that has drawn considerable derision from critics!

Perhaps because it is based upon water, homeopathy fosters a philosophy of healing very different from the conquest of nature that characterizes fire-based allopathy. Allopathic medicine is based on control: killing microbes, dictating hormone levels, cutting out organs and tumors. Whereas allopathic medicine dominates nature, homeopathic medicine sees nature as the body's teacher. The homeopath seeks out the natural substance that can teach the body a healthier pattern of being. Looking within nature instead of seeking to defeat or transcend it, the homeopath approaches healing in the spirit of water instead of fire.

A bit further outside the mainstream is flower essence therapy, which, having been developed primarily through intuitive rather than empirical means, I prefer to call an art rather than a science. As in homeopathy, water serves as a carrier for information originating in flowers or other natural objects, used primarily for emotional or spiritual healing. Here again, each drop of water is unique; here again, even more explicitly than in homeopathy, each drop of water is understood to possess a unique spirit.

Masura Emoto, a Japanese businessman, takes these ideas to their logical extreme. Emoto photographed ice crystals of water that he'd subjected to various influences: electromagnetic, emotional, musical, and so forth. The crystals exhibited striking differences, even when he simply taped different messages onto jars containing samples of distilled water from the same source. For example, the samples shown words like "devil", "you fool!" and so on froze into ugly, amorphous ice, while those shown words like "love", "gratitude", and "cosmos" formed beautiful crystals.[29] Despite an extensive search, I have found no serious refutation of Emoto's findings. Critics typically point to his failure to implement double-blinded controls and his on-line Ph.D., but apparently the substance of his work is beneath them to even address. True, his work is not rigorous, but it isn't meant to be. It is beautiful and, to those to whom it "rings true", suggestive of further directions in research.

Essentially, Emoto's work confirms the metaphorical associations of water as a universal medium, a universal solvent not only for physical materials but for thoughts, feelings, energy, and information. Water carries the imprint of its environment, and since each lake, river, glass, or drop of water is uniquely located on earth, each is subject to a unique combination of influences. At the same time, since this "environment" extends to include the whole planet and beyond, each drop of water contains the informational imprint of the whole. Emoto's work suggests that our every thought and intention affects every drop of water on earth; it's just that the intended target of that thought, along with the water within our own bodies, is most strongly affected.

A primitive hunter-gatherer would not find it difficult to believe that all water had a unique personality, that river water, lake water, rain water, spring water, and water taken from the ground would have differing effects on the body and emotions, and perhaps distinct ceremonial uses as well. I imagine some languages don't even use the same word for these different types of water. Similarly, a hunter-gatherer would find it easy to believe that beloved water would have different properties from despised water. That we believe all water to be a uniform, lifeless "substance" that can be made identical by removing its impurities is a reflection of our ideology of objectivity and reductionism. We once knew better, before we made of the world a thing, before we reduced the infinity of reality to a finitude of generic labels (like "water"). A future technology of water will recover this knowledge, and we will no longer treat water as anything less than sacred.

Emoto's work suggests that we cannot escape the effects of our thoughts, words, and actions. Released into the universe, they leave their imprint there, in effect reconfiguring the reality in which we live. In an Age of Water we will understand this principle. We will understand that, like water, all things eventually cycle back to their source.

An Age of Water will imitate the water cycle in its economics as well. Fire is the epitome of consumption, and it has incinerated social and natural capital for millennia. Today we are seeing the precursors to the cyclical economy of the Age of Water. All of the features of the "restorative economy" I have described—resource recycling, zero-waste manufacturing, full-cost accounting, and non-interest currency systems—equally justify the appellation "water economy". Like demurrage currency, like the energy of the gift, water resists confinement, moves from high places to low, and ultimately circles back to its source.

Perhaps the most profound transformation of the Age of Water will be in our spirituality—how we relate ourselves to the universe. Above, when speaking of animism, I said that each water droplet or other object "has a unique spirit", but that is not quite correct. The conception of spirit as something to be "had", and therefore extrinsic to matter, is a metaphor of separation and of fire. What animism actually implies is that each thing *is* a unique spirit, that matter itself is spiritual, sacred, and special. Spirit can no more be abstracted out from matter than structure can be removed from the water that carries it. The Age of Water, then, is an age in which we treat the earth and everything in it as sacred.

At the same time, water teaches us that the unique spirit of any bit of matter is not discrete and separate from the rest of reality. Like all things including ourselves, water takes on the spiritual qualities of everything that surrounds it; thanks to its ubiquity and receptivity, it is also the medium of this communion of all with all. Unique we are, each one of us, yet no more separate than two drops of water in the ocean. The Age of Separation comes to an end with the dawning of the Age of Water.

Technologies of Reunion

Cyclicity. Abundance. Connection. The money system and economy that embody these qualities of water will give rise to a new mode of technology as well, one that is no longer an agent of separation. Because we have so long associated technology with the qualities of fire—separating, purifying, consuming—some of the new technologies that will define the Age of Reunion we might hesitate to even call by that name. Others will seem mundane or even archaic to us today. There will be a revival of technologies thousands of years old, mixed with new inventions so unexpected as to seem miraculous. A common theme will unify this mélange of low and high technology: all will rise from—and contribute to—the economics, science, and mentality of Reunion that I have described.

Who knows what heights green design and organic agriculture could reach with the full backing of a different currency and a different conception of the human role? What will happen when the trillions of dollars and millions of scientific careers currently devoted toward armaments turn to other purposes? Who knows, when there is an overriding business incentive to do so, how many industrial products and processes will be replaced by ecological alternatives? Who knows what the marriage of tradition and technology will bring to bicycles, to gardening, to sailing ships, hand tools, edible landscaping, fiber production, and farming?

Aspects of the restorative economy are already appearing today, in fits and starts. Wind is the fastest-growing source of electricity. Here and there buildings are going up that are net energy exporters. BMW is building cars that can be disassembled and the parts reused; Matsushita makes washing machines that can be completely disassembled with a single screwdriver.[30] The trend is especially evident in agriculture, from permaculture to the movement of thousands of organic and small farmers away from the factory production model. Some of the technologies of the restorative economy exist already and have existed in some cases for decades. All that is missing is the cultural and economic context: the structural incentives to reward sustainable technologies, and the spiritual revolution that will end our dualistic alienation from nature.

Concurrent with the resurgence of many "low" technologies will be the continued development of certain trends in high technology. For example, the storage, distribution, and manipulation of data will continue to demand less and less of a physical substrate. It will require less and less energy as well. Even nanotechnology, whose inflated promise I derided in Chapter One, will produce new wonders. We will not, however, conceive of nanotechnology as a new level of domination over nature, but as a new arena of intimate, cocreative partnership, motivated by beauty not profit. (Or actually motivated by both, as the two will no longer be at odds.) Again, I do not advocate the abdication of our human gifts of hand and mind. Only the motivation, and therefore the direction and application, of technology will change. We can expect continued "magic and miracles" from technology-as-we-know-it.

Some of the science-fictiony scenarios of future communications technology correspond remarkably to the shift in the human sense of self I have described. We can look at cell phones and instant messaging already as a form of telepathy. These technologies, after all, allow us to share our thoughts with other people over a distance. The dephysicalization of communication media could make such communication even more direct, to the point where we could indeed have access to large parts of each other's minds. Let's go through one of these scenarios, ignoring for now its truly Orwellian implications. Imagine that instead of a keyboard and mouse, implanted data input devices were directly linked to neurons so that we could type (or point and click) at the speed of thought. (Crude prototypes of such devices already exist.[31]) We could also upload to a computer archive any thought we wanted to remember or make available to other people. Meanwhile, we could receive messages with implanted devices that automatically stimulate auditory and optic neurons. Words, voices, and images would flash through our brains from an external source. But what is external? Wouldn't these collectively-accessible data banks be an extension of our own brains, of our own selves—an extension shared with others, a common mind? How would the boundaries of self change, when our private thoughts are no longer private, and the voices and images in our heads might come from somewhere else?

When the linear technologies of certainty and control retreat to develop in their proper sphere, the way will be clear for other kinds of technology. What would a technology look like that were not rooted in the ontology of the discrete and separate self? Like the technologies of water, it would utilize connections that our present delusion of separateness renders invisible to us today. These connections are mediated not just by water but by electromagnetic fields, by DNA, and by vibratory media presently unrecognized or unexplored by science; yet they are foreshadowed in principle by 20th-century science: by the observer-dependency of quantum mechanics, by the spontaneous organization of non-linear systems, and by the cooperation and interconnectedness of all life on earth.

This mode of technology seeks human growth not by dominating an external environment, but by exploring and reifying the true vastness of what we actually are. The next stage of human development will come about because we stop resisting it and allow it to happen, not by engineering reality into some new and improved shape. This is a difference in motivation and conception, but not the end of science. The best of science will remain: the humility of the Scientific Method, the curiosity and wonder that drives it, and the awe and the mystery that sanctifies it. The goal will be different though. In contrast to the ideology of ascent described in Chapter One, which seeks to control what is by knowing what is, future science will seek to know what *should be*. It therefore will require a knowledge of ourselves and our rightful place in nature. In this regard, it will draw from the new paradigms of biology described in Chapter Six, in which environmental purpose helps determine the evolution and behavior of the individual. It will draw as well from the primitive ritual sciences and shamanistic technologies, which sought a reconnection with the natural order and not its domination.

For the natural order is far greater than we imagine it to be. I am not suggesting a retreat to some brute existence where we content ourselves with smelling the flowers and listening to the birds. Wait, let me take that back—there is far more to the scent of flowers and the songs of birds than we realize! What we label "nature" is an important key to unlocking human potential that few people today can imagine.

The scents of flowers… did you know that there are sophisticated, highly nuanced technologies that uses the "vibrational" essence of flowers as an agent for psychospiritual healing and personal development? Dismissed as New-age hooey, flower essence therapy, aromatherapy, and related modalities have tens of thousands of adherents in North America and England whose primary texts exhibit an emotional sophistication and internal logical cogency that defy facile dismissal. They are part of a resurgence in herbal medicine that appeals to a connectedness between the human and plant worlds that goes beyond known biochemical relationships. I especially recommend the books of Matthew Wood, Stephen Buhner, and Eliot Cowan for their placement of herbalism into a vaster, non-reductionistic paradigm. Conventional science, to the extent it recognizes the effectiveness of herbs at all, explains it reductionistically in terms of the direct linear effects of this or that "active constituent". But as Wood, Buhner, and Cowan convincingly demonstrate, plants also work through mechanisms that, for lack of better understanding, we can only call "energetic", "vibrational", or "magical". And this "lack of better understanding" is the hallmark of a science in its mere infancy, hinting at a future science of magic, energy, or vibration that is to present efforts as a supercomputer is to an abacus.

Another technology of Reunion that produces astonishing results through a harmonizing with, and not a domination of, nature is biodynamic agriculture. Derided by some for its seemingly mystical elements, it is another infant technology that even its most ardent proponents are barely beginning to understand. Together with more mundane technologies such as permaculture and more mysterious technologies such as the intuitive gardening techniques developed by Machaelle Small Wright, it holds the promise of a "garden earth". I was about to call it a "cocreative partnership between humans and nature", but the true potential is greater than that. It is the release of the divide between human and nature, between the domestic and the wild. We can envision an agriculture that incorporates all the characteristics of an ecology rather than seeking to resist them.

Besides water, biological chemicals such as DNA, enzymes, and pheromones are also media of interconnectedness. They act together to transmit information from one part of Gaia to another, one part of any ecosystem to another, one part of an organism to another so that each component can fulfill its function in the greater whole. If we human beings are to do the same, it stands to reason that future technology would fully develop the capabilities of each of these media. I am not talking here of "genetic engineering," which attempts to subvert DNA to narrowly-conceived human ambitions—the program of control on the molecular level. I refer rather to the untapped, even unsuspected, genetic potential already present inside us and our symbiotic partners (all of life, including viruses). Sequestered amid our "junk" DNA lies dormant coding for capacities few can imagine today, awaiting the appropriate "switches"—stimuli from the somatic or external environment—to become active.

The biosphere is a treasure-trove of genetic riches that we have barely begun to explore. It contains the means to achieve purposes undreamed of. Paul Stamets did not genetically engineer his mushrooms to detoxify toxic waste—those capabilities existed already. He is merely a student and a steward of a marvelous Nature. His work, and the work of horticulturalists for thousands of years, suggests a different conception of technology—to receive the gifts of the earth and pass them on. That is the "spirit of the gift" applied to science and technology. It is not a passive receiving though. Careful observation, insight, and patient work are necessary to discover these gifts. The difference is a trust in the fundamental providence of nature that makes control-based technologies such as genetic engineering unnecessary.[32]

I just input "DNA activation" into my search engine and got 120,000 listings, a veritable cottage industry utterly outside the bounds of established science. Much of it is in the highly subjective realm of intuition, angelic channeling, and so forth, sometimes dressed up in flimsy pseudo-scientific jargon. I find it more authentic when they don't pretend it is science. I was especially disappointed when I looked into Gregg Braden's "God Code" and found one egregious abuse of mathematics after another in his proposition that our DNA encodes text from the Old Testament of the Bible. "Disappointed" is the word, not contemptuous, because on the level of poetry and myth, Braden is speaking some important truths. He is speaking to the innate sacredness and limitless potential of life. Underneath the New Age dross in this and related areas like biophotonics and bioenergy, there is valid insight and in some cases even sound scientific research.[33]

Pioneers usually get something wrong. Keep in mind that these technologies are in their infancy. Not just DNA but all of our biology harbors untapped potential that can be applied to healing, creativity, and the fuller participation in our cosmic purpose.

Another aspect of our interconnectedness with each other, the planet, and all life is electromagnetic. Once thought to be an insignificant epiphenomenon, the noise of the machine, the electromagnetic fields generated by our brains, hearts, organs, and indeed every living cell are now recognized by a few visionary scientists to carry information to which all life responds. Already, medical pioneers are exploring how to apply this principle to healing through a proliferation of gadgets you can find on the Internet, most of them backed only by anecdotal evidence. Given this unexplored realm of electromagnetic communication that interweaves every living cell, the standard control-based response would be, "How can we take advantage of this knowledge to exert more powerful, more precise control over reality?" In the Age of Reunion, the response will be to become more attuned to the ocean of being in which we are embedded. From that attunement will arise knowledge, such as that we label "intuitive", that would seem supernatural to our present way of thinking. Corresponding to that knowledge will arise undreamed of ways to communicate with each other, with plants, animals, disease organisms, and even our own organs, tissues, and cells. Technologies are already emerging that amplify or clarify this communication; among them are the Feldenkrais Method, EFT, Spinal Network Analysis, applied kinesiology, and numerous "energy" modalities.

While scientific orthodoxy generally rejects the conclusions I've shared above, they are at least plausible in the sense that they appeal to known physical forces. A responsible scientist might say, "Well, it could be true but I doubt that it is": irresponsible conjecture but at least possible. Unfortunately, "known physical forces" apparently cannot account for all the phenomena that are observed in alternative medicine, psi, qigong, and many other fields. For example, effects that might be explained in terms of electromagnetic frequencies often do not obey the inverse-square rule. Others apparently violate causality by projecting effects backward in time. I propose that the mechanisms of water structure, electromagnetism, DNA and so forth described above are just a few of the manifestations of a fundamental principle of interconnectedness or holism that will be the organizing principle of future technology.

With the demise of Cartesian objectivity, the intrinsic inseparability of observer and observed suggests the possibility of influencing reality through focused belief and intention. Experiments at the Princeton Engineering Anomalies Research Lab (PEAR) have shown, over millions of trials, a statistically significant observer effect on supposedly random quantum events. Other experiments pioneered by Helmut Schmidt seem to show that this effect can extend backward in time to affect prerecorded data. The HeartMath Institute has performed experiments documenting a measurable emotional response to disturbing photos randomly placed among a series of banal photographs—several seconds before the photograph is actually viewed. Cleve Backster has spent decades investigating how plants exhibit a galvanic response not only to damage to nearby plants, but even to the intention of doing such damage. Then of course there are the numerous studies on the effects of prayer on healing, including double-blinded studies published in top medical journals. These are but the tip of the iceberg in the world of "anomalies research" which I mention here only to suggest a possibility, not to convince you that it is real. I leave that to others! In the interests of honest disclosure: I believe it is real—with a caveat, however. What does "real" mean? It is very difficult to define real apart from the objective reality "out there" of the Newtonian World Machine. In fact, all the phenomena I've touched upon in this chapter, and many I have not discussed, are notoriously elusive in the laboratory. The more rigorous the controls, the less visible the phenomena, inviting the conclusion that they are but artifacts of sloppy procedure. The most dramatic effects are anecdotal and impossible to verify, experimental results are often not reproducible, the conventional wisdom of thousands of practitioners merges into the placebo effect in the strictest double-blind studies. This elusiveness is a central feature of the phenomena and the key to understanding them. For now, just ponder what "real" might mean in a world that is not objective, where something can happen for me and not for you, where there is no absolute universal Cartesian coordinate system that is the stage for reality.

I mean it: put down this book for a minute and ponder what "real" could mean without an objective backdrop. I've been thinking about this almost daily for ten years now and the concept still gives me vertigo.

Curiously, a similar elusiveness has plagued other controversial areas of science such as cold fusion. In this case, the original researchers, Fleischmann and Pons, announced a dramatic result which other scientists were unable to replicate. Or were they? Many actually did get results; others did not. But did the latter follow the original protocol *exactly*? Some of the laboratories that failed to replicate Pons and Fleischmann's findings had a vested interest in rejecting them—many were part of the "hot fusion" research establishment. Particularly egregious was the definitive M.I.T. rejection of cold fusion when their attempt to replicate the experiment failed to produce the levels of neutrons that conventional theory predicts. It couldn't be happening, they concluded. Only later did it emerge that the M.I.T. experiment actually did produce anomalous heat. The controversy still simmers today, with the general consensus being that the phenomenon is an artifact or a fluke. Much of this comes down to the old standby, "It isn't true because it couldn't be true—there is no known mechanism to explain it." However, a third possibility exists. Perhaps cold fusion only works in an appropriate atmosphere of belief, and only works regularly and generally if that atmosphere of belief pertains to the society as a whole. Perhaps we must understand cold fusion, too, as a technology of mind as well as a technology of matter.

Technotopians fondly dream of a new source of clean, cheap, limitless energy—then all our problems would be solved! But perhaps we can only tap into these sources when we have let go of the ideologies of separation that cut us off from the plenitude of the universe. Perhaps our mentality of lack, want, and scarcity necessarily projects onto our energy technology, and true abundance cannot come until we cease trying to enclose reality into a private human realm of ownership and control. In other words, limitless energy sources will come when we are ready for them. Even if individual researchers in cold fusion (and many other "infinite energy" fields) are getting positive results, an ideological force thwarts their broad application. This will not change until our fundamental worldview, the mentality of scarcity and separation, changes. I for one am glad we don't have infinite energy yet!

Well, actually we do. Even if cold fusion, zero point energy, hydrino energy, over-unity magnetic turbines, and all the other infinite energy technologies are fantasies, the current mounting energy shortage is still the indirect result of our mentality of scarcity. Our universe is awash in limitless energy. Imagine that all the money spent on oil wars in the last half-century had gone toward developing wind, solar, tidal, and geothermal power. Imagine that we had a financial system that encouraged the capitalization of future energy savings. We already have the technology of energy abundance. Most of the uses to which fossil fuel energy is put do little to contribute to human happiness. What proportion is devoted to the manufacture of armaments and the conducting of wars? To shoddy consumer goods that quickly end up in landfills? To energy-hungry McMansions in which people are isolated, lonely, and miserable? To running the car-dependent infrastructure of suburbia, in which people are no happier than in compact villages served by bicycle and rail? The green technologies mentioned earlier in this chapter do not detract from our quality of life, and they needn't await some magical new invention. Yet just as with the more magical "infinite energy" technologies, it is our beliefs that keep their abundance at a distance. At bottom, it is our fundamental belief in the discrete self, separate from nature and from each other, that generates the war, the economic system, the acquisitiveness that consumes the vast part of our energy cornucopia. Many of the ways that beliefs affect reality are quite mundane.

The technologies of mind suggested above come at a price: the loss of certainty. However powerful they are, they cannot be used to impose control over an external universe. Imagine it were all true: that the human mind can in fact bend spoons (and maybe much more), that people can learn to levitate or be two places at once, that cancerous tumors can melt away like snowballs on a hot stove, that people can materialize objects at will, or see into the human body to diagnose disease. And imagine that the scattered stories of such events point to only a tiny fragment of our true potential. A future where the technologies of fire have retreated to their proper place need not be stagnant or dull! But when such technology is possible only through intention and belief, then we must let go of the ideology of the rational observer who sits back and makes observations on which to base beliefs. I mentioned in Chapter Six the idea that it is beliefs that determine observations and not vice-versa as we suppose. If that becomes the foundation of a new science, then the technology arising from that science must also abandon the very same certainty.

In place of certainty, we will have self-exploration, an awareness of choice, and a sense of empowering creativity. Realizing the illusory nature of the self/not-self distinction, the falsity of an independently existing universe "out there", we will abandon the futile program to force life into a mold of security and ease. Futile, because security and ease are at odds with the deep anxiety and insecurity which motivate that program. As we step more and more into the realization of our connectedness, our actions will reinforce that connectedness and bring it even further into our experience. Reality will become more malleable to our beliefs, assuming more of a fluid dreamlike quality. Paradoxically, by releasing the mentality of control, certainty, and proof, we will exercise far greater power to determine our experience of reality, choosing what we shall experience rather than impotently struggling against what we have unconsciously chosen.

Work and Art United

Work is a defining feature of modern life. When we ask someone, "What do you do?" we typically mean, "What is your work," and by that we mean, "What do you do for money?" And work, a defining feature of life, is itself associated with drudgery, tedium, sacrifice, and routine. The daily grind. These features are characteristic of the Machine, and it is no wonder, immersed as we are in a machine civilization, that we view work in contradistinction to leisure and to fun.

The demise of the mechanical view of the universe and the end of Machine civilization holds a new promise—not for the obsolescence of work, but for its transformation. Work in the post-Machine age, in an ecological economy, will take on a new character. This transformation will go beyond the content and nature of work itself to revolutionize the economic relationship of "employment". Work and leisure, job and life, will become one.

For several hundred years, machines have been replacing human beings for rote, laborious tasks. Paradoxically, most human occupations are still tedious and routine. That is because even as machines have taken over more and more functions, new functions have been born of the expanding megamachine, whose human parts—no less than its inanimate parts—must fulfill the machine's requirement of standardization. The very effort to eliminate labor demands more labor than it eliminates, because the effort is itself an intensification of machine methodology. The master's tools can never dismantle the master's house. The transformation of work is coming in a different way than two centuries of futurists have envisioned. Not because all routine functions have been once and for all automated, but because society will require fewer and fewer such functions.

One consequence of local currencies and green technologies will be a decrease in scale in many industries. Today, many mass production processes are only more efficient because they externalize costs. Agriculture is a prime example. Factory farming is more efficient in terms of labor hours, but far less efficient in terms of energy use or pollution production, than organic agriculture. When costs are internalized, many industrial practices will give way to more labor-intensive methods. In many areas, mass production will become obsolete, and along with it the archetype of the assembly line, in which work is routine, repetitive, and highly fragmented, requiring little skill, little personal contact, and many layers of hierarchical management.

These characteristics stand in contrast to repair work, which is local, small-scale, highly skilled, personal, and tangible in its results. The economic logic that today makes it cheaper to buy a new stereo rather than repair an old one will be reversed as raw materials come to reflect their true costs and as designers begin to design for durability and ease of repair. The result will be more labor-intensive (repairs do require significant labor) but the nature of the labor will be vastly different, as will the economic model it entails.

Another impetus toward the passing of the factory system, and along with it the whole psychology of the Mumfordian megamachine, is the increasingly unmanageable complexity of engineering processes and business administration (unmanageable, that is, according to the reductionist paradigm of management). The resulting transformation in business, visible already in the flattening of corporate hierarchies, is not some trendy new business fad but a practical necessity, a consequence of the breakdown of design processes under the weight of their own management. Increasingly, today's complex processes are impervious to traditional cut-and-control engineering, in which the problem is broken down into parts, sub-parts, and so forth, mirroring the traditional business hierarchy. Whether in business or engineering, the old approach to problem solving—break it down and treat each piece separately—is no longer working very well.

In any machine, whether made of human or inanimate parts, the complexity of relationships among parts increases exponentially with their number. The proliferation of variables quickly gets out of control, and at some point traditional reductionistic solutions must give way to solutions that are *grown*, not engineered. This is already happening in many fields such as circuit design, where evolutionary algorithms search for solutions inaccessible to analytic methods. Software engineering is another example in which the product's complexity defies hierarchical management; increasingly, solutions grow from the ground up among legions of independent programmers.

Visionary thinkers such as Michel Bauwens have generalized the peer-to-peer (P2P) revolution in the field of information technology as a new model of economic, social, and political organization. Peer-to-peer networks bear a striking resemblance to the gift economies of ancient times, as well as to the characteristics of a demurrage-based economy. As in a potlatch society, status in such a community comes from how much one contributes, not how much one owns. Resources are shared generally, not hoarded, and the exchange of goods occurs on a gift basis. Such networks have produced a new model of journalism (the blogosphere) that compiles, filters, edits, and organizes enormous quantities of information far more efficiently than traditional news organizations can.[34] Other structures with P2P elements include the Wikipedia and the structures growing up around Amazon.com, eBay, and Google. All of them profit by giving information away for free.[35] One by one, the information providers who clung to the business model of amassing and selling vast quantities of proprietary information have fallen by the wayside (the *Encyclopedia Britannica* is an example). Others in the music, film, and software industries are still struggling to hold on to the information-equals-property model.

Farther out on the margins, a huge shadow economy thrives that is even more explicitly gift-based. The people who participate in file-sharing networks trust fully in the spirit of the gift when they put their music collection or software cracks[36] on line for anyone to download. In some of these subcultures it is considered gauche to ever demand money for data.

The concept of profiting by giving it away is part of a larger shift in the nature of work. Instead of working for money, money becomes a side-effect of doing good work. No longer slave to money, work serves other goals: beauty, service, fun, or self-expression. The worker, in other words, becomes an artist. This is also a consequence of a very different relationship to material objects that will develop in a restorative economy.

Because of the internalization of costs, the mounds of plastic junk we buy today will be far more expensive—so much so that there will no longer be mounds of plastic junk. Consumer goods in general won't be so cheap anymore, and I mean that in both senses of the word "cheap". We consider ourselves to be the wealthiest society in history, but lives full of vast quantities of cheap stuff are themselves poor and cheap. To live among objects that are not cheap, but made with consummate skill, attention, and care, would be true material wealth. Can you imagine a society where each person's talents and gifts were fully expressed in their work, and not suppressed in the interests of machine life? Can you imagine a home in which every appliance, garment, and piece of furniture were so wisely designed, so well-made, so elegant a marriage of beauty and function, that no one would ever miss throw-away things too cheap to really care about? Life will abound with beautifully made objects, because in the restorative economy all of us will be craftspeople, expressing our full talents in our work rather than denying them for the sake of keeping a job. Part of this will be a dramatic revival of traditional handcrafts, as "natural resources" will have become so precious as to merit the best individual workmanship. But even in the high-tech sector the number of narrowly specialized work functions will be far less than it is today, and each person will consider him or herself an artist.

The restorative economy will thus demolish some of our culture's basic dualisms: work and art, work and leisure, utility and beauty. When this happens, the noble ideal at the heart of our dreams of technological Utopia will come to fruition: not in the obsolescence of work, but in the realization of its true nature. When work is seen from the perspective of Mumfordian machine civilization, then it is inescapably narrow and oppressive. That is why the technotopian dreamers looked to the end of labor, each man a king served by machine slaves. Bound by their preconceptions of the nature of work, they could imagine no better than that. But to reunite work and art, to heal the split between work and the rest of life from which arises the selloff of our time, is a much more radical ambition.

It is an insult to our dignity to live off the produce of anyone doing demeaning work because she has to in order to survive. Work done under compulsion is slave work.. A truly wealthy person has no need for such things. Can we envision a life whose objects were all made by people at their best?

From the egalitarian societies of the Paleolithic, humanity evolved into great agrarian civilizations in which the rich were those who owned slaves. In the Machine age, overt slavery disappeared, only to be replaced with a system in which nearly everyone did demeaning work out of survival anxiety. "Do it or you will die!" That's slavery, all right. The great promise of machine technology—Every man a king! Every man a god!—has borne its opposite. Every man a slave. Slaves without human owners, all laboring under the yoke of money. But now, with the end of the age of the Machine, we see the possibility of a return to the original egalitarianism, in which the economy is a flow of gifts within a context of abundance.

If there are still career counselors in the Age of Reunion, they will help us answer the questions, "Which of my creative talents do I enjoy most? Which of the arts would I like to pursue?" No longer will work and art be at odds, but one and the same.

The reunion of work and art accompanies a new kind of materialism. There is a virtue and a pleasure in taking good care of one's things. Before the days of material surfeit and the flood of cheap stuff, people did just that. As recently as the 1930s, people treasured articles like hand tools, fishing rods, tricycles, and children's toys, which received the care needed to last a lifetime, even generations. Today, why bother? Why spend such time and effort on something cheap? A new one only costs twenty dollars. Like TV repairs, caring for our things has become economically inefficient. From one perspective this is a convenience, as we are freed from the burden of having to take care of our things. But from the perspective of Chapter Four it is slavery. Economic exigencies—the exchange of time for money—have rendered us *unable to afford* the virtue and pleasure of taking care of things. Today's flood of cheapness saves time but cheapens life. By undoing the conversion of life into money the restorative economy will free us again to love our things, and will provide us things worth loving.

I am not proposing a disassociation from the material world, or the phony spirituality of overcoming the world and the flesh for the sake of some separate Cartesian soul. To the contrary, I envision a future where we love our material possessions more and not less, so that we care where they come from and whence they go. I envision a future in which we see life in the material world as an opportunity to participate in its constant generation of beauty.

More than just a consequence of our economic system, the cheapness of modern life is a reflection of the devaluing of materiality implicit in the Cartesian abstraction of spirit away from matter. Who cares about the material world, when it is separate from our spiritual selves? The collapse of the Newtonian World-machine will reunite us with the world, and we shall once again fall in love with it. To be in love is to dissolve boundaries, to expand oneself to include an other. Already it is happening. Have you noticed? One by one, we are rejecting our society's priorities and falling in love again with life. That is our true nature, which we can deny only with increasing effort. It is our nature to love life in both senses of the word: biological life and our personal lives. To love the world and to love our time in it. We have long been frightened into rejecting both, accepting as a result their plunder: the reduction of the living world into resources, things, money, and the reduction of our time into commodified hours, jobs, the grim necessity of making a living. The good news is that when we let go of separation and thus fall in love again with life, these results will also give way to their opposites. For the world and for ourselves, we will accept nothing less than lives devoted ardently to creating beautiful things, beautiful music, beautiful ideas. "Could you or I comprehend, even for a moment, how fiercely my great-granddaughter and her friends loved being alive, and that this love was not an evanescent mood, but a never-ending power which pervaded every sleeping and waking moment of their lives?"

I do not deny that there may always be dishes to wash, toilets to clean, buildings to roof—tasks that we consider mindless, laborious, or repulsive.[37] But what is wrong with exerting bodily effort in order to maintain life? Shall we invent chewing machines to save us the "labor" of chewing our own food? Is the goal of humanity's ascent to rest forever in bed, hooked up to various machines that provide us nourishment, pleasure, and excitement all without our own effort? That would be the consummation of the machine's promise, wouldn't it—a servant that not only does all our work for us, but lives our lives on our behalf as well. No, the movements of being alive need not be laborious. What makes work laborious is the sacrifice of variety for the efficiency of repetition and standardization. Herein lies the difference between farming and gardening: the former has been the epitome of drudgery; the latter is such a joy that people do it in spite of its economic irrationality.[38,39] The more a farm resembles a factory (and the less a garden), the more tedious and life-denying farm work is. Conventional agriculture works according to the same principles as any other industrial enterprise. In a garden, no one spends weeks picking cotton, tomatoes, or grapes. On a small mixed farm no one spends weeks, months or years performing over and over again one step in the process of butchering chickens.

What makes a task oppressive is not its content but its duration, motivation, and purpose. Cleaning one's own toilet is neither demeaning nor particularly laborious—quite different from cleaning hundreds of toilets, all day, for strangers. No one was born onto this earth to clean toilets. And no one would submit to such a life who were not broken or coerced by threat to survival. In a society where work is art and money is not scarce, people will be unwilling to sell their time, their dignity, or their integrity for money. Not only assembly-line work and menial labor, but any work that is degrading or life-denying will need to be engineered out of the economy. How often today do industrial design specifications include such a requirement?

The Age of Reunion is not a return to the past nor an abandonment of technology. Rather it is the motivation and organizing principle of technology that will change. When the psychological and economic forces driving the conversion of life into money are reversed, then technology will no longer seek to make that conversion faster and more efficient. The engineers of the future will design for sustainability, for dignity, and for beauty. They will, in other words, be artists, creating technologies for a world of artists on a garden earth. The conflict that all artists encounter between creating for the market and creating for the spirit will cease, when work and art, money and sustainability, come into alignment.

Back to Play

Many teachers think of children as immature adults. It might lead to better and more 'respectful' teaching, if we thought of adults as atrophied children.

-- Keith Johnstone

Early in the industrial revolution, modern schooling was invented in order to acclimate children to work that was tedious, repetitive, degrading, and unfulfilling. The transformation of work into art therefore coincides with a parallel transformation in education that will prepare children for lives as artists. In "The Currency of Cooperation" section I wrote: "Work will no longer be bound to the search for money, but will seek out ways to best serve each other and the world, each according to our unique gifts and temperament." In this context, education will be transformed from a preparation to earn a living into a process of self-discovery. It will seek to help each of us answer the questions, "What do I love to do? What are my unique gifts? How might I serve others and participate in the creation of a more beautiful world?"

That these ideals exist already, both in education and in work, points to our intuitive realization of their validity and the unquenchable longing for a world better than the one we experience today. We know deep down that that is what education *should* be. Educators give lip service to these ideas of self-discovery, but in practice the imperative of control militates against their true expression.

In parallel to the financial crisis that will clear the way for a new system of money and the ecological crisis that will spur the adoption of a new system of industry, an educational crisis is crashing down on us that threatens to obliterate the school system as we know it. Already the schools are hemorrhaging teaching talent, legitimacy, and students as classroom violence escalates and literacy rates plummet. The response, typically, is ever more control: more standardized testing for students, more stringent certifications for teachers, more armed guards, metal detectors, locker searches, and razor wire for the school buildings.

Successful models of schooling exist that rely on the release of control and not its intensification. Their starting point is a faith in the innate curiosity and creativity of the human being. "Man by nature desires to learn," wrote Aristotle, as a child's enormous capacity for uncoerced learning (before schooling starts) demonstrates. Educational philosophers in the Montessori and Waldorf movements have built entire pedagogies around providing resources to meet children's natural curiosity and desire to grow at each stage of development. Their alternatives to coercion, born of trust in the innate curiosity, intelligence, motivation, and wisdom of the human being, go against the institutional requirements of machine civilization and the ideology of conquering (human) nature.

A beautiful unadulterated example of the alternative to control-based education is the Sudbury Valley School of Massachusetts, which takes the principle of trusting the child beyond even Waldorf or Montessori. This is a school with no curriculum, no tests, no grades, and no rules except those legislated by the students themselves. One of the founders writes,

> We wanted [students] to be entirely free to choose their own materials, and books, and teachers. We felt that the only learning that ever counts in life happens when the learners have thrown themselves into a subject on their own, without coaxing, or bribing, or pressure.... In order to be true to ourselves we had to get away from any notion of a school-inspired program. We had to let all the drive come from the students, with the school committed only to responding to this drive.[40]

For example, "At Sudbury Valley, not one child has ever been forced, pushed, urged, cajoled, or bribed into learning how to read." And significantly, "We have no dyslexia. None of our graduates are real or functional illiterates." Despite a lack of any external coercive mechanism or external incentives, Sudbury children display an amazing level of achievement, though not necessarily in traditional academic subjects. Even in the traditional subjects, though, they usually cover the bases, simply because traditional subjects are actually quite simple. One anecdote about the school describes how a self-motivated group of 9-to-12-year-olds learned the entire arithmetic curriculum from first to sixth grade in twenty contact hours.[41] An outside educational expert was not surprised:

> Everyone knows that the subject matter itself isn't that hard. What's hard, virtually impossible, is beating it into the heads of youngsters who hate every step. The only way we have a ghost of a chance is to hammer away at the stuff bit by bit every day for years. Even then it does not work. Most of the sixth graders are mathematical illiterates. Give me a kid who wants to learn the stuff—well, twenty hours or so makes sense.[42]

Even more counterintuitive (to the intuitions of Separation) than self-motivated learning is the complete self-governance of the student body. The school assembly, in which every student age five to eighteen and every staff member gets one vote, makes all the rules and important decisions for the school, in sharp contrast to my own high school voting system, in which we were allowed to choose the school mascot and homecoming queen. At Sudbury, the students have real power: they decide on the disposition of funds, the hiring and firing of teachers, and on rules and penalties for infractions. Rule-breakers are subject to a court system staffed by students. Distrustful of human nature, we imagine such a system could never work, but from most accounts it works beautifully. Sometimes the assembly makes mistakes, of course: At the Circle School, a democratic school in Pennsylvania that my children attend, the assembly once voted to abolish chores (the teachers voted against this proposal but were outnumbered). So, there were no chores—until two weeks later, when the assembly reversed its decision. The students had learned through experience the necessity of chores, very different than the usual model where the reason for doing chores is "because you have to." They learn self-trust rather than obedience to authority, a model that only makes sense if you believe the self is to be trusted. The Sudbury model embodies radical assumptions about human nature and our understanding of self. The school has pioneered a vision of what school will be in the Age of Reunion, when the present effort to hold together the illusion of the discrete and separate self finally becomes unbearable, and collapses. Until that happens, schools like Sudbury Valley, in parallel with local currencies and energy healing, can occupy only a marginal place in our society because they conflict with the dominant institutionalized worldview. Nonetheless, they are extremely important because they provide models which we will naturally and thankfully adopt as we rebuild a society after the convergence of crises.

If they are never "forced, pushed, urged, cajoled, or bribed" into doing anything, what do the students at Sudbury Valley do? So used are we to lives guided by internalized coercion that we imagine, in the absence of self-control, a life of indolence. When I ask my students what would happen if they lost all willpower, they usually say they would stay in bed until noon, lounge around all day in front of the TV, indulging in the nearest pleasures and conveniences, and after that, "a vague never-ending spiral of indulgence, indolence, and apathy."[43] But this is really just a rebellion against "work", when that term is defined, as it has been since the Industrial Revolution if not before, as something unpleasant, degrading, or laborious that we are forced to do in order to survive. It is not human nature. Sloth and indulgence are not human nature. What is human nature? We can see it at Sudbury Valley, because what it is that children do in the absence of coercion is, quite simply, to play.

And here we come up against one of our culture's defining dualities, the distinction between work and play. The last section of Chapter Two, "The Playful Universe," describes how "play purely for play's sake is [seen as] a waste of time, a view based on the purpose-of-life-is-to-survive assumption that underlies modern science and economics. After all, every minute spent playing is a lost opportunity to get ahead in life." However, far from being frivolous or silly, "Play at school is serious business… play is always serious for kids, as well as for adults who have not forgotten how…"[44] Play is nothing less, in fact, than a child's version of what adult life should be as well. Its value is not in the motor skills or problem solving techniques it develops; "What is learned is the ability to concentrate and focus attention unsparingly on the task at hand, without regard for limitations—no tiredness, no rushing, no need to abandon a hot idea in the middle to go on to something else." Play is untrammeled creativity that comes from within. In adult life, play has vanished under the relentless regime of "shoulds" that prevent us from fully devoting ourselves to anything "without regard for limitations". These "shoulds" *are* the limitations: schedules, pressure, guilt, survival anxiety. We are usually looking over our shoulder.

We experience the concentration and unsparing focus of play as a feeling of timelessness. Joseph Chilton Pearce calls this state "entrainment", in which our ability to learn and to create are at a maximum. (Ironically, the coded threats through which we motivate learning actually create a psychological state inimical to learning.) The timelessness of play is almost impossible to reconcile with the busy, scheduled life of the modern adult (or schoolchild). As explained in Chapter Two, primitive societies were timeless societies: hours and minutes, times and dates, clocks and schedules only arose with the complex coordination of human activity necessitated by the division of labor and the Machine. A life at play is a timeless life.

Here is a passage from *Free at Last* that I offer you just because it brings tears to my eyes, so enormous the crime it reveals that our civilization has committed against the human spirit, and so simple is its implied prescription.

> School opens at 8:30 in the morning, closes at 5:00 in the afternoon. It isn't unusual to see someone go into the darkroom at 9:00, lose track of time, and emerge at 4:00 when the work is done.
>
> Jacob seats himself before the potter's wheel. He is thirteen years old. It is 10:30 a.m. He gets ready, and starts throwing pots. An hour passes. Two hours. Activities swirl around him. His friends start a game of soccer, without him. Three hours. At 2:15 he rises from the wheel. Today, he has nothing to show for his efforts. Not a single pot satisfies

> him.
>
> Next day, he tries again. This time, he rises at 1:00, after finishing three specimens he likes.
>
> Thomas and Nathan, aged eleven, begin a game of Dungeons and Dragons at 9:00. It isn't over by 5:00. Nor by 5:00 the next day. On the third day, they wrap it up at 2:00.
>
> Shirley, nine, curls up in a chair and starts to read a book. She continues at home, and the next three days, until it is finished.
>
> Six year old Cindy and Sharon take off for a walk in the woods. It is a lovely Spring day. They are out four hours.
>
> Dan casts his first line into the pond early on Fall morning. Three years later, he is still fishing.

Can you recognize this as a model for the lives we are meant to live? To choose our activities and devote ourselves to them fully until we are satisfied with the results? To be free of any schedule or requirement but our own? So it is in our culture's most powerful archetype of creativity: the Biblical Genesis when, after creating the world, God said, "It is good." Today, instead, we settle not for "good" but for "good enough"—good enough for the grade, the boss, the market—and in so doing deny the creative purpose for which we are here. But like Jacob, age thirteen, something in us desires to create beauty free of these limitations, to immerse ourselves in a creative task for its own sake for as long as it takes so that, in the end, we might look upon it with satisfaction and say, "It is good."

Daniel Greenberg writes, "Time is not a commodity at Sudbury Valley. It is not 'used', either poorly or well. It is not 'wasted', or 'saved'.... The respect the school shows to private rhythms is inviolate." The children there offer us a model, not only for education but for life, that subverts cultural assumptions so deeply rooted that we are hardly aware of them. To no longer think of time as something to use, waste, or save—that would be a revolution far more profound than anything Karl Marx dreamed of.

I have gone into such depth about the Sudbury Valley School because it offers a model for bringing us back to play. And coming back to play is the key to overthrowing the hegemony of measured time that has bound us, tighter and tighter, for thousands of years. Since early childhood, few of us have ever experienced a timeless life, except in brief stolen snatches. We adults have relegated much of our play and creativity to the margins—weekend hobbies irrelevant to our livelihoods—just as ordinary school students must sneak it in between periods or behind the teacher's back.

I do not advocate a return to Stone Age technology as the only means of undoing the bitter consequences of machine ideology. While the outward forms of technology, money, medicine, and education will certainly change, they will do so as a result of a new relationship to the universe that is best summarized, perhaps, as playful. Not only is play timeless, not only is it free of the coercive mechanisms that create ugliness as the price of survival, but it also undoes the artificial self-other distinction that defines the Age of Separation. In the entrainment state of play, a sense of a separate self vanishes as the task at hand absorbs us; instead of a discrete subject manipulating the universe, we become an organic agent of the universe's own creative process. Through us, the universe creates itself. The same goes for another kind of play that instead of creative might be called "exploratory." We play with our boundaries, explore our limits, define who we are. The archetype of this kind of play might be the infant playing with her toes, her hands, and her voice, as she learns to coordinate movement, sense organs, and speech. She is not merely discovering a preexisting reality of her body; her explorations are what stimulates her body's development. In all its aspects, true play is inherently and unavoidably creative.

The Age of Reunion will be an age of play that will redefine every human endeavor, not just education but also work, art, and science. Like an infant playing with yet simultaneously creating her body, science will no longer assume an independently-existing reality "out there" awaiting discovery. But what of the Scientific Method, based on Newtonian assumptions of determinism and objectivity? Articulated by Bacon as a means of interrogating nature and extracting her secrets, the Scientific Method also embodies an intellectual humility, the flip side of its dispassionate objectivity. It represents an intention to hold lightly to preconceived beliefs (hypotheses), to hold lightly onto the existing understanding of the world and our relation to the world (which is really what an experiment is, an exploratory way of relating to the world). The Scientific Method will still be part of future science, but it will be conceived in the spirit of play. Experiment, which is nothing other than the spirit of "Let's see what happens if… " will be a way of playing with nature, not torturing her as Bacon described. We will test hypotheses not only by how they affect the world, but by how they affect us. We will seek not to control nature but to find our place within it. Nature will be our teacher, not the object of our dominion, for we will realize that its beauty and complex wholeness is beyond our rational, reductionistic and hence control-enabling understanding.

Like the body an infant discovers and develops, our place in nature is neither static nor limiting but always unfolding, both discovered and created through our play. The model of science along the lines of exploratory play is consistent with the ecological conception of progress and purpose, in that it seeks to fulfill a role that arises beyond our selves as separate beings. As such, it is much akin to prehistoric ritual, which we, projecting our own ideology, see as an impotent attempt to control nature (bring luck in the hunt, for example), but which actually sought to "restore the earth's balance", to compensate for the damage caused by human beings' already-incipient separation from nature.

One aspect of the Age of Reunion, then, will be the fusion of science and religion, so long sundered. Already we see the portents in the mystical metaphors arising from quantum mechanics, the realization of an organic intelligence pervading nature, and the spiritual awakening that invariably accompanies an ecological understanding of nature. With the crumbling of objectivity, the divide that separates experiment and ritual will crumble as well, for both will seek the same goal: to discover and enact our role and function in the dynamic balance of nature.

The gathering convergence of crises is bringing the Age of Separation to an end, and with it everything that we know as "civilization". Yet the end of civilization-as-we-know-it need not be a return to the past. We equate the ascent of humanity with an escalating domination of nature only because we deny the universe's inherent creative energy, sacredness, and purpose. When we recognize that nature is itself dynamic, creative, and growing, then we need no longer transcend it, but simply participate in it more fully.

The Medicine of Interbeing

I have been wrong. The germ is nothing. The "terrain" is everything.
-- Louis Pasteur

In the opening chapter, "The Triumph of Technology," I described how the triumphs of modern medicine in the 1940s and 1950s in overcoming most infectious disease were never followed by victories over cancer, arthritis, heart disease, and the host of new diseases that have arisen since. Despite vastly increased knowledge of molecular biology, despite the pretensions of nanomedicine and gene therapy, we have made little headway against today's up-and-coming diseases. Certainly there have been no miracle cures, no "magic bullet" drugs like the antibiotics and vaccines of an earlier era.

Instead what we have is an ever-intensifying regime of control, based on killing germs, sterilizing the environment, cutting out organs, replacing hormones, and dictating body chemistry. Like any technological fix, this regime of control comes at a higher and higher price; it is sustained only with increasing effort even as it becomes more and more fragile.

In Chapter Five's "The War on Germs" I asked, "What is the alternative to killing the germs?" What is the alternative to control? Is it to let nature take its course? Yes, it is. However, "let nature take its course" can mean something very different than to just let the body fall apart. Steeped as we are in the ideology of separation, we see nature as a force toward chaos, fundamentally unfriendly. In a survival-of-the-fittest mentality, we are safest, healthiest, and most comfortable when we overcome our competitors—by definition the rest of life—and keep life under control.

New paradigms of biology change all that. The primacy of symbiosis in nature, the genetic fluidity that unifies all life, and the idea of a purpose and beauty written into nature and not accidental to it, all lead to a very different conception of health and healing. No longer is health maintenance a matter of resisting the course of nature; instead, resisting the course of nature generates ill health.

It is almost a cliché that wholeness is a synonym for health, and to heal means to become whole. Throughout this book I have argued that the maladies of our civilization stem ultimately from our misconception of self. Psychologically and biologically, we see ourselves as much less than what we are. In other words, we have defined ourselves as less than our full wholeness. Or, you could say, we have defined ourselves as unhealthy. Illness is built in to our self-definition.

The New Biology understands that no being is separate and distinct from the rest of life; that beyond interdependency, all share an "interbeingness" that admits to no discrete, self-contained unit of life. Cells are alive, organs are alive, organisms are alive, soil is alive, ecosystems are alive, the planet is alive... maybe even the cosmos is alive, and each not only depends but also *exists* only in relation to the others. By warring against germs with our medicine, and by warring against nature in general with our technology, we thereby war against parts of ourselves.

This war against ourselves finds a dramatic enactment in our own bodies. Almost all the new epidemic diseases of the post-modern era involve a dysfunction of the immune system. The healthy functioning of the immune system, at its most fundamental level, depends on the body's ability to distinguish self versus not-self. Otherwise, some integral part of self is rejected as other, a condition called autoimmunity. Diseases such as autism, diabetes, fibromyalgia, multiple sclerosis, Alzheimer's, lupus, asthma, arthritis, and even arteriosclerosis, some of which are increasing by ten or twenty percent per year, are in whole or in part autoimmune diseases. Another way the immune system malfunctions is when it overlooks something that it should reject, treats it as benign, and allows it to proliferate, as with cancer and candida overgrowth. In the case of allergies, the immune system mounts an extreme response to something that is actually harmless. In the case of AIDS it is the immune system itself that comes under attack.[45]

These are the diseases that have reached epidemic proportions in the last two decades and that, not coincidentally, are proving impervious to all the technologies of control at our disposal. The ideology of the Technological Program says that with fine enough understanding, perhaps at the molecular level, we will eventually be able to bring these diseases "under control" too. The evident failure of this program, exemplified by the thirty-year-old "War on Cancer", challenges the whole conceptual basis of control and points toward a dramatically different paradigm of medicine, one that has ancient roots as well as a contemporary basis in the ecological and teleological paradigms transforming science.

The specific etiology of each of the modern immunity epidemics is different: some can be traced to toxins in the environment, others to stress, depletion of intestinal ecology, processed diets, or pharmaceutical medications. All of these immediate causative factors share a common root, though: our separation from nature that compels us to treat it as an object to manipulate, improve, and control. The conventional solution of more control will only exacerbate the source of today's illnesses, while any authentic cure must in one way or another address the root cause of separation.

By definition, anything that brings us more fully to who we are, will bring us more fully to wholeness and therefore to health. In particular, the autoimmune diseases of our era can only be healed through medicine that pacifies the war against the self masquerading as the program of control. A direct example is the body ecology approach to the treatment of autism pioneered by Donna Gates. Her protocol combines probiotic supplements, fermented foods, and a restrictive diet to rebuild the intestinal ecosystem and eliminate systemic fungal overgrowth. The results are extraordinary. Contrary to conventional opinion that autism is incurable, Donna and the people she has trained have helped thousands of children recover from this debilitating autoimmune disease. If the treatment starts early enough, before age five, a complete recovery is possible, but even older children and young adults experience significant improvement.

Donna Gates' approach is holistic. A reductionistic cure would disable whatever element of the immune system is attacking the myelin (and then treat the side effects of compromised immunity with other drugs). Similarly, a reductionistic approach to tree death is to chemically or genetically protect the trees from the proximate cause, for example a fungus, and the reductionistic approach to crop damage is to kill the insects causing it. An ecological approach, instead of trying to fix a nature that is broken or incompetent, sees the problem as a symptom of the disruption of a latent wholeness. What reductionism sees as causes, holism sees as symptoms. On a farm, pest proliferation is a symptom of depleted soil ecology that weakens crop species, as well as depleted plant diversity with its associated loss of insect and bird species that keep pests under control. The reductionistic solution of insecticides is nothing but a technical fix that intensifies the pattern of disharmony of which pest proliferation is a symptom. A holistic solution would be to move away from the linear farm-as-factory model toward an ecological model that incorporates hedgerows, intercropping, and soil recovery. The interrelationships that proliferate among crops, farm animals, insects, wild plants, soil microorganisms, birds, the farm family, and ultimately the community they feed together constitute an ecosystem, an irreducible whole. The ecological solution is the opposite of the reductionistic: instead of trying to isolate variables in a simplified, linear system, it encourages complexity and feedback, eschewing linearity and the control that comes from linearity. An ecological farm, like an ecological body, is not under control.

Control is in fact inimical to health, for it comes from the reduction of non-linear wholes. This is ultimately why we continue to get sicker and sicker as our control of body physiology, plant genes, and soil chemistry grows more and more precise. We are reducing reality, making less of nature. This process started eons ago with the development of symbolic culture, which mediated direct perception of reality with an abstract map of reality. Since then the Fall from wholeness has accelerated to the point where the whole world is rife with illth on every level: planetary, social, bodily. The return to health that is the Age of Reunion is a release of control and a trusting in a wholeness greater than ourselves. So accustomed are we to the control paradigm of technology that it is hard to envision an alternative. The very concept of technology seems to embody control, which is why many radicals conclude that the only healthy society is one that abandons technology and returns to a primitive mode of life. Yet there are modes of technology that are holistic, ecological; they seek not the reduction of nature but to lead us more fully toward nature. For as we have seen, nature is not static but evolves toward greater and greater complexity. Perhaps the emergence of the human species is part of the next upward leap in complexity. That is why I advocate not a return to the past, but to draw from our cultural memory of non-reductionistic, non-linear technologies as models for the future.

In the area of human health, we are fortunate to have at least two such models of holistic technology that are very well-preserved: Ayurveda and Traditional Chinese Medicine (TCM). Both these systems focus not on symptomatic relief (this herb for headaches, that herb for constipation), but on systemic patterns of disharmony that manifest as symptoms. "One does not ask, 'What X is causing Y?' but rather, 'What is the relationship between X and Y?'"[46] Cause and effect are not linear, but interwoven into a whole pattern that cannot be isolated from any other aspect of the patient's body, personality, family life, work, and environment. Ted Kaptchuk writes,

> To Western medicine, understanding an illness means uncovering a distinct entity that is separate from the patient's being; to Chinese medicine, understanding means perceiving the relationships between all the patient's signs and symptoms.... The Chinese method is thus holistic, based on the idea that no single part can be understood except in its relation to the whole. A symptom, therefore, is not traced back to a cause, but is looked at as part of a totality.[47]

How similar this is to the new paradigms of biology, which understand genes and organisms not as discrete and separate entities but in terms of their relationships to each other and the environment. Recall the autocatalytic sets and the emergent phenomena of the Mandelbrot set, that have properties of the whole not isolable in any of the parts. In TCM, a pattern like "Damp Heat affecting the Spleen" is not an isolable cause like a bacteria or virus, but a property of the whole person. It inheres in a relationship among the parts, but cannot be pinpointed in any of the parts themselves. Treatment therefore focuses on altering the whole pattern.

Because the whole pattern involves all of the parts, the right therapy applied to any of the parts can induce a change to the whole. This is the fundamental doctrine of chiropractic. Typically derided as claiming that "spinal misalignment is the cause of all disease," what chiropractic actually says is that spinal misalignment is part of any disease pattern. As with the "signs and symptoms" of Chinese medicine, spinal misalignment is both a cause and a symptom of disease, but it is not only the spine that reflects illness. One chiropractor told me that if we knew how, we could heal by manipulating a single cell—any cell—because the pattern of disharmony projects not only onto the spine but onto every cell, tissue, and organ in the body. Chiropractic focuses on the spine, however, because it is easier to manipulate than a single cell. Other disciplines accomplish the same results focusing on other body parts and systems. Foot reflexology utilizes a "map" of the body on the foot, ear acupuncture on the ear, iridology diagnoses diseases based on their projection onto the iris. Traditional Chinese diagnosis relies on empirical correlations between various disease patterns and corresponding characteristics of pulse and tongue. None of these are "causes"; all are part of an integral pattern. Change any of them and the integrity of the pattern is compromised, offering the possibility of a new pattern emerging.

That each part of the body contains a (distorted but complete) map of the whole resonates with a holographic view of the universe and finds a precise metaphor in the fractal property of self-similarity. The Mandelbrot Set is one such fractal: zoom in on one of the little blips or dots and you'll often find a distorted replica of the entire set, just as complex as the original. The idea that each part contains all the information of the whole deeply challenges reductionism and suggests a very different conception of healing. Holistic healing methods still operate one or another of the parts, but they are not reductionistic; the part is merely a gateway to the whole.

The effectiveness of a given modality depends on how clearly it is able to resolve the pattern of disharmony and whether it has a remedy suitable for that specific pattern. In herbalism, particularly TCM and Ayurveda, a systemic approach identifies generalizable disharmonies involving heat and cold, warmth and dryness, movement and stagnation—the underlying climate of disease—and modifies these basic factors so that a new pattern might emerge. Alternatively, a specific approach matches a single herb with a precise pattern and applies it as a "specific remedy," a "simple", or in Ayurveda, a *prabhava* or "special potency". Classical homeopathy takes this approach as well, using a single remedy matched to the disease pattern.

All of these systems grow out of a very different relationship to the world than underlies Western "allopathic" medicine. Western medicine is based on control: suppressing natural body responses that are uncomfortable (or life-threatening); doing something for the body that it cannot do for itself (kidney dialysis, for example); improving on a flawed natural function (taking out the thyroid and administering thyroxine); or engineering the body (dilating blood vessels to treat hypertension, lowering cholesterol levels with statins, cutting out organs). "The usual biomedical approach is to compensate for the incapacity of the body by introducing a drug that makes the body do what it is supposed to do. The holistic approach is different. Instead of trying to force the body to operate as we want it to, we try to return it to a state of health where it can take care of itself."[48] The holistic approach thus assumes that nature can indeed "take care of itself", that there is a tendency to wholeness built in to nature. The self-emergent, self-regulating properties of complex systems described in Chapter Six lend a new scientific credence to this view. Meanwhile, the idea of "trying to force the body to operate as we want" is the epitome of the Technological Program applied to the human body. This is a dictatorial approach to medicine, a fault shared by alternative modalities whenever they attempt to justify themselves on a conventional biomedical footing. In Matthew Wood's words, they are "aping the mainstream." He comments,

> Herbs can be used in an artificial, suppressive manner. For instance, the killing of bacteria by berberine (found in barberry, Oregon grape, and goldenseal) is an artificial imposition. Bacteria are secondary scavengers, usually following upon a derangement in the tissue state. When the excess in hot or cold, damp or dry, constriction or relaxation is removed, the environment off which the bacteria feeds will be removed. Another method uses echinacea to enhance the body's own method—white cell production—to kill bacteria. Does the practitioner know that the body really needs increased white cell production? Or is this just a convenient

> idea, one little fact torn out of the overall context of Nature, which the practitioner, having no idea how the organism really works, or what it really needs, applies according to his or her status as an outsider, a bumbler, an invader? These forcefully induced activities may temporarily, or perhaps permanently, remove acute conditions, but unless they are working with the body, stimulating or sedating it to its own needs, they are imposing an artificial condition which is never healthy. Killing bacteria is often unnatural and detrimental.[49]

The medicine of the Age of Reunion is not a mere superficial replacement of pharmaceutical drugs with herbal drugs; it embodies a profoundly different understanding of nature and of health. In Chapter Six I wrote of a new approach to science and technology that seeks not the supercession and subjugation of nature, but rather to discover and fulfill our role and function within nature. The medicine of the Age of Reunion thus replaces the dictatorial model with a new paradigm: medicine as resource, teacher, or friend. Used in the general mode, herbs can provide the resources the body needs to right its energetic and biochemical ecosystem. Used in the specific mode, they can teach the body how to be well, even in the vanishingly minute doses of homeopathy. A single molecule might be enough for the body to say, "Ah, here's how to do it"; through a feedback mechanism, the molecule might induce or awaken the processes to recreate it or necessary analogues.

Life does not thrive in sterile isolation. No life form is independent of the rest. We need a continual infusion of information from the rest of life, including bacteria and plants, to maintain a state of health, because our wholeness *includes* the rest of life. When we reduce the infinity of a living plant to a handful of "active constituents," we cut ourselves off from life and therefore from ourselves, exacerbating the opposition to the Other that manifests in our bodies as autoimmune disease, allergies, and cancer. Similarly, when we innoculate and sterilize ourselves into isolation from the microbial world, we cut ourselves off from the genetic plenum that offers the information, genetic and biochemical, we need to maintain a dynamic equilibrium with the rest of life.

The view of medicine as a teacher or a catalyst of a new pattern applies as well to the burgeoning field of "energy medicine". In "The Age of Water" I observed that currently accepted media of information transfer in biology comprise only a tiny fraction of what exists. Just as chiropractic and herbs recognize and utilize different gateways to whole patterns, the various energy modalities act on other gateways less visible to conventional science. And just as in herbology, reductionistic thinking sometimes infects energy medicine too: for example, "repairing torn chakras" with psychic surgery. Now, I'm not saying gifted people like Barbara Ann Brennan should not repair torn chakras! It's just that something fundamentally different is going on than the imposition of control upon another, newly discovered, immaterial organ system.

A common fallacy in the understanding of "energy" in alternative healing is that it is just another constituent of the person, albeit one that mainstream science doesn't recognize. But like spirit, purpose, and consciousness, the "energy fields" that some healers can tune into are emergent properties of the whole organism. The auras that they see are ultimately not objectively existent entities but subjective representations of patterns. They are the way the healer interprets and understands information; they are how the pattern appears to the healer. They may also be inter-subjective to an extent, in that some aspects of the "human energy field" correlate to various electromagnetic properties, and in that there is often substantial agreement among healers about fundamental energy anatomy. They are not, in other words, mere figments of the healer's imagination. The unavoidable subjectivity of energy healing makes it impervious to the program of standardization, mechanization, and the associated education/training paradigms that dominate medicine today. Whereas conventional medicine categorizes illnesses into generic types, holistic medicine recognizes the uniqueness of each patient's condition as well as the uniqueness of each patient-healer relationship. There is thus an inescapable element of uncertainty in each encounter, an element of newness. Something is always missing, healing is always incomplete any time we reduce the individuality of a patient to a set of generic conditions subject to a formulaic treatment. In other words, holistic medicine will never be a "science" in the usual analytic or Baconian sense of the methodical application of objective principles. Certainly, commonalities pertain across a broad spectrum of people, and patterns repeat themselves, but no two are ever identical. Hence, the medicine of Reunion will implement a reversal of the loss of the particular chronicled in "The Origins of Separation", in which labels and numbers obscure the absolute uniqueness of each object.

In the mindset of the Newtonian World-machine, two of the world's components can be considered *the same for all practical purposes.* In an industrial machine this is the principle of standard, replaceable parts. In science it is the idea that we can control variables to create "controlled conditions" under which standard methods will be effective. It is the idea that all variation is merely an illusion arising from different combinations of identical parts—no electron is any different from any other electron. It is the ideology of reducing any given aspect of the world into a data set. A human, then, is for medical purposes just a set of physiological data. The Scientific Program says that when this data set is complete, when it encompasses every relevant body condition, then our medicine too will be complete because we will know the reason for all illness. Until that day, doctors must rely on experience and intuition, but eventually a medical computer might replace them. The collapse of the Scientific Program is thus also the collapse of a deeper program: the abstraction of the world originating with name, number, and symbolic culture.

I imagine that certain conventional medical practices will persist, especially those relating to emergency life-saving technologies, which is where Western medicine excels. Sometimes illness (or injury) progresses to a point beyond the body's self-healing capacity. The way these technologies are practiced, however, will be dramatically different when embedded in a holistic paradigm. The surgeon's knife, like the healer's hands, might be understood as a channel for the communication of a pattern between cosmos and individual. Enlightened healers from all traditions often see themselves as mediums or agents for a power transcending themselves that flows through them and operates their skills. That doesn't mean training, practice, and skill are irrelevant, but that these are the means of attunement to a healing force. That force is the immanent purpose of an intelligent universe seeking always higher and higher levels of fulfillment. It is the emergent order that arises from organic complexity and that is fulfilled only in relationship to the Other, the environment, the planet, the cosmos.

The Spirit of the Gift

Come out of the circle of time and into the circle of love
—Rumi

Chapter Four explained how the conceptual objectification of the world results in an all-consuming regime of money and property. Accordingly, reversing this objectification will bring us back to the economic system that preceded the present compulsion of getting and keeping—the society of the gift. In the realm of human exchange, the demurrage system embodies that basic principle Lewis Hyde identifies with a gift, that it must be passed on or it will stagnate and eventually become a curse. However, gift mentality extends beyond the human realm. It defines a different relationship to nature as well, to the world at large.

It is no coincidence that many of the rituals that pervaded Stone Age culture were conceived in the form of gifts: gifts to the land, to the water, the fish, the trees. When Native American herbalists go to gather herbs, they will typically bring a bit of tobacco or corn meal as an offering, a ceremonially-offered gift to the plants and land from which they have received. Among humans, too, gift-giving is usually accompanied by ritual, which persists to the present in the form of Christmas Day. We instinctively recognize gift-giving as a sacred occasion, from which ritual grows irrepressibly.

The fusion of science and religion, so deeply related to the sacred purpose of the human species, can also be understood in terms of the spirit of the gift. For if science is to seek the fulfillment of our role within a greater whole, then how are we to understand that fulfillment except according to the question, "What have we to give to the world?" Ecology is itself a gift network, in which each organism and each species contributes far more to the environment than the limited Darwinian calculus of "fitness" would permit. If we are no longer to see nature as an object, but to participate in it as ecological beings, then we must join that gift-giving network. No species in nature is redundant and no capability is superfluous. Surely the uniquely human capabilities that we have turned toward world conquest have their purpose too.

I wonder if some readers may be impatient for a more specific statement of human purpose. I have written of a role and function that humanity, no less than any other species, has in the maintenance and evolution of the planetary ecosystem. I have been vague about what that function might be. What I am offering is not so much a program but a different way of thinking. We will create and discover our true role through play. It is not necessary to know at the outset what that role is; what is important is the mentality, the relationship that springs from being in gift consciousness. It is like going shopping for a gift, not having any idea what you will buy, but knowing that you'll find "just the right thing" and that you'll know it when you see it.

The ascent of humanity loses its connotations of domination and separation when we think in terms of "What is our unique gift to the world?" That question will come to define future science and technology. How can we participate in the unfolding beauty of the universe? When the illusion of separation is healed, we will come to define our collective purpose in terms of beauty. Just as individuals will approach work in the spirit of art, so also we will measure a new technology not by whether it will save labor, cut costs, or generate profits, but whether it will contribute to a more beautiful world. And this will not be a salve to the conscience in service of profit; it will imbue the fundamental motivations of science.

Fine, sounds like a nice future, but what about right now? Living in a society based on taking and keeping, is it possible to live today in the spirit of the gift, which is the spirit of abundance, which ultimately means the dissolution of boundaries within the gift-giving circle?

Remember, separation is an illusion. We can choose to live in that illusion or to deny it, but the basic reality that life and the universe is fundamentally provident cannot change. Life itself—our human lives—is a gift. Our lives, our talents, our abilities, our privilege to be human are given to us, and like all gifts they are not to be hoarded. They are not to be devoted, like the capital of classical economics, to the endless increase of me and mine, but must be passed on lest they stagnate and decay. In the ancient circles of gift-giving that defined an identity greater than the skin-encapsulated ego, each individual knew that his or her gifts would be reciprocated someday, in some way. The circle—really a gift *web*—takes care of its own, just as the ecological web of nature sustains every species within it. In other words, each gift eventually finds its way, usually in some altered form, back to the giver. "Our generosity may leave us empty, but our emptiness then pulls gently at the whole until the thing in motion returns to replenish us."

Who then is the Giver of our own personal gifts of life, health, talent and fortune? And how might we reciprocate? What are our gifts for, if not to survive and reproduce, if not for the doomed increase of a delusionary self? The Christian answer to this question is "to glorify God"; unfortunately these days, to glorify God is often interpreted along the lines of singing songs about Jesus. No. To glorify God is to honor and participate in God's most glorious manifestation. God is known, after all, as the Creator, and so to glorify God is to revere and participate in that Creation. Our gifts are creative gifts. The gifts of mind and hand that make us human, the gift of life itself, enable us to participate uniquely in the ongoing process of creation. Unfortunately, for a long time now we have used them with the opposite intent—to reduce Creation, to impose uniformity and linearity upon a world that is neither. That struggle, which since its prehistoric origin in language, number, and time has exacted an escalating price, is almost over. The resources to maintain it are nearly exhausted. Soon we will begin to simply accept nature's gifts rather than try to seize them, to pass them on rather than try to hoard them.

Whether you conceive of the Giver as God or the Universe (and in a wholly enspirited universe, what's the difference?) our lives are a gift, and the way to pass on that gift is to live that life as beautifully as we are able. It does not matter that modern society *appears* to have separated itself off from the gift-giving web. This separation is an illusion. Despite the rational-seeming benefits of keeping and hoarding, in actual fact when we choose beauty over ugliness we find our gifts growing, not shrinking.

The hunter-gatherer's confidence that the forest would always provide is still available to us. "Let us make a feast of all that we have today. Tomorrow we shall eat what tomorrow brings." But a lot of other beliefs must go along with it. To believe that the world is fundamentally provident, to accept the world as a gift-giving web and to enter that web, is to open the boundaries of self. It is to see through the illusion of ourselves as discrete and separate. It is also to trust rather than to control. To fully receive always means to relinquish control; otherwise it is not receiving but merely manipulating the giver, that is, taking. In the spirit of the gift lies the undoing of every manifestation of the regime of separation. When money transactions replace gift transactions, the circle of self shrinks to eventually become the lonely, mercenary domain of John Calvin and Adam Smith. To live in the gift reverses that process, undoing the bonds of the discrete and separate self and all that goes with it. To live in the gift is to relinquish the compulsion to control, the program to label and number the world, the quest for reductionistic certainty, the drive to convert the world into money and property.

Because separation is an illusion, we can "live in the gift" here and now, no less easily than our Stone Age predecessors. The only barrier to doing this lies in our beliefs. The perception that we "cannot afford to" live like this, the perception that it is unsafe, is no more or less true than it ever was. No doubt a hunter-gatherer sometimes did go hungry the next day, needlessly, for not husbanding his food supply or laying up reserves, for not making more of the world his. I am not saying the world is safe. However, the perception that we make the world more safe through our keeping and hoarding is also an illusion. We are no more secure, no better off than we were ten thousand years ago. Then or now, the attitude of trust is still necessary.

Yet it is equally an illusion to think, "If only ours were a hunter-gatherer society, then I would live in the gift, but in modern society it isn't practical." It never was "practical", not in the sense of maximizing the separate self's perception of security.

Part of the gift mentality is the belief "I will be provided for." I will be provided for; therefore it is safe to provide as well. This hunter-gatherer attitude of abundance applies equally today whenever we choose to move into the spirit of the gift, whether in the material, social, or cultural realm. To live in the gift does not require that the whole world change around us first. All anyone need do is to see the world through a different lens. We are so accustomed to thinking in terms of what we can get, how we can benefit, from a given situation. To live in the gift means to approach each person and each choice with the attitude, "What can I create? What can I give?" The modern human, inculcated with survival anxiety, immediately protests, "Well what about me?" When we actually start living in the gift, we find that this protest is founded on an illusion. We find that the universe reciprocates. For example, I urge my students to base their career decisions not on "What career will give me the most money, security, and status?" which is the mentality of taking (and ultimately of agriculture), but rather "What would I most love to give to the world?" Build a career around that and you will be successful in ways you can hardly imagine.

This way of thinking follows naturally when we recognize all of our talents, fortune, and indeed life itself as gifts as well, fostering an attitude of gratitude that impels us to pass on those gifts in whatever form we can. As it was given to me, so shall I give it to the world. In contrast, modern ideology encourages us to see all of our resources not as gifts but as possessions, things that are fundamentally *ours*, devoid of any obligation. Gifts, remember, generate obligation, whether we are speaking of a traditional gift-giving social network, or the gifts of life, fortune, and talent. When we use these gifts otherwise, we experience a disquiet, an ambient anxiety that, completing the vicious circle, fuels even more acquiring, taking, and owning. Unacknowledged obligations weigh on the spirit.

The psychodynamics of living in the gift are particularly apparent in the realm of art and music. The improvisational theater pioneer Keith Johnstone observes that the fountainhead of artistic creativity lies beyond the controlling, rational, planning ego self. To act in improvisational theater, to create a story, to clown around requires that that part get out of the way so that we might become a clear channel for the gift.

> We have an idea that art is self-expression—which historically is *weird.* An artist used to be seen as a medium through which something else operated. He was a servant of God. Maybe a mask-maker would have fasted and prayed for a week before he had a vision of the Mask he was to carve, because no one wanted to see *his* Mask, they wanted to see the God's. When Eskimos believed that each piece of bone only had one shape inside it, then the artist didn't have to "think up" an idea. He had to wait until he knew what was in there—and this is crucial. When he'd finished carving his friends couldn't say "I'm a bit worried about that Nanook at the third igloo", but only, "He made a mess getting that out!" or "There are some very odd bits of bone about these days."[50]

We find that when we try to hold and own the gift, it dries up. To receive we must also be willing to give. Hyde writes, "We are lightened when our gifts rise from pools we cannot fathom. Then we know they are not a solitary egotism and they are inexhaustible. Anything contained within a boundary must contain as well its own exhaustion."[51] Beautiful!

Medieval tradition has it that a sorcerer's power dries up if used for selfish or evil purposes; the Brazilian trance surgeon Jaoa de Deus cites the same reason in refusing to accept money for healing. Acting selfishly, we enforce a separation from the universe from whence our gifts actually come. The same thing happens to artists who "sell out". In trying to keep the fruits of the gift within the limited world of selfish benefit, they exclude from their world the fountainhead of that gift.

Many great artists have recognized that their work comes from a source beyond themselves. The ancient Greeks personified this source as the Muse. The fairy tale of the Elves and the Shoemaker makes the same point: while the shoemaker is asleep (i.e., while his conscious mind is out of the way) magical elves come and make shoes far more beautiful than the shoemaker could himself. Even the word "inspiration" encodes the same understanding, for it literally means take in a spirit. Remember as well the Native American spirit songs, universally claimed to come from an outside source. A similar principle pertains in Eastern traditions. Certain martial arts forms are understood to have been transmitted to human beings from a transhuman source, a phenomenon I witnessed in Taiwan at a Taoist retreat center. There, college students, housewives, and so on would occasionally enter spontaneous, perfectly executed sequences of obscure martial arts forms they had never even seen before. Significantly, people with martial arts training rarely entered this state of receptivity, as if their training got in the way. The same phenomenon is not uncommon in the yoga tradition. The founders of two schools of yoga in the United States, Kripalu and Kali Ray TriYoga, claimed to channel the yoga through their bodies rather than to consciously practice postures, and this spontaneous posture flow is understood to be far more perfect than any conscious approximation of it. Finally, musicians experience the same state when, to paraphrase the Grateful Dead, "the music plays the band."

For the mystic and the artist, something greater than ourselves flows through us. Breath flows through us, food flows through us, matter flows through us, replacing every cell and atom in our bodies repeatedly over the course of a lifetime. Life flows through us. Life lives us, despite our misperception that it is the other way around. To accept and not fight this is to return to the original affluence of the hunter-gatherer.

To enter the Age of Reunion on a personal level means to start living in the gift. Then boundaries begin to dissolve—social boundaries, to be sure, but more importantly the absolutism of the self-other boundary that separates us from the world. Reunion is happening the same way at the collective level. Our species enters the spirit of the gift and ceases the doomed effort to rise above nature when collectively we begin to ask, "What is our proper role and function in the Gaian whole?" Similarly, an individual reunifies with the world when she seeks no longer to triumph over it or control it, but to give to the world and accept its gifts in the full recognition of gratitude. That is what I call "Living in the gift." Harking to the hunter-gatherer, it is a state of abundance without control, a state of creativity and growth without domination, a state of ease that yet fosters exquisite mindfulness. It is available right now. Don't wait until a personal convergence of crises makes the alternative of taking and controlling unbearable.

The entry into the Age of Reunion as individuals is inextricably bound up with our collective entry into that new age. As more of us move more deeply into the spirit of the gift and treat life and the world with gratitude, we will no longer accept the degradation of meaningless work, and no longer choose ever to make the world an uglier place. The results of these choices will eventually reverberate through politics, society, and the economy. Sooner or later it is inevitable that we will reenter the spirit of the gift, if only when disaster forces us to, because that is the nature of reality. The truth will out. Let us stop resisting the truth before it kills us.

Storyteller Consciousness

In the last two chapters I have illuminated something of the magnitude of the changes that will follow the transition to an Age of Reunion. If anything, I have understated them. The end of separation will penetrate far deeper than the forms of money and property, technology and medicine, work and education; it will eventually transform the very psychological infrastructure of the discrete and separate self—symbolic language, number and measure, linear time, and dualistic religion. Already we see how play refuses the linear measure of time and the discrete separation of subject and object: we lose ourselves in play's timelessness and become "an organic agent of the universe's own creative process."

It would seem, based on the arguments of Chapter Two, that representational language inescapably separates us from that world of play and casts us into a divided realm of object and label, self and other. We could perhaps return to the *lingua adamica*,[52] abandoning representational language and all the technology that rests upon it. For let there be no mistake: language is more powerful than, and prior to, any other form of technology. Almost anything we accomplish in today's world, we accomplish through language. We present ideas, we make requests, we describe possibilities, we warn of consequences, we call others into action. Of course, there are some things you can accomplish without language, such as building a tepee, but nothing that requires the coordination of human activity. Activities such as running an airport or building a microchip absolutely require systems of arbitrary symbols, numbers, and the marking of time.

Technology, however, need not be unnatural, nor depend on the doomed maintenance of a separate human realm. The destiny of humankind, in the coming age, is to extend nature into a new realm. We don't need to go back to the Stone Age. As long as waste equals food, as long as it embodies nature's cyclicity and does not pretend to linearity, as long as it enacts the dynamics of the gift rather than the program of control, then the human realm, separate no longer, will itself be natural in every sense of the word. And the communication system of this new realm is the technology of symbol. In it, the role of language is analogous to the role that hormones, neurotransmitters and other signaling molecules play in the body. It is unnecessary beneath a certain threshold of social complexity. It is essential to the physiology of the multicellular metahuman that has sprouted today from the Gaian body. This section will describe an emerging conception of language that, though representational, returns us to its true origin and purpose. Language can be an instrument for humanity's cocreative play with the universe.

The shift in the application of language parallels the shift I have described for the role of technology in our relationship to earth. You see, quite often the purpose of our words is not actually to communicate, but to control. I have witnessed myself in situations where it seemed that every sentence I spoke was part of some devious stratagem to engineer other peoples' perception of me, to get what I wanted, to block their efforts to control me, to project my identity, to maintain my turf, or to create a world in which I could see myself as good and right. Each sentence so contrived was actually a lie—if not in its semantic content, then certainly in its unconscious intention.

The purpose of a lie is at bottom the same as the purpose of the technology of separation: to manipulate and to control. And as we have seen, most words today are some form of a lie. Can we envision, then, a transformation in the technology of language that parallels the transformation in material technologies described in this chapter? The technology of Reunion seeks full participation in the unfolding of a higher natural order. It attempts not the control of nature, but its fulfillment; not the abrogation of natural cycles, but their extension. It brings the separate human realm back into harmony with the rest of reality. How do these qualities translate into the technology of language? Can we imagine a human realm of representation that accords with nature's laws? The Age of Reunion will not mark the demise of representational language, merely its return to its proper place as a conscious play, a device for a marvelous creative game.

Elements of just such a technology are emerging today from diverse sources. Ideas like Brad Blanton's radical honesty, Tamarack Song's truthspeaking, Marshall Rosenberg's non-violent communication, the neurolinguistic programming movement, and Haim Ginott's principles of parent-child communication are coming together to create a language of truth. But the key to understanding the role of language in creating a more beautiful world, a world that rejoins the long-sundered human and natural realms, is to go back to its origin and true purpose. I mentioned in passing in Chapter Two that perhaps the origin and true purpose of language is to tell stories. I would like to unpack this statement now to reveal its full import.

Language is the instrument by which human beings play with time. The *lingua adamica*, by way of contrast, is a communication of the present moment. It is about you and me, right here, right now. It does not and cannot judge, interpret, plan, or speculate. It does not recognize past and future, nor can it coordinate human activity beyond a very small scale. Language is different. Language can do all these things because it creates a separate map of reality that we can manipulate and play with. We create stories, then we act them out. When the stories are held in common, they coordinate our actions and allow us to stamp the stories' image onto physical reality. Herein lies the creative power of language.

Let us not underestimate the power of the new technologies of language. We might be tempted to dismiss them as mere niceties beside the juggernaut of destruction that consumes nature, culture, beauty, goodness, and earth—until we remember that genocide and ecocide alike are ultimately products of perception and communication. No single human being has the physical power to render whole forests to sawdust or whole peoples to slavery. He does so only through language. Our stories in the social realm translate into experiences in the material realm. We wreak our destruction only because we know not what we do. Language has separated us from reality and cut off our love from its natural object.

One possible solution, of course, is to abandon representational language altogether, just as some would return to the Stone Age, and use only a *lingua adamica* that does not distance us from reality and in which it is impossible to lie. I reject that option for the same reasons I reject the abandonment of technology generally. The gifts of hand and mind that make us human exist for a purpose, no different than the gifts of any other animal. Language can be an instrument for healing. I said that no human being has the physical power to render whole forests to sawdust or peoples to slavery, except through language. True. Yet it is equally true that no human being has the power to heal whole ecosystems or free whole peoples, again except through language. Did Gandhi have some superhuman ability to enable him to free India? Did Rachel Carson launch an environmental movement with anything other than the power of her words?

Contrary to the imaginings of the techno-utopians, no new material technology is required to usher in an age of peace and abundance. Scarcity and war are products of our way of relating to each other. The Age of Separation is the result of the story we have built, a story we tell ourselves about ourselves. Can there be any doubt about the creative power of word? This book has explored the "separate human realm" that we have created through our technology and through our symbols. Because the former rests upon the latter, one could say that the entire ascent of humanity is built from symbols. The entire world of modern human experience is built upon a story. Such is the power of word. What is property, for example, but an agreement? What is money, but another agreement about the meaning of yet other symbols, pieces of paper and bits in a computer? What is time, but an agreement as well that something we have constructed *means* something? The coordination of human activity, necessary for anything beyond Neolithic technology, rests on a shared interpretation of symbols. It rests on meaning, and these meanings together comprise our story about the world.

Objects of fantastic unnatural complexity, such as the New York City skyline or an integrated circuit chip, arose in a very brief instant of geological time from a wholly organic matrix. It is primarily to language—shared systems of meaning—that we might attribute these miracles. Today, as I have observed, it will take a miracle to save human civilization. The near-certain future of the planet is plain to see. Only a miracle as great as a microchip can alter our grim course. And the only way I know to generate such a miracle is the same way we generated the first one: by implementing a new story. We have told and retold and endlessly elaborated the story of Separation. Now it is time to tell a different one.

The Age of Reunion, however, entails more than a shift in stories. That has happened before. I have stated that the separate human realm never was really separate. We never were really independent of nature, and our linear consumption never was really linear. We have lived in an unconscious pretense. Similarly, the problem with the language of the Age of Separation is that the storytelling has become unconscious. In our confusion, we mistake the separate realm of words for the reality it is supposed to represent. When we forget that our stories are in fact stories, we end up helplessly lying to ourselves and the world, because the map is always a distortion of that mapped. Words by nature of their abstraction are inexact, a degree removed from the particular objects, processes, and feelings to which they refer, leaving us therefore to infer what the other person really meant, and opening the way for misunderstanding. Moreover, the nature of this abstraction and distortion is not innocent, but infuses our every communication with an unconscious mendacity that abets the regime of separation.

We shouldn't be surprised, then, when the effect of our stories' enactment—the stamp of their image onto physical reality—is the opposite of what we intended. So the economists' stories of economic growth and market development, all told in the language of words and numbers, create a reality of misery and poverty when enacted. Politicians in their war rooms, with all their talk of enemy combatants and collateral damage in pursuit of freedom, security, and the good, create a reality of violence and horror. We have taken the unreal to be real, and then tried as hard as possible to live in the unreality we have created. But reality keeps breaking through, with an intensity now that brooks no denial. Our stories are coming apart.

The story of the Age of Reunion is more than just another story using the same technology of symbol. The fundamental difference is that our storytelling will become conscious. Instead of confusing it with reality, we will use language consciously to *create* reality—to create stories and act them out. In order to do that consciously, we will become acutely aware of what hidden assumptions are embedded in our choice of words. We still might use words like "the environment", but we will do so with full realization that no such thing actually exists separate from ourselves. The same for other words whose hidden assumptions I've observed in this book: the "is" of identity, which says that two things can be the same; the word "exists", which implies an absolute Cartesian reality; distinctions like "matter and spirit" or "human and nature" which artificially divide reality into two parts. In fact, all words encode a lie, and all representation contains misrepresentation. Nonetheless, we will still use words and other forms of representation in the Age of Reunion. We will not, however, delude ourselves into living in that lie. We will never again lose ourselves in the story. We will apply words carefully and consciously, and hold onto their meanings lightly.

Let us call this approach storyteller consciousness. Instead of seeking to describe a reality already out there, we will be aware that we create reality through our story about it. In science, storyteller consciousness means being aware of the creative nature of theories and experiments, whose very language encodes deep assumptions about self and universe. In technology, it is to see our choices as a way to define our relationship with each other and the rest of life. It asks the question, "Whom are we creating ourselves as?" The forms and institutions of politics and government will change most radically of all, as we begin to disbelieve in our labels, categories, and abstractions, and come into contact with human reality. All of these forms and institutions are themselves stories. America is a story. France is a story. Money is a story. The law is a story. Words and symbols, that is all, with no more meaning than what we agree upon. Our mistake has been not in telling stories, only in thinking they are real. When we let go of that, we will be able to play with them consciously and let them go when they no longer serve us.

Over thousands of years, the creative play of storytelling has come to enslave us, and we have lost the storyteller's consciousness. Finally we are awakening, as the effort to maintain the pretense overwhelms us. We cannot maintain the story any more. The story of linearity, the story of separation, the lonely story of a discrete self marooned in a world of other. The story that we are not storytellers, not authors but mere reporters, describing what is, reacting, managing, controlling. We are awakening from that story now, the story that we are not the authors of our world and of our lives.

Indeed it is in our personal lives that the enslavement to unconscious stories has been the most devastating. We live in a fabricated world of interpretation that we mistake for reality. We live in a world of judgments and imposed meanings. Maybe Dad shouted at me a lot, and since I was three I made it mean that I am bad. She left you, and you interpreted it as a betrayal, and made it mean you are unworthy of love, and so you find yourself holding on, manipulating, controlling. We live in our stories, which then create events to justify themselves and strengthen our enslavement.

The origins and multitudinous variations of these stories are beyond the scope of this book; often they are extremely subtle and, because they conform to larger cultural stories of self, wholly invisible. Like the broader, cultural stories, they enslave us only to the extent that they are unconscious. I am advocating the enlightenment and not the abolition of our stories. Yes, we can come back to the present moment, the present experience, and release all judgment if we so choose, just as we can return if we choose to the *lingua adamica.* However, we are not meant to stay there. We are meant to foray into three-dimensional reality, space and linear time, and to create beauty with their tools. We are meant to create meaning and create stories. I am not advocating that we surrender our existence as time-bound material beings, just as I do not propose that we abdicate the gifts of culture and technology that make us human. No longer, though, need we be enslaved to those meanings, to those stories, or to our technology. To enter the Age of Reunion is to awaken to our power as conscious creators.

In this book I have mentioned three cultural stories that many of us have deeply internalized. The first is the Newtonian world of force and mass, which manifests in our personal lives as feelings of compulsion and powerlessness. In language it appears in words like, "have to", "can't", "must", "should", "I will try", and "you made me". The second is the Cartesian split of ourselves into body and soul, a good part and a bad part. It manifests in life as a constant struggle of self-denial and perpetual sacrifice of the present for the future, producing a battle against desire and the imposition of the civilized and conditioned over the natural and the wild. In language, it again manifests as "should" and "shouldn't". The third story is that of separation and scarcity. Manifesting in phrases like "can afford to", it disbelieves in our connection to the universe and all life, a connection that brings our gifts inevitably back to ourselves and makes control and domination unnecessary.

Even naming these stories and observing them in operation already makes them less powerful. However, I have found it useful to deliberately undo them through the way I speak to myself and others. We can use words in ways that deny the stories that enslave us, and thus accelerate our freedom. For example, Marshall Rosenberg suggests rephrasing every "have to" sentence as "I choose to... because..." I used to say, "Even though I hate it, I have to give grades." When I rephrased it as "I choose to give grades because I am afraid I will lose my job if I don't," everything became much clearer. I realized that my job was much less important to me than my sense of integrity, which for me personally was violated by giving grades, and so I decided to leave academia. By thinking in terms of "have to" we surrender our power. The very words carry within them an assumption of powerlessness. Another substitution I've been making is to replace "you should" with "you could", and "I should" with "I can" or "I want to". You can also experiment by abolishing "I will try..." from your lexicon, especially the lexicon of your internal dialog, and replace it simply with "I will..." If you are true to your word, you will think very carefully before agreeing to anything. "I will try" can be a cop-out, a polite way of saying you won't actually do it. It also encodes an assumption of helplessness, a world of external forces that thwart our creativity. A wholly different way of thinking underlies "I can", "I choose to", and "I want to". The story of powerlessness cannot be told with them.

Here is another kind of empowerment, relating to the second internalized story, the Cartesian split into a good part and a bad part. In contrast to my personal age of reason that I described in Chapter Three, I no longer attempt to justify with reasons everything I do. Instead I say, "I did it because I wanted to." What! That's not allowed, is it? We can't follow desire, can we? Resistance to desire is another manifestation of the body-soul division. The good part, the higher part, the spiritual part—the mind and the will—must master the bad part, the fleshly desires. Sacrifice now for a future reward. It is just another variation of the mentality of agriculture, channeled through religion and education, that still dominates us today. Yet Heaven remains forever just around the corner.

Traditional cultures recognized an importance to stories beyond mere reportage or children's entertainment. Storytelling was also a sacred function that carried the spirit of the people and created their world. It is not only the sounds of the *lingua adamica* that have a sacred generative power; our stories do as well. Today we wield that power unconsciously, thus creating unintended effects. We do not know our own power, the power of word. In a way, all speech is a story, because all speech creates a new addition to the world of representation. All speech therefore bears generative power, just as the Native Americans believed, because we enact that world of representation. We live our story, we stamp it onto the world. Why, then, do our words seem so impotent today? It is because, just as our great immersive cultural stories and ideologies are invisible to us, we use words unconsciously too. It is not conscious lying that is weak, it is unconscious lying. A deliberate lie is still a conscious act of world-creation. Many if not all of the disempowering forms of speech described above are unconscious lies. If you would like to restore to your words their generative power, you must treat them as golden. One weakening form of speech is swearing. "Fuck." What are we really saying when we make a sacred life-creating pleasure into a vulgar term of deprecation? "Damn." Do we really wish eternal torment on someone? No, we are speaking unconsciously. Other weakening forms of speech include gossip, small talk, and various forms of negativity. What world-creating story are we telling when we speak like that? For words to truly be powerful, we must align them with our creative intention. Only then can we create the stories of our lives.

Unconscious lying sabotages our credibility to ourselves and others. If we cultivate the habit of speaking truthfully and treating our words as golden, then when we declare great things, they will come to pass. The more we realize the power of our words, the more mindful our speech becomes; the more mindful our speech, the greater the power of our words. We condition ourselves to our words always coming true, and foster a deep confidence in the magical creative power of our speech.

Whether on the collective or personal level, storyteller consciousness is inseparable from the new sense of self that defines the Age of Reunion. It depends on a blurring of the defining distinction of the Age of Separation, between the observer *in here* and the objective world *out there*. It will emerge spontaneously, in tandem with the crisis-induced disintegration of the illusion of separation. The story of powerlessness and separation simply won't be captivating anymore! In its place we will have a story of connectedness, of interbeing, of participation in the all-encompassing circle of the gift. And part of this story is actually a meta-story, a story about stories that invests all of our stories with creative power and motivates us to be conscious in their telling.

The society that may be built around storyteller consciousness centuries hence is so unlike what we have today that I hardly dare describe it on paper. Instead of the present demarcation between drama and real life, future society will consist of stories within stories within stories, plays within plays within plays without any sense that one is "for real" and one is not. Life will be all play, and all play will be in earnest. We might commit to some of these stories as deeply as a human being can commit to anything, as passionately as the greatest artist cares about his greatest masterpiece. Each life will be a masterpiece, and some of our collective projects will span generations and alter the fabric of (what we call) reality. This will be the eventual fulfillment of the Age of Reunion, when we come into full, conscious co-creative partnership with the universe itself. In the meantime, in the next century or so, great storytellers will emerge to inspire us with beautiful and believable stories of what life can be, visions of the world we can create. Those stories will have roles for each of us that draw upon our gifts and develop our potential. It is happening already. Have you heard the casting call? A beautiful life is being offered, if we can only find the courage.

In Love with the World

The last section describes how an emerging language of truth will transform the human drama from an unconsciously scripted tragicomedy into a conscious play, a great work of the storyteller's art. Through language, humanity will become a conscious agent of the universe's unfolding. Pruned of the tangle of excess verbiage that obscures meaning, representational language will be a richer and more meaningful realm than what we have today. Nonetheless, in many important areas of life, language and especially written language will play a reduced role. That is because like reason, like linear technology, like money, the gift of language has exceeded its proper domain. We use words to express many things that words are fundamentally incapable of expressing, and most words express nothing at all.

There are signs that the shrinkage of the domain of language that will accompany the Age of Reunion is already beginning to happen. It is a response, I think, to the "crisis of language" I wrote about in Chapter Two, in which words seem to mean less and less, "forcing us into increasingly exaggerated elocutions to communicate at all"; in which advertising has become ubiquitous and wholly insincere; in which politicians lie routinely and get away with it, and in which we discount nearly all public speech as PR, spin, and hype. Living in a ubiquitous matrix of lies, we naturally respond by discounting all speech. Because of its ubiquity (especially the printed word), we tune it out; because of its insincerity we discount it and second-guess it.

Faced with this crisis, young people in particular are drifting toward other ways of communicating. This may be one of the deep reasons behind the astonishing decline in literacy over the past generation. Ask a retiring college professor about it! When I began teaching at Penn State a few years ago I was shocked to discover that many of my students could not read. They could scan through a page and pull out answers, but they were seemingly incapable of integrating ideas over an entire paragraph; nor could they write coherently. Compared to the standards of a hundred years ago, Americans are shockingly illiterate. Yet we are no less intelligent and no less hungry for communication. To an extent, declining literacy represents another robbery of our spiritual capital, the sell-off of yet another cognitive capacity, but another major cause could be a semi-conscious moral repugnance at the inanity and inauthenticity of the words that inundate us today.

Fed up with the lies, young people in particular turn increasingly to music and other modes of non-textual expression. I think I see the beginning of a trend. One sign is the exploding popularity of websites such as MySpace, Facebook, and YouTube, where, thanks to technology, people share, in addition to words, photographs, music, voice, and video. Meanwhile, the established commercial interests are fighting a desperate battle against the democratization of music production and distribution that technology has made possible. Almost anyone can produce, record, and distribute music (and video) at minimal cost. The technological infrastructure is in place for a return to the gift economy, and only awaits the demise of the extractive, centralizing interest-based financial system we have now. The dominance of text in remote communication may be coming to an end—a momentous development, for it could signal the end of an era that started with the printing press some five centuries ago. Of course there are functions for which text is uniquely suited, and books will surely survive in one form or another. But the written word's monopoly on public and long-distance communication, eroding ever since the age of radio, is coming to an end. Words will reign in a more circumscribed realm. On the Web, podcasts and video streaming may soon eclipse words as the dominant media of communication. Already a lot of people prefer audio books over the written versions. Video-capable cellphones have replaced letter writing, nearly a lost art. I wonder if real literacy will retreat to the three percent or so of the population it comprised in pre-modern times?

Whether or not it is mediated by technology, could we be initiating a halting return to the *lingua adamica* of old? Or will the regime of separation continue to new extremes, a VR world of avatars and voice simulators where nothing is real? I envision a future in which we return to the voice, the song, and the face in communication. The Piraha, whose communication is mostly singing and humming, show us the future as well as the past. At least, they show us a future possibility. I may be wrong, after all; maybe the Age of Reunion will never come, not in our lifetimes, not in a thousand years. Or perhaps both worlds will coexist, and we as individuals shall choose which we live in. In any event, the most important dimension of a return to unmediated communication is direct and personal. There are signs of this as well, transformational technologies such as Authentic Movement, Contact Improv, and the Continuum™ that help people reclaim some of the vocabulary of the

Original Language of body and voice. I encourage you to try them out. An actual experience of the intimacy of communication that is possible without words will be far more effective than this book in persuading you that an Age of Reunion is possible, imminent, right there in front of us. Only our fears, habits, and beliefs keep it hidden.

The *lingua adamica*, the mythic Original Language of humankind, admits no lying and no separation between subject and object. There is no need to infer as to the state of being of the utterer, for her vocalizations are facets of that state. To understand the real voice of another person requires and furthers an intimacy so profound as to dissolve the boundaries between self and other; these days such intimacy typically exists only between mother and infant, sometimes between father and infant, and on occasion between the most trusting of lovers in special moments. I believe that in former times, before the Age of Separation was much underway, such intimacy was the rule in the human relationships of the Stone Age kin band. Language as we know it today was unnecessary. Reports, some by respected anthropologists, abound that describe aboriginal communication abilities that border on the telepathic. Such communication was not limited to human relationships, for people of those times enjoyed a profound intimacy with nature that, by many accounts, allowed them to understand the languages of animals, plants, forests, wind and clouds.

These languages are so unlike the abstract systems of signs that constitute language today that we probably shouldn't even call them by that name. Today's languages are a corruption, a degeneration, at times an imitation but usually an obfuscation of the original *lingua adamica*. To communicate in the Original Language involves a temporary fusion of self and other; it is to *be* the other that is speaking, to understand that other so deeply that its essence is displayed nakedly in its sounds, odors, and motions. It is as impossible to lie in the *lingua adamica* as it is to fake an odor. We can hide it, mask it, or suppress it, but it is always there underneath.

When we bare our true voice to another and really hear the voice behind the words, boundaries dissolve into an unutterably sweet intimacy. This requires total trust, which is impossible when we see the world as competing others, and so happens very rarely today. In the Age of Reunion we will grow gradually less attached to these boundaries. We will be unafraid to relinquish them at will, so that we can live in a profound richness of relationship. To a person living in such intimacy with other people and all Others, life is suffused at every moment with a depth of meaning that we can hardly imagine today. To simply sit and do nothing can be an overwhelmingly sensuous experience. The bliss of such a state is so intense there is hardly reason to do anything but just be. Language or any kind of conceptualization removes us from this state and deprives the world of meaning. (Yet is there a possibility of being simultaneously attuned to both, to word and to voice? Why have we embarked upon the sojourn of separation that began with language?)

The longing for that state of bliss, buried deeply within all of us, drives us never to be satisfied with the lesser world of separation we have created. I know this because I have had the fortune to experience the *lingua adamica*, though just a few times in fullness. Separated by a telephone wire, I and this other merged so completely through voice that there was no distinction between I and thou. In every sound, her total being was manifest to me, and mine to her. Language is utterly insufficient to convey the intensity of this union: words like "mine" and "her" imply a separation that did not exist.

A Homecoming awaits beneath our names, our numbers, our words, and our physical separation; you have been waiting for it your whole life. Sometimes it wells up and we run before it to save our selves, but its trace remains as a promise and a reminder of what life should, can, and will be.

Communication in the *lingua adamica* is an occasion of profound intimacy between the communicants, and it is absolutely specific to them. There are as many Original Languages as there are pairs of communicants. To a third party not sharing their intimacy, the *lingua adamica* would have no more meaning than so many birdsongs—only the grossest emotional states would be apparent.[53] To the intimate, however, a single noise communicates an infinity of nuance. Not only are no two *lingua adamicas* the same, but no two words are ever the same between two people. Each communication is unique, just as each moment is unique. There is no reduction of the infinitude of experience. Speaking the Original Language is much like making love, and indeed it is no coincidence that the vocalizations of sexual passion are among the few outbreakings of the *lingua adamica* in modern adults. But someday we will be in love with the world.

A trace of the *lingua adamica* can be found in poetry and music. Poets and poetically-inclined prose writers know that "the melodic structure precedes and partly determines the structure of the text in poetry."[54] We can only imagine what poetry would be without words—it would be a song, a ululation of non-referential sounds nonetheless laden with meaning, perhaps akin to the spontaneous scatting of jazz vocalists or the spirit songs gifted to Native American vision questers. Spontaneity is a key feature of this sort of communication. It is not a colonization of the other, not an imposition of one's own categories, not a reduction of the world into a finite collection of labels. According to Joseph Epes Brown, Native Americans insist that the songs used in their ceremonies are never composed, but are rather given by spirits.[55]

The *lingua adamica* is not only our distant heritage but also our birthright and our future. The process by which it is learned is wholly different from that of learning any other language. Whereas a language of symbols necessarily involves an equating of A to B and thus a separation and alienation, the *lingua adamica* embodies the uniqueness of each A and B, each part of the whole and each moment in time, and it requires a coming together, not a separation. To learn the *lingua adamica* requires a dropping of the artificial boundaries that keep us apart, a dropping of the conventions of me and mine. To speak it with another means falling in love with that other to the point where it is no longer an other.[56] The more intimate and trusting the love, the more liberty to let down the barriers of self, the more highly developed the understanding of the mutual *lingua adamica* and the less necessary ordinary words become.

If to speak the *lingua adamica* with another is to fall in love with that other, then to understand the languages of plants, animals, and inorganic processes requires nothing less than to fall in love with the world. The *lingua adamica* emerges with the melting of barriers between self and world. As the futility, artificiality, and factitiousness of these barriers becomes increasingly apparent with the breakdown of modern society, we will find ourselves speaking less and coming naturally into our birthright.

To learn the language of the world—what a beautiful and fruitful definition of science that would be! Science, to learn the language of the world so as to fall more deeply in love with it. How different that is from the mentality of mastery and control, from the Baconian inquisition of nature. It begins with the awe of the scientist apprehending the cell, the stars, the complexity of an ecosystem or a metabolism, the miracle of life. It is a return to the presence of the divine in nature. It is the culmination of the journey of science, which chopped nature up into pieces, isolated it and refined it, purified it and reduced it, only to discover at every turn unsuspected new realms of beauty, wonder, and miracles.

The loss of the particular that comes with the generic categorizations of label and number is a cause and a symptom of our falling out of love with the world. Conversely, when we fall in love we see our beloved as a real, unique individual, irreplaceable. The Age of Reunion is nothing more or less than a falling back in love with the world. Nothing, not even an electron, is generic. All are unique individuals, special, and therefore sacred. In physics the quest of four centuries to reduce complex reality to universal laws acting on generic fundamental particles is collapsing, as the variety of "fundamental" particles multiplies, as known physical forces stubbornly resist unification, and as we digest the fact of indeterminacy. Even an electron has an irreducible individuality, an unpredictability to its behavior that is not merely an illusory consequence of our ignorance of more minute causes. Not enslaved to cause, it is an autonomous Chooser. As with the electron, so with the proton, the neutron, and everything made of them. Moreover, this individuality is not a separately existent property, but only intelligible in relation to the rest of the universe. In particle physics we see it in the phenomenon of a "measurement", which is really just relationship manifest; in biology we see it in the web of relationships that define and sustain any organism. In human psychology, the same principle accords us a status as absolutely unique individuals, whose individuality is an infinite set of relationships to the rest of the world.

When we fall in love with the world, we will perceive and treat everything as sacred. Conventional religion sees some things as sacred and others as mundane, inert and interchangeable masses. To be sacred is to be imbued with a specialness, uniqueness, and divine purpose. To see something that way is to fall in love.

The Age of Reunion is a return to the Original Religion of animism. No longer will we divide the universe into the sacred and the profane, the spiritual and the mundane. Everything will be sacred because we know that everything is unique, even each electron, each drop of water, and certainly each human being. We know this already, of course. We have known it for a long time, but relegated it to the realm of Mr. Roger's Neighborhood ("You are special") or half-understood lines of scripture, as we continue to label and judge people and groups of people. Let us imagine a world where it is the basis of life. It is coming, the more beautiful world predicted in so many myths. A world alive with spirit, and a spirituality alive in the world. The core message of New Age spirituality is, "You are divine." From Deepak Chopra to the Gospel of

Thomas, from Louise Hay to the Bhagavad Gita, the teaching of the in-dwelling divinity of all things foretells the spirituality of the Age of Reunion. Instead of spurning the material world for higher things, it loves and embraces the material world and treats all as sacred.

In holistic medicine we will love our patients. Holistic economics will grow from love of people and their work ("each job will be unique"), and holistic science will love the processes of the universe itself. All along, that has been our innate yearning, simply to love and be loved. We feel its stirrings but fear to act on them, so enslaved are we to survival anxiety and the logic of the machine. The former is the feeling of not being able to "afford to" act from our humanity, and we respond to it with the program of control. The latter is the categorization of every person or object as a generic example of one or another type, to be addressed through the professional, dispassionate, and objective application of method.

The convergence of crises will lay bare the fraudulence of the logic of the Machine as well as the futility of the program of control. No longer will fear or ideology keep us from falling back in love with the world. As a result, all the old dualisms will crumble. There will be no distinction between a personal relationship and an economic relationship: economic relationships will *be* personal. There will be no distinction between work and play: *all* work will be playful. There will be no distinction between science and religion: both will lead us to love the world and to understand more fully our role in the unfolding cosmic pattern. There will be no distinction between work and art: both will seek to participate in the creation of beauty.

All of this, and more, will spring from the dissolution of the greatest dualism of them all, that between self and other. That is not to say we will all merge into undifferentiated oneness—nature is not like that, and neither are we. The discrete and separate self will not disappear but will become more fluid, more playful, no longer a permanent, irrevocable feature of our reality. We will feel and define our individuality through relationships to everything else; no longer will we think in terms of *having* relationships, but in terms of *being* relationships. As our relationships change and grow, so will we, thereby participating in and contributing to the growth of the whole.

A psychic once told me of a vision of the "million years of dreaming" that is to follow the collapse of the present world. In a dream, the world is malleable to the mind's reshaping, and the distinction between subject and object fluid. One interpretation would be the fulfillment of the

Technological Program in which nano-technology makes reality malleable to our every whim. But maybe the vision means something quite different. In a dream, the fluidity of perspective allows us to "be" first one character, then another, then to see it all from the outside, then to be the entire dream, which after all is (supposedly) all in our head. The million years of dreaming is not the imposition of our dreams of domination onto reality, but our participation in the universe's dreaming of itself. In love, we identify with another, see and feel the world from its perspective, and so realize our essential unity. So also in a dream. By falling back in love with the world, we enter the cosmic dream, and in releasing our discrete and separate selves, grow into who we are.

CHAPTER VIII

Self and Cosmos

Human Nature Restored

On a country hilltop one fall day, an herbalist challenged me to recall where I'd gotten the belief that I am bad. For in spite of the entire intellectual edifice I've presented in this book, on a deep emotional level I had, like many of us, long been convinced of my inherent unworthiness. "Who first told you you were bad?" she asked.

I couldn't answer her truthfully. If there was a time when I was "first told", or when I first accepted that awful proposition, I cannot remember it. I suppose I could try to pin the blame on mom or dad or teacher, but the fact is that their occasional use of shame, conditional praise, guilt, and so forth was the near-helpless channeling of ambient cultural forces. The message "you are bad" saturates our entire civilization. Relentlessly pounded into us from early childhood, it is bound to our most fundamental beliefs about self and world.

In science, this belief manifests as the selfish gene, the biological discrete and separate self that succeeds by outcompeting the rest of nature. In religion it is the "total depravity of man" or any doctrine that originates in the separation of body and soul, spirit and matter. In economics it is the "economic man", the rational actor motivated to maximize his or her financial "interest". The result is the World Under Control, seeking to rein in the behavior (which we mistake for human nature) that arises from these beliefs. And the apparatus of the World Under Control, the willpower and the coercion and the rules and the incentives, instills and reinforces the message, You are bad.

The message is everywhere.

"No littering—$300 fine." The assumption is that a threat to our self-interest best reins in our natural selfish carelessness.

A teacher: "Without grades, how would we make the students learn?" Unless coerced, they are naturally lazy and content with ignorance.

A parent: "I'm going to make you stay here until you say you are sorry!" People have to be made to feel sorry.

A state law: "Parents must provide a written excuse signed by a doctor for absences due to illness exceeding seven days."

"Johnny how could you!"

You have to. You cannot afford to. You must. You should. How could you? Why did you? You're going to have to try harder. Nature, and human nature, is hostile, uncaring, neither sacred nor innately purposeful, and it is up to us to rise above it, to master it, to control it. Over nature, we exercise the physical control of technology to make it safer, more comfortable, more bounteous. Over human nature, we exercise a psychological technology of control to make it kinder, less selfish, less brutish and bestial. These are the two aspects of control on which our civilization is based.

In this book I have described the inevitable collapse of the program of control, inevitable because it is founded ultimately on falsehoods, and in Chapter Seven I described the world that might arise after the Convergence of Crises has ended it. In this, the final chapter, I will describe an alternative to trying harder to be good (i.e. less selfish, more ethical, less greedy, etc.) based on a faith in nature and human nature. To inspire and sustain such faith in the face of the immense suffering that Separation has brought, I will also describe the dynamics of separation and reunion, so that we may see the cosmic necessity and purpose of our long journey of separation, both as individuals and collectively, and not resist the next stage of our development.

If our ruinous civilization is built on a struggle of good versus evil, then its healing demands the opposite: self-acceptance, self-love, and self-trust. Contrary to our best intentions, we will never end the evil and violence of our civilization by trying harder to overcome, regulate, and control a human nature we deem evil, for the war on human nature, no less than the war on nature, generates only more separation, more violence, more hatred. "You can kill the haters," said Martin Luther King, "but you cannot kill the hate." The master's tools will never dismantle the master's house. The same applies internally. You can go to war against parts of yourself you think are bad, but even if you win, like the Bolsheviks and the Maoists, the victors become the new villains. The separation from self that the campaign of willpower entails cannot but be projected, eventually, in some form, onto the outside world.

Yeah, sure, self-acceptance… the concept is pretty much a cliché these days. In its full expression though, the path to Reunion of self-acceptance, self-love, and self-trust is utterly radical, challenging cherished doctrines of how to be a good person. Let me state it as purely as I can: the path to salvation for us as individuals and as a society lies in being more selfish, not less.

How could this be? Isn't it precisely selfishness and greed that has gotten us into this mess?

No. What we see as selfishness arises from a false view of the self. Our cultural assumptions about who we are have defrauded us of our birthright, yoking us to the aggrandizement of an illusion. As a new understanding of self arises, selfishness will come to mean something quite different.

Already the illusion wears thin. Already we see the bankruptcy of the program of security and success that defines the winners in our society. Already we see how financial independence has cut us off from human community, and how technological insulation from nature has isolated us from the community of life. Increasingly, the program of control fails to benefit even the limited discrete and separate self of our illusions, as health, economy, polity, and environment deteriorate. Ironic indeed, given the ostensible goals of selfishness: security, pleasure, and wealth. That is why the road to the golden future that is possible for us, collectively and as individuals, is not a path of sacrifice and effort, but simply of awakening to what was true all along. In *The Yoga of Eating*, applying this idea to food, I wrote,

> When we deeply examine what we ordinarily think of as selfishness, we find a sad delusion. I imagine a vast orchard, the trees laden with ripe fruit, and myself sitting in the middle of it, warily guarding a small pile of gnarled apples. True selfishness would not be to guard an even bigger pile even more carefully; it would be to stop worrying about the pile and open up to the abundance around me. Without such examination we remain in Hell forever, thinking that our new five-thousand-square-foot house didn't make us happy because what we really needed was ten thousand square feet. On the other hand, very often one must acquire a thing first in order to discover that it doesn't bring happiness after all. That is why even deluded selfishness is potentially a path to liberation,

> and why I urge you to be selfish as best as you are able. Believe it or not, to be genuinely selfish requires courage. When the investment in something is large enough, we dare not ask ourselves if it has made us happy for fear of the answer. After staying in studying throughout high school and college, missing all those fun times, then all those years of med school, and all those sleepless nights as an intern... after all those sacrifices, dare you admit that you hate being a doctor? To be selfish is no easy thing. How many of us, in our heart of hearts, are really good to ourselves?
>
> The realm of food is a way to practice being good to yourself. Think of the greedy eater, eating more than his share, stuffing himself. That's an example of deluded self-interest, of not being good to oneself. The glutton really is getting more food. More more more! But he is hurting himself. If he were more selfish, if he made being good to himself his number one priority, maybe he wouldn't eat so much. It is an irony and a miracle. When you really decide to be good to yourself with food, the end result is a healthier diet, not a less healthy diet, even if the path to that diet might start out with an extra-large helping of ice cream!

When I speak before audiences about radical self-trust, I observe a range of reactions from grateful affirmation ("I've been waiting forever for this—I knew it all along but hardly dared believe it") to outraged protest ("This would wreck civilization as we know it"). Both responses are correct. What would happen to civilization, for instance, if everyone trusted their innate repugnance for any job involving the degradation of themselves and others? I suspect that many people entertain both reactions—gratitude and protest—simultaneously. The conditioned self fears the very freedom it so desperately desires. As on the collective level, to live in self-trust on the personal level is to accept the end of life-as-we-know-it. Anything can happen and everything can change: job, environment, relationships, and more. In exchange for freedom, we must give up predictability and control.

The ideology of control imbues every segment of political and religious belief. Just as religious conservatives believe we must clamp down on our sinful nature, environmentalists tell us to rein in our greed and selfishness, to stop polluting the world and hogging more than our share of resources. And practically everybody believes in "work before play," not permitting ourselves to do as we really want until we have finished what we must—the mentality of agriculture. Anger and blame infuse the writing of crusaders left and right, ideologues as opposite as Derrick Jensen and Ann Coulter, John Robbins and Michael Shermer. Variations on a theme, that is all.

Both sides express the guiding ideology of our civilization, just in a slightly different way. That is why, when one side wins over the other, nothing much changes. Even Communism did not end the domination and exploitation of man by man (let alone woman by man or nature by man). This book proclaims a revolution of a wholly different sort. It is a revolution in our very sense of self and, as a consequence, in our relationship to the world and each other. It will not and cannot arrive through a violent overthrow of the present regime, but only through its obsolescence and transcendence.

Anyone who tells us we must try harder to be good is operating from the same set of faulty assumptions about human nature. Self-trust only makes sense if we are fundamentally good. Looking at human violence and our own failings, we conclude we are not. It appears that the source of violence and evil is ungoverned human nature, but that is a delusion. The source is the opposite: human nature denied. The source is our separation from who we really are.

Does self-trust actually lead to a downward spiral of indolence and greed? It sometimes appears that if we relaxed self-control, we would yell at our children, pig out on junk food, sleep in every day, blow off our schoolwork, have promiscuous sex, quit the bother of recycling, indulge the nearest whim and maximize the easiest pleasure without regard to the consequences for others. But in fact, all of these behaviors are symptoms of disconnection from our true selves, and not our true selves unleashed. We lose patience with children because of our own slavery to measured time—deadlines and schedules—that conflicts with the rhythms of childhood (and with all human rhythms). We pig out on junk food as a substitute for the genuine nourishment so lacking in industrially processed foods and anonymous lives. We want to stay up late and sleep in because we do not want to face the day or live the life scheduled for us; or maybe we are tired from the nervous stress of the constant barrage of a life based on anxiety. We identify with professional athletes whose victories substitute for our own unrealized greatness. We covet financial wealth to replace the lost affluence of connection to community and nature. Perhaps all of our violence and sin is merely a flailing attempt to return to who we are.

In other words, the evils of human nature are actually products of the *denial* of human nature. We are the victims (as well as the perpetrators) of a diabolical fraud that says we must guard against nature and human nature, and ascend beyond both. In fact, as the illusion wears thin, magnificent people appear who show us the results of accepting, loving, and trusting ourselves. Whenever I meet one I am reminded of the intensity of my own limitations and insecurity. There are people who maintain a hunter-gatherer mentality of affluence in the midst of modern society; when meeting them, my own uptightness reminds me of the Jesuit explorer Le Jeune:

> "I told them that they did not manage well, and that it would be better to reserve these feasts for future days, and in doing this they would not be so pressed with hunger. They laughed at me. 'Tomorrow' (they said) 'we shall make another feast with what we shall capture.'"[1]

People like this are never constrained by "Can I afford to?" They have an open hand and an open heart, and somehow, it seems, they are always provided for. Recently I met a man, a shaman and artist, who does not charge for his services. His whole house is furnished with gifts from students and friends. Even without waiting for a restorative economy to appear, we can implement it in our own lives simply by opening up to the gift economy—and the gift ecology—that replaces the money economy. To do that, we need only, simply, to give and to receive. To freely give and receive requires faith that it will be okay. I will be okay. The world will provide. And that will happen when we stop seeing the world as a separate and hostile Other. That is the now-crumbling illusion that sets us in anxious opposition to the world.

We also see the magnificent results of self-trust in the geniuses of our society, the people who believed in themselves enough to devote years to the folly of their passions. I imagine Albert Einstein getting lectured by his boss in the Swiss patent office: "Al, you're never going to get anywhere doodling at your desk—you need good work habits like Mueller over there. Come on, focus!" And maybe Einstein thought, "You know, he's right. I won't play around with Relativity tonight, I'll take home a copy of 'Patents Today' magazine and study up. If I work hard I might even get a promotion." But instead he was drawn to his equations, and his magazine was left unopened. His creative genius didn't come from disciplining himself to do what was prudent, practical, and secure, but from fearless devotion to his passion. So it is with all of us. Earlier I discussed how it is irrational to do anything better than necessary (for the grade, for the boss, for the market), where "rational" means of economic benefit to the separate self. It is only freed of the compulsion of necessity that we can devote ourselves fully to creating beauty. No one will never create anything magnificent if, compelled by anxiety-based limits on time and energy, we make it only good enough for an economic purpose, or to please an authority figure with power over us. Good enough isn't good enough for our own happiness and fulfillment. To do something for someone else because that person or institution holds power over you—the power of threat to your survival—is a good definition of slavery.

Self-trust does not admit to conditions. We are accustomed to channeling our self-determination into safe, inconsequential, or highly circumscribed areas of life. "I will honor my integrity—unless to do so would get me fired." "I will listen to my body—but only if it doesn't want sugar." "I will follow my heart's true desire—but not if it is to get rich." I am not advocating we do without the things we want; I am asserting that the things we really want often aren't what we think they are. Unfortunately, sometimes the only way to find that out is to acquire them. How many people, upon finally achieving fame and fortune, learn that wasn't what they really wanted after all? But they would never have known any other way. Deluded self-interest can be a path toward authentic self-interest.

Perhaps the same is true for our entire civilization. Perhaps nothing less than the collapse of our civilization will be sufficient to awaken us to the truth of who we really are. Perhaps we must fulfill its grand ambition in order to realize its emptiness. True, the Technological Program can never be fulfilled in its entirety, but specific problems indeed succumb to the methods of control, the technological fix. Looked at piecemeal, the Technological Program is a great success. We have attained to a realm of magic and miracles. Godlike powers are ours. Yet somehow, the world around us falls apart. Our confidence in technology is slow to fade, however, because its successes are undeniable within their own limited realm. Perhaps the only experience that can reveal the fraudulence of the technological fix is its irrevocable, undeniable failure on the broadest systemic level.

Recall the metaphor of drug addiction from Chapter One. The drug fix is not impotent to solve the immediate problem. The fix works! I feel bored, I feel uncomfortable, I feel depressed, I feel lonely, and the drug indeed removes these feelings (for the time being), contributing to the lie that the pain is fundamentally avoidable even when its source remains untouched. In the case of technology, the lie is that we can avoid the consequences of our disruption of nature, that instead of bringing it back into balance we can move farther and farther out of balance while covering for the damage already done. It is the lie that our debts need not be paid. It is the lie that there is no inherent purpose to the world beyond that of our own making, and therefore no consequences for disrupting it. It is the delusion that nothing is sacred, so that we may wreck with impunity. Whether drug or technology, it works for a while; hence its allure, so powerful that we imagine that the complications it causes, the further pain it engenders, can likewise be avoided by the same fixes, indefinitely into the future, until the Final Solution.

In the case of drugs, often the addiction does not end until the complications it causes overwhelm its power to mask the associated pain. As the pain from a drug-wrecked life mounts, the power of the drug to numb the pain diminishes; every asset, every recourse is exhausted to keep the gathering problems under control; life becomes unmanageable, and all the postponed consequences emerge to be experienced as a convergence of crises. The addict "hits bottom", life falls apart.

The Technological Program, culminating in the complete elimination of suffering that dreamers think might be possible with the power of coal—I mean, electricity—I mean, nuclear power—I mean, the computer—I mean, nanotechnology—is tantamount to imagining that someday, alcohol or cocaine will not only temporarily remove the pain caused in large part by its previous abuse, but will also solve all the problems causing that pain. An absurd delusion indeed.

At each stage of an addiction there is a possibility of seeing through the lie, not just with reason but with the heart, and abandoning the program of control. It is no solution, not a lasting one, to apply the program of control to the addiction itself, to approach it with the attitude of self-denial. Quitting only works with the heartfelt realization that the fix was a lie, that I am denying myself something I don't want, not something I want. Otherwise, eventual relapse is inevitable.

One purpose of this book is to forestall such a relapse. When crises converge and things fall apart, a new sense of personal and collective self will open up. Let us recognize that and build upon it when the time comes!

Another purpose of this book has been to encourage us not to resist the transition. That is why it is important to describe the dynamics of transformation. In Chapter Five I wrote, "Even worse than the disintegration of the orderly, stable, permanent-seeming life 'under control' is for it to smoothly proceed until time and youth are exhausted." The longer we hang on, the greater the accumulated consequences. Already, the accumulated damage we humans have wrought over the last few thousand years is enough to cause the sixth great extinction in geological history, and the demise of billions of people in the next century due to war, famine, and epidemic. If we continue to deplete our social, spiritual, and natural capital in a desperate gambit to control the consequences of control with yet more control, then the eventual payback will be even worse.

That is why the message, "Be good to yourself as best you know how" must be accompanied by new insight into what it is to be good to yourself. The formula for success in our society is a formula for disaster. Not only on a collective level but individually too, the mortgaging of our life purpose to the demands of security and comfort leads ultimately to bankruptcy, and we are left, lonely and sick, looking back on years wasted in pursuit of a mirage.

Yet those years—and I have wasted many myself—need not be entirely fruitless, not if we learn from them what the surrogate objects of our pursuits were really replacing. All I really wanted was intimacy. All I really wanted was nourishment. All I really wanted was comfort. All I really wanted was to love. All I really wanted was to express my magnificence. The question, then, is what is the true object that humankind, the technological species, is striving toward? For it appears that the Ascent of Humanity is actually a descent, a reduction of the unmediated richness of reality, an abandonment of the original affluence of foraging. But perhaps there is more; perhaps we are groping toward something, a collective purpose or a destiny, and have wrought endless ruin instead in pursuit of a substitute, a sham, a delusion. Maybe it was necessary that our quest take us to the very extremes of Separation; perhaps the Reunion that is to follow will be not a return to a pristine past but a reunion on a higher level of consciousness, a spiraling and not a circling.

What is this transformational process, that it requires such an extreme of Separation? Where might it take us? Could there after all be a purpose, a transformational significance to the crescendo of violence that engulfs the planet today?

The Fall

I have portrayed separation, from ourselves, from nature, and from each other, as an engine of suffering and the cause of the multiple crises facing the planet. Because of this, and because separation has grown throughout history, and because it is in many important respects an illusion, many have conceived of it as an error, a blunder, or a sin, even *the* sin. Hence the age-old thread of returning to an earlier time, casting off the illusions of separateness to return to the original unity.

Yet elsewhere in this book I have hinted of the inevitability of the age-old course of separation, the ways in which it builds upon itself, its inevitable progression implicit in prehuman and even pre-biotic conditions. And intuitively, looking out on the fantastically complex structures of our society, one would like to ascribe to them a purpose, a cosmic reason more than that it was a mistake that should be erased. Yes, the ascent of our self-definition as discrete and separate beings has brought endless destruction, atrocity, and alienation, but it has also brought us to unsuspected points of view, modes of creative expression, and forms of beauty. The wasteland of modern society, its endless strip malls and ravaged ecosystems and dying languages and suicidal youth and empty consumption and lonely people… is it all for nought?

If separation is the root of all evil, then to ask the purpose of separation brings us to the old theological conundrum of the purpose of evil. Why, the theologians ask, did God allow suffering into the world? True, it was Adam and Eve who chose separation from God through eating the fruit of (the knowledge of) good and evil, but it was God who presented them with that choice in the first place, and who must have foreseen what that choice would be. Why, for that matter, did God create the Devil? Why did God create anything at all, why not remain the All and Everything? We can pose the same question in other religions. In Taoism: why did the undifferentiated *hundun*, the original formlessness, differentiate into Yin and Yang, and then the "ten thousand things"? In Buddhism and Hinduism: Why did the non-dualistic Original Mind or Brahman partition itself to create the illusion of self and other?

Through a mythological investigation of these questions, perhaps we can also answer the question in its modern form: what is the ultimate purpose of the descent of humanity "from a place of enchantment, understanding and wholeness" to the separation, alienation, and ruin of the present day?

In the theistic religions the archetypal story of Separation, prior even to the Garden of Eden, is the myth of Lucifer's fall from Heaven. This was the original separation, the template and the essence of all the separation that was to come. The Devil is normally viewed as the epitome of all evil, an identity that makes sense when we understand separation—the illusion of the discrete and separate self—as the source of our present crisis.

Significantly, the Devil as Lucifer is associated with light; that is, with fire, and therefore with the original technology that defined the "circle of domesticity" I described in an earlier chapter. The same function appears in Greek mythology in the figure of Prometheus, whom the gods punished for giving humans the power to be like them, to become, through the technologies of fire, the "lords and possessors of nature", usurping what was once a divine function. The foresight of the Greek mythmakers was truly amazing. How could they have known how close we would come to achieving for ourselves the Olympian powers of flight, rulership over natural processes, maybe even eternal youth and immortality? Just as in Judeo-Christian tradition the role of fire-bringer is combined with that of God's Enemy, the Greeks too understood the doom implicit in the power of fire, the doom that we are experiencing today: "Prometheus, you are glad that you have outwitted me and stolen fire ... but I will give men as the price for fire an evil thing in which they may all be glad of heart while they embrace their own destruction."[2] That evil thing was Pandora's box, appearing—like a drug, like a technological fix—so desirable but filled with the strife, disease, and ruin that we have known since the first technology.

Lucifer's Fall represents the very beginnings of separation from God, just as the technology of fire was humanity's first abrogation and arrogation of nature's cycle of energy flow, marking the beginning of our separation from nature. Then in the figure of Satan, we have the further legend of the Devil at war with Heaven, or trying to set himself above God, a theme echoed in the Babel story, the Tree of the Knowledge of Good and Evil, and countless other myths. This represents our campaign to become the lords and masters of nature; it is the domestication of the wild, the Technological Program, the world under control. And just as in myth, the war against Heaven is ultimately doomed, and the pain endured in this losing campaign is proportional to the intensity with which it is waged. The failure of the technological fix, the failure of the program to bring all the world under control, we address by intensifying the effort to control, sacrificing more and more of Life to War. And so, in the modern era, we have sold away our very lives to the enslavement of time and money.

However, we may take solace in the further observation that the War Against Heaven is itself not outside the divine plan, but part of it. In parallel, the ascent of humanity, the ascent of technology and separation, is not a wrong turn or a blunder but a necessary part of a process. In fact, to believe it a purposeless wrong turn is to reinforce the very assumptions that give it rise. It is to believe ourselves an exception to the rule of nature described in Chapter Six: *no trait evolves accidentally but only in fulfillment of an environmental purpose.* Our separation from nature, our war against nature, our ambition to transcend nature forever is actually in accord with nature's purposes, an evolutionary step not just of the human species, but of the planet as a whole. Our apparent separation is actually a step toward wholeness at a higher level of complexity, a stage in the extension of nature to a new realm.

A mythological narrative can help illustrate this process. Once upon a time there was unity: God but no creation; in Taoism the undifferentiated *hundun*, the original formlessness; in Buddhism the non-dualistic original mind. Then after an eternal interval, this unity decided to experience separation from itself. In other words, God split Godself into trillions of little pieces, which gradually forgot who they were. The reason was to experience what that is like, and eventually to come back together again at a higher level of consciousness. We could even see Lucifer as the first heroic volunteer. "Who shall go?" a voice thundered to itself, and finally part of that voice split off, separated, and therefore descended into Hell, presaging the deepening hell that we find ourselves in when we maintain and widen our separation from nature, other people, and the divine. The mounting crises of today's world are none other than this.

Perhaps separation is not an evil, but rather an adventure of self-discovery. Perhaps technology is an exploration of the illusion of separateness, that the Whole may come to better know itself, achieving reunion, yes, but at a higher level than before. In other words, there is a purpose to the Fall and thus a purpose to the whole course of Separation that is reaching its apogee at the present time. For this purpose to be fulfilled, it is necessary that separation run its full course, to the very extreme we are experiencing today.

For what is the ultimate degree of separation except to have forgotten one's divine nature so completely as to disbelieve in divinity itself; that is to disbelieve in any order, purpose, or destiny other than that which we impose upon reality? The clockwork universe that underlies our present worldview represents the furthest reaches of separation. It is a world in which there is not, *and could not logically be*, any meaning, purpose, or divinity. Today we are exploring these far reaches of Hell, a situation that Rudolph Steiner foresaw as "the war of all against all." How like the Neodarwinian "selfish gene" theory of life that is!

The present-era exploration of the furthest reaches of separation is therefore a cosmic necessity that must and will play out to its final resolution. Despite the near-unanimous recognition by scientists that we are rapidly destroying the basis of human life on earth, the destruction proceeds apace, as if we were helpless to stop it. And helpless we are. Certainly there are isolated victories, roads halted, incinerators closed, forests saved, dams removed, but the overall trend is towards worsening degradation of "the environment". The same applies to various social, political, and medical trends. It is as if some collective will were pushing us toward the full experience of Hell. It is as if we must, like the drug addict, "hit bottom" and render transparently hopeless the campaign to manage life, to control reality. Only then will we surrender to the "higher power" which is nothing other than Nature herself, the uncontrolled, the Wild, and the war against Heaven will be over. No longer holding ourselves away from Heaven, we will be drawn back into it with hardly an effort.

The necessity of exploring the furthest reaches of separation means that none of the long history of separation was an error. Even the Scientific Revolution of Galileo, Newton, Descartes and the rest, which launched the current, most extreme alienation of ourselves from the universe, was a necessity. We had to go through it; it was implicit in everything that came before it. The only error would be not to learn from that experience.

On an individual level too, each breakthrough in our spiritual evolution is preceded by a separation, a dark night of the soul, which can be so thorough as to parallel the Newtonian World-machine in denying not only the existence but the very relevance and possibility of God. When we come back to our true selves, back into Union, it is with greater experience and wisdom. It is almost inevitable that most of us will experience in this lifetime a crisis of faith, which may or may not be explicitly religious in character. It makes little difference what religion we are brought up in, because the crisis will find and obliterate whatever order, purpose, or meaning we assign to the universe, so that we may be reborn into a new understanding of purpose that contains and supersedes the old. Even if we return to the religion of our childhood, it is at a wholly new level of understanding.

So successful are we at holding the crisis of faith in abeyance that it often doesn't strike until old age, when death's imminence renders long-held delusions transparent. Those who have grown up in a religion, or non-religion, and never fully questioned its deepest foundations have as little reason for complacency as a pre-industrial society that has yet to experience industrialism's dislocations, not because they have been transcended but simply because they have not yet arrived. The experience has a developmental necessity; again, the only mistake is not to learn from it, not to integrate it, but to hold to the attitudes and beliefs that generated it in the first place. Whether to cooperate with our birthing into a new concept of who we are, or to fight this process to the last extreme, is up to each of us.

Let us hope we do not wait until the last extreme, the utter ruin of life and world. Let ugliness and wrongness become intolerable before then. Anything you do to spread the knowledge that life is meant to be beautiful and meaningful, anything you do to convince people not to settle for less, will lower our tolerance for ruination and raise the "bottom" from which our addicted society will renew.

I think the best way to spread this message is for us not to settle for less ourselves. I have often been inspired by those who, as Gandhi enjoined, "refuse to participate in anything humiliating," or who dedicate their lives to creating beautiful things, or who are deeply attuned to plants and animals, or who live free of the mentality of what they can afford to do, or who bring people together through the force of their love, who are generous without effort or stint, who see right through me and love me anyway, knowing I am good. These people show us our birthright, what life is meant to be. Seeing their joy, passion, love, generosity, and creativity, we no longer will tolerate a system and an ideology that creates the opposite. Life can be better than this!

The Perinatal Matrix

The tides of separation and reunion repeat on all levels, individual and collective, and in many dimensions of life, taking us back again and again to wholeness, not as a circling back but as a spiraling, each reunion at a higher level of wisdom, consciousness, complexity, and integration. The last section described the mythological template for this process, but perhaps the most intuitive model for the dynamics of separation and reunion is the process of birth. In the same way we are born physically as individuals, so also are we being born collectively by our Mother Earth, and spiritually into a new concept of who we are.

Stanislov Grof has developed a powerful and detailed model of the psychodynamics of birth, which he divides into four "perinatal" stages: uterine bliss, confinement, struggle, and emergence into the light.[3] Since birth is the archetype of the process of separation and individuation, it is useful to apply Grof's model metaphorically to our present condition.

The first stage of the birth process covers the months before the fetus has begun to push up against the limits of the uterus. She lives in a warm, rhythmic, rocking environment where her needs are automatically met, effortlessly. All she does is exist and grow, physically and mentally. The psychological state corresponding to Stage One is one of complete security, complacency, and a feeling of no limits. The world offers endless room for growth.

In the mythic realm, Stage One is represented by the Garden of Eden, where every need is effortlessly met, and in the Golden Age of ancient legend, when people still lived in the bosom of nature and knew not struggle and strife. "The earth herself, without compulsion, untouched by hoe or plowshare, of herself gave all things needful."[4] In terms of humanity and the earth, Stage One was the hunter-gatherer stage. Although life had its occasional tragedies and pain (just as the fetus is sometimes subject to disturbances in the womb) the overall environment was bounteous and nurturing. We were few and the world large. We had not yet begun to test the limits of the environment, or to see the world as limited. This attitude, that there is unlimited room to grow, can be found in the Biblical injunction, "Be fruitful and multiply." Out of habit and inertia, this attitude is still with us today even though it is driving us toward catastrophe.

As a natural consequence of growth, the uterus eventually becomes confining. The fetus loses her freedom of movement as a once-blissful universe turns against her. Because the cervix is still closed, there is literally no way out of this increasingly uncomfortable predicament. When the contractions start, universal, all-encompassing pressure bears down on the fetus from every direction. Stage Two thus corresponds to psychological states of despair, depression, and hopelessness.

In Stage Two, the Eden of the uterus has become a Hell. Just as there is no way out for the fetus, Hell is a place beyond hope of redemption. The basic condition of the universe is that of hopeless suffering. In the language of mystics, this state is known as the Dark Night of the Soul, the feeling of utter abandonment by God. Spirituality seems a cruel joke, faith a delusion. It is the cardboard world. Its meaninglessness is transparent. Existentialism is closely linked to Stage Two of the birth process, in which there is literally No Exit. We are just machines made from meat, and nothing matters and nothing ever could matter.

For several millennia now, the human race has been immersed in the ever-deepening misery of Stage Two. One by one, we have bumped up against the physical and social limits of growth. No longer can we continue to grow as we have for the last ten thousand years; no longer can we continue to expropriate more and more of the environment. Our problem-solving efforts only generate more problems, because they are based on extending our control over the environment, bringing more and more of it into the human realm. They are an attempt at continued growth within the same womb. Any solution which boils down to, "Let's bring more of the universe under human direction" will inevitably exacerbate our condition, bringing us closer to the limits of what the environment can provide. We cannot use the master's tools to dismantle the master's house. At the end of Stage Two, the direction of transformation can no longer be a continuation of the growth in the womb, but must become rather a journey into a new world.

The last century, encompassing two world wars, genocide after genocide, and the accelerating deterioration of the living planet, finally rendered utterly transparent the illusion that control over nature—and its apotheosis in the Machine—could ever fulfill its Utopian promise. It laid bare the bankruptcy of our solutions, and hence our helplessness to improve what we, seeing no alternative, called the "human condition". We were stuck, trapped. The machines our servants had enslaved us; the womb turned poisonous. And we thought it just to be the nature of things, so unable were we to conceive an alternative technology or a kind of growth that was not more *taking*. Meanwhile, our scientific ideology exacerbated those feelings of hopelessness, meaninglessness, and abandonment by putting us in a mechanistic, spiritless universe of impersonal forces and generic masses. We were left alone in a dead, pitiless universe, doomed by nature and human nature to struggle pointlessly for survival in a world rendered ever more horrible by our efforts.

Although we might try to soldier on, the once-bountiful world-womb can no longer sustain the growth of fetal humanity. The effort to squeeze just a few more years out of what remains of the Mother's resources—natural, social, cultural, and spiritual capital—will only poison the womb still further. Yet the exhaustion of these resources for growth does not alter our fetal civilization's built-in imperative to grow. Hence the inevitability of its demise, or its transformation.

It is very simple. A fetus grows; the womb is finite. The limits of growth trigger a birth crisis. Unbearable though it is, Stage Two is a necessary part of any birth process. If the status quo did not become intolerable, there would be nothing to impel change—birth into a new state of being—and we would turn away uncomprehending from the light when it finally presented itself. That is what happens in Stage Three—a way out is finally glimpsed. This way out is not some technological fix that makes the womb inhabitable just a little while longer, much less a technotopian fantasy in which the womb magically becomes infinitely large. That would be a recipe for stagnation and stillbirth. In Stage Three, the enormous pressures on the fetus are revealed to have a purpose, a direction as the cervix opens and a light shines through, promising a new world.

The physical distress of the fetus is even greater now than it was in Stage Two. She is subject to titanic pressures that slowly propel her through the birth canal, a life and death struggle occupying the whole of her being. At this point there is no going back to the womb of the familiar, for that womb is a hell now, and besides, the pressures of birth are too great to resist. While physically more difficult, psychologically a Stage Three state is easier to bear, for the light ahead gives hope and direction. For the human species, it represents our growing knowledge that another way of living is possible. We can see a glimpse of it already, the light at the end of the tunnel—the new modes of technology, money, medicine, education, and so forth I have described.

At this point it may seem that we are resisting the birth process, trying to climb back into the womb, maintaining the delusion of endless linear growth even as it crushes us. This is not unusual in a Stage Three dynamic. Say we enter a new job or relationship, grow within it, then eventually bump up against its limits. The job or relationship becomes increasingly intolerable, but we bear with it, seeing no alternative and hardly daring to believe that the good womb has reached a limit. Then a new possibility presents itself, a new career, a new or transformed relationship, but we may shy away from it in fear, preferring the womb of the familiar even as it grows increasingly intolerable. We crawl back in, but the next contraction is even stronger. Some people go back and forth until their status quo becomes truly intolerable. The early contractions are the gentlest. The first might be a glimpse of an opportunity. Eventually the womb of the old situation becomes a living hell and it is impossible to go back. Forces beyond our conscious control take over, and we are born into a new world.

Collectively, we humans have experienced only the beginnings of the birth pangs that will propel us into the new world we have glimpsed. We are still able to resist, still able to deceive ourselves into thinking that we can expropriate from nature endlessly, that nature has infinite wealth for our taking and infinite capacity for our waste. This illusion is disintegrating rapidly, and the disintegration will accelerate dramatically in our lifetimes. The crises rapidly converging on our species are nothing less than the uterine contractions that will propel us into a new way of being.

We humans have been moving out of Stage Two into Stage Three for nearly a century now. In some areas this transition is more advanced than others. For example, while it is true that we are destroying the environment at an accelerating pace, no longer do we ignore it or unquestioningly assume it is part of the inescapable order of things. On a general level, we know what the problem is and we know what we must do about it. To actually implement that knowledge is the archetypal Stage Three struggle. Solutions such as full-cost accounting, zero-waste manufacturing, community currencies, renewable energy, holistic medicine, and so forth are known—many forces are in denial, but the general solutions are known to us. To many fighting for these causes, the situation looks hopeless. What will spur us to actually implement these necessary solutions? Mother Nature and Mother Culture are having terrible birth pangs now. The contractions take the form of natural disasters, economic crises, famines, epidemics, and soon, environmental catastrophe. It is likely that we will keep trying to return to the womb of unlimited growth until our limits become starkly obvious; it may indeed take a wholesale environmental collapse before we change ourselves.

On the mythological level, Stage Three corresponds to religious archetypes of Armageddon, the final battle between Good and Evil, or Ragnarok, the battle between the gods and the giants in Norse mythology. At Ragnarok, all the worthy warriors who died in battle fight on the side of the gods. This myth refers to the common struggle that we all experience. It means that our personal battles have universal significance. The collective transformation of our species can only be the sum of billions of individual transformations, each driven by the intersection of generalized crises with our individual lives. No longer will we be able to hide from them, no longer will they be something that happens somewhere else, to someone else. In one form or another, they will affect us all personally. Because we are not discrete individuals but exist in relationship to the rest of humanity and the rest of nature, it is impossible to enjoy lasting health amidst an ailing society and a poisoned planet. It is impossible; it is a contradiction in terms.

In Stage Four of birth, the baby is born into a new and unimagined world, where he becomes an anatomically distinct individual. In any kind of birth process, the entity being born cannot imagine what lies beyond the mother's body. In the case of humanity as well, the new society we are being born into is probably beyond imagining. I suspect my halting attempts to describe an Age of Reunion fall woefully short of its true splendor. Whatever form collective humanity and individual life will take, one thing is certain: it will not be a final triumph or mastery over nature. We will be no more independent of nature than an infant, having outgrown the umbilical connection, is independent of her mother.

Birth is a journey that starts with blissful oneness, proceeds through an increasingly unbearable confinement, climaxes in a heroic struggle, and ends with a return to the one, but at a new level of being. In human birth this is the breast, the reuniting with the mother in a new, more highly individuated way. Once upon a time we were enwombed in nature, without the possibility of even an illusory separation. In the Age of Reunion that will follow the present birthing, we will gaze upon Mother Nature's face with the adoring eyes of an infant.

Archeologists and historians are fond of infantile metaphors to describe prehistoric or ancient society: "The cradle of civilization." Perhaps these metaphors are misleading. Even now, humanity is not yet in its infancy. An incredible journey awaits us.

We humans, and even the planet herself, are undergoing a birthing—

to what state we can only speculate. We have passed through the long gestation of the hunter-gatherer, we have grown up against the limits of our environment until we could grow no more, and now the labor is beginning that will likely stretch our capacities to the limit. Some environmentalists, especially those with the most comprehensive knowledge, despair that it is already too late to save our planet—irreversible processes already in motion assure our destruction. But perhaps these conditions are what it will take to turn the long-gathering capabilities of science and technology to their true purpose: the restoration and furtherance of nature's patterns, and the creation of new forms of beauty. Perhaps the impending catastrophe will demand that every facet of our civilization's scientific and technological achievement be turned toward the planet's healing, galvanizing humankind and drawing us together in a way similar to but far, far more powerful than the space race of the 1960s and the last century's wars against cancer, poverty, drugs, and each other ever did. Perhaps nothing less than an all-out struggle will secure the survival of our species; perhaps only in such a struggle can we rise to our potential and our purpose. In healing the ruination of nature, goodness, beauty, and life, we will transcend who we were and be born into something else.

The Gaian Birthing

The mounting destruction, suffering, and catastrophes of the last millennia and especially the last few centuries are the birthing pains of the human race, being born into a new form of relationship to the universe. The question is, if we are being born, who is doing the birthing?

The intuitive answer is Mother Earth, of course, Mother Earth or even Mother Universe. No matter that the birth pangs are manifestly of human origin; they are inescapable accompaniments to separation, which is itself inescapably woven into the fabric of biology. Separation was not a watershed event, a discrete wrong turn as implied in the Eden story, but a cumulative process that started in prehuman times, even prebiotic times. Nature, the universe, even mathematical systems tend toward increasingly complex order and, more significantly, *organization*, which implies a differentiation of roles, an offbudding of the semi-autonomous individuals that we know as life forms. In other words, the present convergence of crises is of transhuman origin, perhaps even an evolutionary inevitability.

Is the crescendo of complexity, from bacteria to nucleated cell to multicellular organism to the human brain, to the tribe to the agricultural civilization to the industrial economy to the noosphere, just a matter of one lucky chance after another? I have explained how a negative answer to this question need not imply divine guidance by an external creator God, not if order and organization is inherent in reality, not if purpose is inherent in the universe. But let's put it more poetically. Let us explore instead the idea that the universe, and in particular the Earth, was *pregnant* with life, with intelligence, with civilization, with the entire course of separation and reunion, from the very beginning.

Does anyone else find it odd that the raw materials for technology are so readily available? That metals are so abundant near the earth's surface, that domesticable animals and plants suitable for agriculture were so accessible, that the earth laid down enormous fossil fuel deposits capable of powering an industrial revolution? What about the fact that many genes expressed only in higher animals were present in the genome hundreds of millions of years before they were ever needed, as if waiting for the time to exercise their function? If we didn't know better, we might think the collection of the earth's organic and inorganic processes that we know as Gaia were consciously executing a plan to bring a technological species into being.

In Chapter Six I speculated that life on earth is of dual parentage, Earth and Sky; that the primordial planet was a fertile womb seeded by genetic material from space. Some people think that this material encodes a program for the eventual rise of technological civilization: a Carboniferous Age to lay down oil and coal deposits, intelligent life forms to utilize them. Because the mentality of manipulation and control—the distancing from nature—is an inseparable aspect of our rise as a technological species, we could also say that the present extreme of Separation is part of that same program.

To what end? Perhaps technology will culminate in some form of space travel, by which we will propagate biological material further throughout the universe and accomplish Gaia's reproduction. Gaia, we might say, is going to seed, devoting every last resource to its production and dispersal. One of the chief criticisms of Gaia theory is that natural selection cannot operate on the planet as a whole, for it has no competitors. But maybe that is untrue, maybe planets (or biospheres, to be more precise) reproduce and compete as well, but on a far grander scale and time frame.

What happens after a plant goes to seed? If it is an annual, it dies; if it is a perennial it goes into a period of hibernation or rest. Is our present despoliation of planet Earth a mechanism for the marshaling of every resource toward reproduction? Is it the poisoning of our womb that will motivate us to search for another? I think not. The birthing that we are undergoing is something far greater than the eventual replication of the same sad pattern in some other place and time. If our planet were to perish—whether or not we sent genetic material, living organisms, or even human colonies into space—I would consider that to be an aborted birth, a stillbirth.

In the environmental movement, the peace movement, the holistic medical movement, and others in education, information technology, psychology, and so on, we have a picture of what humanity could become—the Age of Reunion I have described in this book. But the possibility of stillbirth is also quite real. Perhaps it has already happened before. Perhaps the myths of Atlantis and Babel harbor a kernel of historical truth; perhaps there was a Deluge followed by a new beginning. Sometimes I wonder: What would we do if a total collapse were imminent? Knowing that every written record, maybe every physical record of our civilization would perish, how would we transmit what we have learned to the future? The only way would be to embed it in myths, sacred stories encoding important information to guide us in a future age. It is in part this possibility that has impelled me toward the study of myth. Maybe therein is encoded wisdom that can avert us from being stillborn again.

More likely is that the old myths and legends (Atlantis, the Flood, and so on) embody psychological and not historical truths. Atlantis, for example, represents a submerged higher potential: once glimpsed, now lost, still possible in our personal and collective futures. The mythmakers of old, tapping into a great wisdom, foresaw the inevitable, tragic denouement of humanity's "ascent" to separation, and foresaw as well the possibility of something greater being born from its ashes. This is the Kingdom of God following Armageddon, the renewal of heaven and earth following Ragnarok.

The progression of social and ecological disintegration was written into the future long long ago, at the very dawn of the Separation which took a series of quantum leaps with stone and fire technology, and again with agriculture, and again with modern science and technology. While the catastrophic effects of Separation are flagrantly apparent to many people today, it was not so obvious in the past, when vast amounts of social, natural, cultural, and spiritual capital were yet to be depleted, that disaster was in the works. Who could have guessed, when the first granary was built in prehistoric Mesopotamia and the first forest cut down in Sumer, where it would all lead?

Nonetheless, there have always been visionaries who *have* seen where our separation from self, nature, and other would inevitably lead. Centuries ago, millennia ago, they pointed us urgently in the other direction, even as they recognized the inevitability of the still-ripening catastrophe. All of them did their best to leave us messages, clues and hints that a different way is possible, not so much to stem the tide of history but to teach us how to proceed after the crash of our unsustainable story of self. Their messages are about transcendence, transcending the limited, limiting, and delusory Separated self of our present science, religion, economics, medicine, and psychology.

They spoke their messages in different ways, by whatever means were expedient to transmit them through the ages. Some of them, such as Jesus of Nazareth, were unfortunate enough to have had a religion founded around them that twisted their teachings toward ends diametrically opposed to their original intent; yet even so, the original meanings often still lay encoded in whatever words of theirs have been preserved.

Far more of these teachers quickly became anonymous. We will never know their names, but their messages survive in many forms. Some communicated with the future by creating myths and legends, poems and songs, dances and rituals for receptive people to decode or, more often, that plant a seed in the unconscious mind of the listener or performer. Even if no one can explain the logic behind such rituals, everyone who performs them is changed, quickened, implanted with a knowledge that is only much later, when the time is right, followed by understanding.

Still others communicated with the future through direct personal transmission to disciples. Seeing the hopelessness of stemming the virus of separation until it had consumed all, perhaps seeing also the imminent demise of their own cultures at the hands of encroaching agriculturalists, they founded lineages and hid them within the destroyer civilizations themselves. Perhaps this process started in Neanderthal times, as hinted by the Yurok story of the Wo'gey in Chapter Two. Knowledge was passed down personally through generation after generation of Zen masters, Sufis, shamans, Christian mystics, Kabbalists, Taoists, yogis, wizards, and other individuals, kept disguised within folk religion or completely hidden until times were right for its blossoming. That time is today, and it is no coincidence that many of these formerly secret traditions are making their knowledge public as best they can.[5]

Science fiction writers have envisioned humankind's next evolutionary step as a collective transition to a higher level of organization. Just as a termite mound has an individuality and sentience that transcends each individual termite, just as the brain's intelligence emerges out of the coordinated activity of billions of neurons, so also might a racial intelligence emerge from the tighter and faster linkage of individual human minds. Vernor Vinge has described a future in which people augment their mental capacities and memory with computer add-ons, which make mental processes more accessible from person to person resulting in an übermind.[6] Just as an organism emerges out of the coordinated activity of billions of cells, so might a new kind of metahuman entity arise out of the coordinated activity of billions of humans, especially when communication among them happens as fast as hormonal or bioelectric communication in the body. We might see Gaia, then, as birthing a collective human entity, whose embryonic organs began differentiating with the specialization of labor in the first builder societies, and whose information processing capabilities are finally reaching the speed of thought in the Internet Age.

Vernor Vinge and similar thinkers are firmly encamped in the "Gee Whiz—The Future!" ideology of technological progress. However, their techno-wonderama scenarios arise ultimately from a valid intuition of transcendence in our not-too-distant future. Their mistake lies in assuming that this collective transcendence can happen without a simultaneous and corresponding transcendence on the individual level. What organism could survive whose cells believed themselves to be in constant competition with each other and at war with the environment (the body), in a struggle for resources, without any purpose or significance to life beyond their own replication? That is a pretty good description of cancer cells. No, in a healthy organism the cells seek the perfection and fulfillment of their role. Is this not identical to what I have advocated for us humans? Individually, in the spirit of the gift and the fusion of work and art? Collectively, in the use of science to better understand, and technology to better fulfill, our role and function in nature? It is not our destiny to be, in Agent Smith's words, "a cancer on this planet".

The organismic model of human society has oft been a justification for Fascism, an elevation of the corporate body over the rights, needs, and welfare of the individual. The Fascistic interpretation, however, is itself a product of deeper mechanistic assumptions about both biology and society. For in fact, the specialized cells of a body bear little similarity to the specialized workers in a hierarchical factory system. Recent advances in genetics and neurology, which I touched on in Chapter Six, have forever debunked the top-down control model of morphology. There is no central authority that dictates the development and function of each cell. Of course, each cell responds to signals from the environment, but the environment's intelligence is an emergent property, not centralized in a commanding authority, nor programmed into a collection of genes. Moreover, each cell type responds to the sea of hormones, electrical fields, and other communication media in a unique way; even within a given cell type, individual cells respond uniquely according to their location and other, unknown, factors. Certainly, none of them seek to hoard resources for themselves. Like hunter-gatherers who do not store food, they trust in the ongoing supply of life-sustaining sugar and oxygen. Cells maximize their "self-interest" at the expense of the rest, not unless they are cancer cells.

The imperative of endless growth, like the campaign to maximize me and mine, and like the program to control the world, is founded on an illusion, the discrete and separate self. Now that the illusion is crumbling, now that its scientific foundations are falling apart and its practical consequences destroying us, a possibility is arising for a new kind of society, as well-coordinated as a healthy multicellular organism and as individually creative, fulfilling, and purposeful as the life of a hunter-gatherer.

Our collective transcendence of all we have been is founded on an individual transcendence of what we thought we were. And in a beautiful and necessary synchronicity, it is the same convergence of crises that will bring about both. The individual transcendence I speak of is the spiritual awakening that so often happens in the wake of major life tragedies and transitions, when life falls apart. Usually it follows the model of birth, the perinatal matrix described above. The sudden life event is merely the quickening of a long-gestating process.

The potential for transcendence lies latent in all of us, but in our society is rarely allowed to blossom forth. In fact it is arrested, almost purposefully, by the structures of education, religion, and the "struggle to survive" that define modern adulthood. The arrest of this developmental stage bears striking similarities to that of humanity collectively. Joseph Chilton Pearce describes the natural progression of human development in his books *Evolution's End* and *The Biology of Transcendence*. The growing child goes through several developmental stages—concrete operations, formal operations, and so on—each corresponding to a phase of brain development. Up through early adolescence, each of a child's several brains—the reptilian forebrain, the limbic system, and the cerebral cortex—develops in turn, culminating in the development of rational, analytical thought in the early teen years, which we take as the highest form of cognition.

But Pearce argues compellingly that there is supposed to be another phase of brain organization in the middle to late teens associated with the prefrontal cortex, whose function is largely a mystery to conventional science, and the neurological dimension of the heart, which conventional science ignores altogether. He associates these developments with intuitive, holistic, transpersonal cognitive functions that our society devalues or invalidates. How could we validate them, given that they conflict with fundamental tenets of self and world? Like much learning, adolescents properly develop these functions through modeling, but the models are few indeed. How can someone develop an ability that the ambient ideology says does not exist, and for which there is no model?

Simply put, at about age fifteen human beings are meant to develop empathetic, holistic, transpersonal modes of cognition, abilities that violate the dogma of the discrete and separate self. Primitive societies acknowledged these abilities and fostered their development, providing both models and methods for transcendence of the limited, rational, ego self of the pre-teen. This was the point of many coming-of-age ceremonies, in which the teenager's ego boundaries were temporarily shattered through such means as psychotropic plants, deprivation of food, water, or sleep, isolation from the tribe, pain, or other intense ritual experiences. Returning to the tribe afterward, the young man or woman (no longer a child), would have a profoundly different conception of self. Not by accident, many of these ceremonies incorporate ritual reenactments of the birth process.

The rational, analytical, objective worldview implicit in the discrete Cartesian self is a necessary and proper phase of human development. It is a phase, however, that sets the stage for a further phase, one that we typically no longer experience. We remain therefore stuck in a perpetual adolescence, waiting our whole lives for some momentous happening that never happens. Pearce puts it eloquently:

> What "it" is that is supposed to happen at this age remains a mystery, for though it may linger like Thomas Wolfe's "grape bursting in the throat," it never happens. George Leonard spoke of an anguished longing so acute he knew it could never be assuaged. A university student said that since fourteen she had waited for a momentous happening that didn't happen. (Were the issue sexual it would not be an unknown.) A student wrote his parents that he loved his third year of college but had awakened one night with "the cold hand of terror clutching his heart." Since about fourteen, he reported, he had felt that something tremendous was supposed to happen. Now approaching twenty-one, he had been waiting for seven years and it hadn't happened. Suppose, he had asked himself at that late-night awakening, it never happens, "and I never even know what it was supposed to have been?" a possibility that struck him with despair.[7]

When I read this to my brother John, he replied, "Yeah, and then one day you think, 'Hey, I'm 28 now… I guess it must have happened." And so we live with an emptiness inside, a dissatisfaction stemming from the inner knowledge that there is supposed to be more.

The features of the pre-transcendent adolescent correspond precisely to the worldview of the Newtonian World-machine. That our development remains arrested in the analytic ego phase is necessary for the continuation of society as we know it, for the perpetuation of the World Under Control. A happy corollary, though, is that the ending of this world of property, alienation, and control will coincide with the ending of all the social and ideological conditions that arrest our transcendence. Already it is happening, as the technologies of Reunion build momentum. Even if you are no longer an adolescent, it is not too late. There are more and more models for transcendence now, and our capabilities lie dormant, deeply buried but never dead. All it takes is the equivalent of the coming-of-age ceremony, which is the crumbling of the pre-adolescent world of the discrete and separate self. In the absence of wise elders to implement such a ceremony, the world will do it for us, simply because that pre-adolescent ego world is not sustainable. Like our own society based on the same principles, it inevitably generates converging crises that eventually reveal its fraudulence.

The same dynamic applies to the collective development of the human species. We are poised at the edge of transcendence, having fully developed—to the utmost extreme—the world of the rational ego self. We have fulfilled that stage, in which we are meant to consummate our individuation as separate beings, or in the collective case, as an entity, "Man", separate and distinct from Nature. While we may lament the vast suffering that separation has rendered, another way to look at it is that we are growing up. This process of individuation is necessary to discover who we really are, collectively, as a species.

Throughout childhood and into adolescence our job is to individuate and grow; we take from our mother as much as we need, as much as she can give, with gratitude perhaps but also with a natural, easy selfishness. As a species, this is what we have done to Mother Earth. We have treated her treasures as ours to take, with the same sense of entitlement with which a child demands food from mother. And as we have done so, we have developed as a species an independent identity no longer derived from Mother Earth, from her places, plants, and animals; in some sense, we have been weaned.[8]

In normal development, the adolescent enters true adulthood and no longer takes selfishly from his mother. To be sure, in our society with its lack of coming-of-age rituals we often see adolescence extended for years and decades into what would otherwise be adulthood. (How do you behave when you go home to visit your parents?) This is an abnormal state of affairs. By our late teens or early twenties, we should have become independent of our biological parents and then, desirous of supporting them, releasing the sense of entitlement proper to childhood.

It is time to turn around and begin to cherish and protect the Mother that has given unto near-collapse that we might grow. A mother by nature will continue to give and give, even past her capacity, as the earth is doing now.

What, then, is the coming-of-age ritual for the human race? What is to initiate us into adulthood? Coming-of-age rituals varied across cultures, but they had in common a goal of transcending the childhood sense of self, the individuated ego, separate, objective, rational, maximizing its own aggrandizement, feeding and growing—the very precepts of modern economics. These lower-chakra, limbic, and cerebral-cortex functions are not to be abandoned, but integrated into a higher order of being capable of fulfilling a function in a larger connected whole: the clan, the tribe, the village, the forest. For this to happen, the child's self-definition must be temporarily shattered or released. The adolescent is therefore put in a situation, unprecedented in his experience, to which the old self-conceptions and world-conceptions no longer apply.

And now the coming-of-age ceremony for the human race has commenced. It has been building for millennia, in the vast historical sweep of violence and suffering that has accompanied the ascent of humanity. It is not localizable in space or time, but visits each people, each nation, each individual in a unique way. In the present age, though, a critical mass is building in which the world will crumble for many people all at once, in the space of a few years or decades. Then, the innate spiritual transcendence that is each individual's birthright will once again become the rule, not the exception. Then, the Age of Reunion will have commenced. Then, we will look upon one another, and upon the animals and the trees and the planet, and see not competitors or resources or Others, but extensions of our selves. We will experience a world that is wholly sacred, pregnant with creativity, immanent with purpose, alive with spirit. In love with life, in love with the world, we will humbly join in the ongoing co-creation of beauty. As we are already. But soon it will be with conscious awareness: our individuation complete, we will unite with nature in full-fledged partnership.

Eulogy and Redemption

What, then, of the victims? What shall we say to the men, women, and children whose ruined lives have followed in the wake of our "ascent"? Should we not lament the billions of passenger pigeons whose flocks once darkened the skies? Should we not mourn the dodo, the great auk, the American chestnut, and the millions more now following them to extinction? What of the elder bushes, a century old, keystone species of a fantastic ecology torn up and paved over to build a new road? What of the forests turned to deserts? The native children shot for sport by white settlers? The women tortured and burned alive as witches for practicing herbal medicine? The schoolchildren today, cajoled, coerced, and medicated into spending the Kingdom of Childhood behind a desk, in a room, standing in line? The coal miners of 19th-century England, emerging decades later stunted, broken, and destitute from the mines? The babies deprived of the breast? The women raped, the men tortured and killed, the children watching the soldiers who did it? What can we say of the concentration camps, Auschwitz, the Gulag, the unspeakable hardship of the men sentenced to a lingering death at hard labor? What shall we say to the victims of communist purges, and to the families sent a bill for the executioner's bullet? What of the black man beaten and lynched, and a picture postcard of the event sent to his mother? What shall we say to the starving children, past the point of hunger, bodies falling apart? And what shall we say to their mothers? And what of the children working in toy factories, rug factories, chocolate plantations? The countless assembly line workers, their human creativity reduced to a few rote movements, producing empty consumer junk out of toxic materials destined sooner rather than later for the landfill? The betrayal after betrayal of the Native Americans, people massacred, lands cheated, religion outlawed, culture purposefully destroyed? The cancer victims of a poisoned world? The slaves long ago who labored on the Pyramids? Contrast a life carrying stone to the life of a hunter-gatherer, and the bargain we have made becomes clear. In this, the first monument to the Machine, the folly and the horror of our ascent is clear: an exchange of life for labor to erect a useless edifice.

No authentic peace with the world can be achieved in ignorance of the facts. Read books like *A Language Older Than Words*, *Night*, *Gulag Archipelago*, *The Dying of the Trees*, *The Lost Language of Plants*, *Evolution's End*, *Trail of Tears*, *Rebels Against the Future.* We must be utterly clear about what our civilization has wrought. If we, like the technological Utopians of the Industrial Revolution justifying the mines and mills, maintain that the sacrifices of the victims are a necessary and worthwhile price to pay for our ascent to a higher state, then we must be clear on what that price has been. The price of Separation has been no other, and could be no other, than the furthest possible extreme of evil.

That the Reunion I have spoken of, the rebirth at a higher level of consciousness, could only come through resolving and integrating our age-old course of Separation is not a justification of its evil, no more than a criminal's remorse or a victim's forgiveness justifies the crime. Nonetheless, there is another way to understand the suffering of Separation's victims.

Some years ago, a man I know very well was obliged to dig up a splendid burdock plant that grew outside his home. He had been asked to dig it up before, and in a semblance of compliance had halfheartedly sheared off the leaves, leaving the root intact. This time his wife supervised him to ensure that he did it right, that he dug it up root and all. Every moment his heart was heavy, but his fear of his wife's anger was enough to overcome his reluctance and prevent him from standing up for his integrity. Something changed that day; in his words, "Our marriage has survived many onslaughts because it had a strong, deep root, and now that is gone." The plant had kept coming back, it wanted to grow there, but the man imposed his will, which was not even truly his own, onto nature. He got nature under control. The fate of that burdock plant, the process by which it was destroyed, is really no different in essence from the worst extremes of ecocide and genocide. In both there is a perceived necessity, a fear that overcomes our goodness, and a destruction of the innocent. But later, after the marriage went through a tumultuous period, he realized that the burdock had given him an important teaching that only its self-sacrifice could have delivered. It was a teaching about boundaries, integrity, communication, and change, and he had a clear sense that the plant chose to grow there precisely for that teaching.

Donna Gates, the woman who developed the Body Ecology protocol for curing autism, once told me that she has noticed a similarity among autistic children's households. Beyond the proximate factors of vaccines, mercury, antibiotics and other body ecology disruptions lies a deeper reason for autism—a purpose, not a cause. She believes that these children have in some sense chosen to be born into their circumstances as a way to bring a great gift to their parents and families. Of course, few parents see autism as a gift—having an autistic child is like having a permanent infant who requires intense care and never grows up. In many cases, normal life becomes impossible as the demands of the child consume all leisure time. Life ambitions give way to the demands of caring for another being without thought of recompense.

Reread that last sentence. Isn't that precisely the prescription for joy that the saints have given us through the ages? Perhaps these children are noble spirits choosing their incarnation as a way to help us understand what is important in life. Gates has observed that autism strikes disproportionately in households where life was otherwise smooth sailing, in which the vacuity of modern goals and priorities would otherwise never have become apparent, at least not until time and youth were exhausted. Many autistic and otherwise "mentally retarded" children possess an undeniable spiritual quality about them: the description "special" is not mere euphemism. When they are healed of autism this quality remains: they are often remarkably selfless, content, compassionate, affectionate, and emotionally mature. In one sense, yes they are the innocent victims of modern birthing practices, medical practices, pollution, and dietary ignorance, but from a higher perspective they fulfill a noble purpose in our society's healing.

We often think of misfortune as some kind of punishment for past evil, a theme which runs through religious thought both East and West. In the East it is the idea that present suffering represents the negative karma generated through past misdeeds; in the West we have the image of Yahweh striking down the cities of Sodom and Gomorrah for their sins, threatening Ninevah for its "wickedness". However, the self-evident fact that it is often the innocent who suffer the most demands all kinds of theological contortions, from past lives to original sin, from future rebirth to Heaven and Hell.[9] How else to explain the sweet, innocent babies in the children's cancer wards? If we are not to resort to blind, pitiless, purposeless chance, we need another explanation for the innocence of our victims. Perhaps they are great souls, meeting the huge necessity for innocent victims that our civilization has wrought. "I will go," they say. "I am big enough. I am ready for this experience."

We might look on whole peoples, cultures, or even species in the same way. While we might understand the decay of our civilization as a just dessert for the violence it has perpetrated, how can we explain the destruction of the beautiful indigenous cultures of North America (or any continent you care to choose)? What sin against God, man, or nature could justify their violent extinction? True, some were supposedly warlike and unfriendly to outsiders, but many amazed their first European contacts with their childlike trust and easy generosity. None of them, not even the most warlike, perpetrated anything comparable to the human and environmental ruin that are the handiwork of Machine civilizations. We might just as well try to explain how three-year-olds dying of cancer actually deserved it. No, if we are to believe in a purposeful universe, we must look elsewhere for an answer.

The answer I offer you is that all of the people, cultures, species, and ecosystems that we have destroyed constitute, together, a medicine for the great disease of our civilization, the disease named Separation. It is in the nature of the disease to destroy what is beautiful: to convert reality into a data set and life into money, with all the violence that such reduction of life implies. In the process of separation and eventual reunion at a higher level, selfless beings who already live in non-separation are structurally necessary. They, like the burdock root, like the autistic children, have taken on a noble and magnanimous role. The cultures, species, and people we have extirpated have delivered to us a teaching and a medicine.

The cultures we have destroyed have not vanished without a trace. Anything we destroy leaves its imprint on our own spirit, whether on the personal or cultural level, automatically becoming a future medicine when it emerges into conscious experience. Please do not misunderstand: I do not mean to exculpate the victimizers because, after all, the victims volunteered for it. Nor do I mean to depreciate the magnitude of the crime or the tragedy. Nonetheless, their sacrifice was not in vain.

Because all acts of violence leave their imprint in the perpetrator, the perpetrators ultimately will suffer violence equal to that they have inflicted. Perhaps that is why Jesus, facing his tormentors, said, "Forgive them Father, for they know not what they do." He understood what they were in for, the oceans of remorse they would need to traverse before arriving at peace. Etymologically, remorse means to bite back. Jesus saw that what they did to him, they were doing to themselves. Studies of soldiers with post-traumatic stress disorder find that the most seriously disturbed are not those who have witnessed or suffered violence, but those who committed it. "Soldiers who were in low-intensity battles but had killed someone suffered higher rates of PTSD than soldiers who experienced high-intensity battles but did not kill anyone."[10] As researcher Rachel MacNair puts it, "Despite all the killing we've done, the human mind is not designed to kill. Portions of us get sick when we kill. Killing is against our nature."[11]

Enormous forces must be applied to render a human being into a killer, someone who could cut down forests, tear up land, or kill innocent people. To do the things we do requires that we be removed from our natural-born state of wholeness, enchantment, connectedness, and biophilia. To commit the heinous violence of our culture, even in its muted, indirect forms such as consumerism, we must first be mangled ourselves. We perpetrators are the end products of a spirit-wrecking machine thousands of years in the making, that has battered and wounded us almost beyond recognition. Our healing happens through our victims, just as my friend's healing required that he destroy and mourn the burdock plant.

Our Separation, our ruined wholeness, our Fallen state leads inevitably to acts of violence. Violence is a symptom of a wounded spirit. And the medicine for this disease is precisely the consequences of that violence. The process of acknowledging and mourning what we have done is itself healing. To simply withhold opportunities for violence from a wounded person is not a sustainable solution. Something has to bring it to the surface, and something eventually will.

Does this mean that I can excuse myself from all the hurt I've caused in my life, thinking, "Well, my wound drove me to it, and I needed to do that to recover"? No. The healing comes only through the realization, "My God, what have I done?" It is the remorse that is healing. On a cultural level, then, it is healing for us to face up to the crimes of our civilization, the dirty secrets of our past. Living in denial of the bitter facts only perpetuates more violence and prolongs our state of separation and suffering. The truth is coming to light now, as we acknowledge what we have done to our planet, its cultures and people. This is another sign that the Age of Reunion is nigh. Yes, many segments of our society are still in denial, choosing to live with the imprints of the wounds they have inflicted upon the world, not knowing that world and self, I and thou, are not really separate and that no amount of control can keep the consequences from eventually seeking out the perpetrator. The denial cannot last forever. The continuing pain of the festering wounds, which cannot be hidden forever, will eventually make the truth impossible to ignore or deny.

Once upon a time the Great Spirit spoke to the world. The Great Spirit said, "The world is sick. Millions of people have separated themselves off from life, and their suffering grows with each passing year. Soon they will utterly destroy themselves and all that is good. They need medicine, but I warn you, most of the medicine they take they will destroy most horribly. Who is ready to be the medicine?" And the Spirits of the Tribes said, "We are ready." And the spirits of the forests said, "We are ready." The spirits of the frogs said, "We are ready." The Earth herself said, "I am ready."

Martin Prechtel once said, "The redwoods are perfectly happy to go extinct." I know another man who described a long conversation he had with redwoods under the influence of LSD. The redwoods told him they were sad for the people chopping them down, and hoped they would stop doing that before they destroyed themselves. The redwoods know the enormous, inescapable price to be paid for destroying such a magnificent being for the sake of mere money.

All the life and beauty that has been destroyed, cut down, paved over, exterminated, raped, imprisoned, and enslaved has given the world a great gift. I sometimes ponder the Trail of Tears, so named not for the tears of the Cherokee, for there were none, but for the tears of the crowds of whites that gathered to watch them pass dignified and unbroken. That image is burned indelibly onto the national psyche, and it will never let us rest until we have healed our own separation, softened the callusing of the soul that enabled us to commit such a crime. Nations and cultures, not just individuals, bear the self-inflicted wounds of their collective crimes; karma is not just an individual phenomenon. Collective salvation will only come when we face up to the ugliness of our own past and feel the mirror image of the pain of every slave lashed, every man lynched, every child humiliated. One way or another, we must weep for all of this.

The suffering of Separation's victims is never in vain. From separation comes violence, which then reverberates in the soul of the perpetrator to form the seed of the separation's healing.

I hope this is of some consolation to those of you who are among the victims (and that is all of us; we are all among the victims and perpetrators both). Usually the eventual healing and redemption is invisible to us; part of the suffering, in fact, is that it seems purposeless. The victims too experience a complete alienation, a loneliness intrinsic to all suffering. The image comes to mind of Christ on the cross: "Father, why hast thou forsaken me?" In this archetypal story, the Redeemer experiences the same extreme of separation from God—separation, that is, from all that we are and can be—that drove his tormentors. Do not think that the redwood, the burdock, the murdered, enslaved, and ruined, go without agony. Let us not underestimate the suffering of this world. Each of them partakes in Christ, making the ultimate sacrifice so that we might become whole. And you, dear reader, are no exception. Love and appreciate yourself as a noble being, born into this vale of tears for a sacred purpose. None of your hurts were in vain.

One way or another, we must weep for all this. What goes for the crimes of humanity, nations, and cultures goes as well for us individually. Even as we appreciate our nobility and tend gently to our wounds, so also must we lament the violence, the scarring, the ruin of the Other that has sprung from our separation, if ever we are to become whole. The Bodhisattva Path, to remain in samsara until all beings are free, is more than a noble sacrifice—it is an organic necessity.

The full integration of the pain from the life of separation is what impels us back toward wholeness. One way or another, the pain will be felt. We can either wait for it to come to us, like an addict determined to get a fix at any cost, or we can go to it. Perhaps if we can see the futility of control, the futility of perpetually postponing the consequences, then we will have the courage to face them. It is said that no addict truly enters recovery until he or she hits bottom; however, it is equally true that "bottom" is different from person to person. At some point the addict decides no longer to evade the pain of a shattered life, wrecked family, sick body, or ruined career. He feels the accumulated agony, mourns what is lost, tries to make amends. Sometimes he succeeds in doing so before all is lost, before all friendship, all wealth, all health has been converted into money for the fix. Perhaps we humans will do so as well, and begin making amends to the world we are ruining before all beauty, all goodness, all wealth, all life is consumed.

At Play Beside The Tower

In the reduction of reality to number and name, in the program of owning and controlling the world, we have wrought a Tower of Babel, seeking with our finite tools to take the infinite by storm. To do this we have so specialized and separated, and so reduced and exhausted the world, that the coherency of the vast megamachine that makes possible our ascent to the heavens is threatened. Our tools of control are insufficient to manage the chaos we have unleashed. Our ascent, even the *illusion* of our ascent, slows to a standstill as the effort merely to hold everything together grows to consume all resources. Now, as the Tower totters under its own weight, now as each attempt to shore up its crumbling sections adds to the instability of the whole edifice, perhaps we can see more clearly, from amid the ruins of our civilization, what the collective purpose that we have yearned for might be.

The supreme irony in our Babelian quest of attaining the infinite through finite means is that we are actually enacting precisely the opposite. We are liquidating all that is infinite, sacred, and unique, converting it into the finite, the controlled, the generic, standard, and measurable. Think of the redwood forests reduced to furniture, possessions, and ultimately to money: numbers in a computer. We are cashing in the earth, selling off our lives, reducing reality to data. Soon there will be nothing left to convert, as all social, cultural, natural, and spiritual capital is exhausted.

The Tower totters and sways. Once proudly leading the vanguard in a glorious conquest of nature, science and technology—and the whole regime of management and control—are now consumed in an ever more-desperate attempt to simply hold things together. Having cut ourselves off from nature, from wholeness, and therefore from health, we try frantically to manage the consequences with one technological fix after another. Like an addict trying to hold his life together, we shift debts, create rationalizations, and generate long-term consequences to solve short-term problems, pretending all the while that everything is under control. "Science will find the solution," we think, as we manage problems by putting them off until a future day of reckoning.

Like the addict's increasingly unmanageable life, such an effort is destined eventually to collapse completely, bringing us face to face with the realization that we can only recover, only heal, by relinquishing the entire mindset of control implicit in existing systems of technology, science, medicine, money, property, and education. At that point we will be open to the healing power of nature, the wild, the inherent purposefulness of the universe, the beneficence and fecundity of something beyond rational understanding and control. Something greater than our selves as we have conceived them.

On one level, this means modeling our industrial, social, educational, and economic processes after those of nature, replacing the metaphor of the machine or computer with that of the ecosystem, where there is no such thing as waste that is not also food, no place that is external, where there is no centralized organization, where each part is dependent on all the others, where the most successful are those who best fulfill their function in meeting the needs of the whole.

More than that, nature can also bestow upon us a model of creativity that does not entail the reduction of life and world. I have long felt that the disappointments of the environmental movement stem from its failure to articulate anything more positive than "sustainability". The creative gifts of humanity, culture and technology, having enabled us to destroy so much, are understandably seen as an arrant scourge that must be reined in. In the introduction I asked, "Can the gift be separated from the curse?" My heart says yes. Does the salvation of humanity lie in the denial of the very essence of our species? My heart says no, but for a long time I could see no other option. Now, finally, I can envision a more-than-sustainable future that accommodates the exuberant expression of the gifts that make us human. Like the indigenous artist who understood his work to be the revealing of the form already contained within the carving block, we can see ourselves collectively as an agent of nature's continuing self-revelation. We will no longer impose, but discover and reveal. To get back to nature doesn't mean passivity, to desist in the effort of creativity. We are creative beings, made "in the image", it is said, of the Creator, part of Nature which is nothing less than Creativity itself. No longer need we see nature as the passive and inert substrate of an external creator's art, nor as the empty, arbitrary result of random interactions of forces. Reflecting upon nature as it really is—endlessly creative of new forms and systems of beauty—we will understand that our highest purpose is to actualize unmanifest realms of beauty too. Indeed we have been doing this for millennia already. In addition to ruin, technology has afforded us new modes of creative expression that would have been unimaginable just a few centuries ago. The difference is that soon it will become our conscious collective purpose. The forces that today pressure us to "sell out" will disappear along with the illusion of separateness that gives them rise. In their place, new forces of the civilization of Reunion will arise which will reward wholeness, beauty, sustainability, discovery, and art. Freed from the anxiety inherent in the manage-and-control mentality, we will also be free to create beauty rather than to sacrifice it to the apparent necessity of survival.

The human gifts that have empowered us to bring the planet to the brink of catastrophe are not intrinsically evil, demonic powers to be spurned, but are, in the end, sacred means to take the creation of beauty to a new level. The problem is that we have not respected them as sacred. We have prostituted our gifts. We have been stuck in the delusion that their purpose is to gain us comfort, security, and pleasure, which follows from the idea that there is no real purpose to life but to survive, which follows from our deeply held Newtonian ontology, which itself is just the culminating articulation of separation.

When I say the purpose of technology is to take nature and the creation of beauty to the next level, I do not echo the common view that now that technology has essentially solved the problem of survival, it is time to halt the destructive spiral of materialism and turn our attention to art, music, literature, pure science, and aesthetic enjoyment. In other words, it is time to retire! This view goes back at least to the Age of Coal, when the terrible suffering of industrial laborers was justified on the grounds that it was a temporary sacrifice necessary to usher in the Golden Age of plenty. Now, the thinking goes, it is here, or at least it could be here if only we weren't so greedy, if only we didn't spend a trillion dollars on weapons, if only the economic system weren't so skewed. This view ignores that fact that *there never was* a "problem of survival".

The Age of Plenty is not the fruit of technology; it is the mindset of technology that has led us away from plenty to a world of anxiety, scarcity, and alienation. But if the misconception of self and world that has driven our technology were to change, its function as a force for separation would change as well.

The creation of beauty I speak of is not limited to the traditional aesthetic arts, which, isolated in their museums, have become a category largely separate from life. Every industrial process, every social institution, every relationship of our lives is a suitable object of our art. Humanity's turning to art is not the hobby of a retiree, it is the fusion of life and art, art and work, work and play.

Instead of focusing on survival ("making a living"), our interaction with the world will be our play. After all, our purpose is to understand, appreciate, and participate in nature's ongoing creation of new realms of beauty, and how do we do that? It is through play. Isn't that how a child learns to "understand, appreciate, and participate" in the world? In a sense, the entire course of separation has been nothing but a cosmic play; the difference will be that we will no longer be lost in the game, no longer oblivious to the illusory nature of our separation. With this consciousness, our play will become again playful.

The parallel with storyteller consciousness, described in Chapter Seven, is significant, and in fact play and storytelling are deeply connected. Play is an enactment of a story, a provisional reality with its own rules and agreements. As we become conscious creators of our stories, so we become conscious players in the cosmic game. All the accouterments of the separate human realm—label and number, images and machines, technology and culture—become our playthings and the instruments of our art. No longer unconsciously lost in that separate human realm, we are free to reunite it with the natural. We reunite its linearity with the rest of the cycle from which we tried to separate it. We reunite its symbols and stories with our conscious creative intentions. We reunite its technology with the purposes and processes of nature. Wielding our gifts consciously now, we can create a human realm no longer at odds with the natural.

To reunite with nature, to reconceive self and world, may sound like an unachievable ideal, but actually it is closer than close, as available at any instant as nature is herself. A Chinese proverb goes, "As far away as the horizon, yet right in front of your face." On the one hand, no matter how far we travel, we can no longer find pristine, perfectly undisturbed "nature" anywhere on this planet. There is no escape from the sounds, the chemicals, the lights, and all the other signatures of technology. All ecosystems are disturbed. Moreover, we take ourselves, our thoughts, and our being with us anywhere we go, and by our very presence as sojourners from civilization adulterate the purity of our destination. Like the horizon in the proverb, pristine nature recedes as we approach it.

On the other hand, nature is also right in front of our face, in us and all around us. It only recedes as we approach it when we conceive it as something separate from ourselves. We could, I suppose, attempt the reunion of the human and natural realms by willfully abandoning technology and returning to the Stone Age, but I suspect that the yearned-for state of purity would recede before us like the horizon in the proverb. The origins of separation go back beyond the Stone Age. Shall we overthrow the dictatorship of the eurkaryotes? Fortunately it is unnecessary. A return to nature, as the proverb implies, is as easy as a shift of perception. I will conclude this book by offering a few thoughts on how to reunite the human and natural realms on the individual level.

Going back to nature can be as simple as crawling on your hands and knees for a few minutes sniffing the dandelions. The healing power of even this tiny action is amazing. No matter how doubtful you are, however reason denies it, sniffing the dandelions just for the experience of it, watching a bug just to see what it does, looking at the clouds for five minutes, will have a noticeable effect. As Tom Brown Jr. put it, "A five-minute walk through a vacant lot or the park will have regenerating qualities about it. They'll be able to see more and feel more and, therefore, realize their aliveness." It seems trite, but even the most conventional ways of "reconnecting to nature" can erode the illusion of separateness, which is so much easier to maintain in the boxes of our houses, cars, and computers.

Nature is also in our bodies. The Cartesian mind-body split, which located the self apart from the body, can be healed through various practices that render that split experientially absurd. Yoga, Taichi, martial arts, the Feldenkrais Method, Authentic Movement, Contact Improv, the Continuum, and various hands-on healing modalities can reveal to us that body is an aspect of mind, and mind is an aspect of body.

But is signing up for a yoga class or taking more walks in the park going to heal the planet? Obviously not, except that, imperceptibly at first, such minor changes begin to erode the illusions of separation. The process of Reunion on a personal level often starts with a persistent disquiet before it erupts into a full-blown convergence of crises. It is the sense that something about life and the world is just not right—and this feeling alone can constitute a crisis in the sensitive. One's job, one's plans, one's way of life doesn't make sense anymore in light of a dawning truth. Eventually, all aspects of life undergo a thorough transformation.

Many books on our environmental and social crisis offer nothing but despair, either in the form of "These problems are too vast for you to do anything about," or in the form of tepid, palliative suggestions like buying "green" products and recycling your beer cans. In a way, the despair is justified. Everyone knows that even if you reduce your ecological footprint to zero, your individual action is as nothing compared to the colossal forces that are inexorably destroying our planet. A life that makes sense in the full realization of the tragedy of the human condition cannot be achieved by switching brands or buying a Zen meditation kit. Eventually we realize that the transformation must reach the pith of life, core issues of relationship and work. By degrees, spiritual realizations take on a material character. The despair comes from realizing that life as we know it cannot go on. If this realization is unconscious—no matter—the unconscious mind will engineer crises that propel birth into a new state of life.

In this core transformation, we tap into a power that makes the aforementioned despair irrelevant. It comes again from a "return to nature", but on a much more subtle and profound level than the beginning steps of reconnecting with the body and the outdoors. It parallels a primary theme of this book, which is the transformation rather than the abandonment of the separate human realm, so that it is no longer unnatural. On the individual level, it comes through the power of word, storyteller consciousness, and living in the gift. These are what enable us to realize our full potential as world-creating beings. It is no coincidence that these concepts contain within themselves nearly all of the world's great spiritual teachings: non-attachment, love, opening to something beyond our separate selves. More than in the outdoors or the body, nature is in these things. Living in the gift: rejoining the gift circle of ecology in which purpose lies in the fulfillment of our role and function in an ever-blossoming, ever-transforming whole. Storyteller consciousness: assuming a conscious role in the universe's ongoing play of self-creation.

Not only does going back to nature free us from the chains of survival anxiety and teach us our purpose as creators of beauty—that is, artists—it also teaches us how best that beauty might be created. By observing the grand pattern greater than anything the manufactured and artificially separated self could contrive, we become aware of the unique role we have to play in that pattern. This understanding comes from the simple fact that we *are* part of the pattern. It is not distinct from us. By observing nature, we observe ourselves; by learning about nature we learn about ourselves. The function of "I", "this provisionally self-aware part of the pattern" becomes apparent.

Sooner or later, whether driven by a crisis from within or without, the normal lives we have known are going to end. The mad scramble of technology and self-improvement seeks to continually find ways to maintain normalcy just a little longer. What is seldom recognized is that normality is at the root of the problems, and contains the seeds of its own demise. It can be maintained only with constant and growing suffering—the very suffering that now consumes our world. Less often acknowledged is that the normality, even if it were sustainable, is not worth sustaining. We have grown accustomed to enormously impoverished lives. Yet a buried memory remains of what life can and should be, a memory sometimes brought to the surface in those lucid moments of joy and connection I described in the Introduction. I speak to this memory and this knowing. I wish to remind myself and everyone that a far more beautiful world and life is possible, and that this possibility demands a revolution in human beingness. The call is urgent. Live a life that makes sense in light of all the truths you are awakening to. The social and planetary crisis, the illusion of separation, the impermanence of your discrete and separate self, the futility of the program of infinite control, the robbery of our spiritual capital, the reduction of the world to money, the selloff of time and life... and for what?

All we can do and all we need do is to live a life that makes sense. The danger is that even after seeing these truths, we continue to pursue an illusion anyway. Old habits are hard to change. A saying goes, "The truth will set you free." Nothing more is needed; nothing less will suffice. The eons of striving to transcend human nature are over; now we are learning we need only come more fully into who we are. Part of the coming to wholeness that I have described as the Age of Reunion is to no longer hide away parts of the world and parts of ourselves whose existence makes life-as-usual inconvenient. We will not heal our hurting planet, nor will we help any living soul, by denying our selves. Quite the opposite: our true nature denied—separation from who we are—is what has caused our present crisis to begin with. For centuries the message has been to be less: to overcome human nature with self-discipline, just as we overcome the rest of nature with technology. But today, the conception of self and world upon which the ideology of control is founded is obsolete. The war on nature and human nature is over. It is time to step into what we truly are, and so assume our divine purpose in nature's evolution to its next level of beauty.

Herein lies the self-acceptance and self-trust with which I opened this chapter. These do not lead to the destructive, narrow greed of the discrete and separate self, because this is not really who we are. Ultimately, it is the path of self-love that will necessarily bring us back into love with the world. This path is not without its pain; indeed, it encompasses all the pain that there is. But on the other side of the pain and sadness is understanding, wholeness, and therefore freedom. By integrating the sad truth of what we have made of life and the world, the sad truth of our millennia-long reduction of reality into label and number, money and property, we regain a vision of what we can be, should be, and actually are; we reclaim our birthright as whole, creative beings, in love with life and life in love with us.

The infinity we seek is here already, and it always has been. The collapse of the Tower—the world under control, the quest for certainty in science—is laying bare the fraud that has enslaved us for ten thousand years. Yet we must remember that this fraud too has its purpose. We must remember the playful origins of separation, this exploratory game we have lost ourselves in and from which we are now awakening. Our quest, our journey to the farthest reaches of separation, is now nearly complete. However hard the birthing pains, a light beckons us, a Reunion with that place of enchantment, understanding, and wholeness. Let that light sustain us through the coming darkness.

Notes and Sources

Chapter I

---

1 Quoted by Kirkpatrick Sale. *Rebels Against the Future*, Addison Wesley Publishing, 1996, p. 59

2 For example, the U.S. Department of Agriculture and others boasted a belief that soon DDT and other miracle chemicals would enable us to eliminate all insects from the earth—a worthy goal, it was assumed. Then would we be able to conduct agriculture without the messy uncontrolled variables that insects and other life forms represent.

3 Lewis L. Strauss, Speech to the National Association of Science Writers, New York City, September 16, 1954.

4 Magalhães, João Pedro de. Nanotechnology. http://www.jpreason.com/science/nanotech.htm

5 Pearce, David. The Hedonistic Imperative. http://paradise-engineering.com/heav22.htm

6 Mohawk, William. Nano-Economomics. http://www.geocities.com/computerresearchassociated/NanoEconomics.htm

7 Moore's Law says that the complexity of integrated circuit chips doubles approximately every 18 months. The naïve interpretation is that computers are getting exponentially smarter.

8 Farley, Patrick. The Guy I Almost Was. http://www.e-sheep.com/almostguy/

9 As of 2006, 12.5% of American babies are born premature, an increase of 13% since 1992 and 30% since 1982. See "Preterm Birth: Causes, Consequences, and Prevention", Report by the Institute of Medicine, July 13, 2006. and "Premature Birth Rate in U.S. Reaches Historic High; Now Up 29 Percent Since 1981", http://www.marchofdimes.com/aboutus/10651_10763.asp

10 Centers for Disease Control and Prevention, National Center for Health Statistics, National Vital Statistics System. http://www.cdc.gov/nchs/fastats/lifexpec.htm

[11] Ibid

[12] That is my personal estimate, but already some mainstream researchers are claiming that the new generation will be the first in 200 years to have a lower life expectancy than their parents. In a paper published in the March 17, 2005 issue of the *New England Journal of Medicine*, a team of scientists claims that the obesity epidemic could shorten lifespans by up to five years in the next few decades. (Olshanky, et al. "A Potential Decline in Life Expectancy in the United States in the 21st Century".) Other authorities disagree, citing "advances in modern medicine" that will offset these losses.

[13] Juliet Schor. *Overworked American: The Unexpected Decline of Leisure*. Basic Books, 1993.

[14] "Grain Stocks Continue to Shrink, Despite Record Production", USDA Foreign Agriculture Service Circular Series, May 2004.

[15] From *The Excursion*, Book Eighth, starting on line 150. Kirkpatrick Sale quotes the same passage in *Rebels Against the Future*, which is how it came to my attention.

[16] "51% Of U.S. Adults Take 2 Pills or More a Day, Survey Reports" (Scripps Howard News Service). *San Diego Union-Tribune*, Bowman, L., Jan. 17, 2001:A8.

[17] "Seniors and Medication Safety", Minnesota Poison Control System, http://www.mnpoison.org

[18] Tainter, John. *The Collapse of Complex Societies*. Cambridge University Press, 1988.

[19] Francis, Eric. "Conspiracy of Silence", *Sierra magazine*. Sept./Oct. 1994. The claim that PCBs are present in every living cell appears in a 1998 introduction to that article (http://www.planetwaves.net/silence.html)

[20] Sardar, Ziaduddin "Cyberspace as the Darker Side of the West". *The Cybercultures Reader*. Routledge, 2000, p. 742.

[21] Hodgkinson, Tom. "A Philosophy of Boredom", New Statesman, March 14, 2005. This is a review of Svendsen, Lars Fredrick. *A Philosophy of Boredom*. Reaktion Books, 2005. Translation by John Irons.

[22] Hinduism is similar to Buddhism in its explanation of suffering. As for Taoism, suffering could be said to result from ignorance of the Tao; that is, resisting the natural flow of life. This too is a form of separation.

[23] Stephen Harrod Buhner, *Sacred and Healing Herbal Beers*. Siris Books, 1998.

[24] Lee, Richard B. *The Dobe !Kung*. Holt, Rhinehart and Winston, New York, 1979. p. 50-55

[25] Norberg-Hodge, Helena. *Ancient Futures: Learning From Ladakh*. Sierra Club Books, 1992.

[26] Derrick Jensen, *A Language Older Than Words*, Context Books, 2000. p. 85-86.

[27] Sahlins, Marshall, *Stone Age Economics*, Aldine-Atherton, 1972. pp. 30-31

[28] Price's findings appear in his classic work, *Nutrition and Physical Degeneration*. Price-Pottenger Foundation, 1970.

[29] Lee, Richard B. *The Dobe !Kung*, Holt, Rhinehart and Winston, 1984. p. 81, p. 91

[30] Wrangham, Richard, and Dale Peterson. *Demonic Males: Apes and the Origins of Human Violence*. New York: Mariner, 1996. p. 76

[31] Chagnon, Napoleon A. "Life Histories, Blood Revenge, and Warfare in a Tribal Population", *Science*, Feb 26, 1988 v239 n4843 p985(8)

[32] Thomas Melancon, *Marriage and Reproduction among the Yanomamo Indians of Venezuela.* PhD dissertation, UMI, 1982, p 42. Cited in a Brazilian Anthropological Association letter to Anthropology News, 1989.
[33] Gregory, Juno. "Macho Anthropology", *Salon* , Sep. 28, 2000.
[34] Tierney, Patrick. "The Fierce Anthropologist" *The New Yorker,* Nov. 6, 2000.
[35] Power, Margaret. *The Egalitarians: Human and Chimpanzee: An Anthropological View of Social Organization.* Cambridge University Press 1991
[36] I spend time watching bees' nests outside my home. Often ten or fifteen bees (wasps, actually) congregate "on the doorstep" outside the hive. Occasionally they will crawl around, touch feelers, and groom, but much of the time they really do nothing at all except wave their antennae or be still. Of course, even then they may be gathering data about their environment, but it is not a stress-induced effort. We need not be driven. Merely by being, we can live in the world.
[37] Gusinde, Martin. *The Yamana,* Human Relations Area Files, 1961. p. 27, cited by Sahlins, p. 28.
[38] Quoted by Michael Shermer in *Scientific American,* February 2002, p. 35.
[39] I mean "atoms" in the sense of indivisible minimum units, in this case sub-"atomic" particles. The most recent version of the atom would be the vibrating strings of String Theory.
[40] http://www.apsa-co.org/ctf/pubinfo/ask/archive/askarchiveG.html

Chapter II

[1] Diamond, Jared. "The Worst Mistake in the History of the Human Race." *Discover Magazine,* May, 1987. pp. 64-66. However, note that Diamond's later writings tend to the opinion that agriculture was inevitable, not a bad choice as the word "mistake" implies.
[2] Diamond, Jared, *Guns, Germs, and Steel,* New York, W.W. Norton & Co. 1997, p. 105
[3] McKee, Jeffrey K. *The Riddled Chain: Chance, Coincidence, and Chaos in Human Evolution.,* New Brunswick, NJ, Rutgers University Press, 2000, p 107. Note that various authorities offer widely varying dates for the origin of fire use.
[4] Kosseff, Lauren, "Primate Use of Tools", http://www.pigeon.psy.tufts.edu/psych26/primates.htm#monkeys
[5] Goodall, Jane and H van Lawick, "Use of Tools by the Egyptian Vulture (*Neophron porenoptemus*)", *Nature,* vol. 212, 1966, pp. 1468-1469.
[6] Gee, Henry, "The Maker's Mark", *Nature Science Update,* May 6, 1999, http://www.nature.com/nsu/990506/990506-9.html.
[7] Precellular life might be an exception to this homeostatic principle, except for the fact that there is no evidence that non-cellular life has ever existed. The so-called "naked" replicating ribozymes of the proto-biotic soup exist in theory only, and this theory is more a projection of our culture's notions of self than it

is a plausible account of biogenesis. See Chapter Six.

[8] Even this form of separation can be considered an illusion when we speak of combinatorial rather then thermodynamic entropy. It gets at deep questions about the how an initially low-entropy universe came to be in the first place. At some point, usable energy had to be created out of nothing; either all at once in big-bang cosmology, or ongoingly in steady-state cosmology. The deep question here is, "Do we live in a universe of scarcity or of abundance?"

[9] For a lucid exposition of lateral gene transfer in bacteria, see W.J. Powell, *Molecular Mechanisms of Antimicrobial Resistance*, Food Safety Network, February 2000.

[10] Zerzan, John. *Elements of Refusal.* CAL Press, 1999.

[11] During the witch hunts of Europe, the primary repositories of plant knowledge—the female herbalists—were belittled, demonized, and even exterminated. Although these and other crimes against woman and indigenous cultures certainly had the effect of eliminating competitors to the dominant fire-based technology, I am referring instead to the deeper imperative underlying and unifying such campaigns.

[12] Mead, Margaret, *Male and Female: A Study of the Sexes in a Changing World*, William Morrow Publishing, 1949. p. 20

[13] Pearce, Joseph Chilton, *The Biology of Transcendence*, Park Street Press, 2002. p. 111. Pearce fixes the blame for this decline mostly on television—another very likely culprit!

[14] Usually, the word "true" in this passage is rendered in English as "eternal". However, the Chinese word *chang* has a very complex meaning, with connotations of permanency, perdurance, and therefore of being real or true.

[15] These sounds are mentioned in some of the very earliest works on *qigong*, at least back to the "Frolic of the Five Animals" of 200 BCE, and are still practiced today.

[16] Brown, Joseph Epes, *Teaching Spirits*, Oxford University Press, 2001. p. 42.

[17] Id., p. 43-44.

[18] Id., p. 45.

[19] In Sanskrit, prana means both breath and spirit. In Chinese, qi also refers both to breath and to a spiritual energy. The same word is used in Japanese and Korean. I've been told the same identity exists in ancient Hebrew and Arabic. Even in English, the word "respiration" literally means to re-spirit oneself.

[20] Johnston, Jack, *Male Multiple Orgasm* (audio CD), Jack Johnston Seminars, 1994

[21] Fonagy, Ivan, *Languages within Language: An Evolutive Approach*. John Benjamins Pub., 1994. p. 18, 87-106

[22] Id., p. 5

[23] Thoreau, Henry David, *A Week on the Concord and Merrimack Rivers,* 1849. p. 88

[24] Diamond, A.S., The History and Origin of Language. New York, 1959. Cited in Zerzan, p. 33

[25] Interview with Derrick Jensen, Published in *The Sun*, April 2001. Reprinted at http://hiddenwine.com/indexSUN.html.

[26] Oppenheimer, Stephen, *Out of Eden*, Constable and Robin, 2004. p. 30

[27] Id., p. 25

[28] Projecting our present selves backward into a prelinguistic setting would be misleading: the frustration and inconvenience we would experience may be a product of our atrophied direct perceptions. Atrophied, but still present in vestigial form and capable of development.

[29] Zerzan, p. 36

[30] Burke, James and and Robert Ornstein, *The Axemaker's Gift*, Jeremy P. Tarcher/Putnam, 1997. p. 68

[31] Interestingly, the first Chinese dictionary was compiled in the first century C.E.

[32] Zerzan, p. 32

[33] Thoreau, p. 218.

[34] Levy-Bruhl, Lucien, *How natives think*, New York: Humanities Press, 1926. p. 181

[35] Aristotle, *Metaphysics*. Cited by J.B. Wilbur and H.J. Allen, *The Worlds of the Early Greek Philosophers*, p. 86.

[36] Wilbur, J.B. and H.J. Allen, *The Worlds of the Early Greek Philosophers*, p. 87.

[37] Jensen, Derrick, *A Language Older than Words*, Context Books, 2000. p. 41

[38] Zerzan, p. 48

[39] Burke and Ornstein, p. 45.

[40] The original definition of the meter was abandoned when it was realized that the circumference of the earth is not absolute and unchanging. Now we are beginning to suspect the same may be true of the physical "constants" of the universe. This might be another example of the doomed quest for certainty described in Chapter Three.

[41] Zerzan, p. 57

[42] David Deutsch demonstrates convincingly the impossibility of perfect reality-simulation on a classical (non-quantum) computer; however on a quantum computer the very uncertainties of uncontrolled reality creep back in in subtle ways.

[43] Brin, David, "Three Cheers for the Surveillance Society!" *Salon* , 8/4/2004. www.salon.com/tech/feature/ 2004/08/04/mortal_gods/index1.html

[44] Mumford, Lewis, *Technics and Civilization* , Harcourt, Brace & Co. 1963, c1934. p. 14, cited by Zerzan, p. 23

[45] Pynchon, Thomas, "Nearer, my Couch, to Thee," *New York Times Book Review*, June 6, 1993.

[46] Campos, Paul F., *Jurismania: The Madness of American Law*, Oxford University Press, 1998. p. 32

[47] Mumford, *Technics and Civilization,* p. 15

[48] Zerzan, p. 24

[49] Thoreau, p. 255

[50] Greenberg, Daniel, "When does a Person Make Good Use of His Time?" *The Sudbury Valley School Experience*, Mimsy Sadofsky and Daniel Greenberg eds., Sudbury Valley School Press, 1992. p. 105.

[51] Marshal Sahlins, "The Original Affluent Society." Excerpted from *Stone Age Economics*.

[52] Oppenheimer, p. 120

[53] Brown, p. 71.

[54] Saul, John Raulston, *Voltaire's Bastards*, The Free Press, 1992. p. 427

[55] Ibid
[56] Zerzan, p. 22.
[57] Saul, p. 430.
[58] Id., p. 435
[59] Shainin, Jonathon, "Politics-a-palooza", *Salon* , May 12, 2005.
[60] Saul, p. 439.
[61] Zerzan, p. 68
[62] Brown, p. 61
[63] Id., pp. 61-62
[64] Berry, Wendell, *Sex, Economy, Freedom & Community*, Pantheon, 1993. p. 112-113
[65] Everett, Daniel L., "Cultural Constraints on Grammar and Cognition in Pirahã: Another Look at the Design Features of Human Language" *Current Anthropology*, Aug-Oct 2005.Vol.46, No. 4
[66] Gordon, Peter, "Numerical Cognition without Words: Evidence from Amazonia," *Science*, August 19, 2004. pp. 496-499
[67] Video footage by Peter Gordon, available on his Columbia University website.
[68] Everett, p. 628
[69] Id., p. 626
[70] Id., p. 633
[71] See for example Sage, R.F., 1995, "Was low atmospheric CO2 During the Pleistocene a Limiting Factor for the Origin of Agriculture?" *Global Change Biology*, 1, 93-106
[72] Diamond, *Guns, Germs, and Steel*, p. 99. The inevitability of agriculture (that it developed every place where there were domesticable species) is a main theme in Diamond's book.
[73] Quinn, Daniel. *Ishmael.* Bantam Books, 1995.
[74] Gusinde, Martin. *The Yamana,* Human Relations Area Files, 1961. p. 27, cited by Sahlins, p. 28.
[75] Mumford, Lewis *The Myth of the Machine: Technics and Human Development*, Harvard/HBJ Book, 1971. p. 191
[76] Mumford, *The Myth of the Machine: Technics and Human Development*, p. 193.
[77] Id., p. 196.
[78] Ibid
[79] Id., p. 212
[80] Sale, Kirkpatrick, *Rebels Against the Future*, Addison Wesley Publishing Company, 1996. p. 59
[81] Id., p. 36
[82] Kay, Jane Holtz. *Asphalt Nation.* University of California Press, 1998, p. 55.
[83] Oppenheimer, p. 101
[84] The lack of standardization, referred to by Diamond (p. 38) is highly significant, because standardization implies a division of labor and the possible commodification of products.
[85] Brown, p. 17
[86] Campbell, Joseph, *Myths to Live By*, The Viking Press, 1972. p. 74
[87] Id., p. 76
[88] Zerzan, John, "Future Primitive", http://www.primitivism.com/future-

primitive.htm.

89 Pearce, Joseph Chilton. *Evolution's End.* Harper-Collins, 1992. pp. 154-172

90 Brown, p. 16.

91 John 1:1

Chapter III

1 Miller, Henry, "The World of Sex", 1940.

2 I'm not actually sure if Wudka originated this phrase, because it is all over the Internet. It matters little: the belief it articulates is centuries old.

3 The verb is *facere* (to make, do, or perform). The root of "fact" might also be the participle *factus*, "made" or "done".

4 Modern depictions of the Minotaur show the head of a bull atop a human body, but in many classical drawings it was reversed.

5 Cowan, Tom, "The Fourfold Approach to Cancer", Nov. 13, 2005 presentation at the Wise Traditions Conference, Chantilly, VA.

6 Mumford, Lewis, *The Myth of the Machine*, v.2: *The Pentagon of Power.* Harvest/HBJ, 1974. p. 53

7 Hume, David, *An Enquiry Concerning Human Understanding*, Part II. 1748.

8 Descartes, Rene, *Discourse on Method*, Part 6, 1637.

9 Johannes Kepler actually predicted a moon landing in his book "Dream", written at the outset of the 17th century, and even accounted for the various difficulties he foresaw.

10 These quotes and many more appear on http://www.evolutionarychristianity.org/view.html

11 Actually, except in simple linear systems, as a practical matter neither predictability nor control follows from determinism, but this was not widely understood until the advent of chaos theory in the last half-century.

12 This passage created a very strong impression on me when I first read it in a college philosophy class. It is not insignificant that disputes can and most certainly do arise between accountants! Propositions can be proven within a formal system, but the correspondence of that system to reality cannot be proven mathematically but only argued for empirically.

13 Mumford, p. 72

14 Actually, the ascription of an atomistic ontology to the Chinese is somewhat of a misnomer. The five elements (wu xing) are better translated as the "five phases"; they are interdependent and, like yin and yang, do not have a separate and independent existence.

15 The reason that they lead back to animism is that these hidden variables are unknowable. They are a mathematical artifice for deriving the results of quantum mechanics, but can never be experimentally isolated. Thus, while the hidden variable theory seems to advance the reductionist program, it actually buries it.

---

[16] Kervran's work is hard to find in English, but I refer the reader to *Biological Transmutations,* published by Happiness Press, 1989. I also searched for a convincing refutation of his work, but mostly what I found were accusations of elementary errors, based on the logic: "The result could not be true, so he must have failed to account for XYZ."

[17] Mumford, p. 84

[18] Campbell, Joseph, *Myths to Live By*, Viking Press, 1972. p. 76 of the 1993 Compass reprint.

[19] Mumford, Lewis, *Technics and Civilization*, Harcourt, Brace & Co., 1934. p. 51

[20] Although contemporaries such as Leibnitz and Berkeley disputed Newton's idea of "absolute space", insisting that position could be defined relative to stars and other objects, the absolute Cartesian coordinate system is very hard to get rid of whether or not it is seen as physically real. That is because it is encoded into the Euclidean mathematics of Newton's theory. Even if there is no physically real absolute space, the fact that properties such as position, length, and time are invariant across frames of reference allows one to mathematically construct an absolute coordinate system.

[21] Velmans, Max, *Understanding Consciousness,* Routledge, 2000. p.264.

[22] Cited by Anthony D'Amato. *Whales: Their Emerging Right to Life* (with Sudhir K. Chopra), 85 American Journal of International Law 21 (1991)

[23] See for example Andrew Newberg, Eugene G. D'Aquili, Vince Rause, *Why God won't Go Away*, Ballantine Books, 2002

[24] Acknowledgement for tracing Vitalism through Aristotle back to Thales goes to Johnjoe McFadden (*Quantum Evolution*, p. 7-9).

[25] Contrary to what is taught in medical school, the mechanical pump model does not successfully explain the circulation of the blood at all. I refer the reader to Craig Holdrege's *The Dynamic Heart and Circulation*, or Tom Cowan's *The Fourfold Path of Healing*. Notice that the traditional heart-driven model of circulation, in which the blood vessels are passive carriers of blood, is more at home in a dualistic worldview than the alternative hydraulic ram model in which circulation is an organic property of the whole system.

[26] Lynn Margulis, [interviewed in] *The Third Culture* by John Brockman, Simon and Schuster, 1995. p 133. Cited by Brig Klyce http://www.panspermia.org/neodarw.htm#%202txt.

[27] The foregoing is actually an account of gradualism, the classical form of Neodarwinism still espoused by many biologists such as Ernst Mayr (see his recent book, *What Evolution Is*, for a pure exposition of this now-crumbling orthodoxy). However, the tide seems to have turned toward Stephan Jay Gould and Niles Eldridge's theory of punctuated equilibrium, which argues that the fossil record shows not gradual evolution but sudden, dramatic jumps followed by long periods of relative stasis. While some evolutionary biologists try to explain these jumps as statistical artifacts arising from the fossil record's incompleteness, most accept at least some version of punctuated equilibrium. Their attempts to reconcile it with the gradual accretion of mutations usually involve some way of accumulated mutations resting unexpressed in the genome until triggered by a mutation in a coordinating gene. These attempts are highly

problematic, but necessary in order to preserve the non-purposefulness of evolution.

28 Talk to the Green Gathering conference, September 2003.

29 To be sure, game theoretic models of cooperation demonstrate that it is *consistent* with self-interest, but they are subject to the same "bootstrap problem" that plagues every step of evolution that requires a significant jump in complexity. See Chapter Six for more on the bootstrap problem.

30 The flip side of the coin is the prevalence of economic terminology in talking about ecosystems: the "profits" and "costs" of various adaptations.

31 For example, in *The Descent of Man* (p. 201) we read regarding the future progress of humanity, "The break will then be rendered wider, for it will intervene between man in a more civilized state, as we may hope, than the Caucasian, and some ape as low as a baboon, instead of as at present between the Negro or Australian and the Gorilla."

32 Quoted by Michael Shermer in *Scientific American*, February 2002, p. 35.

33 Monod, Jacques, *Chance and Necessity*, Vintage Press, 1972. pp. 145-6

34 Id., p. 172-3

35 From "A Free Man's Worship", published in 1903.

36 I am indebted to Wendell Berry for the use of this quote in this context: see "Christianity and the Survival of Creation" from his book *Sex, Economy, Freedom, and Community*. Pantheon Books, 1992.

Chapter IV

1 We might become friends with our professional colleagues, especially to the extent we are devoted to a common cause. However, the possibility of workplace friendship is poisoned by the preeminence of money as the main motivating factor in our work. And personal gain is not a common cause. On the contrary, all too often workers must compete for promotion, tenure, or other benefits. Since money is the primary motivation sending people to work, to make money naturally supersedes any joint creation as the primary goal of that work. Workplace friendships are therefore often highly qualified.

2 Hyde, Lewis. *The Gift*. Vintage Books, New York, 1979. p. 17.

3 By "specialist" I mean someone paid to perform a specific function, in this case preparing food. It does not imply that the work requires a high level of training or technical expertise.

4 The figure was 50% in 1998 as reported by the USDA President's Council on Food Safety, August 4, 1999; recently on NPR I heard a figure of two-thirds but I cannot find a citation for it. Other sources state a figure of 50% for meals *eaten* outside the home, so I suspect the addition of ready-to-eat takeout meals would bring that figure even higher.

5 Sale, Kirkpatrick, *Rebels Against the Future*, Perseus Books, 1995. p. 38

6 Cited by Charles Siegel, *The End of Economic Growth*, The Preservation Institute,

1998.

7 Some vestiges of the old days remain at Yale, where students still gather at Mory's, a private club associated with the university, and sing old songs. Such behavior is exceptional, however.

8 For example, in Sweden, the right of *Allemansrätt* allows individuals to walk, pick flowers, camp for a day or two, swim, or ski on private land (but not too near a dwelling).

9 When I most recently asked this question to my class, I was nonplussed to find that a surprising number of students don't feel the slightest twinge of guilt over shoplifting either. Keep this in mind as you read the discussion of intellectual property. Could it be that on some unconscious level, they realize that indeed, "Property is theft?"

10 Jefferson, Thomas, letter to Isaac McPherson, August 13, 1813. This quote is widely cited.

11 These are the terms of the Copyright Term Extension Act passed in 1998 and upheld by the Supreme Court in 2003—just as Mickey Mouse and other iconic characters were to have passed into the public domain.

12 Lawrence Lessig, *The Future of Ideas*, Random House, 2001, excerpted at http://cyberlaw.stanford.edu/future/excerpts/.

13 I have read that images such as commercial storefronts and the New York City skyline are also proprietary. In the human world, that covers the entire landscape. I imagine that as the natural landscape turns increasingly toward proprietary crop and animal varieties, non-urban landscapes might require rights clearance too.

14 Based on comments of Dr. Jonathan Kind, Professor of Genetics at MIT, quoted by Bernard Lietaer in *The Future of Money*.

15 "Trademark Litigation Hall of Fame," Overlawyered.com, April 2001, http://www.overlawyered.com/archives/01/apr1.html, and "Michigan Lawyer's Demand: get your case off my website", Overlawyered.com, June 2001, http://www.overlawyered.com/archives/01/june2.html

16 Lewis Mumford, *Technics and Civilization*, p. 142

17 Berry, Wendell. "Christianity and the Survival of Creation." *Sex, Economy, Freedom, and Community*. Pantheon Books, New York, 1993. P. 101

18 Roland Wall, "Erosion: Wind and Water, Food and Money", The Academy of Natural Sciences, http://www.acnatsci.org/education/kye/nr/kye82002.html

19 Several of these USDA composition tables comparing vitamin and mineral content between 1975 and 2001 are laid out in Alex Jack, "The Disappearing Nutrients in America's Orchards". December 14, 2004, published on line by the National Health Federation. http://www.thenhf.com/articles_56.htm

20 Lester Brown and Brian Halweil, "Populations Outrunning Water Supply as World Hits 6 Billion", Worldwatch Institute press release, September 23, 1999.

21 World Medical Association Statement on Water and Health, WMA General Assembly, Tokyo 2004

22 To be sure, some forms of agriculture dispense with some of these steps, but the general principle remains valid that some work is required to lift the land beyond its natural carrying capacity (for human beings).

[23] Hyde, p. 121
[24] Brown Jr., Tom, *Tracker*. Prentice Hall, 1978. p. 56
[25] Quoted in *Adbusters* Magazine, Vol. 12, No. 5, September/October 2004. No page numbers are used—yay!
[26] Berry, Wendell. "Christianity and the Survival of Creation." *Sex, Economy, Freedom, and Community*. Pantheon Books, New York, 1993. P. 97
[27] Id., p. 98
[28] Pearce, Joseph Chilton, *Evolution's End*, p. 162
[29] Id. p. 169
[30] To experience the mind's craving for constant stimulus, try standing in a supermarket checkout line without allowing your attention to skip to various tabloid headlines, magazines, products, and other pictures and text. It isn't easy.
[31] After some years of controversy, the phenomenon of falling sperm counts was confirmed by a massive meta-study led by Shanna Swan of California's Department of Health Services ("Sperm Count Decline Confirmed" by Magie Fox, Reuters, November 24, 1997). No one agrees on the reason, but I think the main culprits are toxic estrogen-mimicking chemicals like PCBs, excessive soy in the diet, hormones in industrial meat animals, and tight underpants. Just kidding about the last one.
[32] See for example, "India's New Outsourcing Business – Wombs", by Sudha Ramachandran. The Asia Times, June 16, 2006.
[33] These were the official words of the Nobel Foundation when it issued Becker's prize in 1992. For more details see, http://nobelprize.org/economics/laureates/1992/press.html.
[34] To see the hyperrationalism and abstraction of this approach, see Gary Becker and Judge Richard Posner's blog, http://www.becker-posner-blog.com.
[35] Lewis Mumford, *Technics and Civilization*, p. 104
[36] Paul Hawken, *The Ecology of Commerce*, p. 76
[37] Coase, Ronald H. The Problem of Social Cost. J. Law & Econ. 3, p. 1 (1960)
[38] Thomas Carlyle, "Gospel of Mammonism", Past and Present, Book 3, Chapter 2, quoted by Kirkpatrick Sale in *Rebels Against the Future.*
[39] See Schenk, Robert. "From Commodity to Bank-debt Money" http://ingrimayne.saintjoe.edu/econ/Banking/Commodity.html, for a basic yet thorough description of the process by which money is created.
[40] Lietaer, p. 47
[41] Lietaer, p. 52
[42] Hyde, p. 139
[43] Hyde, p. 23
[44] The reason that an infinite amount of money can have a finite "net present value" is that it comprises a converging series.
[45] These quotes are taken from Adbusters Magazine, "Let us Eat Cake", Issue #55, September/October 2004.
[46] Greider, William, *One World Ready or Not*, Simon & Schuster, 1998.
[47] This refers to the fact that oil product will soon reach, or has already reached, its peak level of production. Petroleum geologists are nearly unanimous in asserting that discoveries of new reserves cannot possibly keep pace with the depletion of

old reserves.

Chapter V

1 The original esoteric teachings of all religions say quite the opposite; I am talking here about institutional religion.

2 The doctrine of total depravity has been expounded upon by generations of Protestant theologians. I refer the reader to Arthur Pink's *The Total Depravity of Man* for an articulate exposition. Given the atrocities perpetrated by Luther, I wonder if he was merely projecting his understanding of his own self onto reality.

3 There are esoteric interpretations of these concepts that do not depend on Original Sin, however.

4 Maslow, Abraham, *Religions, Values, and Peak Experiences.* Penguin Books, 1970. p. 38

5 See Pinker,Stephen, *The Blank Slate: The Modern Denial of Human Nature.* Viking Penguin, 2002.

6 Another way that fundamentalist Christianity dovetails with the Scientific Program is in the literalist interpretation of the Bible, which accords to words an absolute meaning and reified status. Their goal is to discover an inerrant standard by which to determine truth, an absolute reality "out there" that is beyond the subjectivity of cultural construction—how similar indeed to the goal of the scientic method. Superficially very different, Fundamentalism and Science share many of the same ontological assumptions and goals.

7 Dawkins, Richard, *The Selfish Gene*, Oxford University Press, 1976. p. 71

8 This quote is usually attributed to Dostoevsky, but David Cortesi (http://www.infidels.org/library/modern/features/2000/cortesi1.html) contends that these words actually do not appear in *The Brothers Karamazov*, as commonly attributed. I have seen other versions of this sentence quoted as well. Perhaps it is a matter of translation. In any event, the sentiment is surely there.

9 This quotation is all over the Internet. I have not bothered to track it down in Columbus's actual journal, which is typically the source cited.

10 See Mauss, Marcel, *The Gift*, W.W. Norton & Company, 2000 (originally published in French in 1925)

11 Posner, Richard. "A Theory of Primitive Society, With Special Reference to Law", *Journal of Law and Economics*, April 1980

12 Jensen, Derrick, *A Language Older than Words*, Chelsea Green, 2004. p. 212

13 Mumford, Lewis, *The Myth of the Machine: Technics and Human Development*, Secker & Warburg, 1967. P. 206

14 See Chapter One for a discussion of the history of human lifespan. While it may be true that lifespans in classical and Medieval societies were much shorter than they were today, much of their mortality was due to civilization itself: epidemics, famine, war, etc. Remote agrarian and foraging societies live much longer.

[15] Here are a couple of children's tales' endings, from *Tales of the Brothers Grimm*: Rumpelstiltskin: "Then, in his passion, he seized his left leg with both hands and tore himself asunder in the middle." Ashenputtel: "Afterwards, when they were coming out of church, the elder was on the left, the younger on the right, and the doves picked out the other eye of each of them. And so for their wickedness and falseness they were punished with blindness for the rest of their days." And Snowdrop (also known as Snow White): "But iron slippers were heated over the fire, and were soon brought in with tongs and put before [the queen]. And she had to step into the red-hot shoes and dance till she fell down dead." A little different from the Disney version!

[16] I do not mean here to dismiss the concept of an afterlife. It is just that we smuggle into our understanding of the afterlife the same misconception and misdefinition of self that we apply to our current existence.

[17] Hanna, Thomas, *Somatics: Reawakening the Mind's Control of Movement, Flexibility, and Health*. Da Capo Press, 2004. p. 90

[18] Hanna, Thomas, *The Body of Life: Creating New Pathways for Sensory Awareness and Fluid Movement*, Knopf, 1980. p. 58-59

[19] Eisenstein, Charles, *The Yoga of Eating*, New Trends, 2003. p. 144

[20] Jensen, p. 102

[21] I remember reading this once in some Sixties' manifesto, but I cannot find the source.

[22] Gelineau, Kristin, "Baby Workouts Touted to Ward off Obesity" *Salon Magazine*, June 13, 2004.

[23] Illich, Ivan, *Medical Nemesis*, Pantheon, 1982. p. 85

[24] Erdmann, Jack with Larry Kearney, *Whiskey's Children*. K Publishing, 1998. P. 198. I recommend this book to anyone who thinks he is in any way superior to an addict.

[25] Gatto, John Taylor, *The Underground History of American Education*, Odysseus, 2001. p. 53.

[26] For an insightful discussion of this trend, see "Abridged too Far", Hillary Flower http://www.salon.com/books/feature/2004/03/29/willows/index.html.

[27] Bloom, Harold, *The Western Canon*, Riverhead Books, 1994. p. 520

[28] Gatto, p. 38

[29] Illich, Ivan, *Deschooling Society*, Perennial, 1972. p. 2-3

[30] Harris, William Torrey , *The Philosophy of Education*, quoted by Gatto p. 105

[31] General Education Board, *Occasional Letter Number One*, 1906, quoted by Gatto p. 45

[32] Gatto, p. 156

[33] Gatto, p. 173

[34] Statistics from Habitat for Humanity, "Creeping Affluence", http://www.lawrencehabitat.org/About/Library/creepingaffluence.pdf

[35] The details in this paragraph and the quotations following it are taken from Linda Baker, "Why don't we do it in the road?" May 20, 2004, http://archive.salon.com/tech/feature/2004/05/20/traffic_design/index.html. However, anyone who has lived in a third-world city has already seen this model in action.

---

[36] Campos, Paul F., *Jurismania: The Madness of American Law.* Oxford University Press, 1999. p. 5

[37] Campos, p. 29-30.

[38] I have simplified the statement of the theorem by neglecting the distinction between a formal system (or theory) and the interpretation (or model) of that system. The correct statement would be that there are unprovable sentences that are true in an interpretation of that system. The difficulties inherent in the distinction between mathematical theory and model also pertain to the law, and are partly responsible for the endless elaboration and repetition that characterize legal writing.

[39] The exception is for axiom systems that are not consistent, that contain an embedded contradiction, from which it is possible to prove anything. This is actually closer to the state of the law, which embodies many contradictory principles arising from contradictory social values. An example is free speech versus restriction of hate speech. If both these principles are written into law, certain actions will be legally justifiable according to one and unjustifiable according to the other. And this contradiction seems to persist no matter how fine a distinction is made.

[40] Illich, p. 16

[41] For a typical example, see "The Antibiotic Arsenal", http://www.microbeworld.org/htm/cissues/resist/resist_2.htm

[42] Buhner, Stephen Harrod, *The Lost Language of Plants*, Chelsea Green, 2002. p. 139

[43] One way this happens is through the introduction of simian viruses into humans via the vaccine culturing medium. See for example *Journal of Infectious Disease*, September 1999;180:884-887. Many also blame the large quantities of mercury used as a preservative in many vaccines.

[44] Autism, Autoimmunity and Immunotherapy: a Commentary by Vijendra K. Singh, Ph.D, http://libnt2.lib.tcu.edu/staff/lruede/singhfeature.html

[45] The parts of the Third World that still harbor such epidemics are still "in the past" in the sense that their alienation from nature has not reached the phase it has in the West.

[46] "Fowl play: The poultry industry's central role in the bird flu crisis," http://en.groundspring.org/EmailNow/pub.php?

[47] For an in-depth discussion of the origins and faulty science of pasteurization, see *The Untold Story of Milk* by Ron Schmidt.

[48] Illich, *Medical Nemesis*, p. 73

[49] Id., p.47

[50] For a similar analysis of psychiatric disorders like ADD, depression, and so on read *Commonsense Rebellion* by the renegade psychologist Bruce Shapiro.

[51] Illich, *Medical Nemesis*, p. 135

[52] Id., p. 128

[53] Gatto, p. 129. Chautauqua refers to an ideal of social engineering that you can read about in his book.

[54] See Pearce's *The Biology of Transcendence* for a remarkable exposition of the necessity and means of transcendence and the consequences of its frustration.

[55] For an eloquent and ardent overview of this phenomenon read *Commodify your*

Dissent: Salvos from The Baffler, by Thomas Frank and Matt Weiland eds. W. W. Norton & Company, 1997

[56] Prechtel, Martin, speech to the Green Nations Gathering, September 2003

[57] Jensen, p. 320

Chapter VI

[1] While David Bohm's "hidden variables" interpretation of quantum mechanics attempts to restore determinism, it does nothing to rescue the program of complete understanding/control, because these hidden variables are fundamentally unknowable.

[2] For examples of the intentional use of observation to affect reality, I suggest reading about "null measurements" or "quantum Zeno effect".

[3] This was demonstrated once and for all in the Aspect experiment, which demonstrated an observer effect when the observation happened outside the space-time light cone of an affected part of the system.

[4] And even then you may never know for sure. It could very well be that running the Turing Machine for a billion iterations tells you only that it does not halt for the first billion iterations.

[5] Basically what that means is that given a random list of arbitrary mathematical truths, there is in general no shorter way to characterize that list: the list itself is its own shortest description. These results are presented in depth in Chaitin's controversial classic, *Information, Randomness and Incompleteness: Papers on Algorithmic Information Theory*.

[6] Chaitin, Gregory, *Meta Math! The Quest for Omega*, Pantheon, 2005. p. 20

[7] Borwein, Jonathon and David Bailey *Mathematics by Experiment*, A.K. Peters, 2003. p. 4-5. There is a recent trend in mathematics toward "experimental mathematics" which forgoes the certainty of traditional analytic proof, and seeks insight instead through the use of computers. There may be certain basic truths that are inherently unprovable in present axiom systems; for example, the conjecture that all irrational algebraic numbers are Borel-normal. This has been computationally confirmed for some numbers up to trillions of digits. While at first glance experimental mathematics may seem just another form of empiricism applied to mathematics, and thus consistent with the Baconian assumption of an objective universe "out there", the matter is actually extremely tricky. If the digits are in some sense random, in what sense do they exist or in what sense are they necessarily what they are, before they are calculated? This question is not trivial, but more involved ruminations are beyond the scope of this book. I refer the reader to the works of Gregory Chaitin for a philosophical discussion of related issues.

[8] A detailed technical overview of these constants is offered in the classic *The Anthropic Cosmological Principle*, by Frank Tippler and John Barrow. Their explanation, which is really an anti-explanation, is that these constants have to

be what they are because if they weren't we wouldn't be here to even ask about them. Most people find their reasoning sound but deeply unsatisfying. Why are you reading this book? Well, obviously everything in your life that brought you to this book *had* to be that way, else you wouldn't be reading this book.

9 I have vastly oversimplified a complex issue here. If the hyperbolicity conjecture for the M. set is true, then the set (actually its intersection with the algebraic complex plane) is in a certain sense computable. However, that conjecture remains unproven despite enormous efforts to prove it.

10 By random here I mean not recursively enumerable. Philosophers might quibble with the equation of randomness and non-recursive enumerability, so let me note in passing that this is essentially equivalent to the definition of randomness used in algorithmic information theory (AIT). In AIT a set of numbers is random if there is no computer program shorter than the set itself that can generate the set.

11 Kauffman, Stuart, *At Home in the Universe: The Search for the Laws of Self-Organization and Complexity*, Oxford University Press, 1995, and *The Origins of Order: Self-Organization and Selection in Evolution*, Oxford University Press, 1993.

12 Provided the initial setup contains a finite number of white squares or a finite number of black squares.

13 Even if such a proof were to be found, there are other emergent properties of cellular automata that are formally undecidable. John Conway's "Game of Life" is one example. Conway proved that this cellular automaton can be configured into the equivalent of a universal Turing Machine, which implies via the Halting Problem that there is no finite, general way to decide whether a given starting configuration grows without bound.

14 Pfeiffer, Dale Allen. Drawing Lessons from Experience: The Agricultural Crises in North Korea and Cuba. http://www.fromthewilderness.com/free/ww3/111703_korea_cuba_1_summary.html. Attempts to plan out artificial societies have been uniform failures. The more determined the effort to exercise centralized control, the more dysfunctional the society becomes: Stalinist Russia and present-day North Korea are two excellent examples. On a smaller scale, communities such as the Owenite experiments of the 19th century disintegrate swiftly unless the foundational "plan" is allowed to coevolve with the society it underlies.

15 Actually, Darwin, who was a humble and modest man, did not himself contend that this was the *sole* mechanism of evolution, just that it could explain a lot. In that sense, Darwin was not himself a Darwinist.

16 Quoted by Michael Shermer in *Scientific American*, February 2002, p. 35.

17 The same non-dualistic understanding which is central to animism, the mother of all religions, has been shared by the world's great spiritual teachers over the millennia, despite interpretations to the contrary. It can be found at the heart of all modern religions.

18 Nijhout, H.F., "Metaphors and the role of genes in development." BioEssays, vol.*12* (1990) p.444-446.

19 Lipton, Bruce. *The Biology of Belief.* Mountain of Love Productions, pre-publication draft, no page numbers.

20 Ibid.

21 From the Preface to Shaw's play *Back to Methuselah*. It is worth reading the whole of the remarkable essay, which eloquently lays out many of the spiritual ramifications of the Darwinian paradigm.

22 Larmarck, Jean Baptiste, *Philosophie zoologique, ou exposition des conside'rations relatives a` l'histoire naturelle des animaux*. (Zoological Philosophy. An Exposition with Regard to the Natural History of Animals), 1809. Translated by Hugh Elliot Macmillan, London 1914, Reprinted by University of Chicago Press, 1984

23 Studies supporting this assertion are almost too numerous to mention. Many have been done with bacteria, for example Loewe, L., Textor, V., and Scherer, S., "High deleterious mutation rate in stationary phase of Escherichia coli." *Science* vol. 302 (2003), pp. 1558-1560

24 Rosenberg, Susan M. and P.J. Hastings, "Genomes: Worming into Genetic Instability", *Nature* vol. 430, August 5, 2004, pp. 625 - 626

25 Cairns, J., J. Overbaugh, and S. Miller, 1988 "The Origin of Mutants." *Nature* vol. 335, pp. 142-145

26 B.G. Hall, "Activation of the bgl operon by adaptive mutation", *Molecular Biology and Evolution*. Jan. 15, 1998 pp. 1-5.

27 Rosenberg, Susan. "Evolving Responsively: Adaptive Mutation" *Nature Reviews Genetics* , vol. 2, 2001, pp. 504 -515. This is a comprehensive review of the state of research on adaptive mutation.

28 T. Beardsley, "Evolution Evolving," *Scientific American*, September 1997, pp. 15–18

29 McFadden, Johnjoe, *Quantum Evolution*, Norton, 2001.

30 Darwin, Charles, 1876 letter to Moritz Wagner, referenced in *Obituary Notices of the Proceedings of the Royal Society*, vol. 44, 1888.

31 I did the research for this section several years ago. Now it seems that the RNA world is falling rapidly out of favor.

32 Joyce, Gerald F. and Leslie E. Orgel, "Prospects for Understanding the Origin of the RNA World," from *The RNA World*, Second Edition, Cold Spring Harbor Laboratory Press, 1999. p. 68

33 Joyce, G.F., Visser G.M., van Boeckel C.A.A., van Boom J.H., Orgel L.E., and van Westrenen J. "Chiral selection in poly(C)-directed synthesis of oligo(G)". *Nature* , vol. 310, 1984, pp. 602-604.

34 Joyce and Orgel, p. 51

35 Joyce and Orgel, p. 62

36 Bartel, David P. "Recreating an RNA Replicase." *The RNA World*, p. 143-159

37 Kauffman's more recent book, *At Home in the Universe*, will be more accessible to the lay reader. It applies the same concepts far beyond biogenesis and evolution, and is an excellent resource to help develop non-dualistic intuitions about the origin of order and beauty in the universe.

38 Displayed at www.nd.edu/~networks/gallery.htm.

39 Also keep in mind that even in the world of masses and forces, there is much we are ignorant of.

40 The emergent property of soul might be associated with a bound energy that contributes to the mass of a living being. Death brings a decoherence of

innumerable processes of life, an enormous loss of embodied information and energy, and thus an equivalent loss of mass. I realize that standard physics admits no way for the mass—measured at several ounces—to just disappear without being converted to energy on the order of $9x10^{16}$ joules. Where does it go? I'm not going to try to answer that question right now.

[41] Wickelgren, I. Immunotherapy: Can Worms Tame the Immune System? *Science* vol. 305, 2004, pp. 170-171.

[42] Buhner, Stephen Harrod, *The Lost Language of Plants*, Chelsea Green, 2002. p. 163

[43] Just for fun, I shall note here that some renegade geologists think the planet *does* grow, and cite this as an alternative explanation for continental drift.

[44] Buhner, p. 160

[45] The lack of new, more complex organisms is hidden by the fact that complex and unforeseen patterns of behavior indeed arise among the seed creatures and their variants, which do indeed evolve, only not toward leaps in complexity. See Richard E. Lenski, Charles Ofria, Robert T. Pennock and Christoph Adami, "The evolutionary origin of complex features", *Nature*, vol, 4238, May 2003, p 139-144 for some interesting examples.

[46] Liangbiao Chen, Arthur L. DeVries and Chi-Hing C. Cheng. "Evolution of antifreeze protein from a trypsinogen gene in Antarctic notothenioid fish" vol. 94, *Proceedings of the National Academy of Sciences, USA*, April 1997, p 3811-3816.

[47] Klyce, Brig, "The Origin of Antifreeze Protein Genes", http://www.panspermia.org/neodarw.htm

[48] Thanks to Brig Klyce's close reading of the paper for spotting this teleological phrase.

[49] Mckee, Jeffrey, *The Riddled Chain*, Rutgers University Press, 2000. p. 196.

[50] Charles Darwin, *Origin of Species*, p. 154, cited in Behe, Michael, *Darwin's Black Box*, Free Press, 2006.

[51] Debbie Lindell et al., "Transfer of photosynthesis genes to and from *Prochlorococcus* viruses", *Proceedings of the National Academy of Sciences, USA,* vol. 101, July 27, 2004, p 11013-11018.

[52] Jason Raymond et al., "Whole-Genome Analysis of Photosynthetic Prokaryotes", *Science*, v 298, Nov 22 2002, p 1616-1620.

[53] "Photosynthesis Analysis Shows Work of Ancient Genetic Engineering," Arizona State University, 21 Nov 2002. http://clasdean.la.asu.edu/news/blankenship.htm

[54] Grillot-Courvalin, Catherine, Sylvie Goussard, and Patrice Courvalin, "Bacteria as Gene Delivery Vectors for Mammalian Cells", *Horizontal Gene Transfer*, ed. Michael Syvanen & Clarence Kado.

[55] Hartl, D.L., Lohe, A.R. And Lozovskaya, E.R., "Modern Thoughts on an Ancyent Marinere: Function, Evolution, Regulation." *Annual Review of Genetics*, vol. 31, 1997, p. 337-358.

[56] *Nature Reviews Genetics,* vol. 5, 2004, p. 638-639

[57] Svitil, Kathy A. "Did Viruses Make Us Human?" *Discover*, Nov 2002.

[58] Hughes, Jennifer F. and John M. Coffin, "Evidence for genomic rearrangements mediated by human endogenous retroviruses during primate evolution", *Nature Genetics*, vol. 29, Dec 2001, p 487-489, n. 4

59 Herbert, Alan, "The four Rs of RNA-directed evolution" *Nature Genetics*, vol. 36, Jan 2004, p 19-25, n. 1
60 Nicole King et al., "Evolution of Key Cell Signaling and Adhesion Protein Families Predates Animal Origins", *Science*, v 301, July 18, 2003, p 361-363.
61 Klyce, Brig, "Neo-Darwinism: The Current Paradigm", http://www.panspermia.org/neodarw.htm
62 Westphal, Sylvia Paga'n, "Life goes on without 'vital' DNA", *New Scientist*, June 3, 2004
63 For you sticklers out there, I understand that what Darwin implies is not "The purpose of life is to survive" but "The purpose of life is to survive long enough to pass on my genes and ensure their survival and further replication." Don't you think the first sentence has a better ring to it though? Anyway, the upshot is the same.
64 http://www.panspermia.org/whatsne33.htm
65 See Arp, Halton, *Seeing Red*, Aperion, 1998, and Lerner, Eric, *The Big Bang Never Happened*, Vintage, 1992.

Chapter VII

1 The root is not the Greek eu-, meaning "good", but rather ou-, meaning "not".
2 Fox, Matthew, *The Coming of the Cosmic Christ*, HarperSanFrancisco, 1988.
3 Farley, Patrick, "Chrysalis Colossus", www.e-sheep.com/chrysalis
4 Gesell, Silvio, *The Natural Economic Order*, 1906. Trans. Philip Pye. I consulted an on-line version: http://www.systemfehler.de/en/neo/
5 Today this could all be done electronically of course.
6 Bernard Leitaer gives a nice discussion of the history of demurrage in *The Future of Money*
7 Gesell, ch. 4.1
8 Sahlins, Marshall, *Stone Age Economics*, Aldine, 1972. p. 209
9 Lee, Richard. *The Dobe !Kung*. P. 101
10 Silvio Gesell, *The Natural Economic Order*, 1906. Trans. Philip Pye. Ch. 5E. Gesell also advocated the abolition of land ownership.
11 Gesell, Ch.4.4
12 For example, in a deflationary depression the scarcity of money leads everyone to hoard it, exacerbating the scarcity.
13 Gesell, ch.4.4
14 Hyde, Lewis, *The Gift*, Vintage, 1979. p. 23
15 The waste of food is at all stages of production and consumption, from the barn to the dinner plate. At the production stage, economic efficiency trumps solar efficiency in the conversion of sunlight into food, with the result that labor-intensive farming that can achieve higher nutritional yields cannot compete with machine agriculture (especially with its hidden subsidies). There is further waste at the processing stage, for example in the failure to use organ meats and

imperfect fruits and vegetables. At the distribution stage there is enormous waste at supermarkets, which must throw away everything that spoils or expires. As for the consumption stage, simply go to a university cafeteria and observe the window where students bus their trays.

16 Prentice, Jessica. Stirring the Cauldron – New Egg Moon, April 13, 2005. www.wisefoodways.com

17 Most of the "efficiency" is actually due to their bargaining power, not their productive efficiency. A large purchaser can demand lower prices from producers without necessarily being more efficient in any other way.

18 Gesell, ch.4.4

19 The quote and general outline of this historical incident are from Lietaer, Bernard, *The Future of Money*, Century, 2002. p. 156-7

20 Income tax reinforces the regime of control in another way too, by requiring that one keep records of all income and allowable deductions. As life becomes increasingly economic, these means that more and more of life goes on record and thereby becomes data.

21 Leitaer, p. 156

22 "Non-toxic" is not an absolute category. Substances must be produced in quantities small enough for the biosphere to utilize, as many substances that are beneficial in small quantities are destructive in large quantities. It is not the chemical constitution of the substance, it is whether it contributes to a cyclical flow. That means that no action can be understood in isolation from its place and time. The same substance that is poison in one location might be food in another.

23 The discussion of the "intelligent product system", green taxes, and pollution permits draws its basic facts from Paul Hawken's *The Ecology of Commerce*, as well as *Natural Capital* by Hawken and Lovins.

24 I am not exaggerating when I say "amazing". Your heart will probably leap when you read "Mushroom Power" by Paul Stamets, *Yes!,* Spring 2003.

25 Hawken, Paul, The Ecology of Commerce, Harper, 1993. p. 21

26 Hawken, p. 20

27 Interview with Derrick Jensen, *The Sun*, April 2001.

28 Many of the ancient world's great agricultural civilizations eventually destroyed their ecosystems. The deforestation of the Greek islands and the desertification of Mesopotamia and North Africa illustrate the destructive capacity of even low levels of technology. On the other hand, I have read that certain areas of China have been under continuous cultivation for five thousand years.

29 I first discovered Emoto's work in a Chinese translation of his original Japanese book. Some of them are available on line: see http://www.wellnessgoods.com/messages.asp.

30 Hawken, p. 72

31 At the present writing, most or all of these are for medical applications. For example, Cyberkinetics Inc. has built an implantable device that allows paralysis victims to control prosthetic devices. And implantable microchips are now commonplace for a variety of medical and security applications.

32 The deep problem with genetic engineering is that it tries to impose linear

control over something that is highly non-linear. All kinds of unexpected consequences arise from such bumbling. We know not what we do. I think we are thousands of years away from having the knowledge and wisdom to use this kind of technology.

[33] See for example the work of the HeartMath Institute, the International Institute for Biophysics, and the Qigong Institute.

[34] This is a complex topic. The blogosphere and traditional journalism are growing into a symbiotic relationship from which something entirely new will emerge. Neither will swallow up or replace the other. A similar process is beginning to transform the tottering scientific journal system.

[35] In the case of Amazon, the free information comprises the vast database of user reviews, reviews of user reviews, and so on. Of course that is not the sole reason for Amazon's profitability, but giving something away for free is certainly one way in which the company profits.

[36] A software crack is a procedure for making illegal copies of commercial software operable.

[37] Note, however, that much of the work of modern life is an artifact of our compulsion to maintain separation from nature. Woodchucks and hunter-gatherers do not clean toilets or wash dishes. Most do construct and maintain dwellings however. To undo even that form of separation, we must go back to a pre-mammalian state.

[38] When I speak of a garden, I have a concept in mind of cooperation with nature and not control. A garden can embody either. On one extreme there is the Victorian garden of exotic species, each precisely positioned in a total human imposition on the land. On the other extreme, there is simply the altered ecosystem that arises from any animal's interaction with its herbal environment.

[39] As for the issue of economic irrationality, calculate how much money you save by growing your own lettuce instead of buying it at the supermarket. Then factor in the time you spend. I doubt you'll save more than fifty cents an hour, even if you buy organic produce.

[40] Greenberg, Daniel, *Free at Last*, Sudbury Valley School Press, 1995. p. 3

[41] Greenberg, p. 15-18

[42] Ibid

[43] This quote is from my earlier book, *The Yoga of Eating* (New Trends, 2003) which has an extensive discussion of the fallacy of willpower.

[44] Greenberg, p. 80

[45] I mean a self-attack, not an attack by the HIV virus. Irrepressible evidence is mounting that HIV is a symptom and not a cause of AIDS. For an exhaustive demonstration of this, see Henry Bauer's series in The Journal of Scientific Exploration, Fall 2005 through Summer 2006.

[46] Kaptchuk, Ted J., *The Web that Has no Weaver*, Contemporary Books, 2000. p. 4

[47] Kaptchuk, p. 6-7

[48] Wood, Matthew, *The Practice of Traditional Western Herbalism*. North Atlantic Books, 2004. p. 115

[49] Wood, p. 116

[50] Johnstone, Keith, *Impro: Improvisation and the Theatre*. Routledge, New York, 1979.

pp. 78-9

51 Hyde, p. 20

52 The *lingua adamica* is humanity's hypothetical original language, the "language of Adam". It is described in Chapter Two.

53 Of course, birdsongs have far more meaning than most of us realize, and indeed are part of the Original Language of Nature. The metaphor refers to the modern listener.

54 Fonágy, Ivan, *Languages within Language*, John Benjamins Publishing, 2001. p. 181

55 Brown, Joseph Epes, *Teaching Spirits*, Oxford University Press, 2001. p. 35

56 The forced, contrived, premature opening of boundaries in the absence of love does not bring one any closer to the *lingua adamica*, but only invites violation.

Chapter VIII

1 Le Jeune, le Pere Paul. 1897. "Relation of What Occured in New France in the Year 1634", in R. G. Thwaites (ed.), *The Jesuit Relations and Allied Documents*. Vol. 6. Cleveland: Burrows. (First French edition, 1635) Quoted by Marshall Sahlins.

2 Spoken by Zeus in Hesiod, *Works and Days* 55

3 Grof, Stanislov, *Realms of the Unconscious*, Viking, `1975. Reprinted by Condor Books, 1995.

4 Ovid, *Metamorphoses* 1.89

5 But be careful to distinguish this from the commodification of spirituality, the attempt to market the teachings of the various spiritual traditions and thus convert this form of cultural or spiritual capital into yet more money. The "knowledge" I speak of here is not in the form of information, secret teachings, and so on. It is a way of being.

6 See Vinge, Vernor, *A Fire on the Deep* for a very accessible portrayal of the emergent nature of intelligence. It's a great story too!

7 Pearce, Joseph Chilton, *Evolution's End*, HarperCollings, 1992. p. 190

8 Of course, we are not yet and will probably never be at the point of complete material independence of Mother Earth; we are not even close. Our dependence on the geophysical and organic processes of Gaia is far greater than most people imagine. We are, for example, a long way from being able to manufacture an artificial atmosphere capable of sustaining human life; nor is synthetic food yet a viable alternative. Much of our technology-derived independence from nature is really an illusion, especially when it comes to the food supply. Yet such independence is not a prerequisite for adulthood, just as the tribal youth is not expected to live independent of the tribe. The key difference is a change of attitude, away from "It is mine to take" toward "It is ours to honor."

Perhaps someday we will become independent of Mother Earth, in some science-fiction scenario of space-roaming biospheres or consciousness uploaded onto computers, or alternatively, a New Age vision of spiritualization, the shedding of material bodily needs or even of the body itself, to live in the realm of pure spirit. Yet I suspect that even if such scenarios come to pass, it will not be as an escape from a planet we have ruined, but as the fruiting of a healed and vibrant Earth moving into the next stage of its development.

[9] Actually, I believe in all of these, with the qualification that concepts such as reincarnation and the afterlife mean something quite different outside the context of linear time and the discrete, separate self.

[10] This finding is ascribed to Rachel MacNair, *Yes!* Magazine, Winter 2005, p. 22.

[11] Id., p. 23

Index

LaVergne, TN USA
04 January 2010
168779LV00002B/4/P

9 780977 622207